APPLICATIONS OF FRACTURE MECHANICS TO REINFORCED CONCRETE

This volume is based on the papers presented at the International Workshop on the Applications of Fracture Mechanics to Reinforced Concrete—held in Turin, Italy, 6 October 1990.

APPLICATIONS OF FRACTURE MECHANICS TO REINFORCED CONCRETE

Edited by

Alberto CARPINTERI

Department of Structural Engineering
Politecnico di Torino, Italy

CRC Press
Taylor & Francis Group
Boca Raton London New York

CRC Press is an imprint of the
Taylor & Francis Group, an **informa** business
A TAYLOR & FRANCIS BOOK

CRC Press
Taylor & Francis Group
6000 Broken Sound Parkway NW, Suite 300
Boca Raton, FL 33487-2742

First issued in paperback 2019

ISBN-13: 978-1-85166-666-9 (hbk)
ISBN-13: 978-0-367-86461-3 (pbk)

Visit the Taylor & Francis Web site at
http://www.taylorandfrancis.com

and the CRC Press Web site at
http://www.crcpress.com

British Library Cataloguing in Publication Data

Applications of fracture mechanics to reinforced
concrete.
I. Carpinteri, A. (Alberto)
620.1366

ISBN 1 85166 666 4

Library of Congress Cataloging in Publication Data

International Workshop on the Applications of Fracture Mechanics to
Reinforced Concrete (1990: Turin, Italy)
Applications of fracture mechanics to reinforced concrete/edited
by A. Carpinteri.
p. cm.
"Based on the papers presented at the International Workshop on the Applications of Fracture Mechanics to Reinforced Concrete, held in Turin, Italy, 6 October 1990" Prelim. p.
Includes bibliographical references and index.
ISBN 1-85166-666-4
1. Reinforced concrete—Cracking—Congresses. 2. Fracture mechanics—Congresses. 3. Reinforced concrete—Testing—Congresses. 4. Reinforced concrete construction—Testing—Congresses.
I. Carpinteri, A. II. Title.
TA445.I68 1990
620.1′376—dc20 91-22614
CIP

Preface

I was very glad and honoured that the Politecnico di Torino could host the International Workshop on the Applications of Fracture Mechanics to Reinforced Concrete, a one-day conference that was held on 6 October 1990, just after the Eighth European Conference on Fracture (ECF8), with the contribution of the leading experts of the international community. To the original eighteen lectures, five additional contributions were added by invitation, so as to form the twenty-three chapters of the present volume, with the authors coming from twelve different countries.

The purpose of the Workshop and, therefore, of the present volume, was to emphasize the most recent advancements of fracture mechanics applied to steel-reinforced concrete. For several years technical committees of the most important international civil engineering associations (RILEM, CEB, ACI, SEM) have been successfully working on the application of fracture mechanics concepts to plain and fibre-reinforced concrete. It is now time to extend these concepts to steel-reinforced and prestressed concrete design. This is especially true for high strength concrete, which is a very brittle material, and in the case of large structural members the application of fracture mechanics appears to be very useful for improving the present design rules, in a compromise between structural reliability and economical aspects.

Four specific topics were selected, for which a rational approach could introduce variations in the concrete design codes. These constituted the session themes of the Workshop:

1. size effects
2. anchorage and bond
3. minimum reinforcement for elements in flexure
4. shear resistance

The chapters of the present volume are ordered in a logical sequence, according to the themes defined above. However, the first theme permeates all the chapters of the volume and represents the most original and revolutionary contribution of fracture mechanics. No partitioning of the chapters is performed, since the single topics are strongly correlated and it is difficult to class them objectively.

The first five chapters are devoted to the size effects in plain concrete structures and to their description through the size effect law, a 'best fit' method proposed by Z. P. Bažant. While Bažant and Yunping Xi extend such a method to include Weibull statistical strength variations, Eligehausen and Ožbolt give confirmation to the original method by notched tension specimens, three-point bend specimens, pull-out headed anchor specimens and beams loaded in torsion. A similar investigation regarding high strength concrete is presented by Duda and König, while Barr and Tokatly raise some doubts about the applicability of the size effect law to a limited number of tests. Van Mier deals with the size effect in compression and emphasizes how the internal length scale, introduced by the heterogeneous nature of concrete, must be considered only an average of a continuously decreasing crack band width.

Chapters 6–8 propose and illustrate the resistance curve method (R-curve) as a useful mean to describe ductile crack propagation, especially in fibre-reinforced cement composites. Ouyang, Mobasher and Shah define the R-curve as an envelope of fracture energy release rate of specimens with different sizes but the same initial notch length. Elices and Planas review the concept of equivalent elastic crack and show that the R-curve approach is a special case of this equivalence. They define a size-dependent R-curve which is tendentially in agreement with the bell-shaped curve proposed by the Editor of the present volume (*Eng. Fracture Mechanics,* **16** (1982) 467–81). Mai makes the assumption of a power law strain-softening characteristic of the damage zone and a linearized crack profile. In this way, the stable crack growth in fibre-reinforced cement composites can be readily modelled with the R-curve.

Chapters 9–13 deal with the problems of anchorage pull-out and the bond between steel bar and concrete. Bocca, Carpinteri and Valente illustrate the results of an extensive pull-out investigation on axial-symmetrical steel anchor bolts embedded at variable depths in slabs of concrete and mortar. Elfgren and Ohlsson provide similar size-dependent results related to a planar geometry. Rots presents two applications of softening fracture mechanics to bond and anchorage in reinforced concrete, emphasizing brittle snap-back instabilities. Mazars, Pijaudier-Cabot and Clement, on the other hand, apply the damage theory to the same problems, showing the ability of the model to predict a size effect consistent with the experimental observations. Reinhardt and van der Veen investigate the dependence of the maximum bond stress on the thickness of the concrete cover. A strain-softening constitutive law provides realistic results, whereas analyses with elastic or plastic laws lead to lower or upper bounds, respectively.

The problem of the minimum amount of reinforcement in bending is discussed in Chapters 14–17. Bosco and Carpinteri interpret the results of very extensive experimental investigations through a crack opening displacement congruence condition and the brittleness number concept. The minimum percentage of reinforcement appears to be size-dependent, and, more precisely, decreases with the beam depth. Hawkins and Hjorteset confirm such a trend, using as a fracture parameter the ratio of the bending strength to the tensile strength of concrete. Baluch, Azad and Ashmawi reconsider an early approach of the Editor (*J. Structural Eng.* (ASCE), **110** (1984) 544–58), removing the assumption of steel yielding at incipient fracture. Rokugo and Koyanagi introduce the concept of fracture toughness in compression, which is size-dependent and provokes the decrease with size of the so-called 'rupture limit reinforcement ratio'.

The problem of the shear resistance is addressed in Chapters 18–21. Tassios and Vintzeleou use a model previously developed for the prediction of the behaviour of sheared reinforced concrete interfaces. The total energy produced up to failure as well as the critical slip are calculated. Hillerborg applies the fictitious crack model so that crack development and shear resistance are analysed. He concludes that the shear resistance decreases as $d^{-1/4}$, where d is the beam depth, and this is in full agreement with a comprehensive summary of test results. Similar results are obtained by Li, Ward and Hamza for fibre-reinforced mortar and concrete, putting in evidence the shear resistance variation with the beam span and using the ratio of the bending strength to the tensile strength as fracture parameter. The transition from flexural failure to shear failure by increasing the amount of reinforcement, is analysed by Karihaloo, according to fracture mechanics models proposed by different authors. The loading capacity appears to decrease by increasing the size with a power which can be estimated between $-1/4$ (numerical result by Hillerborg) and $-1/2$ (stress-singularity power of linear elastic fracture mechanics).

The last two chapters deal with two topics not directly related to the preceding. Andrea Carpinteri considers cyclic actions on reinforced concrete beams. A numerical procedure following the fatigue crack growth and the hysteretic energy dissipation, is confirmed by some experimental results. Swartz, Kan and Hu present an expert system tutorial approach able to introduce the majority of concepts described in the present volume to the design engineering community.

My most sincere thanks and acknowledgements are due to RILEM (International Union of Testing and Research Laboratories for Materials and Structures) and to CEB (Comité Euro-International du Béton) for the prestigious scientific support to the Workshop; to CNR (Consiglio Nazionale delle Ricerche) for the generous financial support; to UNICEM-Torino for the welcome sponsorship; to the session Chairmen Professor L. Elfgren (Chairman of the RILEM Committee 90-FMA), Professor F. Levi (CEB Honorary President), Professor S. P. Shah (Chairman of the RILEM Committee 89-FMT), Professor T. P. Tassios (CEB Honorary President) for the fruitful and brilliant discussions they provoked during the Workshop; to the distinguished lecturers and authors for their precious contributions; to all the participants to the Workshop for their confidence in the organizer. On the other hand, the Workshop organizer was valiantly helped by the members of the Local Organizing Committee: Professor P. Bocca, Dr C. Bosco, Dr C. Scavia and Dr S. Valente.

Finally, I wish to express my warmest thanks to: Professor R. Zich (Rector of the Politecnico di Torino), Professor E. Antonelli (Dean of the Faculty of Engineering), Professor D. Firrao (President of the Italian Group of Fracture), for attending the Workshop and addressing their welcome to the participants, and also to Professor E. Giangreco (RILEM Past-President), who was prevented from participating by illness.

Alberto CARPINTERI
Torino, Italy, December 1990

Contents

List of Contributors

W. ASHMAWI
Department of Civil Engineering, King Fahd University of Petroleum & Minerals, KFUPM—Box 1469, Dhahran—31261, Saudi Arabia

A. K. AZAD
Department of Civil Engineering, King Fahd University of Petroleum & Minerals, KFUPM—Box 1469, Dhahran—31261, Saudi Arabia

M. H. BALUCH
Department of Civil Engineering, King Fahd University of Petroleum & Minerals, KFUPM—Box 1469, Dhahran—31261, Saudi Arabia

B. BARR
Department of Civil & Structural Engineering, School of Engineering, University of Wales College of Cardiff, PO. Box 917, Cardiff CF2 1XH, UK

Z. P. BAŽANT
Department of Civil Engineering, Northwestern University, 2145 Sheridan Road, Evanston, IL 60208, USA

P. BOCCA
Politecnico di Torino, Department of Structural Engineering, 10129 Torino, Italy

C. BOSCO
Politecnico di Torino, Department of Structural Engineering, 10129 Torino, Italy

Al. CARPINTERI
Politecnico di Torino, Department of Structural Engineering, 10129 Torino, Italy

An. CARPINTERI
Istituto di Scienza e Tecnica delle Costruzioni, Facoltà di

Ingegneria, Università di Padova, Via Marzolo 9, 35131 Padova, Italy

J. L. CLEMENT
Laboratoire de Mecanique et Technologie, Ecole Normale Supérieure de Cachan, 94235 Cachan, France

H. DUDA
Technische Hochschule Darmstadt, Institut für Massiv-bau, Alexanderstrasse 5, 6100 Darmstadt, Germany

L. ELFGREN
Division of Structural Engineering, Luleå University of Technology, S-95187 Luleå, Sweden

M. ELICES
Departamento de Ciencia de Materiales, Universidad Politécnica de Madrid, ETS Ingenieros de Caminos, Ciudad Universitaria, 28040 Madrid, Spain

R. ELIGEHAUSEN
Institut für Werkstoffe im Bauwesen, Universität Stuttgart, Pfaffenwaldring 4, 7000 Stuttgart 80, Germany

A. M. HAMZA
Advanced Civil Engineering Material Research Laboratory, Department of Civil Engineering, University of Michigan, Ann Arbor, MI 48109-2125, USA

N. M. HAWKINS
Department of Civil Engineering, University of Washington, 223 More Hall FX-10, Seattle, WA 98195, USA

A. HILLERBORG
Division of Building Materials, Lund Institute of Technology, PO Box 118, S-22100 Lund, Sweden

K. HJORTESET
Department of Civil Engineering, University of Washington, 223 More Hall FX-10, Seattle, WA 98195, USA

K. K. HU
Department of Civil Engineering, Kansas State University, Seaton Hall, Manhattan, KS 66506, USA

Y. C. KAN
Department of Civil Engineering, Kansas State University, Seaton Hall, Manhattan, KS 66506, USA

B. L. KARIHALOO
School of Civil & Mining Engineering, University of Sydney, NSW 2006, Australia

G. KÖNIG
Technische Hochschule Darmstadt, Institut für Massiv-bau, Alexanderstrasse 5, 6100 Darmstadt, Germany

W. KOYANAGI
Department of Civil Engineering, Gifu University, Yanagido, Gifu 501-11, Japan

V. C. LI
Department of Civil Engineering, University of Michigan, Ann Arbor, MI 48109-2125, USA

Y.-W. MAI
Centre for Advanced Materials Technology, Department of Mechanical Engineering, University of Sydney, NSW 2006, Australia

J. MAZARS
Laboratoire de Mecanique et Technologie, Ecole Normale Supérieure de Cachan, 94235 Cachan, France

B. MOBASHER
NSF Center for Science and Technology of Advanced Cement Based Materials, Northwestern University, 1800 Ridge, Evanston, IL 60208, USA

U. OHLSSON
Division of Structural Engineering, Luleå University of Technology, S-95187 Luleå, Sweden

C. OUYANG
NSF Center for Science and Technology of Advanced Cement Based Materials, Northwestern University, 1800 Ridge, Evanston, IL 60208, USA

J. OŽBOLT
Institut für Werkstoffe im Bauwesen, Universität Stuttgart, Pfaffenwaldring 4, 7000 Stuttgart, Germany

G. PIJAUDIER-CABOT
Laboratoire de Mechanique et Technologie, Ecole Normale Supérieure de Cachan, 94235 Cachan, France

J. PLANAS
Departamento de Ciencia de Materiales, Universidad Politécnica de Madrid, ETS Ingenieros de Caminos, Ciudad Universitaria, 28040 Madrid, Spain

H. W. REINHARDT
Institut für Werkstoffe im Bauwesen, Universität Stuttgart, Pfaffenwaldring 4, D-7000 Stuttgart, Germany

R. ROKUGO
Department of Civil Engineering, Gifu University, Yanagido, Gifu 501-11, Japan

J. G. ROTS
TNO Building and Construction Research, Delft University of Technology, P O Box 49, 2600 AA Delft, The Netherlands

S. P. SHAH
NSF Center for Science and Technology of Advanced Cement Based Materials, 1800 Ridge, Evanston, IL 60208, USA

S. E. SWARTZ
Department of Civil Engineering, Kansas State University, Seaton Hall, Manhattan, KS 66506, USA

T. P. TASSIOS
National Technical University of Athens, 42 Patission Street, GR 106 82 Athens, Greece

Z. Y. TOKATLY
Department of Civil & Structural Engineering, School of Engineering, University of Wales, College of Cardiff, PO Box 917, Cardiff CF2 1XH, UK

S. VALENTE
Politecnico di Torino, Department of Structural Engineering, 10129 Torino, Italy

C. van der VEEN
Delft University of Technology, Division of Mechanics and Structures, Stevinweg 1, NL-2628 CN Delft, The Netherlands

J. G. M. van MIER
Delft University of Technology, Department of Civil Engineering, Stevinlaboratory, PO Box 5048, 2600DA Delft, The Netherlands

E. VINTZELEOU
National Technical University of Athens, 42 Patission Street, GR 106 82 Athens, Greece

R. WARD
Cygna Consulting Engineers, Boston, MA 02210, USA

YUNPING XI
Department of Civil Engineering, Northwestern University, 2145 Sheridan Road, Evanston, IL 60208, USA

Chapter 1

Size Effect in Quasi-brittle Micro-heterogeneous Structures: Deterministic and Statistical Theories

ZDENĚK P. BAŽANT & YUNPING XI

NSF Center for Science and Technology of Advanced Cement Based Materials, Northwestern University, Evanston, Illinois 60208, USA

ABSTRACT

The classical applications of Weibull statistical theory of size effect in quasi-brittle structures such as reinforced concrete structures are re-examined in the light of recent test results. The classical Weibull-type approach ignores the stress redistributions and energy release during the stable large fracture growth before failure, which causes a strong deterministic size effect. Furthermore, the classical theory does not agree with recent test data. Therefore, the failure probability of structures must be calculated from the stress field that exists just before failure, rather than the initial elastic field. Accordingly, fracture mechanics stress solutions are utilized to obtain the failure probabilities and formulate an amalgamated theory that combines the size effect due to fracture energy release with the size effect due to random variability of strength having Weibull distribution. For the singular stress field of linear elastic fracture mechanics, the failure probability integral diverges. Convergent solutions, however, can be obtained with the nonlocal continuum concept. This leads to nonlocal statistical theory of size effect. According to this theory, the asymptotic size effect law for very small structure sizes agrees with the classical power law based on Weibull theory, while the asymptotic size effect law for very large structure sizes coincides with that of linear elastic fracture mechanics of bodies with similar large cracks. For very large structures, the failure probability is dominated by the stress field in the fracture process zone, while the stresses in the rest of the structure are almost irrelevant. The size effect predictions agree reasonably well with the existing test data. The failure probability can be approximately calculated by applying the failure probability integral to spatially averaged stresses obtained, according to the nonlocal continuum concept, from the singular stress field of linear elastic fracture mechanics. More realistic is the use of the stress field obtained by nonlinear finite element analysis according to the nonlocal damage concept.

1. INTRODUCTION

Traditionally, the size effect in failure of concrete structures has been explained by Weibull's statistical theory [1], [2]; see Freudenthal [3], Mihashi & Zaitsev [4]; Carpinteri [5]; Mihashi [6]; Mihashi & Izumi [7]. In Weibull's theory, the failure is determined by the minimum value of the strength of the material, and the statistical size effect is due to the fact that, the larger the structure, the smaller is the strength value that is likely to be encountered in the structure. This explanation is certainly valid for one-dimensional structures such as a long chain or a long fiber, but extension to multidimensional structures depends on certain simplifying hypotheses which do not have to be satisfied for all types of structures. These hypotheses appear to be applicable to most metal structures, but not to quasi-brittle structures, such as concrete structures, because of their ability to develop large fractures in a stable manner prior to failure.

The central idea in Weibull-type statistical analysis of failure and size effect is that the survival probability of the structure is the joint probability of survival of all its elementary parts. However, implementation of this idea is clear and simple only for a long fiber or a long chain, but is difficult for two-dimensional and three-dimensional structures.

This paper will examine the limitations of the classical Weibull theory of size effect from the viewpoint of quasi-brittle structures, such as reinforced concrete structures, and will present a new formulation which overcomes the main limitations. The basic idea of the present formulation has been briefly outlined in a previous conference paper by Bažant [8].

2. BASIC IDEAS OF WEIBULL THEORY

Weibull statistical theory tries to predict the failure probability P_f of a large structure on the basis of the failure probabilities $P_j(\sigma_j)$ of small elements, which may be called the representative volume of the material, as customary in the statistical theory of heterogeneous materials and in nonlocal theories; subscript $j = 1, 2, \ldots$ labels the structural elements. Probability P_1 is a function of the stress, given by the well-known Weibull distribution [1], [2],

$$P_1 = \begin{cases} (\sigma - \sigma_u)^m / \sigma_0^m & \text{for } \sigma > \sigma_u \\ 0 & \text{for } \sigma \leq \sigma_u \end{cases} \tag{1}$$

in which σ_0 and σ_u are empirical material parameters; m = shape parameter (Weibull modulus), σ_0 = scale parameter, σ_u = strength threshold (datum parameter). As an acceptable approximation, one may assume $\sigma_u = 0$, in which case the results of direct tensile tests of concrete indicate approximately $m = 12$ [9].

The survival probability of one representative volume is $1 - P_1$. The central idea in Weibull theory is that if the whole structure should survive, all of the elements must survive. This means that the probability of survival of the structure is the joint probability of the survival of all its representative volumes of the material. According to the theorem of joint probability, $1 - P_f = [1 - P_1(\sigma_1)][1 - P_1(\sigma_2)] \ldots [1 - P_1(\sigma_N)]$ or $\ln(1 - P_f) = \sum_{j=1}^{N} \ln[1 - P_1(\sigma_j)]$, where N = number of elements. Now probability P_1 is always very small in practical situations, and so $\ln[1 - P_1(\sigma_j)] \approx -P_1(\sigma_1)$. Therefore,

$$\ln(1 - P_f) = -\int_V \sum_{i=1}^{n} \left\langle \frac{\sigma_i(\mathbf{x}) - \sigma_u}{\sigma_0} \right\rangle^m \frac{dV(\mathbf{x})}{V_r} \tag{2}$$

in which we replaced σ by the principal stresses σ_i ($i = 1, 2$ or $1, 2, 3$) to take into account in an approximate manner the triaxial nature of stress; n is the number of dimensions (2 or 3); V = volume of the structure; and $\mathbf{x}$ = coordinate vector of material points in the structure; $\langle\ \rangle$ = Macauley brackets, $\langle X \rangle = X$ if $X \geq 0$ and 0 if $X \leq 0$.

In many papers dealing with probabilistic failure of concrete structures, the stress field $\sigma_i(\mathbf{x})$ has been taken as the elastic stress field. The stress redistributions due to large fracture growth near the maximum load have been ignored. In view of the experimental and theoretical research on the deterministic size effect in concrete structures, it is now clear that this simplification causes a large error and is inadmissible. The stress field substituted in eqn (2) must be the stress field just before failure, reflecting the stress redistributions due to stable fracture growth.

The incorrectness of using the elastic stress field can also be demonstrated from the fact that, when the elastic stress is used, every structure is equivalent (from the viewpoint of failure probability) to a uniaxially stressed bar of variable cross-section. This is obviously unreasonable. Therefore, a revision of this classical solution of size effect based on Weibull type reasoning is required.

Another questionable aspect of classical Weibull-type theories based on eqn (2) is the neglect of spatial correlation. This might be justified for the links in a chain, but not for continuous bodies of concrete cast at one time. If the strength value realized in one small material element is on the low side of the average strength, the strength value realized in the adjacent material elements is more likely to be on the low side than on the high side of average strength. The standard way to deal with spatial correlation would be to introduce a spatial autocorrelation function for strength, but that approach would be rather complicated for the present purpose. There is nevertheless another way.

3. NONLOCAL CONCEPT IN FAILURE PROBABILITY

The lack of spatial correlation between survival probabilities of adjacent material elements can be eliminated by introducing a nonlocal concept, in which failure probability at any point depends on the average state of a certain region (representative volume) surrounding the point. But there also is another, mechanics-based argument for the nonlocal concept.

The stress distribution function to be used in the integral for failure probability of the structure must be the stress distribution at incipient failure, rather than some stress distribution that exists long before failure. This distribution must reflect localization of strains and stresses that occurs prior to reaching the maximum load. In the extreme case of complete localization of cracking, a sharp crack develops upon reaching the maximum load, as illustrated in Fig. 1. The stress distribution is then singular and has the form:

$$\sigma_i = \sigma_N \rho^{-1/2} \phi_i(\rho, \theta) \tag{3}$$

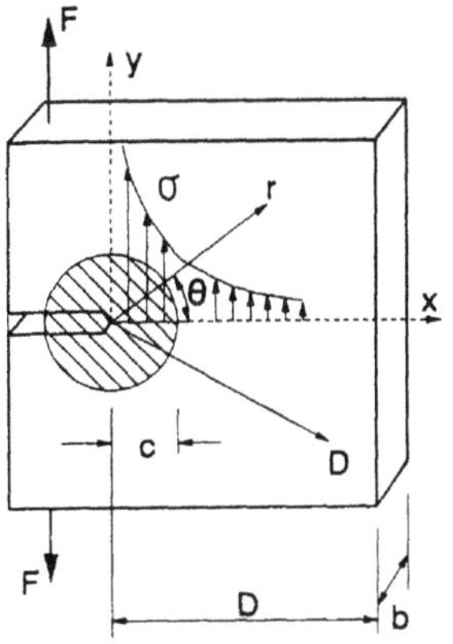

Fig. 1 Stress distributions and crack process zone in specimens.

in which σ_i $(i = 1, \ldots, n)$ are the principal stresses, $\rho = r/D$, D = characteristic dimension of the structure; ρ, θ = polar coordinates centered at the tip of the crack, and $\sigma_N = F/bD$ where b = structure thickness, and F = applied load; σ_N represents the nominal strength (nominal stress at failure); $\phi_i(\rho, \theta)$ is a non-singular function that is continuous and smooth (except at concentrated loads and at boundary corners).

Stress singularity is an abstraction that does not exist in reality. The stresses near the tip of a sharp crack are blunted by a fracture process zone that is large due to large material inhomogeneities. One effective way to take the material heterogeneity into account is the nonlocal continuum concept ([10], [11]), which is properly applied only to the variables associated with failure or damage [12]–[15]. Thus, recognizing that failure at a point of a heterogeneous material must depend not only on the continuum stress at that point but also on the stress resultant or average stress within a certain representative volume V_r of the material, we realize that the probability of failure should not depend on the local stresses $\sigma_i(\mathbf{x})$ but on the average stresses

$$\bar{\sigma}_i(\mathbf{x}) = \int_{V_r} \sigma_i(\mathbf{s})\alpha(\mathbf{x}-\mathbf{s})\,\mathrm{d}V(\mathbf{s}) \tag{4}$$

in which $\alpha(\mathbf{x}-\mathbf{s})$ is a given empirical weight function, which must satisfy the normalizing condition, $\int_{V_r} \alpha(\mathbf{x}-\mathbf{s})\,\mathrm{d}V = 1$. When the representative volume V_r protrudes across the body boundary or the crack boundary, the protruded part must be chopped off and the weights α must be scaled up so as to satisfy the normalizing condition. The weight function introduces the nonlocal material properties. For the special case that α is the Dirac δ-function, one has $\bar{\sigma}_i(\mathbf{x}) = \sigma_i(\mathbf{s})$, which is the case of local continuum. For zero strength threshold ($\sigma_u = 0$), eqn (2) may now be rewritten as

$$-\ln(1-P_f) = \int_V \sum_{i=1}^{n} \left(\frac{\bar{\sigma}_i(\mathbf{x})}{\sigma_0}\right)^m \frac{dV}{V_0} \tag{5}$$

Equations (4) and (5) are too difficult for an analytical solution. Therefore, one must take the nonlocal aspect into account in the simplest possible manner. As is known from previous studies, the nonlocal averaging (eqn (4)) is unnecessary in regions of small damage, i.e. far from the fracture front (large ρ). However, within volume V_c of the fracture process zone (shaded in Fig. 1), some form of nonlocal averaging is necessary. As an approximation, we may consider a constant average stress value $\bar{\sigma}_i$ through this entire zone, which is physically justified by the stress limit posed by inelastic deformation (similar to a yield limit). Equation (5) may thus be simplified as follows

$$-\ln(1-P_f) = \sum_{i=1}^{n} \left(\frac{\bar{\sigma}_i}{\sigma_0}\right)^m \frac{V_c}{V_0} + \int_{V_e} \sum_{i=1}^{n} \left(\frac{\sigma_i}{\sigma_0}\right)^m \frac{dV}{V_0} \tag{6}$$

in which $V_e = V - V_c$ = volume of the rest of the body outside the fracture process zone.

One may now be tempted to consider the value of $\bar{\sigma}_i$ as a constant yield stress. However, regardless of the absence of a clearly defined yield limit in the material, this would be incorrect since the $\bar{\sigma}_i$-value must be considered as a random variable and must be determined also on the basis of an extreme value distribution (such as Weibull's) in relation to the random nominal strength σ_N. Therefore, it is proposed to approximate $\bar{\sigma}_i$ as the value of the elastically calculated stress σ_i at a point that lies on the crack extension line ($\theta = 0$) at a certain fixed distance $r = c$ from the tip of the ideal sharp crack. Thus, for the region $V_c(\rho \leq c/D)$, we introduce the approximation

$$\bar{\sigma}_i \simeq k[\sigma_i]_{\substack{r=c\\ \theta=0}} = \sigma_N \left(\frac{c}{D}\right)^{-1/2} k\phi_{ci} \qquad \phi_{ci} = \phi_i\left(\frac{c}{D}, 0\right) \tag{7}$$

in which D = characteristic dimension (size) of the structure; k and c are empirical constants and c may be interpreted as the effective radius of the fracture process zone, which in turn is related to the characteristic length of the nonlocal continuum model approximating the heterogeneous material [12], [13]. The value of $\bar{\sigma}_i$ is random because σ_N is random.

4. APPROXIMATE NONLOCAL ANALYSIS

In the integral in eqn (6), we have $dV = bD^2\rho\, d\theta\, d\rho$, where $b =$ thickness of the body. This expression applies for the case of two-dimensional similarity. To be more general and cover also the case of three-dimensional similarity, one may generalize the foregoing expression as follows:

$$dV = b_0 D^n \rho\, d\theta\, d\rho \tag{8}$$

in which $b_0 =$ non-dimensional constant and $n =$ number of spatial dimensions. For two-dimensional similarity ($n = 2$), we have $b = b_0$, and for three-dimensional similarity ($n = 3$), $b = b_0 D$. In the case of axisymmetric fracture situations, b_0 is a constant without any meaning of thickness. The effect of the number of dimensions on the first expression in eqn (6) is a more difficult question. Strictly speaking, the volume of the fracture process zone is $V_c = \pi c^2 b_0$ ($b_0 = b$) for two dimensions ($n = 2$) and $V_c = \pi c^2 b_0 D$ for three dimensions ($n = 3$), which in general may be written as $V_c = \pi c^2 b_0 D^{n-2}$. For three dimensions, however, this would mean that the cracks would propagate independently in various parts of the fracture process zone throughout the thickness of the fracture specimen, which is impossible except if the fracture process zone is very thick ($b \gg c$). It seems more reasonable to assume that once the crack forms or propagates, it must do so simultaneously throughout the whole thickness b of the specimen, so that the probability of survival depends on the fracture process zone area rather than the volume. Consequently, we will assume that, for the purpose of survival probability in eqn (6),

$$V_c = \pi c^2 b_0 D^p \tag{9}$$

where $p = 0$. Later, however, we will also explore the case $p = n - 2$.

Substituting eqns (7) and (9), we may rearrange eqn (6) as follows

$$-\ln(1 - P_f) = \left[A_0\left(\frac{D}{c}\right)^{(p+m/2)} + A_1\left(\frac{D}{c}\right)^n H_D I\left(\frac{c}{D}\right)\right]\sigma_N^m \tag{10}$$

in which $A_0 = \pi b_0 c^{p+2} k^m \sigma_0^{-m} \sum_i \varphi_{ci}^m / V_0$, $A_1 = b_0 c^n \sigma_0^{-m} / V_0$, which are size-independent constants, H_D = parameter to be defined later, and

$$I\left(\frac{c}{D}\right) = \int_{-\pi}^{\pi} \int_{c/D}^{\hat{\rho}(\theta)} \sum_{i=1}^{n} [\rho^{-1/2} \phi_i(\rho, \theta)]^m \rho \, d\rho \, d\theta \qquad \text{except if } c \ll D \quad (11)$$

Here function $\hat{\rho}(\theta)$ represents the relative radial distances to the boundary points and defines the geometry of the structure. For geometrically similar specimens, the value of $I(c/D)$ is constant.

Furthermore, it has been necessary to introduce in eqn (10) parameter H_D which takes into account interaction of the fracture process zone with the specimen boundaries. For the extreme cases of very large and very small specimens relative to the fracture process zone size c, one must obviously have

$$\text{for } \frac{c}{D} \ll \hat{\rho}(\theta): \qquad H_D = 1 \qquad (12a)$$

$$\text{for } \frac{c}{D} \approx \hat{\rho}(\theta): \qquad H_D = 0 \qquad (12b)$$

Equation (12a) pertains to the case where the fracture process zone size is negligible compared to the structure size (dimension) D. Equation (12b) pertains to the case where the fracture process zone extends roughly over the entire structure (or at least the entire ligament length). In that case the domain V_e of the integral in eqn (6) is vanishing, and the vanishing of the integral is ensured by setting $H_D = 0$. For the intermediate body sizes, for which the fracture process zone is neither small nor large, parameter H_D must be considered to be a smooth function of the relative size of the process zone, c/D, i.e. $H_D = H_D(c/D)$. We leave it as an empirical function, although it could conceivably be calculated after solving the boundary value problem in its nonlocal form (eqns (4) and (5)).

From eqn (11), one can easily demonstrate that the singular stress distribution of linear elastic fracture mechanics cannot be used in the integral for failure probability. For c/D approaching zero, the integral in eqn (11) converges if and only if $m \leq 4$. But this is not a realistic situation. The typical value of Weibull modulus for concrete is $m \approx 12$ (with $\sigma_u = 0$). Therefore, it is inevitable to consider c as finite, as proposed in this paper.

The preceding consideration shows that the stress distribution in eqn (4), with σ_i proportional to σ_N, is unrealistic for the fracture process zone when the body is very large, i.e. $D \gg c$. For that case the stresses at the ideal crack tip become very large and would exceed the intrinsic strength f^* of the material without any flaws or microcracks. Therefore, the stress must be limited as $\sigma_i \leq f^*$ and must be taken as $\sigma_i = f^*$ when eqn (4) exceeds this value. From this consideration it follows that, for $c \ll D$, the integral in eqn (6) should be proportional to f^{*m}, i.e. a constant. This may be achieved by setting

$$\text{for } c \ll D \qquad I\left(\frac{c}{D}\right) \simeq k_0\left(\frac{f^*}{\sigma_N}\right)^m \tag{13}$$

in which k_0 is some constant.

If the failure probability P_f is fixed, eqn (10) yields for the size effect the following relation

$$\sigma_N = \left\{\frac{1}{-\ln(1-P_f)}\left[A_0\left(\frac{D}{c}\right)^{(p+m/2)} + H_D A_1 I(c/D)\left(\frac{D}{c}\right)^n\right]\right\}^{-1/m} \tag{14}$$

For $P_f = 0{\cdot}5$, eqn (14) yields the median strength, which is known from experiments on concrete to be close to the mean strength.

Let us now discuss the limiting cases, assuming that $p = 0$ and $m/2 > n$ ($n = 2$ or 3). For very small D/c, $(D/c)^{m/2} \ll (D/c)^n$, while for very large D/c, $(D/c)^{m/2} \gg (D/c)^n$. Also note that $I(c/D)$ tends to a finite value for very small sizes (eqn (13)) because σ_i is bounded. So the first term in the bracket of eqn (14) must dominate for very large sizes, while the second term must dominate for very small sizes. Thus we obtain the following asymptotic size effect laws (with $p = 0$):

$$\text{for small } D/c \qquad \sigma_N \propto D^{-n/m} \tag{15}$$

$$\text{for large } D/c \qquad \sigma_N \propto D^{-1/2} \tag{16}$$

In the foregoing asymptotic results, it is particularly noteworthy that for very large structures (eqn (16)) the size effect is the same as that obtained by deterministic analysis on the basis of energy release, i.e. $\sigma_N \propto D^{-1/2}$. The reason is that, for large structures, the stress peak in the fracture process zone dwarfs the stresses in the rest of the structure, thus making the statistical variability of the strength in the rest of the structure irrelevant. Only the statistical variability of strength within the fracture process zone matters. But the size of this zone is essentially fixed, independent of the body dimensions. Consequently, the statistical part of the size effect disappears and the only size effect left is deterministic—namely that due to the increase in the energy release rate as the structure size (along with the fracture length at imminent failure) is increased. (The fact that for large structures the statistical size effect should approach the deterministic size effect of linear elastic fracture mechanics was stated in [8].)

Equation (16) for large D/c is based on the hypothesis that in the fracture process zone a crack must form or propagate simultaneously within the entire specimen thickness b. For very thick specimens, however, it is conceivable that a crack might form or propagate only in a part of thickness and be stationary in the remaining part of the thickness. In that case, as already explained, it might be more appropriate to use some value of p between 0 and $p = n - 2 = 1$ in eqn (9) for three-dimensional similarity ($n = 3$). Equation (16) would then be replaced by:

$$\text{for large } D/c \qquad \sigma_N \propto D^{-(p/m+1/2)} \quad \text{(3D-similarity only)} \tag{17}$$

Assuming $m = 12$, we would have $0 < p/m < 1/12$, which is a relatively small correction. But it is noteworthy that a size effect slightly stronger than $D^{-1/2}$ cannot be ruled out. Experimentally, though, it has not been observed in tests with three-dimensional similarity (torsion, pull-out, punching shear), and so we will not consider this possibility any further.

To calculate the mean strength $\bar{\sigma}_N = E(\sigma_N)$, in which $E =$ expectation, one may use the following well-known relation

$$\bar{\sigma}_N = E(\sigma_N) = \int_0^1 \sigma_N \, \mathrm{d}P_f = \int_0^\infty (1 - P_f) \, \mathrm{d}\sigma_N \tag{18}$$

which follows from the fact that the cross-hatched area in Fig. 2 can be obtained by integrating either over the horizontal strips or over the vertical strips. Now, denoting the expression in the square brackets in eqn (14) as Z, one obtains (from eqn (14)), $1 - P_f = \exp(-Z\sigma_N^m)$. So eqn (18) yields:

$$\begin{aligned}\bar{\sigma}_N &= \int_0^\infty \exp(-Z\sigma_N^m) \, \mathrm{d}\sigma_N \\ &= Z^{-1/m} \frac{1}{m} \int_0^\infty t^{1/m-1} \mathrm{e}^{-t} \, \mathrm{d}t = Z^{-1/m} \Gamma\left(1 + \frac{1}{m}\right)\end{aligned} \tag{19}$$

in which $\Gamma =$ Gamma function. Substituting the bracketed expression in eqn (14) for Z, one gets for the mean strength the following size

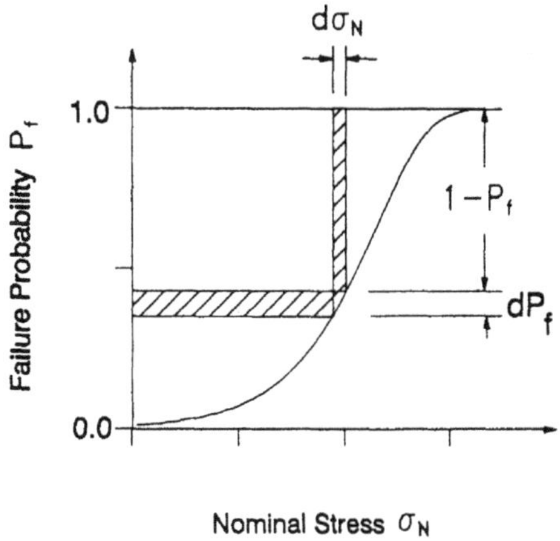

Fig. 2 Cumulative distribution of failure probability of the structure.

effect law

$$\bar{\sigma}_N = \Gamma\left(1 + \frac{1}{m}\right)\left[A_0\left(\frac{D}{c}\right)^{m/2} + H_D A_1 I\left(\frac{c}{D}\right)\left(\frac{D}{c}\right)^n\right]^{-1/m} \tag{20}$$

As we see, except for a factor, this law is formally identical to the size effect law in eqn (14) for a fixed failure probability P_f. The limiting behavior for very small and very large sizes is, for the mean strength, again the same as that in eqns (15) and (16). By a similar procedure, one can easily also calculate the variance of σ_N.

5. EMPIRICAL INTERPOLATION BETWEEN THE ASYMPTOTIC SIZE EFFECTS

The asymptotic size effect laws for very small and very large sizes in eqns (15) and (16) are very simple. More importantly, they are also independent of the geometry of the structure. The complete size effect law described by eqn (14) is of little value because the intermediate variation of coefficients I and H_D between the asymptotic cases is difficult to determine, and is sure to depend on the geometry of the structure. Therefore, eqn (14) cannot be proposed for practical use.

From extensive experimentation (diagonal shear [16]; pull-out [17]; torsion [18]; punching shear [19]) it appears that, as a good approximation relative to the scatter of the test results, the form of the size effect law can be considered as shape-independent through the entire range of sizes up to about 1:20. This has been confirmed by the success of the approximate size effect law

$$\sigma_N = \frac{Bf_t'}{\sqrt{1+\beta}} \qquad \beta = \frac{D}{D_0} \tag{21}$$

proposed in [20], with Bf_t' and D_0 as empirical constants (f_t' is the direct tensile strength). In view of this observation, it is not inappropriate to obtain the complete statistical size effect law for bodies with large fractures by empirical interpolation between the asymptotic cases in eqns (15) and (16). A simple empirical interpolation formula with the correct asymptotic properties (eqns (15) and (16)) is

$$\sigma_N = \frac{Bf_t'}{\sqrt{\beta^{2n/m} + \beta}} \tag{22}$$

6. DETERMINATION OF MATERIAL PARAMETERS

As for eqn (21), it is possible to identify parameters D_0 and B by linear regression of σ_N-data for geometrically similar specimens of various sizes. To this end, we may algebraically rearrange eqn (22) to the linear plot $Y = AX + C$ in which

$$\begin{aligned} X &= D^{1-(2n/m)}; \qquad Y = \left(\frac{f_t'}{\sigma_N}\right)^2 D^{-2n/m}; \\ A &= B^{-2}D_0^{-1}; \qquad C = D_0^{-2n/m}B^{-2} \end{aligned} \tag{23}$$

(n is the given number of dimensions). Thus, if m is known, A and C, along with the coefficients of variation of A and of the deviations of the data points from the regression line, can be determined from the plot of Y versus X by linear regression; B and D_0 follow from eqn (23). The linear regression previously introduced for the deterministic size effect law (eqn (21)) is the special case of eqn (22) for $m \to \infty$.

Similar to the previous use of eqn (21) ([8], [21]) the regression results for B can be used to determine the fracture energy, G_f, defined as the critical energy release rate in an infinitely large fracture specimen. According to LEFM, one has $G_f = F^2 g(\alpha)/E'b^2 D$ where $E' = E/(1-\gamma^2)$ for plane strain, $E' = E =$ Young's modulus for plane stress ($\gamma =$ Poisson ratio); $\alpha =$ relative crack length; and $g(\alpha) =$

nondimensional energy release rate for the given specimen geometry, which can be easily determined by elastic analysis. Inserting $F = \sigma_N bD$, one gets $\sigma_N = [E' G_f / g(\alpha) d]^{1/2}$. From eqn (22), the asymptotic behavior for large D is $\sigma_N = Bf_t'/\sqrt{\beta}$, while $\alpha \rightarrow \alpha_0$ (initial crack or notch length). Setting these two expressions equal, one obtains

$$G_f = \frac{B^2 f_t'^2 d_0}{E'} g(\alpha_0) \tag{24}$$

This happens to coincide with the expression obtained previously from eqn (21). This means that the statistical strength effects on G_f are nil, which is not surprising.

Similar to Bažant & Kazemi [16], one could also deduce from eqn (22) the effective length c_f of the fracture process zone length, but this value is affected by the Weibull statistical parameters m and σ_0.

7. COMPARISON WITH TEST RESULTS

To be able to check the asymptotic trend defined by eqns (15) and (16) and the empirical interpolation formula in eqn (22), test data of a rather broad size range are needed. Although no known data are perfect in this regard, the recent data in Fig. 3(a) are of a sufficiently broad range. They show the optimum fits by eqn (21) (nonstatistical) and eqn (22) (statistical) for the test results of Bažant & Kazemi [16] on diagonal shear failure of longitudinally reinforced concrete beams without stirrups. The size range of these data, 1:16, was quite broad. We see that the statistical eqn (22) fits these data somewhat better than the deterministic eqn (21). The difference between these two size effect laws is discernible only for small specimen sizes, and the small-size asymptotic slope $-n/m$ seems to be acceptable.

In evaluating these tests, there is a question with regard to the failure mode of the smallest specimen, which was of a different type: it was essentially a bending type failure, while all the large beams exhibited typical diagonal shear failure [16]. For the present phenomenologic analysis of failure, this would hardly be a sufficient argument for excluding the smallest beam data, which justifies keeping the smallest beam data in Fig. 3(a). But if one speaks strictly of diagonal shear, these data must of course be excluded. For that case, the optimum fit by eqns (21) and (22) is shown in Fig. 3(b), which is however not too different from Fig. (3a).

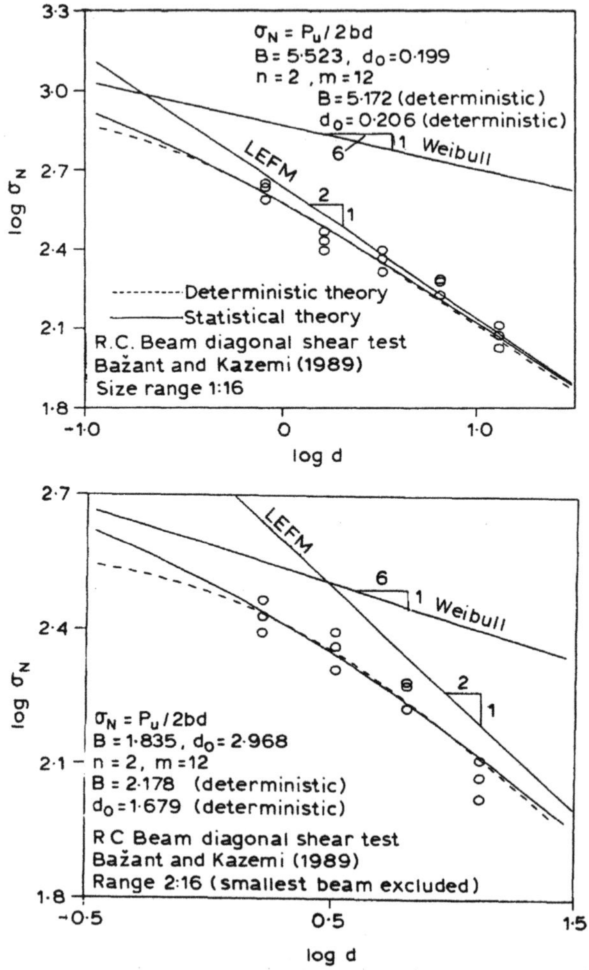

Fig. 3 Test result on diagonal shear failure of reinforced concrete beam without stirrups, and the optimum fits by eqns (21) and (22). (Fig. 3(a) is for 5 beams and 3(b) for 4 beams).

8. CONCLUSIONS

1. The Weibull-type integral that yields the failure probability must be based on the stress field at imminent failure, which reflects the stress redistributions due to large fracture. However, the singular

stress field obtained according to linear elastic fracture mechanics cannot be used, since the probability integral diverges.

2. Divergence of the probability integral is avoided in general by the nonlocal concept, in which the failure probability at any point depends not only on the stress at that point but also on the stresses in a small neighborhood of the point.

3. Nonlocal statistical analysis of failure probability indicates that the classical Weibull-type size effect law (applicable only to bodies without large fractures) is valid asymptotically for very small specimen sizes, while the size effect law of linear elastic fracture mechanics is approached asymptotically for very large sizes. The failure probability for a very large size of a structure with a macroscopic crack proportional to the structure size is dominated by the stress field in the fracture process zone and is nearly independent of the stress field in the rest of the structure.

4. The aforementioned asymptotic behavior as well as the presently proposed empirical interpolation curve agrees reasonably well with the existing test data, although the agreement cannot be regarded as a proof.

5. Finally, the reasons that, for very large specimens, the present statistical theory yields a nonstatistical size effect, the same as linear elastic fracture mechanics (conclusion 3), are basically two: (1) The mechanics of localization causes the failure to depend in the limit of infinite size solely on the properties of the fracture process zone; (2) the size and state of this zone, in an infinitely large specimen, becomes independent of the structure size (as well as shape).

ACKNOWLEDGEMENT

Partial financial support under NSF Grant MSM-8815166 to Northwestern University is gratefully acknowledged.

REFERENCES

[1] Weibull, W., A statistical theory of the strength of materials. *Royal Swedish Academy Eng. Sci. Proc.*, **151** (1939) 1–45.

[2] Weibull, W., A statistical distribution function of wide applicability. *J. Applied Mech.*, **18** (1951) 293–7.

[3] Freudenthal, A. M., Statistical approach to brittle fracture. In *Fracture—An Advanced Treatise*, ed. H. Liebowitz, Vol. 2, Chapter 6. Academic Press, New York, 1968, 591–619.

[4] Mihashi, H. & Zaitsev, J. W., Statistical nature of crack propagation. Report to RILEM TC 50-FMC, Chapter 4, 1981, 1–21.

[5] Carpinteri, A., Decrease of apparent tensile and bending strength with specimen size: two different explanations based on fracture mechanics. *Int. J. Solids and Structures,* **25** (1989) 407–29.

[6] Mihashi, H., A stochastic theory for fracture of concrete. *Fracture Mechanics of Concrete,* Chapter 4.3, ed. F. H. Wittmann. Elsevier Science Publishers, Amsterdam, 1983, 301–39.

[7] Mihashi, H. & Izumi, M., Stochastic theory for concrete fracture. *Cement and Concrete Research,* **7** (1977) 411–22.

[8] Bažant, Z. P., Fracture energy of heterogeneous materials and similitude. In *Proceedings of the SEM/RILEM International Conference of Fracture of Concrete and Rock,* ed. S. P. Shah & S. E. Swartz. Springer, 1989, pp. 390–402.

[9] Zech, B. & Wittmann, F. H., A Complex Study on the Reliability Assessment of the Containment of a PWR, Part II. Probabilistic approach to describe the behavior of materials. In *Trans. 4th Int. Conf. on Structural Mechanics in Reactor Technology,* ed. T. A. Jaeger & B. A. Boley. European Communities, Brussels, Vol. H, J1/11, 1977, pp. 1–14.

[10] Kröner, E., Elasticity theory of materials with long-range cohesive forces. *International J. Solids and Structures,* **3**(5) (1967) 731–42.

[11] Eringen, A. C., Linear theory of nonlocal elasticity and dispersion of plane waves. *International J. Engineering Science,* **10**(5) (1972) 425–35.

[12] Bažant, Z. P. & Pijaudier-Cabot, G., Measurement of characteristic length of nonlocal continuum. *J. Eng. Mech.* (ASCE), **115**(4) (1989) 755–67.

[13] Bažant, Z. P. & Pijaudier-Cabot, G., Nonlocal continuum damage localization instability and convergence. *J. Applied Mech.* (ASME), **55** (1988) 287–93.

[14] Bažant, Z. P. & Lin, F. B., Nonlocal smeared cracking model for concrete fracture. *J. Structural Engineering* (ASCE), **114**(11) (1988) 2493–2510.

[15] Bažant, Z. P. & Lin, F. B., Nonlocal yield limit degradation. *International J. Numerical Methods in Engineering,* **26** (1988) 1805–23.

[16] Bažant, Z. P. & Kazemi, M. T., Size effect on diagonal shear failure of beams without stirrups. Report No. 89-8/498S, Center for Advanced Cement-Based Materials, Northwestern University, 1989.

[17] Bažant, Z. P. & Sener, S., Size effect in pullout test. *ACI Materials J.,* **85**(5) (1988) 347–51.

[18] Bažant, Z. P., Prat, P. C. & Mazen, R. T., Antiplane shear fracture tests (Mode III). *ACI Materials J.,* **87**(1) (1990) 12–19.

[19] Bažant, Z. P. & Cao, Z., Size effect in punching shear failure of slabs. *ACI Structural J.,* **84**(1) (1987) 44–53.

[20] Bažant, Z. P., Size effect in blunt fracture: concrete, rock, metal. *J. Eng. Mech.* (ASCE), **110**(4) (1984) 518–35.

[21] Bažant, Z. P. & Pfeiffer, P. A., Determination of fracture energy from size effect and brittleness number. *ACI Materials J.,* **84**(6) (1987) 463–80.

Chapter 2

Size Effect in Concrete Structures

ROLF ELIGEHAUSEN & JOŠKO OŽBOLT
Institut für Werkstoffe im Bauwesen, Universität Stuttgart, Pfaffenwaldring 4, 7000 Stuttgart 80, Germany

ABSTRACT

The size effect for notched-tension specimens, three-point bend specimens, pull-out headed anchor specimens and beams loaded in torsion are calculated using a 2D and 3D finite element program. The program is based on the nonlocal microplane model. The calculated failure loads are compared with previously obtained experimental results. Test results and calculated data are compared with the recently proposed size effect law. Results of tests and analysis exhibit significant size effect that should be taken into account in design practice. It is demonstrated that the nonlocal microplane model used in a 2D and 3D finite element code can correctly predict failure loads for similar specimens of different sizes.

1. INTRODUCTION

The size effect in concrete structures is a well known phenomenon. For example, the bending strength decreases with increasing specimen height. Another example is the shear strength of concrete beams without shear reinforcement. Kani [1] was one of the first to demonstrate that the shear strength of identical concrete beams decreases with increasing beam depth and that the shear design provisions used at that time were unsafe for larger beams. This size effect can well be explained by fracture mechanics, because the fracture in a concrete structure is driven by the stored elastic energy that is released globally from the entire structure. However, before failure, microcracking in the concrete causes deviations of the size effect from the geometrical size effect known from linear elastic fracture mechanics (LEFM), because for normal geometrical sizes the fracture process zone is relatively large with regard to the geometry of the structure and therefore the size effect can only be correctly calculated using nonlinear fracture mechanics (NLFM).

In numerical analysis it is very difficult to model damage and fracture processes in materials such as concrete correctly. At present, three different material models for strain-softening damage exists:

(1) Continuum models used together with fracture mechanics.
(2) Random particle model, in which the microstructure is imagined to consist of randomly arranged rigid aggregate pieces with elastic-softening interactions between them.
(3) Micro-finite element models, in which the matrix as well as the aggregate pieces in concrete are subdivided into many finite elements, whose inelastic behaviour and cracking as well as interface bond failures are taken into account.

The last two material models automatically take into account the structural size effect but they are still extremely demanding of computer time and cannot be used in structural analysis. Therefore, the continuum material models must be formulated in such a way that they are capable of describing fracture of the structure in a correct way.

In the present study a number of finite element analyses are performed to investigate the structural size effect. A continuum material model, called the *nonlocal microplane model,* based on the smeared crack approach is used. Numerical studies of the structural size effect are presented and discussed for plain concrete specimens loaded in centric tension, three-point bending, axisymmetric pull-out of headed anchors and beam torsion. The calculated results are compared with available test data as well as with the size effect law recently proposed by Bažant [2].

2. SUMMARY OF NONLOCAL MICROPLANE MODEL FOR CONCRETE

The microplane models were initiated by Taylor [3], who suggested the principle for the modelling of plasticity of polycrystalline metals.

In that approach, developed in detail by Batdorf & Budianski [4] and others, the plastic slips were calculated independently on various crystallographic planes based on the resolved shear stress component, and were then superimposed to obtain the plastic microstrain. Later this approach was extended, under the name multilaminate model, to the modelling of non-softening plastic response of soils or rocks [5]. Recently [6]–[9], this approach was extended to include strain softening of concrete, and was renamed more generally as the *microplane model,* in recognition of the fact that the approach is not limited to plastic slip but can equally well describe cracking and strain-softening damage. To prevent instability due to strain softening, the microplanes must be constrained kinematically rather than statically, in which case the use of the principle of virtual work must replace the direct superposition of the plastic strains as used in the slip theory.

In the present study the microplane model originally developed by Bažant & Prat [9], is slightly modified and implemented into a 2D and 3D finite element code. The basic hypotheses used in the model are as follows:

Hypothesis I—The strains on any microplane represent the resolved components of the macrostrain tensor ε_{ij} (kinematic constrains).

Hypothesis II—Each microplane resists not only normal strain ε_N, but also shear strain ε_T. Shear strain is split into two mutually perpendicular in-plane components and as a consequence the shear stress vector is not parallel with the shear strain vector. This is the main modification in contrast to the originally proposed microplane model [9], where stress and strain vectors are parallel.

Hypothesis III—The normal microplane strain is split into volumetric and deviatoric components, i.e. $\varepsilon_N = \varepsilon_V + \varepsilon_D$.

Hypothesis IV—The stress–strain relation for each microplane is path independent as long as there is no unloading on this microplane for that component. During each unloading and reloading, which is defined separately in each microplane, the curves of the stress and strain differences from the state at the start of the unloading are also path independent. Thus, all the macroscopic path dependence is produced by various combinations of loading and unloading on all the microplanes.

Hypothesis V—The volumetric, deviatoric and shear responses on each microplane are mutually independent.

These five hypotheses were shown to allow an excellent representation of nonlinear test data for concrete in 1D, 2D and 3D stress–strain states [9].

A basic requirement for a continuum model for a brittle heterogeneous material such as concrete is that it must correctly display the consequences of heterogeneity of the microstructure. A continuum constitutive model lumps the average response of a certain characteristic volume of the material (Fig. 1). In essence, one may distinguish two types of interactions among the particles or damage sites in the microstructure, which must be somehow manifested in the continuum model: (1) Interaction at a distance *among various sites* (e.g. between A and B, Fig. 1); and (2) interaction *among various orientations* (see angle α in Fig. 1).

The interactions at a distance control the localization of damage. They are ignored in the classical, local continuum models but are reflected in nonlocal models [10]. The nonlocal aspect is a requisite for a realistic description of the size effect, as well as for the modelling of fracture propagation in the form of a crack band.

According to the nonlocal concept, the stress at a point depends not only on the strain at the same point but also on the strain field in a certain neighbourhood of the point [11]–[14]. For strain-softening behaviour, this concept was introduced by Bažant *et al.* [15]. In the current study, an effective form of the nonlocal concept, in which all variables that are associated with strain softening are nonlocal and all other variables are local, is used. The originally proposed nonlocal concept [10], is here modified by introducing additional weighting functions that control averaging into the directions of the main principle stresses. An important advantage of this formulation, called

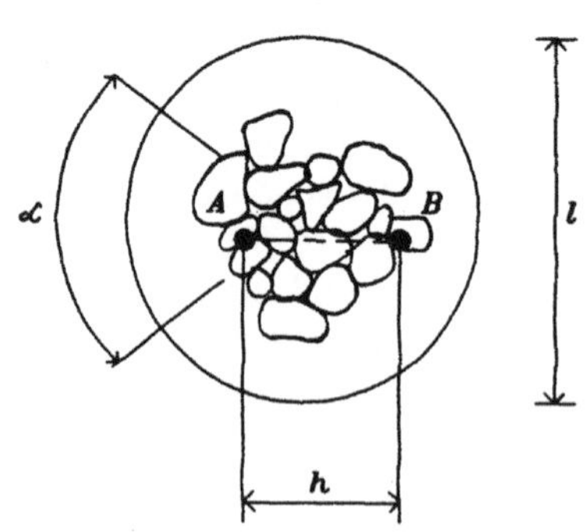

Fig. 1 Interaction among the various orientations and interaction at a distance.

nonlocal damage or nonlocal continuum with local strain, is that the differential equations of equilibrium as well as the boundary conditions are of the same form as in the local continuum theory, and that there exist no zero-energy periodic modes of instability.

The key parameter in the nonlocal concept used is the characteristic length l_c over which the strains are averaged, because it has a significant influence on the results of the analysis. Bažant & Pijaudier-Cabot [16] assumed that this length is a material parameter which can be correlated with G_F and approximately taken as $3d_a$ (d_a = maximum aggregate size) in the case of uniaxial macroscopic stress–strain state. However, presently in general 3D stress–strain situations, the concrete fracture property cannot be measured and correlated with the characteristic length. As a consequence, l_c is difficult to interpret as a material parameter depending on the concrete mix only, but may be influenced by other parameters as well. Further studies are needed to clarify whether this length depends only on the concrete composition or also on other parameters such as strain conditions in the failure zone. Therefore in the current study, in general, the characteristic length was determined such that together with the assumed tension strain-softening relationship, the failure load of a certain type of specimen with given size was correctly matched. Then, in the analysis of the specimens of different sizes, this characteristic length was taken as constant. Only in the analysis of the pull-out specimen was the characteristic length taken arbitrarily as $l_c = 12$ mm.

In a preceding paper [17], the nonlocal microplane model as well as an effective numerical iterative algorithm for the loading steps, that is used in the finite element code, is described in detail.

3. NUMERICAL STUDIES

To demonstrate that the nonlocal microplane model, implemented in the 2D and 3D finite element code, can correctly predict the failure load of plain concrete specimens of different sizes, numerical studies of four cases are presented (Fig. 2):

(1) notched tension specimen;
(2) three-point bend specimen;
(3) pull-out specimen with headed anchors;
(4) torsion of short beams.

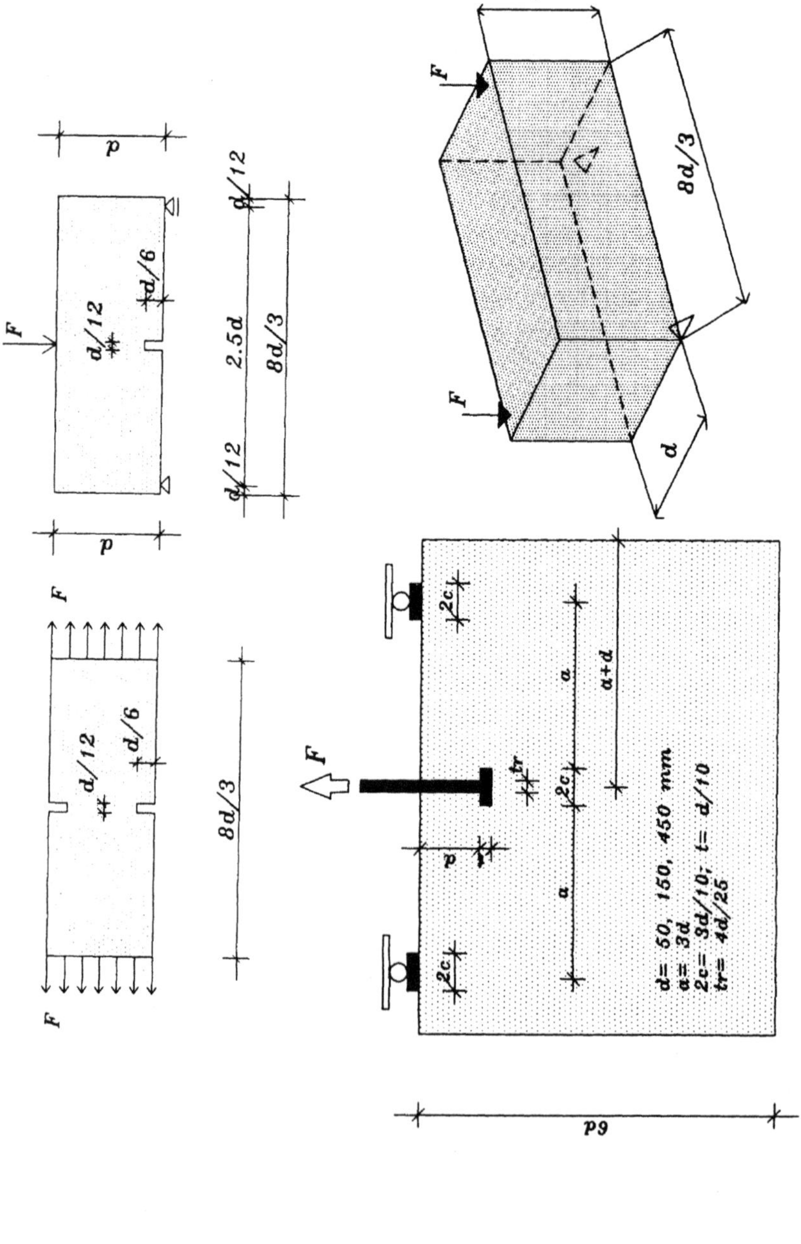

Fig. 2 Geometry of the specimens analysed in the present study. (a) Notched tension specimen; (b) 3-point bend specimen; (c) headed stud anchor; (d) torsion specimen.

In all examples, specimens of three different sizes, in the size ratio 1:2:4, with geometrically similar shapes are used. Exception is made in example (3) where a size ratio of 1:3:9 is studied. Cases (1) and (2) are analysed using four-node plane stress isoparametric finite elements with two by two integration rule. The analysis of the pull-out specimen is made using four-node axisymmetric finite elements with two by two integration rule. Finally, in the torsion problem, eight-node finite elements are employed with two by two by two (smallest specimen) and three by three by three integration rule, respectively. All specimens in the analysis are loaded by prescribing displacement increments in each loading step.

Microplane model parameters in cases (1), (2) and (4) are determined so that they represent the concrete properties used in the experiments. In all the examples the shape of the tension stress–strain curve is determined so that the fracture energy G_F for the unit area of a specimen of length 360 mm is approximately 0·1 N/mm.

Example (2) has been analysed by Bažant & Ožbolt [17], but was re-analysed using slightly different material parameters. Example (3) has been analysed by Eligenausen & Ožbolt [18], and results are shown from this work.

Example (1)—The notched tension specimen shown in Fig. 2(a) is considered. This type of specimen was tested by Bažant & Pfeiffer [19] using concrete with maximum aggregate size $d_a = 12{\cdot}7$ mm. The depth of the smallest specimen was $d = 38{\cdot}1$ mm, the depth of the notch was always 1/6 of the depth of the specimen and the thickness of the specimen was constant for all sizes, $b = 19$ mm. In the analysis only one half of the specimen is modelled. The finite element meshes are shown in Fig. 3 in the deformed state. The characteristic length is taken as $l_c = 3d_a = 38{\cdot}1$ mm. The microplane model parameters are taken so that the calculated tension strength is $f_t = 2{\cdot}70$ MPa. In the experiment the estimated average tension strength, calculated on the basis of the measured uniaxial compressive strength, was $f_t = 2{\cdot}69$ MPa. The characteristic length is chosen such that together with the microplane material parameters, the average failure load of the specimen with $d = 76{\cdot}2$ mm is matched.

In Fig. 4 the nominal stresses at failure related to the total area $\sigma_N = F_U/bd$ (F_U = peak load), obtained in the numerical analysis and in the experiments (average values) are compared with the size effect

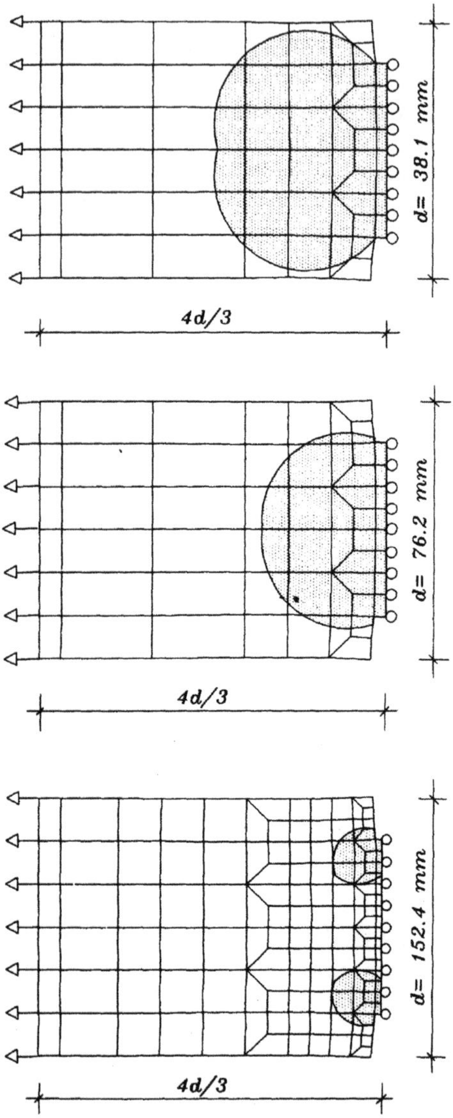

Fig. 3 Deformed finite element meshes and fracture process zone (shaded areas) at peak load for the notched tension specimen.

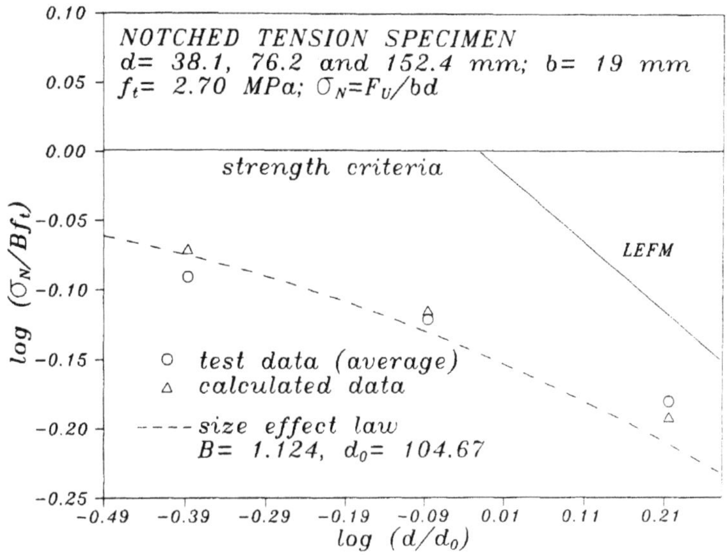

Fig. 4 Comparison between calculated and measured failure loads with size effect law for the notched tension specimen.

law is proposed by Bažant [2]:

$$\sigma_N = Bf_t(1+\beta)^{-1/2} \qquad \beta = d/d_0 \tag{1}$$

The optimum values for the parameters B and d_0 are obtained by linear regression of the numerical results (Fig. 5). In Fig. 6 the nominal stresses at failure (numerical and experimental results) are plotted as a function of the specimen depth in normal scale.

It can be seen from Figs 4–6 that the numerical and experimental results indicate a size effect: the nominal stresses at peak load decrease with increasing specimen depth, that means the absolute failure loads increase approximately by a factor of 1·5 when doubling the depth, much less than the increase in failure area.

In Fig. 3 the shaded areas indicate zones where the tensile stress at peak load exceeds approximately 75% of the uniaxial tensile strength. According to the assumed stress–strain relationship, the concrete starts to exhibit nonlinear behaviour for stresses $\sigma > 0{\cdot}75f_t$. Therefore the shaded area can be assumed as the size of the fracture process zone at peak load. Note that the scale for the specimens of different sizes is inversely proportional to the specimen depth. It is evident from

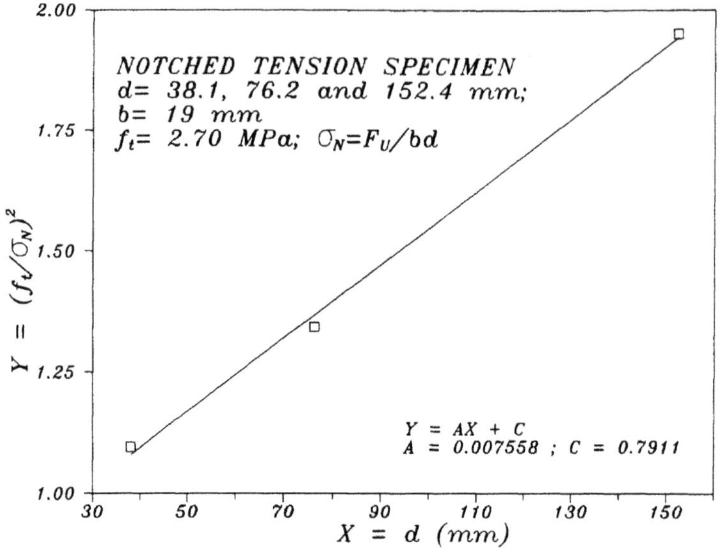

Fig. 5 Linear regression analysis of the calculated peak loads for the notched tension specimen.

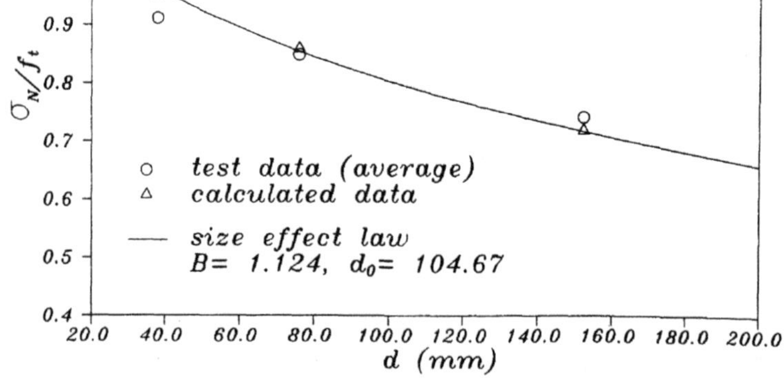

Fig. 6 Comparison between calculated and measured failure loads with size effect law for the notched tension specimen, shown in normal scale.

Fig. 3 that, relative to the specimen size, the fracture process zone decreases with increasing specimen depth. This is a consequence of the fact that the volume of the nonlocal continuum over which the strains are averaged is constant and therefore this volume is, relative to the specimen size, smaller if the size of the specimen is larger.

In Fig. 7 axial strain profiles across the symmetry line of the specimen at the start of the analysis and at peak load are plotted for all sizes. This figure clearly indicates that the strain distribution over the cross-section is more uniform if the size of the specimen is smaller. Therefore with decreasing depth, the stresses in the critical section are more uniformly distributed and the average stress increases.

Summarizing, the size effect can be explained by two effects: (1) The size of the fracture process zone relative to the specimen size decreases with increasing specimen depth; (2) because of (1) the strain and stress distribution becomes less uniform with increasing member depth, resulting in a decrease of the nominal stress at peak load.

Example (2)—The three-point bend specimen shown in Fig. 2(b) was tested by Bažant & Pfeiffer [19], using concrete with maximum

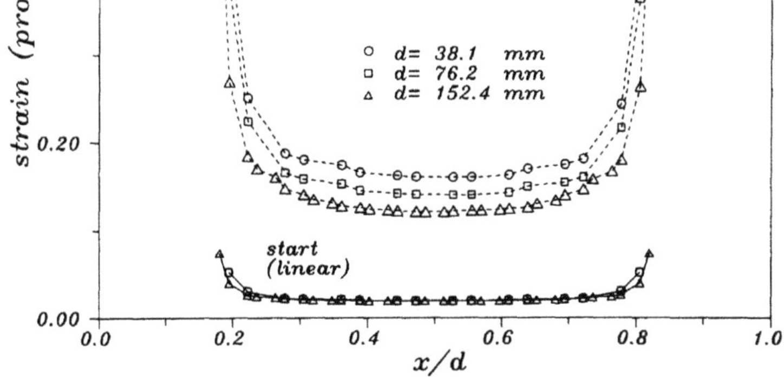

Fig. 7 Strain redistribution in the critical cross-section of the notched tension specimen.

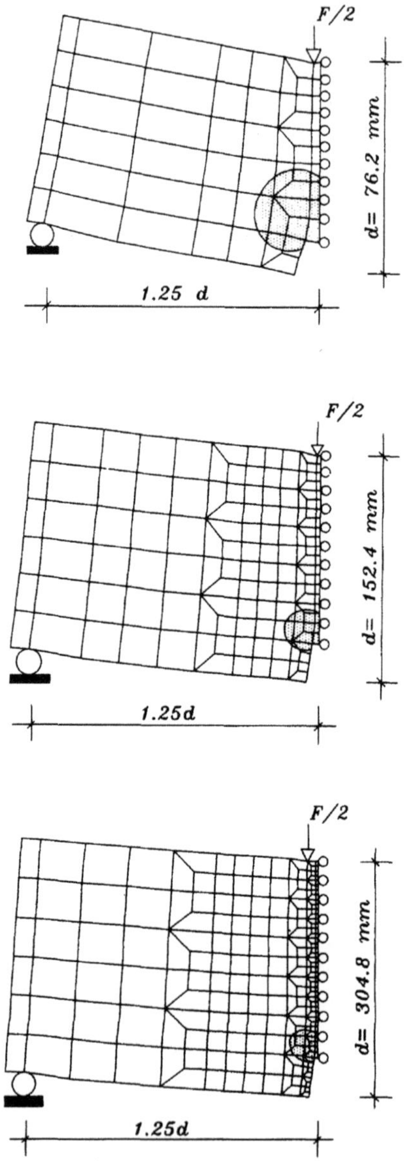

Fig. 8 Deformed finite element meshes and fracture process zone (shaded areas) at peak load for the three-point bend specimen.

aggregate size $d_a = 12{\cdot}7$ mm. The geometry of the specimens basically was the same as in the case of Example (1), except that the depth of the smallest specimen size was $d = 76{\cdot}2$ mm and the thickness of all specimens was $b = 38$ mm. In the analysis, again only one half of the specimen is modelled. The finite element meshes are shown in Fig. 8 in the deformed state. The characteristic length is taken as $l_c = 3d_a$, the microplane model parameters are chosen so that the tension strength is $f_t = 2{\cdot}74$ MPa. In the experiments the average estimated tension strength was $f_t = 2{\cdot}90$ MPa. The material parameters were taken such that the average failure load of the specimen with $d = 152{\cdot}4$ mm is matched.

In Figs 9–11 the nominal bending stresses at peak load according to the theory of elasticity, related to the total depth d, $\sigma_N = 15F_U/(4bd)$, obtained numerically and experimentally are compared with each other and with Bažant's size effect law. Again, calculated results and experimentally measured data exhibit a very strong size effect, well known for bending specimens [20], [21]. According to Fig. 11, the bending strength for a specimen with $d = 76{\cdot}2$ mm is $\sigma_N \simeq 1{\cdot}5f_t$. This relatively small bending strength is due to the notch, because the

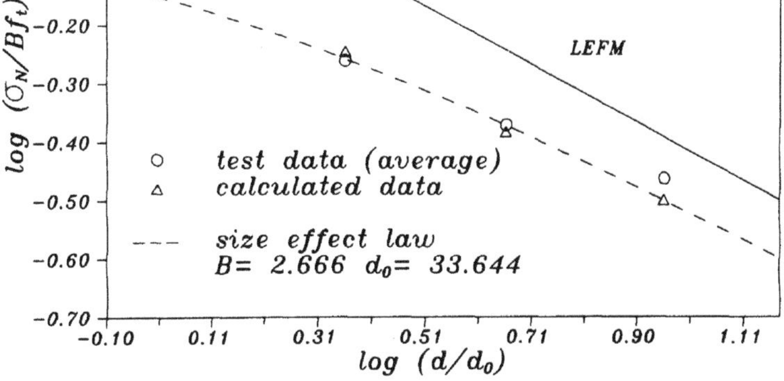

Fig. 9 Comparison between calculated and measured failure loads with size effect law for the three-point bend specimen.

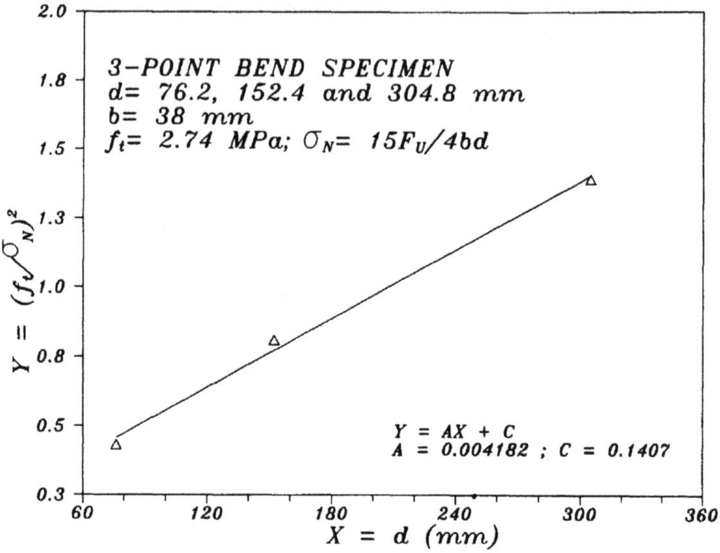

Fig. 10 Linear regression analysis of the calculated peak loads for the three-point bend specimen.

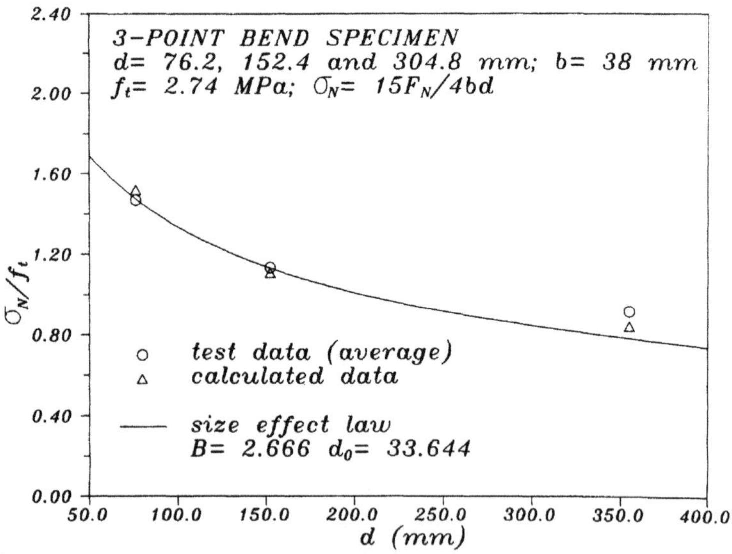

Fig. 11 Comparison between calculated and measured failure loads with size effect law for the three-point bend specimen, shown in normal scale.

strength related to the net area is $\sigma_N \simeq 2{\cdot}15 f_t$ which agrees with the value expected for unnotched beams.

The shaded areas in Fig. 8 indicate the size of the fracture process zones. As in the previous example, the relative size of the fracture process zone decreases with increasing member depth. When, in addition, the strain and stress distribution over the critical cross-section is analysed one comes to the same explanation for the size effect as in the case of the tension specimen. However, the size effect is much more pronounced than in notched tension specimens, because the size of the fracture process zone relative to the member depth is smaller.

In Fig. 12 the nominal bending strengths related to the value for $d = 100$ mm are plotted as a function of the member depth. The numerical values compared are calculated for the net member depth $d_1 = 5/6d$, with predictions according to different proposals valid for unnotched specimens. The test results by Heilmann [20] and Malcov & Karavaev [21], agree rather well. The bending strength decreases from $\sigma_N \simeq 2f_t$ for $d = 100$ mm to $\sigma_N \simeq 1{\cdot}1f_t$ for $d = 1000$ mm. According to the CEB Model Code [22] the bending strength is only $\sigma_N \simeq 1{\cdot}5f_t$ for $d = 100$ mm but approaches $\sigma_N = 1{\cdot}0f_t$ for larger specimens. The

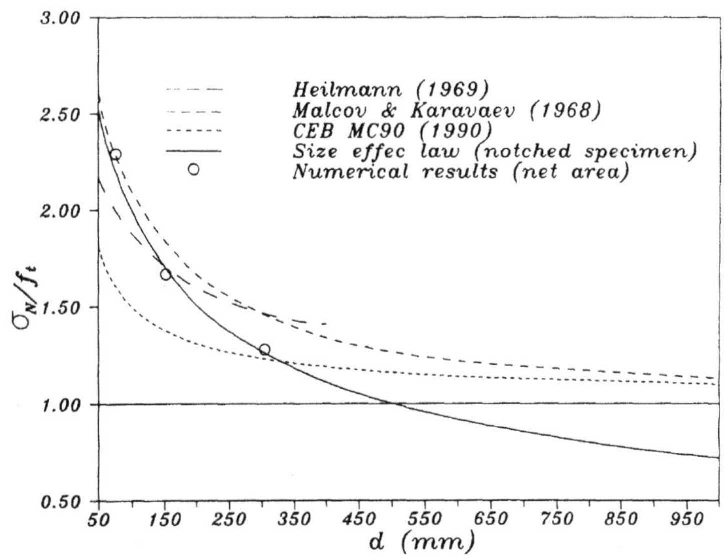

Fig. 12 Relative bending strength as a function of the member depth.

numerical results for notched specimens agree roughly with the other predictions; however, when extrapolating them by the size effect law to larger specimens the nominal bending strength is much lower than the centric tension strength. This is in contradiction to the experimental results for unnotched specimens. This is probably due to the fact that the size effect law was adjusted to fit the results of notched specimens. Therefore unnotched specimens of different sizes should be analysed and the resulting size effect law should be compared with test results.

According to the CEB Model Code 1990 (MC90) [22], the ultimate bending moment of large specimens ($d \lesssim 1$ m) increases in proportion to d^2. In contrast to this, the size effect law and linear elastic fracture mechanics predict an increase of M_u in proportion to $d^{1\cdot5}$. This means that the failure moment calculated according to MC90 might be unconservative for large specimens.

Example (3)—The concrete cone failure load of headed·anchors embedded in a large concrete block is studied. The geometry of the specimen is shown in Fig. 2(c). It is correlated with the embedment depth d. The smallest embedment depth is $d = 50$ mm. The distance between support and anchor is $3d$, so that an unrestricted formation of the failure cone is possible. The axisymmetric finite element mesh, shown in Fig. 13 (deformed shape), is constant in all analysed cases, i.e. the elements are scaled in proportion to d. Contact between anchor and concrete in the direction of loading exists under the head of the anchor only. To account for the restraining effect of the embedded anchor, the displacements of the concrete surface along the anchor in the vicinity of the head are fixed in the direction perpendicular to the load direction. Except at supports, all other nodes at the concrete surface are supposed to be free. Microplane model parameters are taken so that the calculated tension strength is approximately $f_t = 3$ MPa and the uniaxial compression strength is $f_c = 40$ MPa. The characteristic length of the nonlocal continuum is taken as $l_c = 12$ mm. Pulling out of the anchor is performed by prescribing displacements at the bottom of the head.

According to Eligehausen & Ožbolt [18], the concrete cone failure load can be calculated with Bažant's size effect law

$$F_U = F_N B(1+\beta)^{-1/2} \qquad \beta = d/d_0 \tag{2}$$

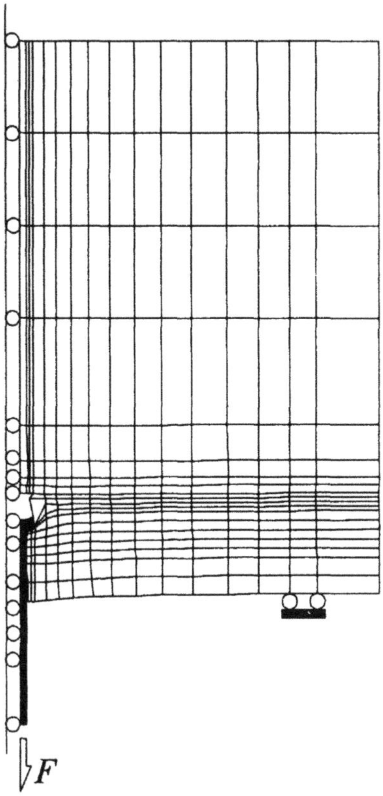

Fig. 13 Finite element mesh for the headed stud specimen, shown in deformed shape at peak load.

where F_U represents load at failure including size effect, F_N a failure load without size effect, and d is embedment depth. B and d_0 are again obtained using linear regression analysis of the numerical results (Fig. 14). F_N, the ultimate load with no size effect, is calculated using the formula

$$F_N = a\sqrt{f_c}\, d^2 \tag{3}$$

where f_c represents the concrete compression strength, a is a factor to calibrate calculated failure loads with measured values and to ensure the dimensional correctness of eqn (3). Equation (3) is proposed by ACI 349, Appendix B (1978) [23], for the prediction of the concrete cone failure load.

In Fig. 15 the results of the analysis are plotted and compared with the size effect law (eqn (2)). The coefficient a in eqn (3) is fixed such

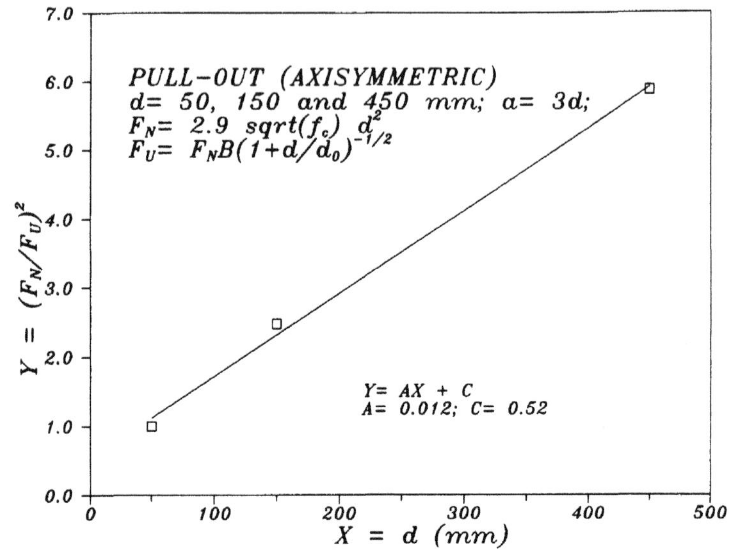

Fig. 14 Linear regression analysis of the calculated peak loads for the headed stud specimen.

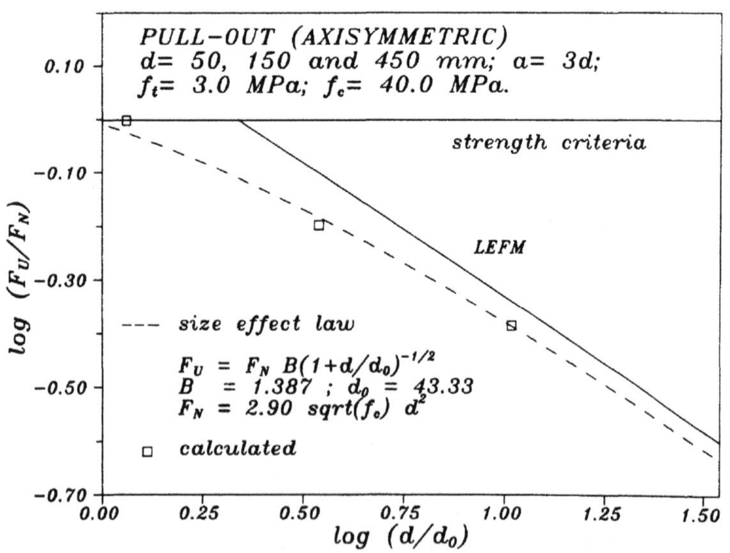

Fig. 15 Comparison between calculated peak loads and size effect law for the headed stud specimen.

that the numerically obtained failure load for anchors with an embedment depth $d = 50$ mm is predicted correctly. As can be seen from Fig. 15, the concrete cone failure loads exhibit a strong size effect, because the numerical results are close to the LEFM solution.

In Fig. 16 the results of the analysis are compared with different failure load equations. The relative failure loads are shown as a function of the embedment depth. The failure load for an embedment depth $d = 150$ mm is taken as the reference value. Plotted are the relative failure loads according to the size effect law (eqn (2)), a formula that neglects the size effect (eqn (3)), and a formula derived on the basis of linear elastic fracture mechanics (eqn (4)) [24]:

$$F_U = a_1\sqrt{EG_F}\, d^{3/2} \tag{4}$$

In eqn (4), a_1 is a constant and E is Young's modulus. The fracture loads predicted by eqn (4) agree rather well with test results [24]. Assuming no size effect, the failure loads should increase in proportion to d^2, that means by a factor of nine, when tripling the embedment depth. The results of the analysis show that the increase of the failure load is much less (approximately by a factor of 5·7). Therefore the size effect should be taken into account in the design of anchorages, otherwise the failure loads are underestimated for small embedment depths (Fig. 16(a)) and are overestimated for large embedment depths (Fig. 16(b)). The agreement between the size effect formula and the formula based on linear elastic fracture mechanics is good in the entire embedment range. This could be expected on the basis of Fig. 15. The size effect has also been observed in tests by Bode & Hanenkamp [25] and by Eligehausen *et al.* [26]. According to these authors, the failure load increases in proportion to $d^{1\cdot5}$.

The relative shapes of the fracture cone for three different embedment depths, estimated from the numerical analysis at peak load, are plotted in Fig. 17. In Fig. 18 the distribution of the tensile stresses perpendicular to the failure cone surface are shown as a function of the ratio l_h/l_{hmax} where l_h represents the distance from the anchor and l_{hmax} is the failure cone radius taken from Fig. 17. These distributions are estimated from the results of the numerical analysis.

From Figs 17 and 18 the size effect can be explained as follows. With increasing embedment depth the ratio of the diameter of the failure cone to embedment depth decreases, i.e. the effective relative cone surface area decreases as well. Furthermore, the average stress over the failure surface also decreases with increasing embedment depth

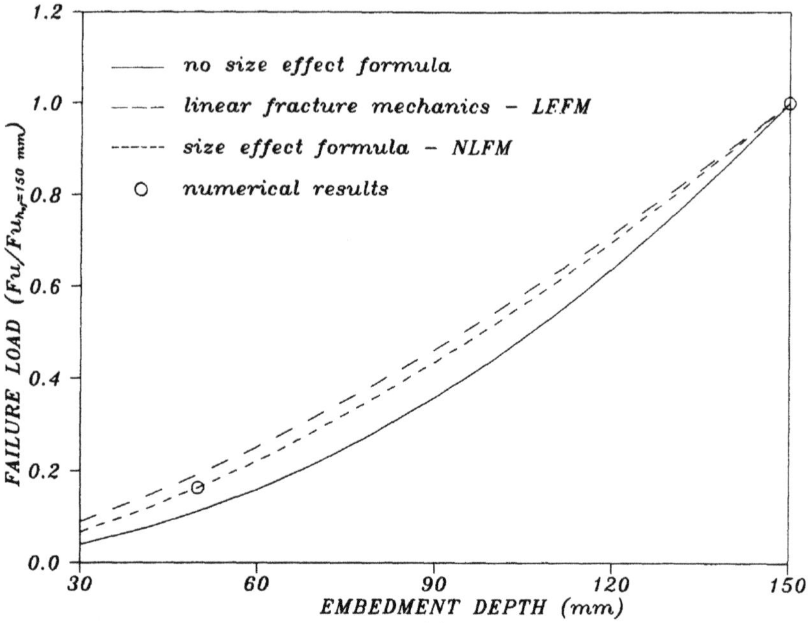

(a)

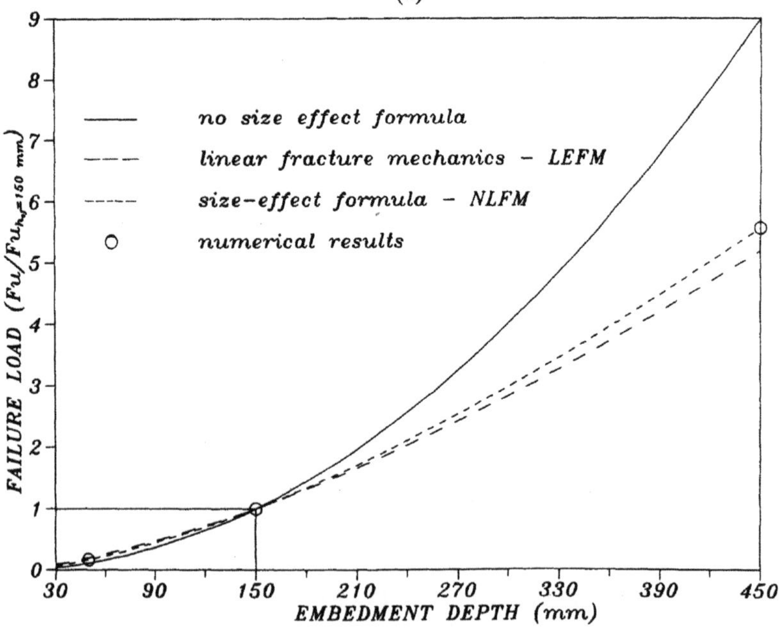

(b)

Fig. 16 Prediction of the failure loads for the headed stud specimen according to different proposals.

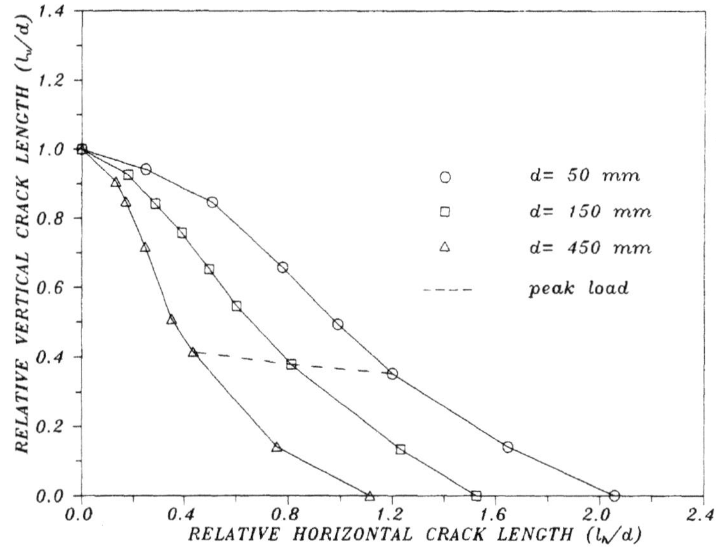

Fig. 17 Shape of the failure cone surface area in axisymmetrical pull-out.

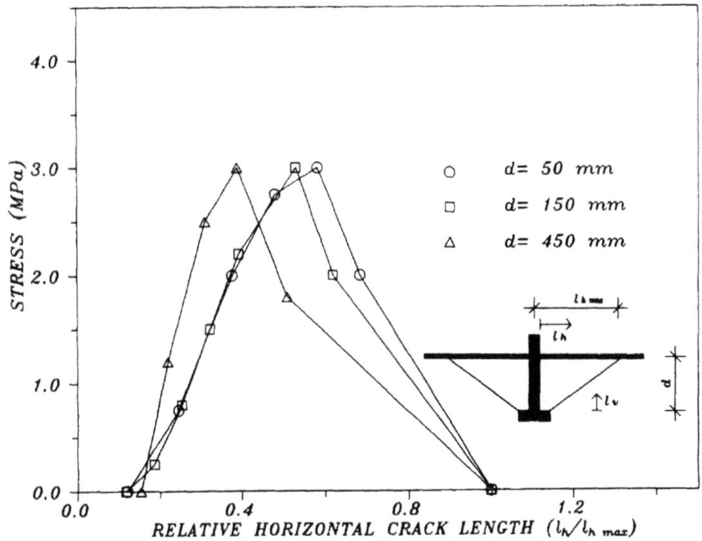

Fig. 18 Tensile stress distribution along the cone surface at peak load in axisymmetric pull-out.

because the stress distribution is more triangular as in the case of a large embedment depth and more parabolic in the case of smaller embedments.

Example (4)—The short beams loaded in torsion (Fig. 2(d)) were tested by Bažant *et al.* [27] using concrete with maximum aggregate size $d_a = 4{\cdot}8$ mm. The depth of the smallest specimen was $d = 38{\cdot}1$ mm. The finite element meshes are plotted in Fig. 19. The same mesh is used for the small and middle-sized specimens (72 finite elements), while for the largest specimen the number of finite elements is increased (176 finite elements). For the beam with $d = 38{\cdot}1$ mm, eight integration points are used in each finite element, while in the middle-sized and the largest specimen 27 integration points are used. To avoid localization due to concentrated loads imposed at the beginning and at the end of the specimen, the first and last cross-sections of the finite element mesh are supposed to behave linear elastically. The characteristic length is taken as $l_c = 15$ mm, the microplane model parameters are chosen so that the calculated tension strength is $f_t = 2{\cdot}60$ MPa and the uniaxial compression strength is

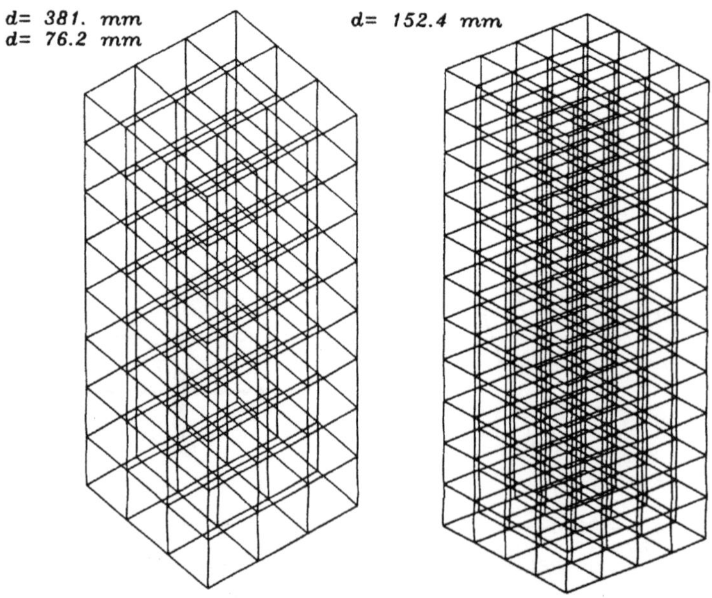

Fig. 19 Finite element meshes used in the analysis of the torsion specimen.

$f_c = 43$ MPa. The average estimated tension strength in the experiments was $f_t = 2{\cdot}70$ MPa. Material model parameters are obtained on the basis of fitting the average experimental failure load for the smallest specimen.

In Fig. 20 the nominal torsion stresses at peak load calculated on the basis of linear elastic theory, $\sigma_N = M_t/(0{\cdot}208d^3)$ with M_t = peak torsion moment, are compared with the average experimental values and the size effect law. The optimum values for the parameters B and d_0 are found by linear regression analysis of the numerical results (Fig. 21). Figure 22 represents a similar comparison in nonlogarithmic scale. As in the previous examples, experimentally and numerically obtained failure loads exhibit significant size effect.

To explain in detail the reason for the size effect in this complicated stress–strain state, further studies are required.

In the present numerical analysis and the tests, the concrete composition was constant. However, note that in practice the maximum aggregate size is not constant, and that for larger structures coarser aggregates are often used. In this case the size effect should be less pronounced than that found in this study. This can be seen from

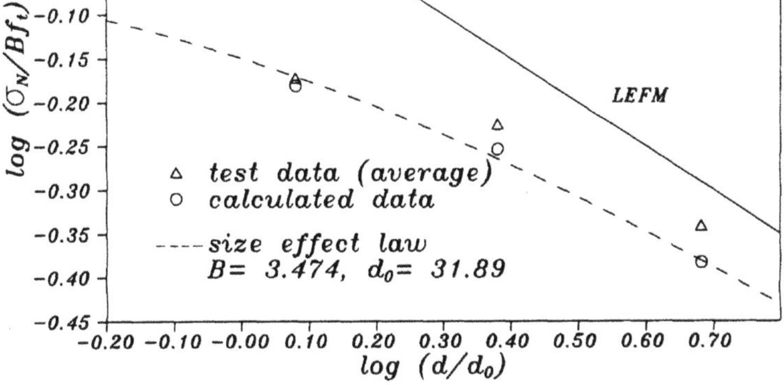

Fig. 20 Comparison between calculated and measured failure loads with size effect law for the beam loaded in torsion.

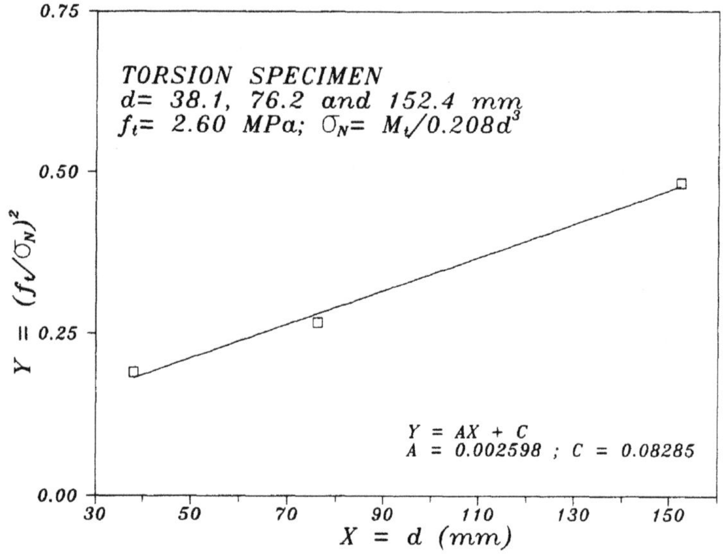

Fig. 21 Linear regression analysis of the calculated peak loads for the beam loaded in torsion.

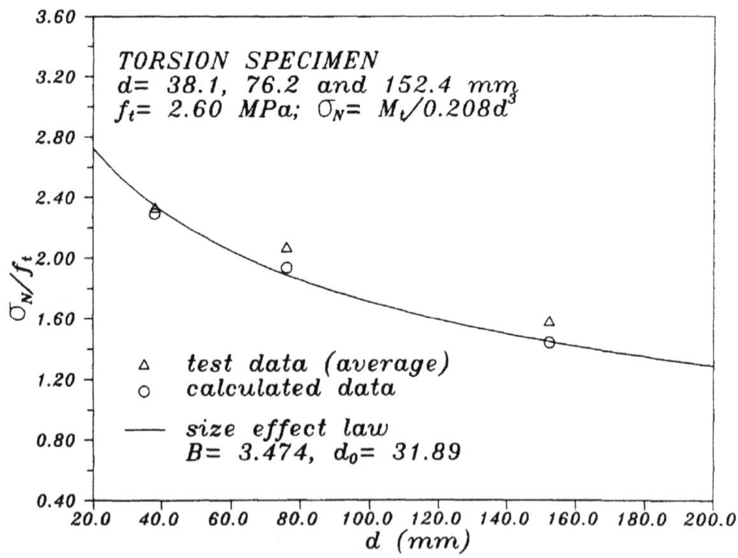

Fig. 22 Comparison between calculated and measured failure loads with size effect law for the beam loaded in torsion, shown in normal scale.

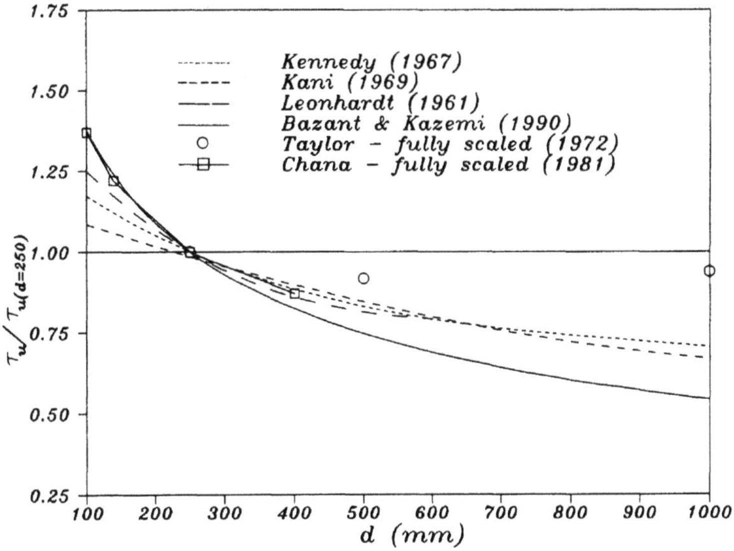

Fig. 23 Relative shear strength of beams without shear reinforcement as a function of the member depth.

Fig. 23 which shows the relative shear stresses at peak load (shear failure) of beams without shear reinforcement as a function of the member depth. The shear strength for slabs with $d = 250$ mm is taken as a reference value. In Fig. 23 test results of Leonhardt & Walter [28], Kani [1] and Kennedy [29] and the size effect law, as proposed by Bažant & Kazemi [30], are plotted. In these investigations the concrete mix was constant. As can be seen, the relative shear strength decreases significantly with increasing member depth. Taylor [31] tested fully scaled specimens that scaled all parameters, including the aggregate size. The shear strength did not decrease significantly with increasing specimen size. However, Chana [32] who also tested fully scaled specimens found that influence of the member depth on the shear strength was almost the same as in the investigations with constant concrete mix.

4. CONCLUSIONS

The results of the present numerical study on the behaviour of plain concrete structures under different loading conditions demonstrate that the peak loads exhibit a significant size effect. Therefore, the increase of the failure load is much less than the increase of the failure surface area. This is in accordance with experimental evidence. Similar results can be expected in other cases where the concrete tension strength plays a dominant role, such as a bond between deformed reinforcing bars and concrete, frame corners, punching, etc.

The analysis demonstrates that the microplane material model based on the nonlocal strain concept is capable of correctly predicting the behaviour of concrete structures in respect of fracture processes, peak load and size effect. Since the microplane model is a fully 3D material model it can be effectively used in 2D and 3D finite element codes. The fact that in the numerical analysis the size effect is calculated correctly is due to the nonlocal strain concept.

Bažant's size effect law or a suitably simplified formula can predict size effect rather well in a small range of dimensions. But to check this law in a broader range, tests of very large structures are required.

Further studies are needed to clarify the influence of the concrete mix on the size effect. Furthermore, design provisions should be evaluated, which take the practical conditions into account, and which should be incorporated in codes.

The size effect in concrete structures is significant and should be taken into account in the design codes.

REFERENCES

[1] Kani, G. N., How safe are our large concrete beams? *ACI Journal*, Proceedings, **64** (1967) 128–41.

[2] Bažant, Z. P., Size effect in blunt fracture: Concrete, rock, metal. *J. Eng. Mechanics* (ASCE), **110**(4) (1984) 518–35.

[3] Taylor, G. I., Plastic strain in metals. *J. Inst. Metals*, **62** (1983) 307–24.

[4] Batdorf, S. B. & Budianski, B., A Mathematical Theory of Plasticity Based on the Concept of Slip. NACA TN1871, April, 1949.

[5] Zienkiewicz, O. C. & Pande, G. N., Time-dependent multi-laminate model of rocks—a numerical study of deformation and failure of rock masses. *Int. J. Num. Anal. Meth. in Geomechanics*, **1** (1977) 219–47.

[6] Bažant, Z. P. & Gambarova, P. G., Crack shear in concrete: crack band microplane model. *J. Struc. Eng.* (ASCE), **110**(10) (1984) 2015–35.

[7] Bažant, Z. P., Microplane model for strain-controlled inelastic behaviour. In *Mechanics of Engineering Materials*, ed. C. S. Desai & R. H. Gallager. John Wiley & Sons, Chichester and New York, 1984, Chap. 4, pp. 45–59.

[8] Bažant, Z. P. & Oh, B.-H., Microplane model for progressive fracture of concrete and rock. *J. Eng. Mechanics* (ASCE), **111**(4) (1985) 559–82.
[9] Bažant, Z. P. & Prat, P. C., Microplane model for brittle–plastic material—Parts I and II. *J. Eng. Mechanics* (ASCE), **114**(10) (1988) 1672–1702.
[10] Bažant, Z. P. & Pijaudier-Cabot, G., Nonlocal continuum damage, localization instability and convergence. *J. Applied Mechanics* (ASME), **55** (1988) 287–93.
[11] Kröner, E., Interrelations between various branches of continuum mechanics. In *Mech. of Generalized Continua*, ed. E. Kröner. Springer, W. Berlin, 1968, pp. 330–40.
[12] Eringen, A. C. & Edelen, D. G. D., On nonlocal elasticity. *Int. J. Eng. Sci.*, **10** (1972) 233–48.
[13] Krumhansl, J. A., Some considerations of the relations between solid state physics and generalised continuum mechanics. In *Mech. of Generalized Continua*, ed. E. Kröner. Springer, W. Berlin, 1968, pp. 298–331.
[14] Levin, V. M., The relation between mathematical expectation of stress and strain tensor in elastic micro-heterogeneous media. *Prikl. Mat. Mehk.*, **35** (1971) 694–701 (in Russian).
[15] Bažant, Z. P., Belytschko, T. B. & Chang, T. P., Continuum model for strain softening. *J. Eng. Mechanics* (ASCE), **110**(12) (1984) 1666–92.
[16] Bažant, Z. P. & Pijaudier-Cabot, G., Measurement of characteristic length of nonlocal continuum. *J. Eng. Mechanics* (ASCE), **115**(4) (1989) 755–67.
[17] Bažant, Z. P. & Ožbolt, J., Nonlocal microplane model for fracture, damage and size effect in structures. Report 89-10/498n Center for Concrete and Geomaterials, Northwestern University, Evanston, 1989, 33 pp. Also *J. Eng. Mechanics* (ASCE), (in press).
[18] Eligehausen, R. & Ožbolt, J., Size effect in anchorage behaviour. Paper presented at Proceedings of the Eighth European Conference on Fracture—Fracture Behaviour and Design of Materials and Structures, Torino, Italy, 1–5 October 1990.
[19] Bažant, Z. P. & Pfeiffer, P. A., Determination of fracture energy from size effect and brittleness number. *ACI Materials Journal*, **84** (1987) 463–80.
[20] Heilmann, H. G., Beziehungen zwischen Zug—und Druckfestigkeit des Betons. *Beton*, **2** (1969) 68–72 (in German).
[21] Malkov, K. & Karavaev, A., Abhängigkeit der Festigkeit des Betons auf Zug bei Biegung und ausmittiger Belastung von den Querschnittsabemssungen. *Wissenschaftliche Zeitschrift der Technischen Universität Dresden*, **17**(6) (1968) 1545–7.
[22] Comité Euro-International du Beton, CEB-FIP Model Code 1990, First Draft. *Bulletin d'Information Nos 195 and 196*, CEB, Lausanne, March, 1990.
[23] ACI 349-76: Code Requirements for Nuclear Safety Related Concrete Structures. *ACI Journal*, **75** (1978) 329–47.
[24] Eligehausen, R. & Sawade, G., A fracture mechanics based description of the pull-out behavior of headed studs embedded in concrete. *Fracture*

Mechanics of Concrete Structures—RILEM Report, ed. L. Elfgren. Chapman and Hall, London, 1989, pp. 281–99.

[25] Bode, H. & Hanenkamp, W., Zur Tragfähigkeit von Kopbolzen bei Zugbeanspruchung. *Bauingenier,* **60** (1985) 361–7 (in German).

[26] Eligehausen, R., Fuchs, W. & Mayer, B., Tagverhalten von Dübelbefestigungen bei Zugbeanspruchung. *Betonwerk + Fertigteil—Technik,* **12** (1987) 826–32, and **1** (1988) 29–35, (in German and English).

[27] Bažant, Z. P., Sener S. & Prat, P., Size effect test of torsional failure of plain and reinforced concrete beams. *Materials and Structures,* **21** (1980) 425–30.

[28] Leonhardt, F. & Walter, R., Beiträge zur Behandlung der Schubprobleme im Stahlbetonbau, *Beton und Stahlbeton* (Berlin), **56** (12) (1961) and **57** (2, 3, 6, 7, 8) (1962) (in German). The Stuttgart Shear Tests 1961. C. & C. A. Library Translation No. 111, Cement and Concrete Association, London, UK.

[29] Kennedy, R. P., A Statistical Analysis of the Shear Strength of Reinforced Concrete Beams. Technical Report No. 78, Department of Civil Engineering, Stanford University, USA, April 1967.

[30] Bažant, Z. P. & Kazemi, M. T., Size Effect on Diagonal Shear Failure of Beams Without Stirrups. Internal Report, Center for Advanced Cement-Based Materials, Northwestern University, Evanston, USA, 1990.

[31] Taylor, H. P. J., Shear strength of large concrete beams. *J. Struct. Div. Proc. ASCE,* **98** (1972) 2473–91.

[32] Chana, P. S., Some aspects of modelling the behaviour of reinforced concrete under shear loading. Cement and Concrete Association, London, July 1981.

Chapter 3

Stress-Crack Opening Relation and Size Effect in Concrete

H. Duda & G. König

Koenig und Heunisch, Beratende Ingenieure, Oskar-Somner-Str. 15–17, D 6000 Frankfurt, Germany

ABSTRACT

In the first part of this chapter the size effect in concrete is studied by means of finite element analysis. The rheological material model developed at TH Darmstadt was used for a parameter study. The strength of the concrete and the shape of the stress-crack opening relation was varied. The increasing flexural strength of concrete with decreasing depth of the beams is described. It is caused by the strain-softening behaviour of concrete. The effect of a normal force acting simultaneously with a bending moment is also taken into account. The outcome is that size effects also depend on eccentricity and material properties. In the second part the rheological material model was used to analyse three-point bending specimens of four different sizes. The measured crack-mouth-opening curve was used to recalculate the stress-crack opening relation (σ–w relation). It is shown that the same σ–w relation is able to describe the specimen behaviour independent of the size. With the unique σ–w relation found for the concrete used, the peak loads of the whole range between very small and very large specimens were calculated and compared with a yield criterion and with linear elastic fracture mechanics.

NOTATION

A_0	Net cross-section
b	Specimen width

CMOD	Crack mouth opening displacement
d	Specimen depth
d_{max}	Maximum aggregate size
e	Eccentricity
E_c	Modulus of elasticity
f_{ct}	Tensile strength
f'_{c14}	Average 14-day compression strength
$f(\alpha)$	Specimen geometry function
f_{cf}	Flexural tensile strength
G_F	Fracture energy
K_{Ic}	Critical stress intensity factor
l	Specimen length
l_{ch}	Characteristic length
M_u	Ultimate bending moment
N_u	Ultimate normal force
P_u	Peak load
w	Crack width
w_a	Parameter of the primary bearing mechanism
w_b	Parameter of the secondary bearing
W_0	Area under the load–deflection curve
α	Relative notch length
σ	Stress
σ_a	Parameter of the primary bearing mechanism
σ_b	Parameter of the secondary bearing
σ_{ct}	Tensile stress dependent on w
σ_n	Nominal peak stress

1. THE SIZE EFFECT—A FINITE ELEMENT STUDY

1.1 Introduction

The idiom 'Size Effect' is used to describe two totally different effects. First, the effect that the element size of finite-element-nets may have on the results of the calculation. Secondly, the effect that specimens of different sizes, though geometrically similar, do not behave similarly.

1.2 The Element Size Effect

The post-failure behaviour of tensile loaded concrete is governed by the development of a local crack or crack-band. Transmission of stresses over the crack is possible due to aggregate interlock. The relation between stress and crack width is the *σ–w relation*. Using a stress–strain relation (*σ–ε relation*) instead of the stress–crack width relation, the crack width has to be distributed continuously over a certain length, which in most cases is the element size. The additional strain due to the crack, which is the result of dividing the crack width by the element width, depends, therefore, on the element width.

If a FE-program ignores this dependence and uses stress–strain relations for strain-softening, independent of the element size, the results of the calculation will depend on the element size.

Using 'reasonable' element sizes and stress–strain relations the results will be independent of the element size. Problems with the definition of 'reasonable' will occur when the crack is not in parallel with the element borders. This is true also in the case where the locality of cracking is unknown or free or if more than one crack may occur. In all these cases a FE-calculation may be element size dependent, but these element size effects are mere mathematical problems and not material behaviour.

1.3 The Specimen Size Effect

If geometrically similar specimens do not behave similarly for different sizes, this is called a size effect. Size effects occur anywhere in concrete. They are most significant in cases of tensile or shear load and less significant in the case of compression. A very well-known size effect is the dependence of the flexural strength of plain concrete on the depth of the beam.

The increase in flexural strength is caused by the strain-softening behaviour of concrete, which is the relation between the crack width and the transmissible stress (*σ–w* relation). A finite-element study is presented where the dependence of the flexural strength on the depth, the eccentricity, and the *σ–w* relation is analysed. A four-point bending beam was used as shown in Fig. 1. The crack was described by contact elements. The depth *d* was varied between 5 cm and 2 m. The other dimensions were varied proportionally.

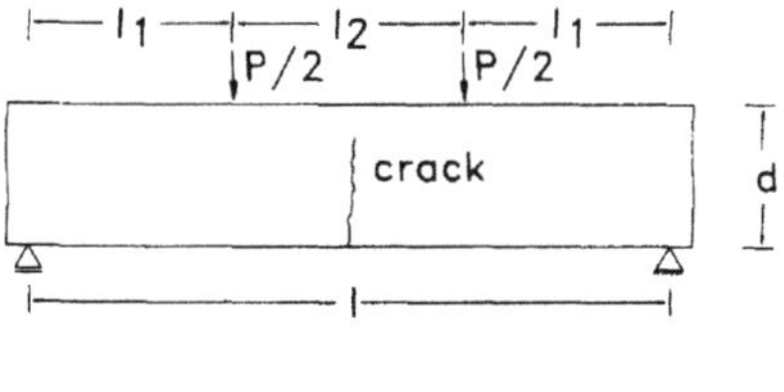

Fig. 1 Four-point bending beam.

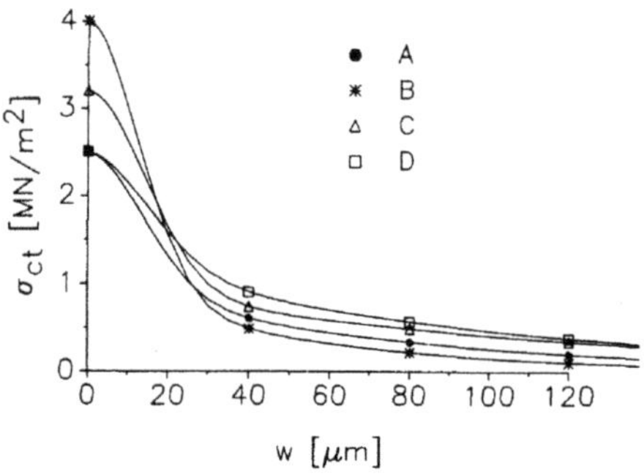

Fig. 2 σ–w relations A–D.

Four different σ–w relations were assumed. They are shown in Fig. 2. The envelope of the σ–w relation is described by eqn (1) (see also [1]).

$$\sigma_{ct}(w) = \sigma_a e^{-(w/w_a)^2} + \sigma_b e^{-w/w_b} \tag{1}$$

The parameters of the σ–w relation are listed in Table 1.

As Fig. 2 shows four different relations were used. The fracture energy of type A and B is $G_F \approx 100$ N/m, and of type C and D, $G_F \approx 150$ N/m. The fracture energy is defined as the area under the σ–w relation:

$$G_F = \int_0^\infty \sigma_{ct}(w)\, dw \tag{2}$$

The tensile strength of type A and C is $f_{ct} = 2{\cdot}5$ N/mm^2. The characteristic length of type A and C is $l_{ch} \approx 480$ mm. The characteris-

TABLE 1 Parameters of the σ–w relations

	Unit	*Type A*	*Type B*	*Type C*	*Type D*
f_{ct}	MN/m^2	2·5	4·0	3·2	2·50
G_F	N/m	100	100	150	150
E_c	MN/m^2	30 000	30 000	33 000	33 000
l_{ch}	mm	480	187	483	792
σ_a	MN/m^2	1·5	3·0	2·2	1·2
w_a	μm	20	18	20	22
σ_b	MN/m^2	1·0	1·0	1·0	1·3
w_b	μm	73	52	111	97

tic length is defined as (see also [2]):

$$l_{ch} = \frac{E_c G_F}{f_{ct}^2} \tag{3}$$

The parameters of the four σ–w relations are not chosen arbitrarily. Type A is a normal concrete (approximately $C25$). The aggregate is gravel with top size diameter of aggregate $d_{max} = 8$ mm. Type B is a high-strength concrete (approximately $C80$) with crushed limestone as aggregate. Type C is a normal concrete (approximately $C35$). The aggregate is gravel ($d_{max} = 32$ mm). Type D is $C25$ with crushed basalt ($d_{max} = 16$ mm). These four types of concrete represent the concrete most commonly used.

The flexural tensile strength of concrete f_{cf} is defined as the ratio between the ultimate bending moment M_u and the moment of resistance.

$$f_{cf} = \frac{M_u 6}{bd^2} \tag{4}$$

The ratio of flexural tensile strength f_{cf} to axial tensile strength f_{ct} is plotted in Fig. 3 over the depth of the beam. It is obvious that the increase in flexural strength depends on the depth of the beam and on the kind of σ–w relation.

A normal force acting simultaneously with a bending moment influences the increase in flexural strength. In Fig. 4 the relative flexural strength is plotted over the depth d for different relative eccentricities e/d. The flexural strength for normal force N and bending moment $M_u = eN_u$ is defined as:

$$f_{cf} = \frac{N_u}{bd}\left(\frac{e6}{d} + 1\right) \tag{5}$$

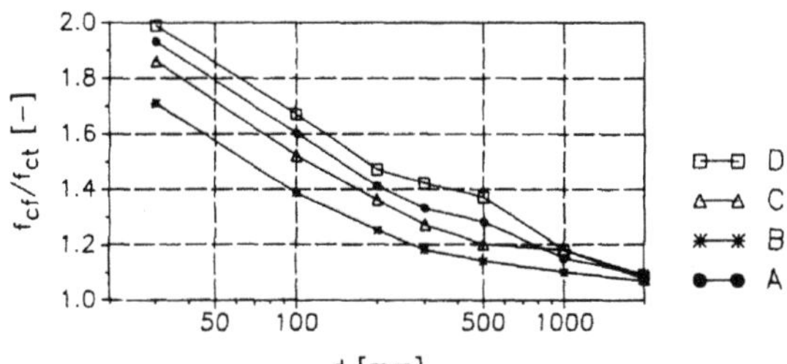

Fig. 3 Relative flexural strength of concrete under pure bending.

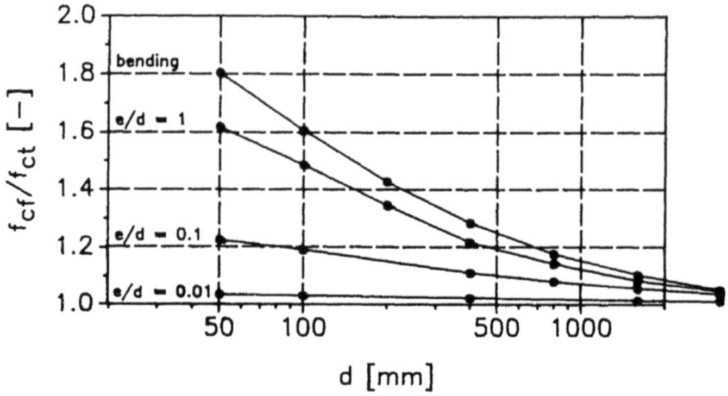

Fig. 4 Relative flexural strength of concrete under bending and normal force.

The increase of flexural strength occurs due to the redistribution of the tensile stresses in concrete. The stress distribution under ultimate bending moment is plotted in Fig. 5 for the depths $d = 10\,\text{cm}$, $d = 40\,\text{cm}$ and $d \rightarrow \infty$.

The extension of the redistribution zone depends on the σ–w relation. The relative size of this zone is small for large beam depth and for small eccentricities. The critical cases where no redistribution occurs are:

- large (infinity) beam depth under bending;
- axial tensile load ($e = 0$);
- no transmission of stresses through the crack.

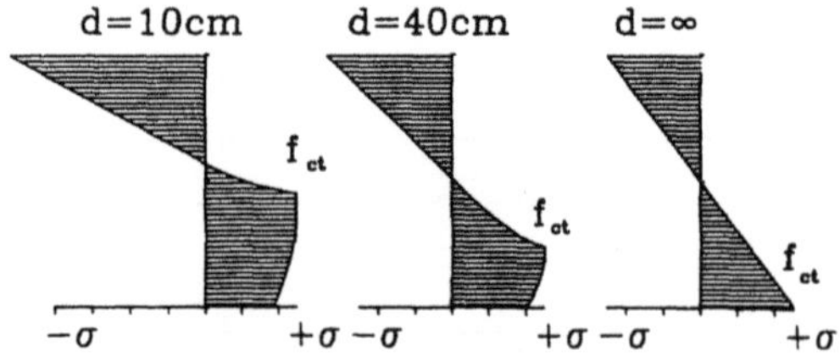

Fig. 5 Stress distribution for different beam depths.

2. STRESS-CRACK OPENING RELATION AND SIZE EFFECT IN HIGH STRENGTH CONCRETE

2.1 Introduction

The tests analysed in this part were carried out at Northwestern University, Evanston, by R. Gettu *et al.* [3]. Geometrically similar three-point bending specimens of four different sizes were tested.

Concretes of strengths exceeding 80 MPa are now commonly used in the construction of high-rise buildings and offshore structures. The utilization of concrete of such high strength has been spurred on by the superior mechanical properties of the material and the cost-effectiveness it provides. Typical high strength concrete has a very high strength matrix, is more compact and possesses well-bonded aggregate mortar interfaces.

In the past, research on high strength concrete has primarily concentrated on increasing the strength of the material. In the last decade, however, considerable efforts have been spent in studying its mechanical properties and structural behaviour. Nevertheless many aspects such as the fracture behaviour need much more detailed investigations.

In most cases concrete exhibits a behaviour which is somewhere in between ductile and brittle. The fracture process is distributed over a certain zone. The size of the zone depends on the geometry of the specimen, the loading conditions and the material behaviour. If the size of this zone is of the same order as the specimen size the specimen behaviour approaches the yield criteria; but this is only true for very small specimens—in the case of pure bending, depth less than $\approx$5 cm. If the size of the fracture zone is very small compared with the specimen size the specimen behaviour approaches the criteria of linear elastic fracture mechanic (LEFM). This is true for very large specimens, in the case as above—depth greater than $\approx$1–5 m. The upper and lower boundary value shows that for the dimensions of most common structural members neither a yield criteria nor LEFM can be used to describe the behaviour. Nonlinear models like cohesive crack models, e.g. the rheological TH Darmstadt model, the size effect law introduced by Bažant or extensions of LEFM have to be used to achieve a reasonable description of specimen behaviour.

TABLE 2 Concrete composition for about 0·8 m^3 (1 yd^3)

Dundee cement (ASTM Type I)	363 kg
Sand (FA2)	490 kg
Gravel (CA15) (crushed limestone)	880 kg
Water	127 kg
Fly ash (Class C)	91 kg
Micro-silica	16 kg
Retarder (naphthalene-based)	1 litre
High-range water reducer	5·3 litres

2.2 Experimental Investigation

In order to study typical high strength concrete used in the industry, the investigation was conducted on material obtained directly from a batch mixed for the construction of a high rise building in downtown Chicago. The concrete mix was designed to exceed a 28-day compressive strength of 83 MPa. Silica fume and fly ash were used as mineral admixtures. The maximum aggregate size (d_{max}) was 9·5 mm. The details of the mix composition are given in Table 2. The average 14-day compression strength (f'_{c14}) (cylinder 102 mm × 204 mm) was 85·5 MPa. Geometrically similar beams of four different sizes were cast from the same batch of concrete. The geometry is shown in Fig. 6. All specimens were 38·1 mm thick. The depths were $d = 304{\cdot}8$, 152·4, 76·2 and 38·1 mm. For more details see [3].

The peak load of the largest specimens were measured by loading them under stroke control in a 534 kN load frame. The other specimens were tested in an 89 kN load frame under crack mouth

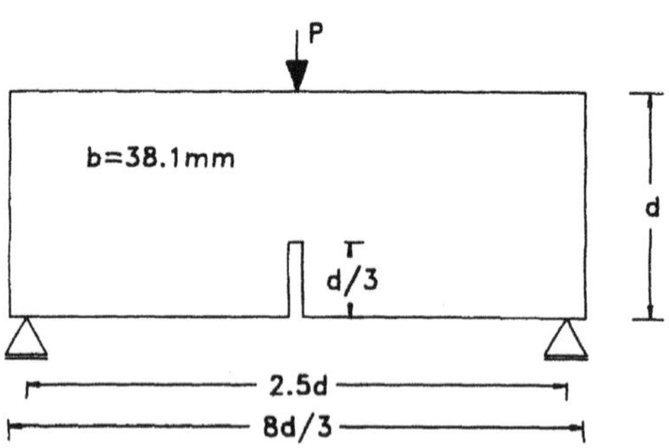

Fig. 6 Geometry of the three-point bending specimen.

opening (CMOD) control. For these specimens the complete load–deflection and load–CMOD curves were recorded. The load-point displacements were measured between the tension faces of the beams and the cross-head of the loading frame, i.e. the 'deflection' is the real beam deflection, the deformation of parts of the load frame and most likely some plastic deformation at the supports.

2.3 Review of the Rheological Material Model

The material model allows a realistic description of the complex behaviour of concrete loaded by tension. The model consists of a parallel arrangement of springs and friction blocks. It describes the stress-crack opening relation (σ–w relation) for monotonic and cyclic loading. In general a σ–w relation can be divided into the first steep branch and the second considerably flatter branch. The first branch may be interpreted as failure of the primary bearing mechanism, that is the bond between grain and matrix: the second branch as secondary bearing mechanism, that is friction between grain and matrix after cracking. For the simple case of monotonic loading both branches can be described by exponential functions:

$$\sigma_{ct}(w) = \sigma_a e^{-(w/w_a)^2} + \sigma_b e^{-w/w_b} \tag{6}$$

where σ_{ct} = tensile stress dependent on w
w = crack width
w_a; σ_a = parameters of the primary bearing mechanism
w_b; σ_b = parameters of the secondary bearing mechanism
$\sigma_a + \sigma_b = f_{ct}$ = tensile strength.

The fracture energy G_F is defined as the area under the σ–w relation. This leads to

$$G_F = \int_0^\infty \sigma_{ct}(w)\,dw = \sigma_a w_a \sqrt{\pi}/2 + \sigma_b w_b \tag{7}$$

2.4 Analysis of Test Results

The test results were analysed carrying out a recalculation procedure using the finite element code SNAP [4]. Due to symmetry only half of the beam has to be modelled (see Fig. 7). In the symmetry-axis, contact elements were used to model the σ–w relation (eqn (6)).

The recalculation was conducted using the whole measured load–crack-mouth-opening curve (load–CMOD curve). The modulus of

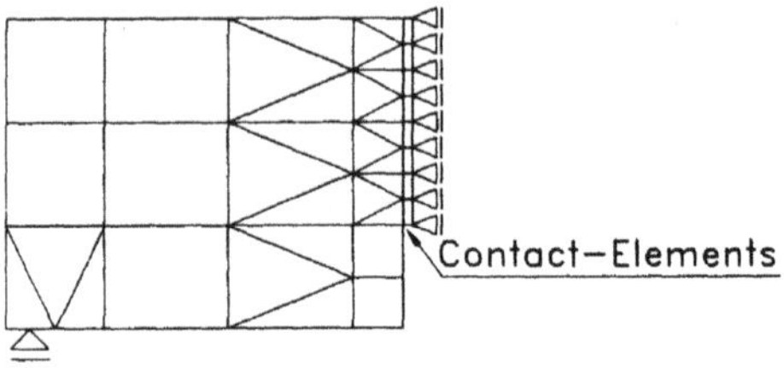

Fig. 7 Finite element net.

elasticity (E_c) was calculated by the initial slope of the load–CMOD curve. In the first step values for the four parameters of the monotonic σ–w relation were assumed. Within the next steps the parameters were varied until a reasonable fit between the calculated and measured load–CMOD curve was reached. The measured and calculated load–CMOD curves are plotted in Figs 8, 9 and 10.

As the differences between analysis and test are of the same order as the scatter within the test results the parameters used can be regarded as the average parameters of the concrete used. The descending branches of the largest specimens (Fig. 10) exhibit a big scatter. This is most likely due to out-of-plane bending which could not be prevented for a specimen with a depth/width ratio of 4/1. Out-of-plane bending was not taken into consideration recalculating the test.

The result of the recalculation was a unique σ–w relation for the three specimen sizes. The σ–w relation is plotted in Fig. 11, the parameters are given in Table 3, column 3.

The σ–w relation of the HSC can be compared with the relation of

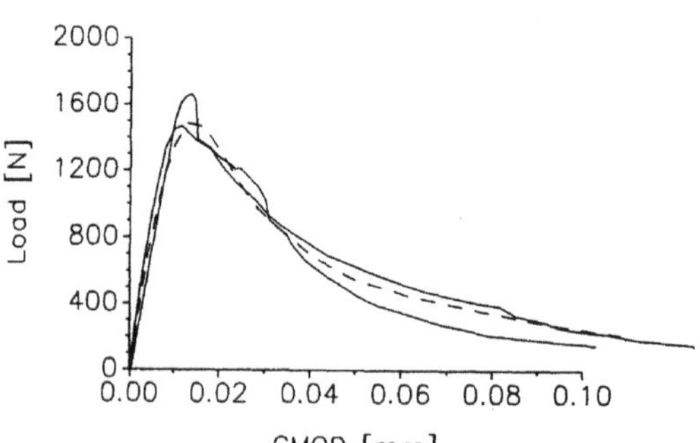

Fig. 8 Load–CMOD curves, $d = 38{\cdot}1$ mm.

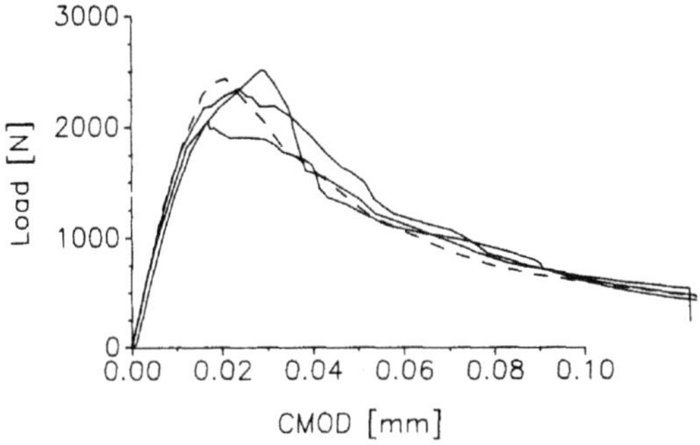

Fig. 9 Load–CMOD curves, $d = 76{\cdot}2$ mm.

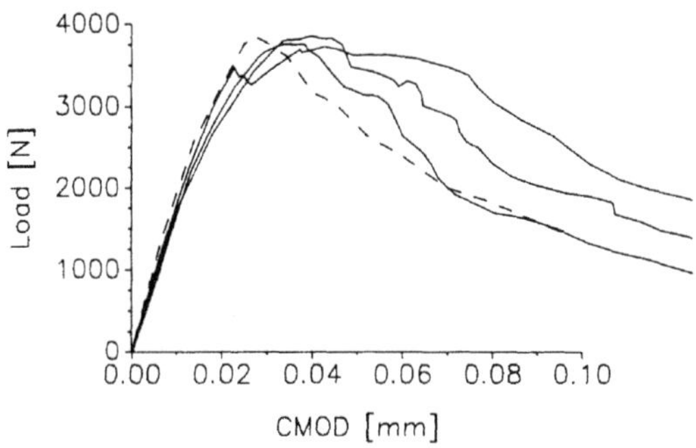

Fig. 10 Load–CMOD curves, $d = 152{\cdot}4$ mm.

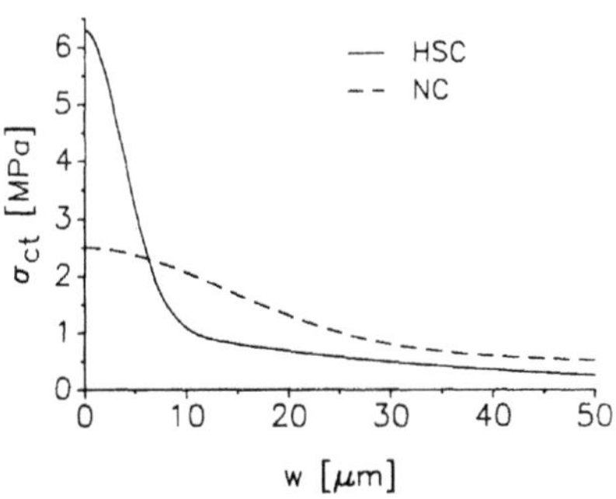

Fig. 11 σ–w relation of the HSC used.

TABLE 3 Parameters of the σ–w relation

1	2	3 HSC	4 NC
E_c	MN/m^2	43 000	
G_F	N/m	62·5	102
σ_a	MN/m^2	5·0	1·8
w_a	μm	5·3	20
σ_b	MN/m^2	1·3	0·7
w_b	μm	30	100

an average normal concrete. For this comparison, test results of a *C*25, (aggregate: water worn gravel; $d_{max} = 16$ mm) was chosen as 'normal concrete'. The σ–w relation of this concrete is plotted in Fig. 11 as a dashed line. The parameters are given in Table 3, column 4. The major difference between high and normal strength concrete is due to the primary bearing mechanism. HSC starts with high tensile strength, but the decay in bearing capacity is very steep. In this particular case already after a crack opening of 6·5 μm the bearing capacity of normal concrete exceeds the bearing capacity of HSC. In this particular case—in spite of the higher strength—HSC produces less fracture energy than normal concrete. This demonstrates that a concrete should not be judged by its strength only. The brittle response of HSC is caused by the very steep decay in bearing capacity after the first cracking. Due to this only minor stress redistribution is possible.

Figure 12 shows the calculated and measured mid span deflection of the specimens with $d = 152{\cdot}4$ mm. Obviously there is a big difference

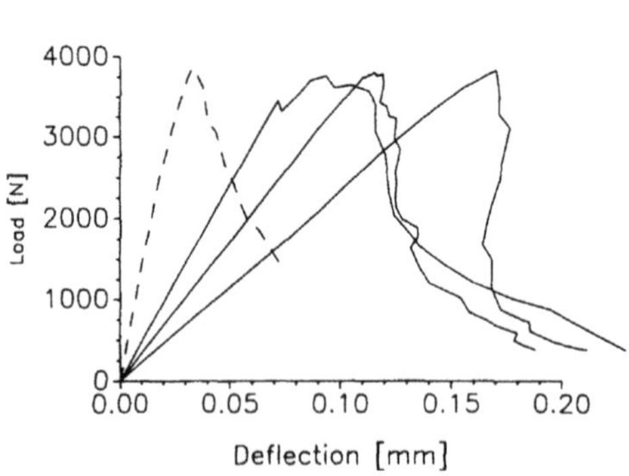

Fig. 12 Measured and calculated mid-span deflection.

between the different experimental measurements and between the calculation and the measurements. From the initial slope—which is independent of the chosen σ–w relation—it is indisputable that a major part of the measured deformation is the deformation of the load frame and/or plastic deformation under the supports. Therefore use of this measurement is very doubtful without knowledge about the two influencing phenomena. This consideration may also explain the difference in the fracture energy found from the integration of the σ–w relation (see Table 3) and the value calculated from the area under the load–deflection curve (see Table 4). As the elastic deformation of the load frame does not influence the area under the load–deflection curve it seems that there are some other energy consuming deformations outside the fracture zone.

2.5 The Size Effect

The size effect which was studied in this research is the dependency of the nominal peak stress σ_n on the structural size. The nominal stress is defined as the ratio between peak load and cross-section ($\sigma_n =$

TABLE 4 Summary of test results

Specimen dimensions[a] b/d [mm]	*Peak load* [N]	$G_F = W_0/A_0$ [N/m]
38·1/101·6	1 668	n.r.
	1 490	76·5
	1 735	n.r.
76·2/203·2	2 357	86·3
	2 091	73·9
	2 535	73·4
152·4/406·4	3 825	93·5
	3 736	107·9
	3 825	84·2
304·8/812·8	7 072	n.m.
	7 442	n.m.
	7 695	n.m.

[a] All specimens 38·1 mm wide.
W_0 Area under the load–deflection curve.
A_0 Net cross-section.
n.r. Values not recorded due to failure of instruments.
n.m. Values not measured.

$P_u/(bd)$). Using LEFM this dependency can be written as (see [3])

$$\sigma_n = \frac{P_u}{bd} = \frac{K_{Ic}}{\sqrt{d}f(\alpha)} \tag{8}$$

where P_u = peak load
b = width
d = depth
K_{Ic} = critical stress intensity factor
α = relative notch length
$f(\alpha)$ = specimen geometry function.

The critical stress intensity factor can be calculated from the fracture energy G_F and the modulus of elasticity E_c as $K_{Ic} = \sqrt{G_F E_c}$. With the values given in Table 3 this leads to $K_{Ic} = 51{\cdot}8\,\text{MPa}\sqrt{\text{mm}}$. For the specimen geometry used, the geometry function is given in [3] as

$$f(\alpha) = 6{\cdot}647\sqrt{\alpha}(1 - 2{\cdot}5\alpha + 4{\cdot}49\alpha^2 - 3{\cdot}98\alpha^3 + 1{\cdot}33\alpha^4)/(1-\alpha)^{3/2} \tag{9}$$

The initial relative notch length is $\alpha = 1/3$, this leads to $f(1/3) = 3{\cdot}76$. In a double logarithmic plot the $\sqrt{d}$ dependency of σ_n is the straight line plotted in Fig. 13.

An upper limit of the nominal stress may be calculated using a yield criteria. Assuming a stress distribution as shown in Fig. 14 the

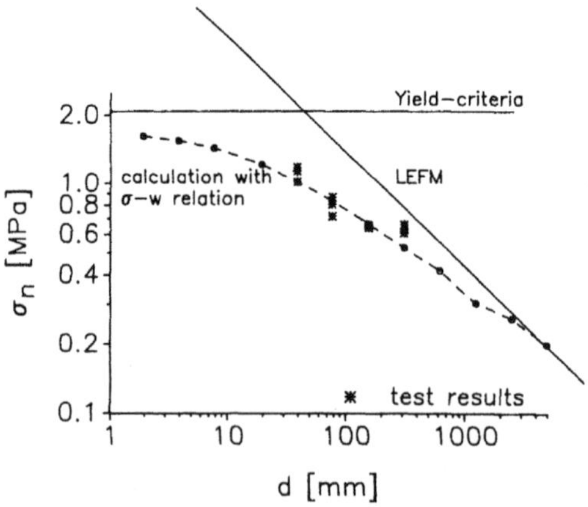

Fig. 13 Size effect curve.

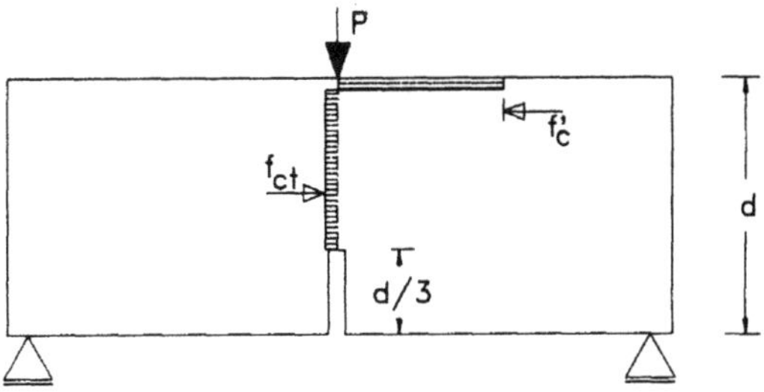

Fig. 14 Stress distribution for yield criteria.

equilibrium conditions lead to

$$\sigma_n = \frac{16}{45} \frac{f_{ct} f_c'}{f_{ct} + f_c'} \tag{10}$$

With $f_c' = 85{\cdot}5$ MPa as given by the compressive tests, and $f_{ct} = 6{\cdot}3$ MPa resulting from the recalculation of the bending tests, this leads to $\sigma_n = 2{\cdot}086$ MPa; which is the horizontal line plotted in Fig. 13.

The test results given in Table 4 are marked with crosses. The circles connected by the dashed line are the results of the recalculation of the tests and the extension of the calculation for both very large and very small specimens.

The test results prove that HSC exhibits a considerable size effect, but also for HSC—which is usually judged to be very brittle—the behaviour is not in coincidence with LEFM.

The finite element calculation using the cohesive crack model fits not only with the test results—where it was calibrated—but also with LEFM for very large dimensions and with the yield criteria for very small dimensions.

3. CONCLUSIONS

The specimen's real size effect must be distinguished from the element size effect of FE calculations. The first influences the behaviour of structural members, the second is only a mathematical problem.

One specimen size effect—the increase in flexural strength—was examined when varying the depth, the eccentricity and the material properties. The outcome was that the increase in flexural strength depends not only on the depth but to the same extent on the eccentricity and on the material properties. Even in this very simple case it is impossible to describe the size effect by a simple equation depending only on a geometrical parameter, e.g. the depth of the beam. Generalized for other cases, it may be concluded that equations describing the size effect only with a geometrical parameter like the depth can approximate the real specimen behaviour only very roughly.

The σ–w relation does not depend on the structural size. This was proven by analysing geometrically similar specimens of three different sizes. This supports the view that the softening curve is a real material property.

It is a well known fact that high strength concrete behaves with more brittleness than normal concrete. This is explained by the very steep decay in bearing capacity after the first cracking.

Although HSC behaves with more brittleness, it is not brittle enough to permit the use of LEFM. The peak loads are significantly lower than those that LEFM would predict.

With the use of cohesive crack models together with finite element programs the whole range between very large and very small dimensions can be described. The finite element solution approaches the yield criteria for small sizes and the LEFM criteria for large sizes. The cohesive crack model seems to be the most promising model for general civil engineering application.

ACKNOWLEDGMENTS

The second part of this work was carried out during a three-month visiting scholarship of the first writer at Northwestern University supported by Deutsche Forschungs Gemeinschaft. The writer thanks Prof. S. P. Shah and his group for hospitality at the Center for Advanced Cement-Based Materials. Special thanks to Mr R. Gettu for providing the test data and discussion.

REFERENCES

[1] Duda, H., Bruchmechanisches Verhalten von Beton unter monotoner und zyklischer Zugbeanspruchung. PhD thesis, TH Darmstadt, Germany, 1990.

[2] Petersson P.-E., Crack Growth and Development of Fracture Zones in Plain Concrete and Similar Materials. Report TVBM-1006, Thesis, Div. of Building Materials, Univ. of Lund, Sweden, 1981.

[3] Gettu, R., Bažant, Z. P. & Karr, M. E., Fracture properties and Brittleness of High Strength Concrete. Report No. 89-10/B627f, Center for Advanced Cement-Based Materials, Northwestern University, Evanston, USA, October 1989. (Accepted for publication, ACI).
[4] König, G., Rothe, D. & Schmidt, Th., SNAP–Ein nichtlineares Finite-Element-Programm, Bericht zum Schlusskolloquium des DFG-Schwerpunktprogramms 'Nichtlineare Berechnungen im Konstruktiven Ingenieurbau' am 2–3 March 1989 in Hannover. Springer-Verlag, New York, Berlin, Heidelberg, London, Paris, Tokyo, 1989, pp. 687–700.

Chapter 4

Size Effects in Two Compact Test Specimen Geometries

B. BARR & Z. Y. TOKATLY

Division of Civil Engineering, School of Engineering, University of Wales College of Cardiff, PO Box 917, Cardiff CF2 1XH, UK

ABSTRACT

Size effect studies are reported for torsion test specimens prepared from cylinders and/or cores and compact compression test specimens. Since both geometries are compact test specimens they allow varying sizes to be readily investigated. In the torsion tests, the approximate Size-Effect Law proposed by Bazant is shown to be only reasonably applicable to the experimental results. The variation between actual experimental results and those determined from the application of the Size-Effect Law is significant. In the compact compression tests, the results show that this specimen is an excellent test geometry to study size effects and that in the case of specimens with shallow notch depths, the type of failure approaches that which corresponds to LEFM. Finally a new brittleness number, based on strain energy density calculations is proposed which can be used to compare the relative brittleness of different test specimens as well as the effect of notch depth.

1. INTRODUCTION

Most of the initial theoretical work on brittle fracture has been developed on the basis of the concepts introduced by Griffith in 1920 [1]. His writings on the theory of rupture introduced two basic ideas.

The first idea is that of the presence of flaws in all real materials. Griffith wrote that 'the weakness of isotropic solids, as ordinarily met with, is due to the presence of discontinuities, or flaws'. The measured strength of real materials are 10^2–10^3 lower than their theoretical strengths arrived at on the basis of material homogeneity. The second idea introduced by Griffith was the relationship between the work to spread a crack and the surface energy of the new surfaces formed.

As the strength of brittle materials is affected by the presence of imperfections as first suggested by Griffith, it seems reasonable to expect that the value of the ultimate strength will depend upon the size of specimens. As specimens increase in size the strength should decrease since the probability of having weak spots or links is increased. This effect is referred to as the size effect and an explanation of the fact on a statistical basis was furnished by Weibull [2]. He showed that if two series of tensile tests are performed on geometrically similar specimens of two volumes V_1 and V_2, the corresponding values of ultimate strengths (σ_{ult}) will be in the ratio

$$\frac{(\sigma_{ult})_1}{(\sigma_{ult})_2} = \left(\frac{V_2}{V_1}\right)^{1/m}$$

where m is a constant of the material. By today the work of Weibull permeates most areas of composite materials design.

The above statistical size effect is not the only size effect observed during the fracture process in concrete. The second size effect has been referred to as the fracture-type size effect by Bazant *et al.* [3]. This fracture-type size effect is approximately described by Bazant's size effect law [4], [5] which has been shown to agree well with test data. This chapter deals specifically with the fracture-type size effect.

The main difficulties associated with the application of fracture mechanics theory in concrete technology are the effects of slow crack growth that may occur during testing and the large size of the test specimen required to ensure valid fracture toughness results. The influence of slow crack growth under Mode I loading conditions has received considerable attention. Most researchers agree that slow crack growth cannot be neglected in any fracture toughness test. The present authors believe that the importance of slow crack growth varies with the relative stiffness of the test specimen and the testing arrangement and, in particular, with the strain energy stored in the test specimen prior to crack initiation. The effect of strain energy is considered in detail later.

The second potential difficulty associated with the application of fracture mechanics in concrete technology is the minimum size requirement for the test samples. The test specimens should be sufficiently large to ensure that the plastic zone which may develop at the crack tip is not large in comparison with the crack length, specimen thickness, uncracked ligament, etc. This potential problem is enhanced with FRC materials when the aggregate size and/or fibre length may be relatively large compared with the uncracked ligaments length. This effect has traditionally been investigated by varying the overall size of the test specimens and introducing various notch depths into the specimen.

For example, the effect of specimen size on the fracture energy of concrete has been investigated by Mindess [6]. He carried out tests to examine the effect of changing the size (but not the relative dimensions) of the test specimen recommended by RILEM. The fracture energy was obtained from test results on notched beams subjected to three-point bending. The specimen dimensions ranged from 100 mm × 100 mm × 840 mm to 400 mm × 400 mm × 3360 mm. For the large beams, the self-weight was considerable and was taken into account in determining the fracture energy. The main conclusion from the work reported by Mindess was that the fracture energy increased considerably in the case of the large test specimens. These results were based on only nine tests and only one notch depth ratio (notch depth was one-half of depth of beam in all tests).

The effect of specimen size on the fracture toughness of concrete has also been reported by Nallathambi *et al.* [7]. Again notched rectangular test specimens subjected to three-point bending were used in the study. The specimen dimensions ranged from 40 mm × 51 mm × 200 mm to 80 mm × 300 mm × 1800 mm (width × depth × span). This study included an investigation of the over-all specimen dimensions, the notch–depth ratio, and also the maximum aggregate size used in the mix.

Nallathambi *et al.* determined the fracture toughness of concrete using both the energy (to give G_{Ic}) and the stress intensity factor method (to give K_{Ic}). The G_{Ic} results showed that the fracture toughness increased substantially with increasing specimen size and increasing uncracked ligament length. On the other hand, the K_{Ic} results were relatively independent of notch depth effect but did show a slight increase with increasing specimen size. However, the G_{Ic} results were an order of magnitude greater than the K_{Ic} results. Nallathambi and his colleagues argued that the energy method is the more appropriate for determining the fracture toughness of concrete since it better approximates the process of energy dissipation and consumption.

Most research workers have isolated the size effect from other factors by testing geometrically similar test specimens. However, there is another method which can be used to overcome the size effect problem. Bazant & Pfeiffer [8] considered three distinct test specimen geometries—three-point bend, edge-notched tension and eccentric-compression specimens. In the case of the notched tension specimen, the entire ligament is subject to tension. In such a case the fracture process zone can be very large and in extreme cases extend over the entire ligament length. At the other end of the scale, the eccentric compression specimen has a very limited amount of the ligament in tension. Hence the fracture process zone must be limited in extent. The three-point bend specimen is somewhere between these two extreme cases.

The eccentric compression specimen reported by Bazant & Pfeiffer was used some time ago by one of the present authors [9] to determine the fracture toughness of plain and polypropylene fibre reinforced concrete. The modified cube and testing arrangement is shown in Fig. 1. Since the testing system had some similarity with the traditional compact tension test specimen used in Fracture Mechanics, the test specimen was designated the compact compression test specimen.

The results given in Ref. [9] were all obtained from modified cubes and in all the tests carried out on the compact compression test specimens the two notches were symmetrical. This work was extended for the general case of a compact test geometry as shown in Fig. 2.

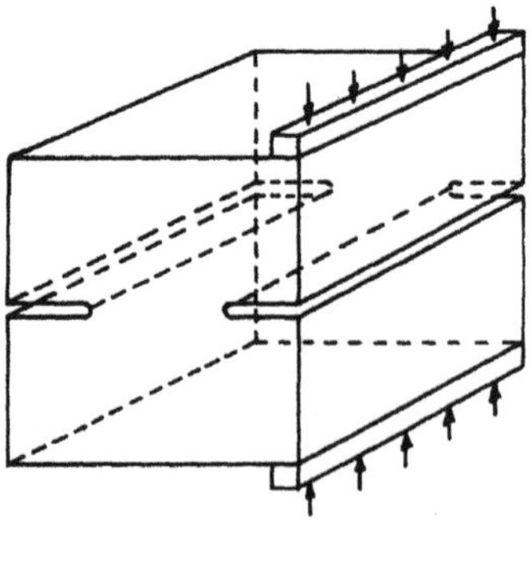

Fig. 1 Compact compression test specimen (cube).

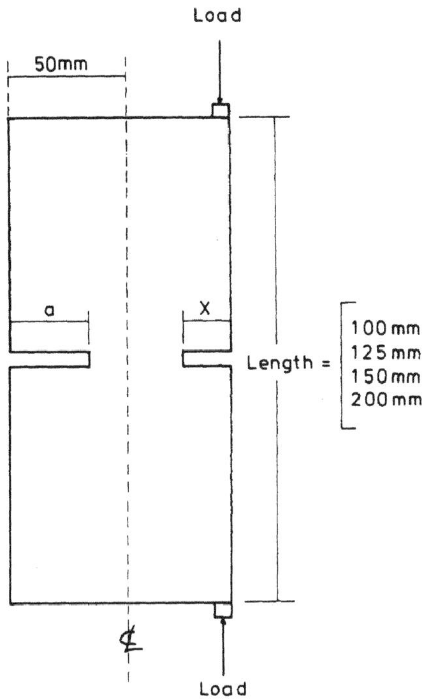

Fig. 2 General compact compression geometry.

The compact compression test specimen was modified as follows:

(a) to include unsymmetrical notches; and
(b) to include various lengths of specimens.

Both experimental and numerical results have been obtained by Barr & Sabir [10]. One of the main conclusions of this work is that there is no benefit in increasing the over-all length (height) of the compact compression test specimen. Increasing the height simply increases the amount of concrete used in the test sample. The most important geometrical value is the uncracked area at the point where crack propagation occurs, i.e. the notch depth and the uncracked ligament are the most significant lengths in the test specimen.

It is the view of the authors that not enough thought has been given to the choice of test specimen geometries in the past. Fracture test specimens have been developed without consideration being given to the existing standard test specimen geometries, i.e. cubes, beams and cylinders. The essential requirements of a successful new test specimen geometry should include a simple compact geometry, ease of preparation, simple loading system and being able to be performed in non-specialist laboratories. These requirements can generally be met by developing test specimen geometries based on the standard control specimens.

In general, it is more likely that cubes will be more readily available than beams and that concrete cores will be more readily available than cylinders. Both cubes and cores are eminently suitable for being developed as compact fracture test specimen geometries. Two compact geometries are reported in this chapter—compact torsion test specimens prepared from cylinders or cores and compact compression test specimens. In both cases it has been relatively easy to investigate size effects in these test specimens.

2. EXPERIMENTAL STUDIES

Two test specimen geometries have been used in the work reported here. The first test specimen geometry was developed from earlier work on torsion testing of cores and was first reported in Ref. [11]. The system of loading is illustrated in Fig. 3 and shows two supports providing upward reactions, a third support providing a downward reaction and the fourth support being the point of application of the applied load. Although Fig. 3 shows a solid cylinder being subjected to a torque, the same loading arrangement can be used to test circumferentially notched cylinders. In the study on notched cylinders,

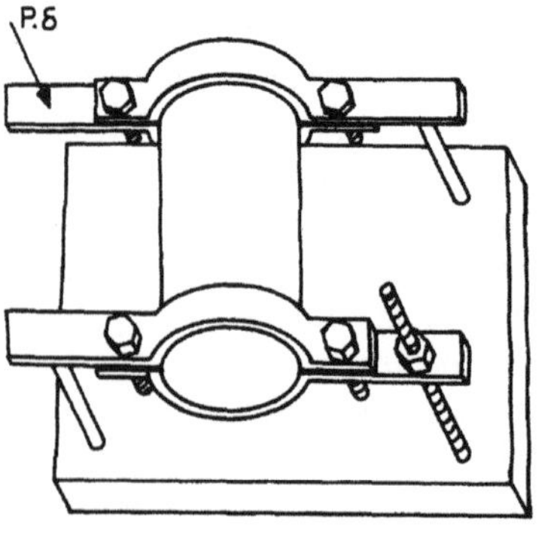

Fig. 3 Torsion rest rig.

standard 100 mm diameter cylinders were initially notched circumferentially to a depth of either 20 mm or 25 mm. The test specimens were subjected to opposing couples via a pair of split collars. Full details of the sizes of the solid and notched test specimens used in this part of the study are given later.

The second test specimen geometry used in this study is the compact compression test specimen. The experimental details have already been reported elsewhere [12]. This test specimen geometry was developed initially from modified standard cubes. The two critical dimensions of any concrete fracture test specimen are the crack length and the uncracked ligament. These two dimensions can be varied readily in the compact compression test specimens. The overall dimensions (thickness × width × length) of the test specimens reported in the study are as follows:

(a) 100 mm × 200 mm × 200 mm specimens;
(b) 100 mm × 300 mm × 300 mm specimens;
(c) 100 mm × 400 mm × 400 mm specimens.

As the compact compression test specimens increase in size, their self-weight becomes significant. It can be shown that the effect of self-weight (due to the large eccentricity of the test specimens in the testing machine) can be of the order of 9%, 18% and 29% of the failure load for 200 mm, 300 mm and 400 mm cubes, respectively. Hence the compact compressive test specimens used in the work reported here were loaded as shown in Fig. 4, i.e. they were loaded

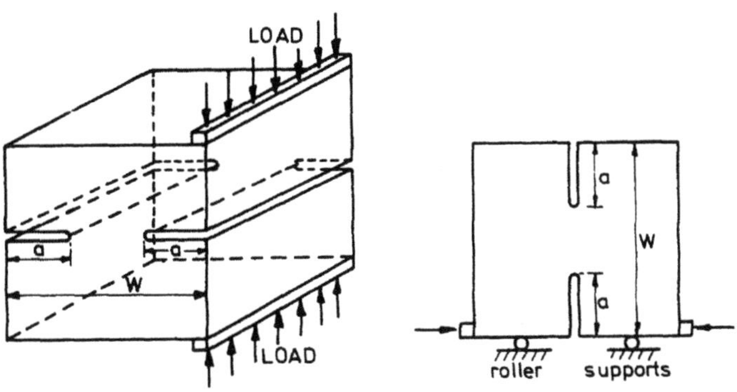

Fig. 4 Testing arrangement for large compact compression test specimens.

with the applied loads acting horizontally and the self-weight of the specimens was supported on rollers.

A nominal Grade 50 concrete having proportions by weight of 1:1·8:2·8 of cement:fine aggregate:coarse aggregate was used for both test specimen geometries. The water–cement ratio was kept constant at 0·5 giving a compressive strength in the range 50–55 N/mm^2. The cement was ordinary Portland cement, the fine aggregate was a local sea-dredged sand and the coarse aggregate was crushed limestone. In the case of the torsion tests only 10 mm maximum size coarse aggregate was used. However, in the compact compression tests three maximum sizes of crushed limestone coarse aggregate were used—5 mm, 10 mm and 20 mm.

At the start of the study on the effect of the maximum size of the coarse aggregate, the authors considered designing the mix to be used for each maximum size of coarse aggregate to give a constant compressive strength. This exercise was carried out by Nallathambi *et al.* [7] but, as the results show, there was still some variation in the slump and compressive strength values. The present authors decided to use exactly the same mix proportions for all three maximum sizes of coarse aggregate and to take particular care in noting the variation in workability and compressive strength results.

The mixes were prepared in a pan mixer and the specimens, including the control cubes, were compacted by means of a vibrating table. All specimens were cured under water for 26 days with any necessary notches (e.g. circumferential notches or notches in the compact compressive specimens) being introduced at the end of this curing period. The specimens were allowed to dry overnight and the collars added (where necessary) on the following day. Testing was carried out at 28 days.

The torsion tests on the smaller diameter specimens were carried out in an Instron machine under deflection-controlled conditions. All the tests were carried out at room temperature (20°C) and the load–deflection graphs (see Fig. 5) were plotted autographically. The load–deflection response is linear up to the cracking load, P_c, at which point the test specimen shows signs of softening. The maximum load, P_m, is generally some 10% greater than P_c. Once the maximum load is reached the test specimen shows a rapid reduction in the load of some 50% or more, and thereafter, the load reduces gradually as the diagonal tensile crack opens up.

The larger diameter torsion specimens could not be accommodated

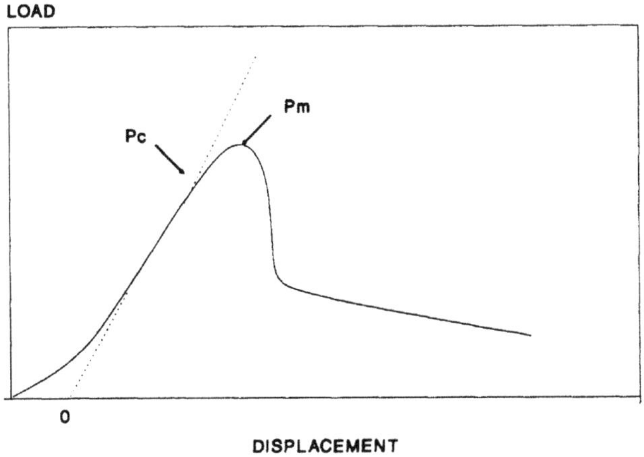

Fig. 5 Typical load–displacement curve (torsion tests).

in the Instron machine. The large specimens were therefore tested by means of an Avery-Denison machine. The loading rate was chosen so that the time taken by the load to reach its maximum value (for all sizes) was approximately 5 mins.

All the tests on the smallest compact compression test specimens (100 mm cubes) were carried out by means of an Avery-Denison machine. The tests were carried out under deflection-control at a strain rate of approximately 10×10^{-3}/sec. The test specimens were loaded as shown in Fig. 1, the load being applied through two square steel bars located along the edge of the test specimens. Typical load–deflection graphs have not been included since all the curves showed a simple linear relationship following an initial bedding of the specimens at the loading points. A distinct maximum load was recorded at the point of fracture in all cases. There was no evidence of a change in the compliance of the test specimen at the point of maximum load—all failures occurred suddenly with a large amount of noise being produced, i.e. a typical brittle type of fracture occurred.

The larger compact compression test specimens were loaded in a specially prepared test rig (Fig. 4). This testing arrangement was used to overcome the problem of the self-weight of the large test specimens. All the dimensions were kept geometrically similar including the steel loading strips used to transfer the loads into the test specimens.

3. TORSION RESULTS

Torsion tests, illustrated in Fig. 3, have been carried out on a range of cylindrical test specimen geometries. Approximately half of the tests were carried out on solid cylinders and the other half on circumferentially notched cylinders. Four test specimen diameters have been investigated in the study (80 mm, 100 mm, 150 mm and 200 mm) together with three notch-depth ratios ($D/4$, $D/5$ and $D/6$). Only the results for the test specimens with a notch-depth ratio of Diameter/5 are presented here. In all cases the length of the test specimens were twice the length of the diameter.

The experimental results obtained from the study have been used to investigate the applicability of the Bazant Size-Effect Law to concrete. The Size-Effect Law is illustrated in Fig. 6. For small test specimens the strength at failure is proportional to the material strength and there is no size effect. For large test specimens, the strength at failure is proportional to a characteristic dimension to the power $-\frac{1}{2}$ which represents the maximum possible size effect. The first case corresponds to plastic limit analysis and is represented by the horizontal line in Fig. 6 and the second case corresponds to classical linear elastic fracture mechanics and is represented by the inclined line in Fig. 6. In practical experimental work on most concrete test specimens, the results lie in the transition zone between these two extreme cases.

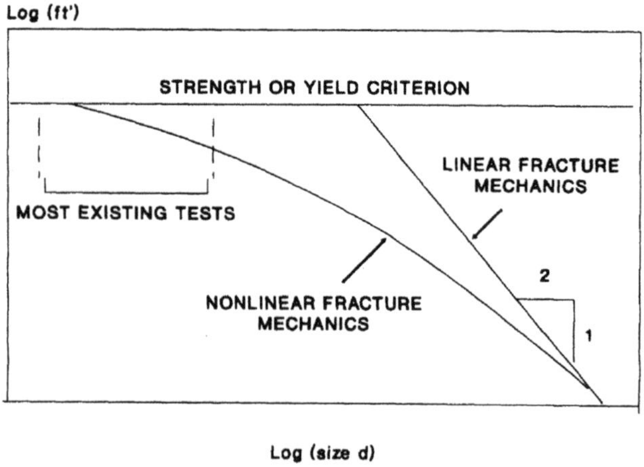

Fig. 6 Effect of specimen size.

A full description of the Bazant Size-Effect Law is not presented here. It can be shown that the nominal stress at failure (T_N) approximately obeys the following Size-Effect Law

$$T_N = Bf_t'[1 + \lambda/\lambda_0]^{-1/2}$$

where $T_N = C_N \,.\, P/d^2$ for three-dimensional similarity, P = load at failure, d = characteristic dimension of the specimens and C_N = arbitrary non-dimensional constant; $\lambda = d/d_a$ = relative structural size where d_a = maximum aggregate size; B and λ_0 are empirical parameters characterizing the fracture energy of the material and f_t'= tensile strength of concrete. The above formula, which was derived by dimensionless analysis and similitude arguments, represents a gradual transition from the failure strength criterion and limit analysis to the linear elastic fracture mechanics failure as illustrated in Fig. 6.

In order to determine the two parameters B and λ_0, the above equation can be transformed into the following form:

$$\left[\frac{f_t'}{T_N}\right]^2 = \frac{1}{B^2\lambda_0} \cdot \frac{d}{d_a} + \frac{1}{B^2}$$

This transformed equation gives a linear relationship between $[f_t'/T_N]^2$ and d/d_a and may be rewritten in the form $Y = AX + C$ where $Y = [f_t'/T_N]^2$, $X = d/d_a$, $C = 1/B^2$ and $A = C/\lambda_0$. Thus the constants A and C may be obtained by linear regression of the test results from which $B = 1/C^{1/2}$ and $\lambda_0 = C/A$. The regression also yields statistics of errors, i.e. the coefficient of variation and the correlation coefficient. The tensile strength f_t' was determined from the cylindrical compressive strength results.

Bazant & Prat [13] have applied the Bazant Size-Effect Law to results obtained from notched cylinders. The same procedure is followed here. Although three notch-depth ratios have been studied as described earlier, only one notch-depth is reported in detail, i.e. all the results are for a notch-depth ratio of Diameter/5. The experimental results for the four diameters tested are reported in Table 1. Three test specimens were tested for each diameter and the coefficients of variation were generally below 10%.

According to the size effect formula, $T_N = C_N P/D^2$ where $T_N = 16T/\pi D^3$. Therefore $C_N = 16R/\pi D$, which is a constant since R/D = constant for geometrically similar specimens ($T = PR$ where R is the lever arm of the applied load). The plots of the linear regression and

TABLE 1 Maximum torque results for the notched cylinders (notch depth = $d/5$)

Cylinder diameter (d) (mm)	*Notch depth (mm)*	*Maximum torque (N . m)*			*Average maximum torque (N . m)*	*Coefficient of variation (%)*
		T_1	T_2	T_3		
80	16	263·8	225·8	258·0	249·2	8·2
100	20	411·8	389·3	408·3	403·1	3·0
150	30	1170·0	1332·0	1231·2	1244·4	6·6
200	40	2857·8	2996·4	2871·0	2908·4	2·6

size effect are shown in Fig. 7. The results show that they are consistent with the proposed Bazant Size-Effect Law.

Since four experimental results are available, by taking three of the four experimental results and obtaining the Size-Effect Law, it should be possible for the fourth result to be determined. Thereafter the fourth result (obtained from the size effect study) can be compared with the fourth experimental value. This comparison between the real fourth experimental result and the prediction from the Size-Effect Law is very interesting. The regression plots and size effect plots for the four combinations of three results are shown in Figs 8–11. (Figure 8 shows the results from the 80 mm, 100 mm and 150 mm diameter test specimens, Fig. 9 the results from the 80 mm, 100 mm and 200 mm diameter test specimens, Fig. 10 the results from the 80 mm, 150 mm and 200 mm diameter test specimens and Fig. 11 the results from the 100 mm, 150 mm and 200 mm diameter test specimens.)

A comparison between the values obtained from the experimental results and the corresponding results obtained from the application of the Size-Effect Law is given in Table 2. The variation between the real experimental results and the corresponding results obtained from the Size-Effect Law together with the high coefficients of variation obtained from the application of the Size-Effect Law relative to the coefficient of variation obtained from the experimental results suggests that the Size-Effect Law cannot be used to predict unknown experimental results. The relatively high spread of results obtained from the Size-Effect Law will extend over a wide range of values and therefore the experimental results will invariably lie within that range.

Xu & Reinhardt [14] have also examined the validity of Bazant's Size-Effect Law when applied to the results obtained from circum-

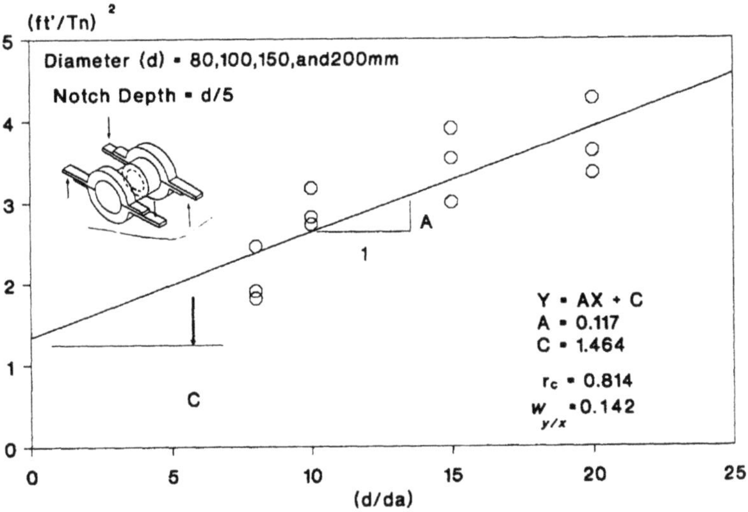

(a)

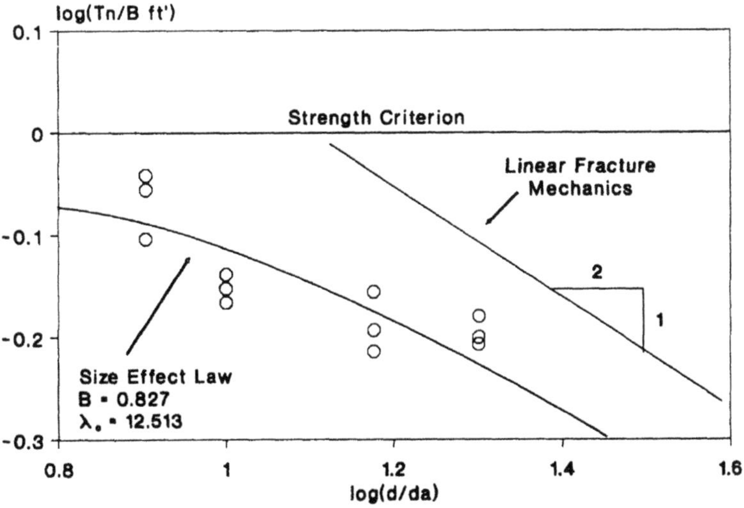

(b)

Fig. 7 (a) Linear regression and (b) size effects plots for notched cylinders (all four diameters).

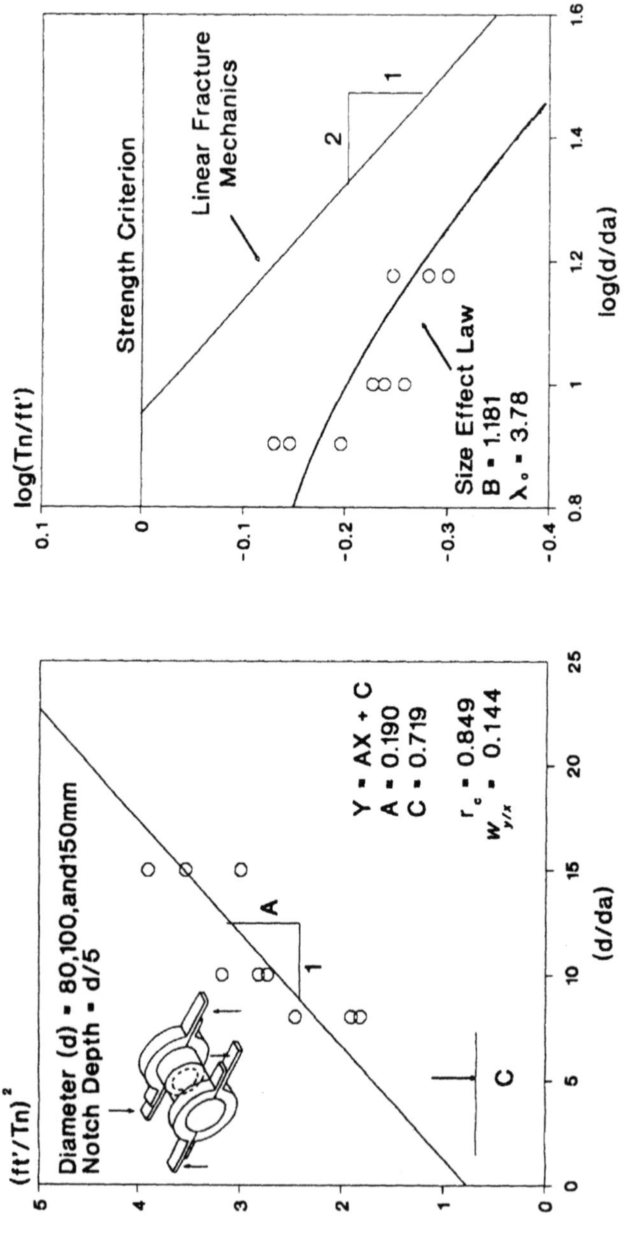

Fig. 8 (a) Linear regression and (b) size effect plots for notched cylinders ($d = 80$ mm, 100 mm, 150 mm).

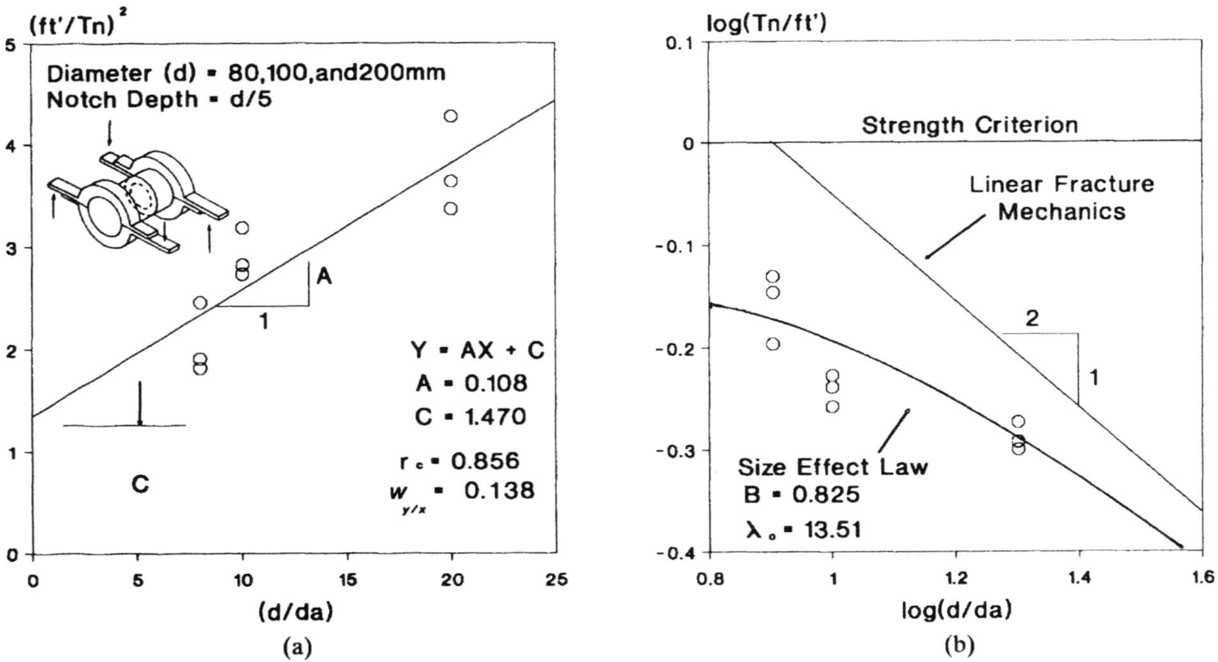

Fig. 9 (a) Linear regression and (b) size effects plots for notched cylinders ($d = 80$ mm, 100 mm, 200 mm).

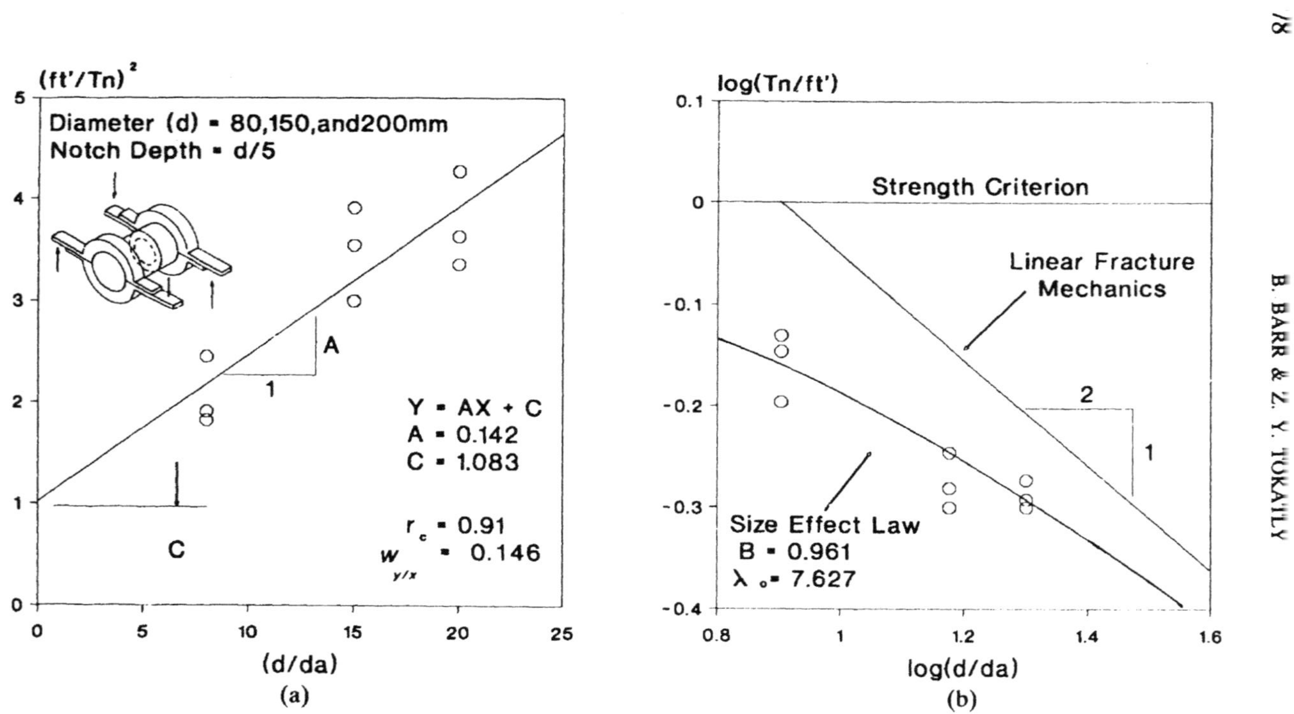

Fig. 10 (a) Linear regression and (b) size effects plots for notched cylinders ($d = 80$ mm, 150 mm, 200 mm).

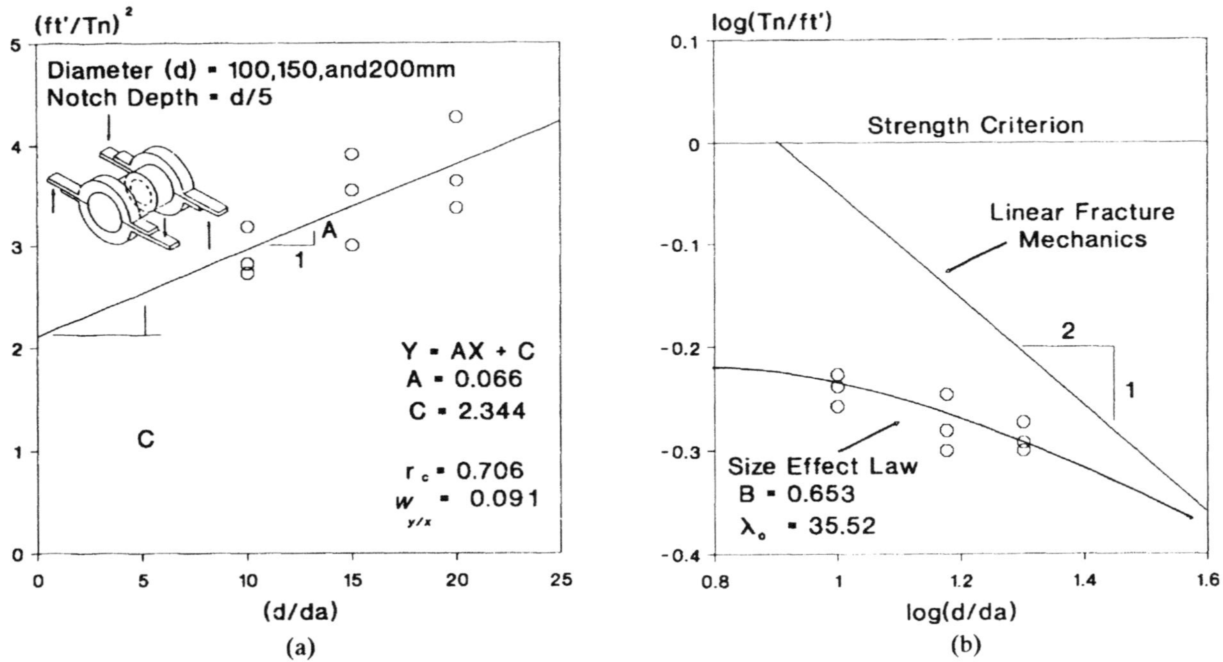

Fig. 11 (a) Linear regression and (b) size effects plots for notched cylinders ($d = 100$ mm, 150 mm, 200 mm).

TABLE 2 Comparison between the exact experimental results with those obtained from Bazant Size-Effect Law

Cylinder diameter (d) (mm)	*Experimental results*		*Bazant Size-Effect Law*	
	Average nominal strength at failure (N/mm^2)	*Coefficient of variation* (%)	*Critical strength at failure* (N/mm^2)	*Coefficient of variation* (%)
80	2·48	8·1	2·12	9·1
100	2·05	3·1	2·31	14·6
150	1·89	6·5	2·01	13·9
200	1·85	2·7	1·70	14·4

ferentially notched concrete cylinders tested under Mode III type of loading. They concluded that Bazant's Size-Effect Law does not apply to the results obtained in their study. This could possibly be due to the fact that their specimens were all of the same length but with different diameters. It should be noted that one of the main requirements of the Size-Effect Law is that only geometrically similar specimens be used—this requirement was not met in the work of Xu & Reinhardt.

A series of solid cylinders (i.e. un-notched test specimens) were also used in the experimental study. The results from the solid cylinders are presented in Table 3. The same test specimen diameters were used as in the case of the notched specimens, i.e. $D = 80$ mm, 100 mm, 150 mm and 200 mm. In all cases the length of the test specimens was twice the length of the diameter.

The regression and size effect plots are illustrated in Fig. 12. The results indicate that the Bazant Size-Effect Law can be applied to the solid cylinders. The coefficient of variation of the results presented in Table 3 is also within 10%.

TABLE 3 Maximum torque results for the solid cylinders

Cylinder diameter (d) (mm)	*L/d*	*Maximum torque (N . m)*			*Average maximum torque (N . m)*	*Coefficient of variation (%)*
		T_1	T_2	T_3		
80	2	389·3	381·0	393·8	388·0	1·7
100	2	665·3	689·8	611·8	655·6	6·1
150	2	2069·1	2442·9	2229·3	2247·1	8·4
200	2	5075·4	5173·2	4980·0	5076·2	1·9

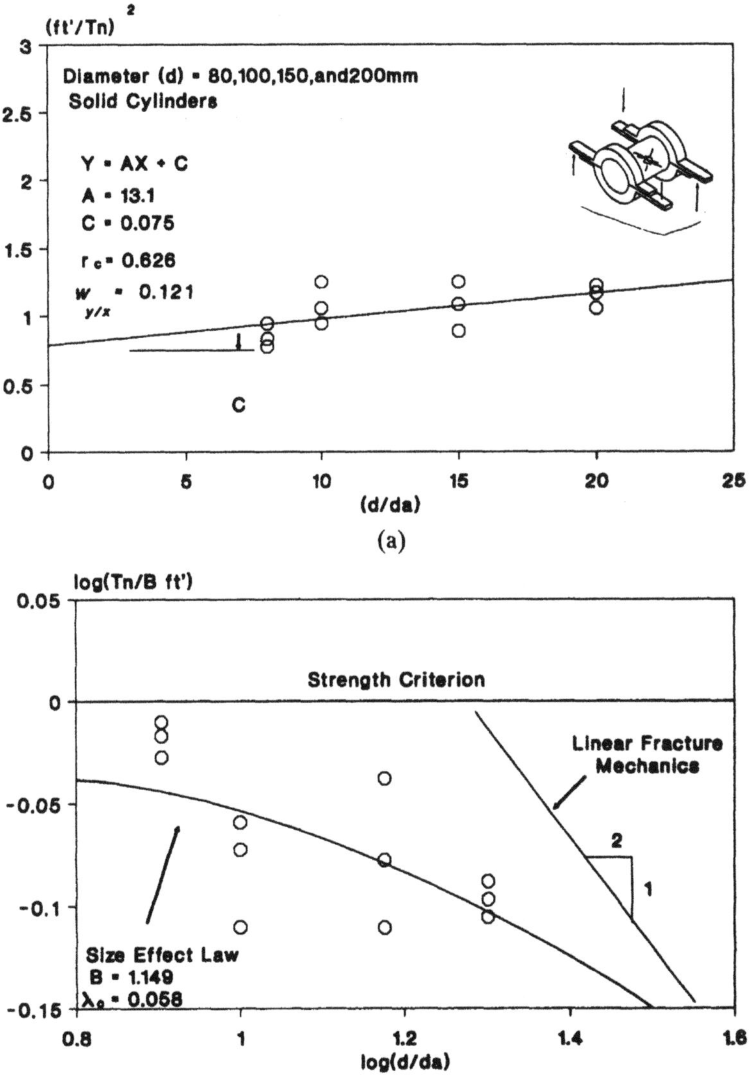

Fig. 12 (a) Linear regression and (b) size effects plots for solid cylinders (all four diameters).

4. COMPACT COMPRESSION RESULTS

Compact compression tests, illustrated in Fig. 4, have been carried out on a range of geometrically similar test specimens. In all tests the specimen width was kept constant at 100 mm—this value was considered to be sufficiently large not to influence the test results. The test results are represented in Table 4. The first series had a notch-depth ratio of 0·25 and the overall size of the specimens were 100 mm × 200 mm × 200 mm, 100 mm × 300 mm × 300 mm and 100 mm × 400 mm × 400 mm. The second series had a notch-depth ratio of 0·30 and the overall size of the specimens in this series were 100 mm × 100 mm × 100 mm, 100 mm × 200 mm × 200 mm and 100 mm × 300 mm × 300 mm.

The experimental results obtained from the compact compression tests have also been used to investigate the applicability of the Bazant Size-Effect Law. The results presented in Table 4 can also be used to investigate the effect of the maximum size of coarse aggregate used in the mix. In this study three maximum sizes of coarse aggregate were

TABLE 4 Compact compression test specimen results

Notch depth (mm)	*Specimen size (mm)*	*Aggregate size (mm)*	*Cracking load (kN)*	*V (%)*	K_{Ic} *(N/mm$^{3/2}$)*
0·25	100 × 200 × 200	20	17·5	6·6	20·3
		10	19·1	2·2	22·1
		5	19·0	2·2	22·0
	100 × 300 × 300	20	23·4	3·9	22·0
		10	24·9	3·4	23·5
		5	24·4	3·7	23·0
	100 × 400 × 400	20	28·2	3·1	23·1
		10	29·3	4·1	24·0
		5	28·8	4·0	23·6
0·30	100 × 100 × 100	20	5·7	9·2	16·0
		10	6·1	4·5	17·1
		5	6·0	4·5	16·8
	100 × 200 × 200	20	9·6	6·3	19·0
		10	11·8	1·6	23·4
		5	10·8	5·7	21·4
	100 × 300 × 300	20	14·3	6·8	23·1
		10	14·7	4·2	23·7
		5	14·5	3·3	23·4

used—5 mm, 10 mm and 20 mm. The fracture toughness results given in Table 4 are shown graphically in Fig. 13 and are reported in detail in Ref. [12]. The results for all three mixes are in close agreement, although the highest fracture toughness values are given in each case by the 10 mm aggregate mix and the minimum by the 20 mm aggregate.

Unfortunately, the required mode of failure could not be achieved for the 100 mm cubes with a notch-depth ratio of 0·25—for this geometry, shear failure occurred at the point of application of the load. The results for the fracture toughness for the cubes with a notch-depth ratio of 0·30 are much lower than expected. Previous work in the same laboratory would suggest that the fracture toughness from the cubes should be of the order of 20 N/mm$^{3/2}$.

Disregarding the results for the 100 mm cubes, Fig. 13 shows that the fracture toughness results are increasing gradually with increasing test specimen size. These results confirm the trend reported by Mindess [6] and Nallathambi *et al.* [7], although the increase in this study is much less than that reported in the other two studies. The results presented in Fig. 13 also suggest that the fracture toughness

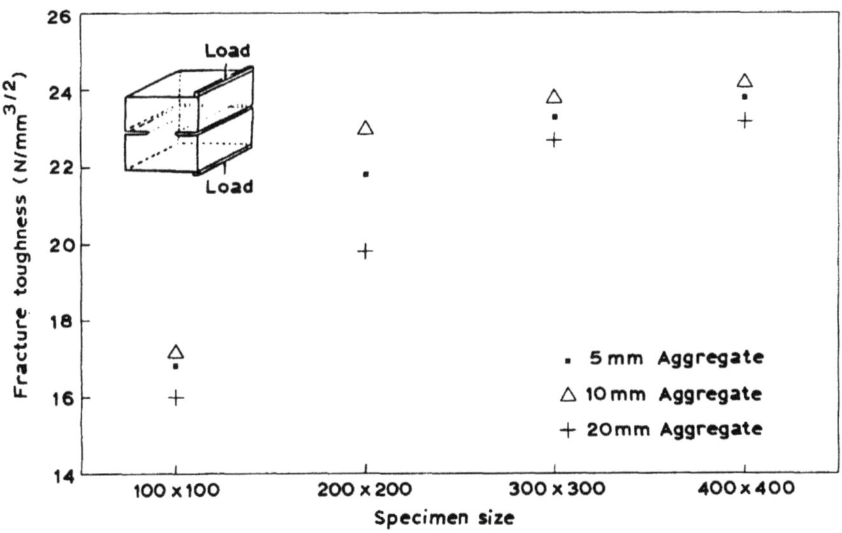

Fig. 13 Variation of fracture toughness with specimen size for the compact compressive specimens (for three aggregate sizes).

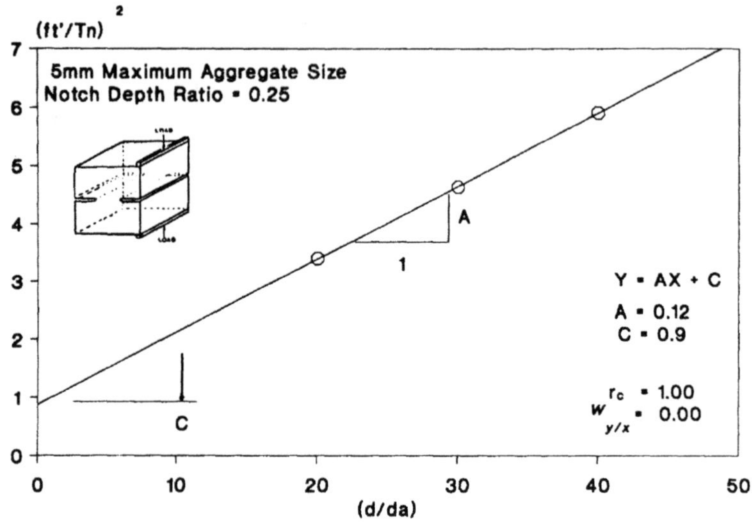

(a)

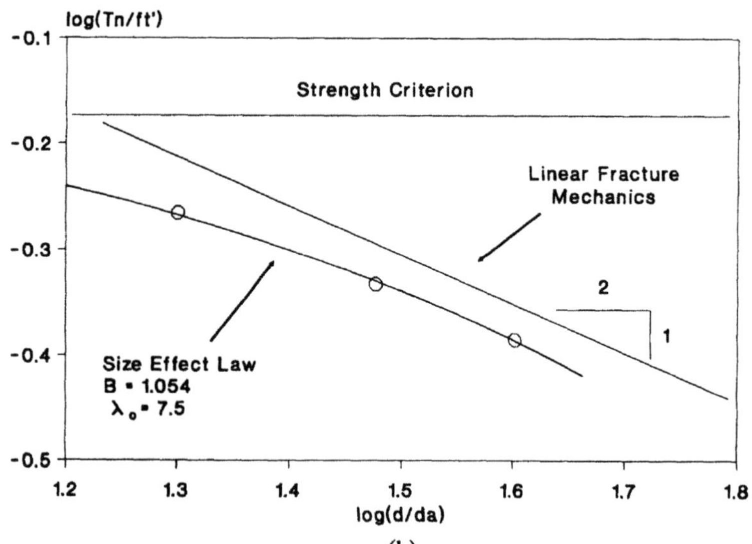

(b)

Fig. 14 (a) Linear regression and (b) size effects plots for compact compression specimens (5 mm aggregate).

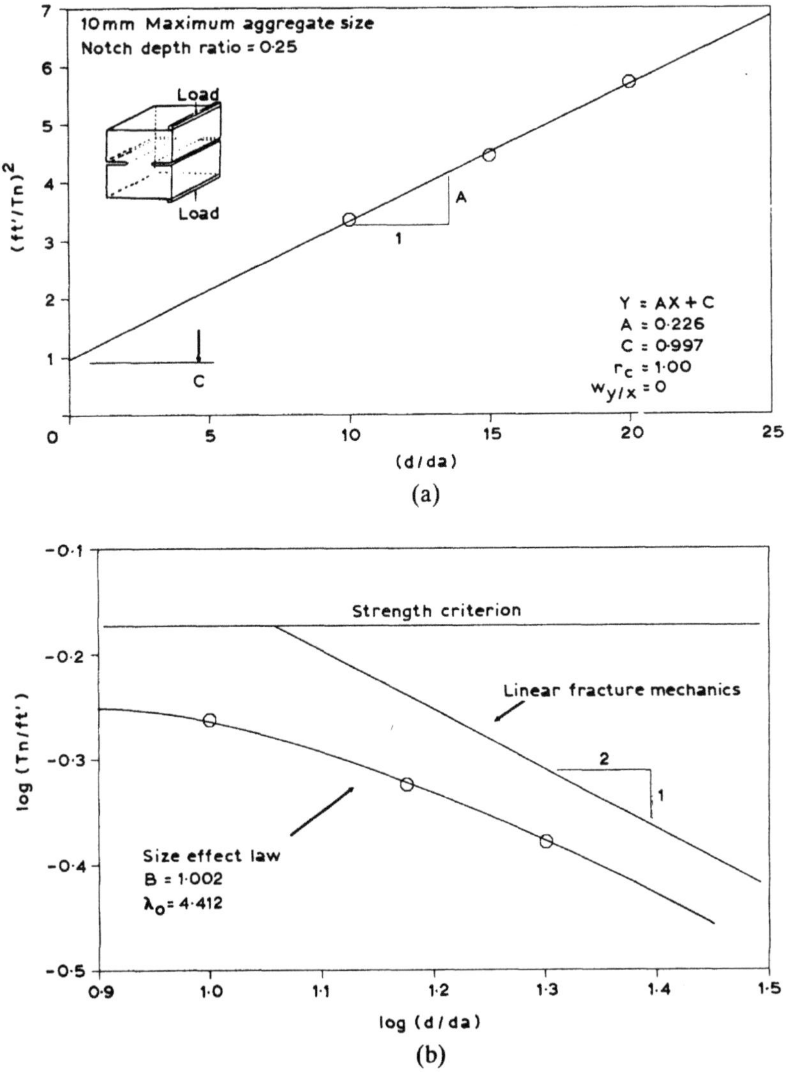

Fig. 15 (a) Linear regression and (b) size effects plots for compact compression specimens (10 mm aggregate).

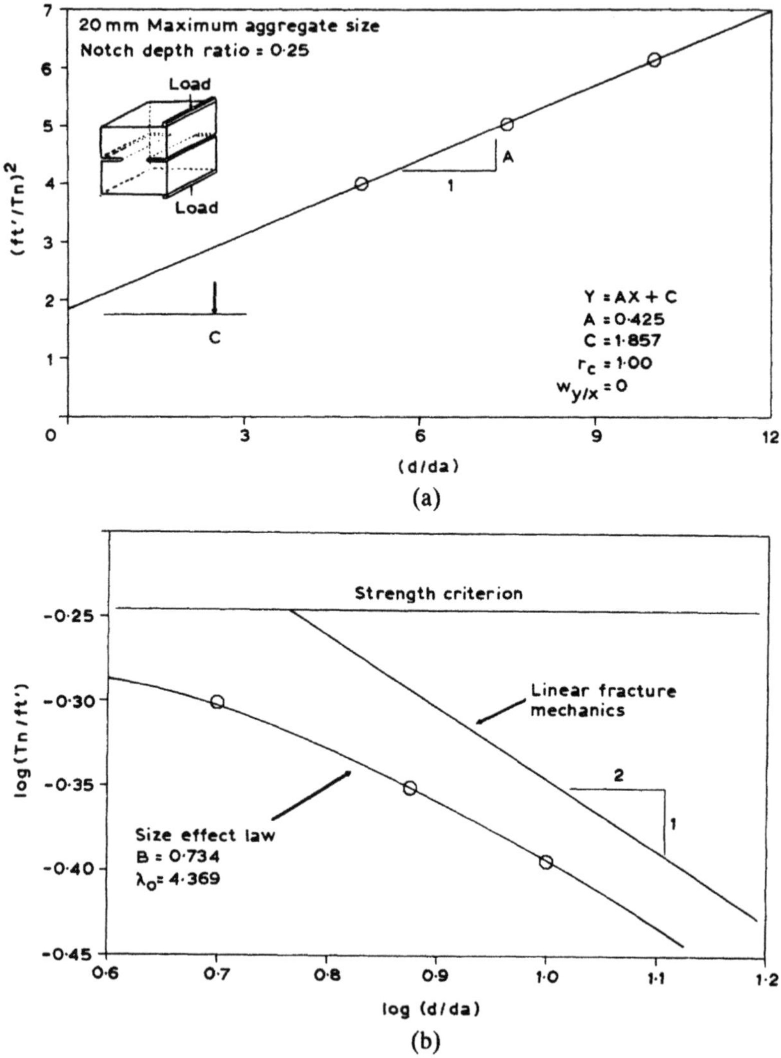

Fig. 16 (a) Linear regression and (b) size effects plots for compact compression specimens (20 mm aggregate).

values are approaching a limiting maximum value since the results appear to be levelling off at the maximum size tested. This would suggest that the maximum size of test specimen used in this study does in fact approach the necessary size required for linear elastic fracture mechanics to apply. This is in keeping with the observations made during the testing programme. Fracture of the large test specimens with relatively small notch-depth ratios was sudden with a loud noise being produced during the fracture process, i.e. a typical brittle type of fracture was observed.

The Bazant Size-Effect Law has been applied to the experimental results presented in Table 4. Since the more brittle type of behaviour was observed for the shallower notch-depth ratios, the Size-Effect Law has been used initially to consider the results for the 0·25 notch-depth ratio test specimens. The same procedure as that described earlier for the torsion tests was followed. In this case C_N was equal to unity, the characteristic length d was equal to the width of the specimen (b = thickness = 100 mm) and the tensile strength was again determined from the cylinder compressive strength. The linear regression and size-effect plots for the various aggregate sizes are shown in Figs 14–16. The results suggest that the size of the maximum aggregate used has little effect on the fracture behaviour. The results shown suggest that only the notch-depth ratio has an effect on the fracture behaviour in the compact compression test.

5. BRITTLENESS NUMBERS

A number of research workers have tried to gain a better understanding of the probable material behaviour during the fracture process by means of various brittleness numbers. Brittleness numbers can serve as a basic indication of the type of behaviour, i.e. brittle or ductile fracture. In this sense the brittleness number is analagous to the Reynolds number used in fluid mechanics.

One of the first attempts at introducing a brittleness number was made by Hillerborg *et al.* [15]. By the combination of G_F, f_t and E, Hillerborg *et al.* introduced a characteristic length for concrete and other cement-based materials, denoted l_{ch}, which was equal to EG_F/f_t^2 (where E = Young's Modulus, G_F = fracture energy and f_t = tensile strength). From the shape of the stress-crack opening displacement curve, the values of f_t and G_F can be completely defined. Hillerborg and his colleagues proposed that the characteristic length, l_{ch}, is a material property, which has no physical interpretation, but can be calculated from the values of E, G_F, and f_t.

Carpinteri [16], [17] characterized the effect of structure size on its brittleness by the non-dimensional ratio of $S_E = G_f/bf_t$ in which G_F and f_t are the same as defined in [15] above and b = beam depth. According to Carpinteri's definition of S_E (which is similar to l_{ch}), the system is brittle for low S_E values.

Bazant & Pfeiffer [8] developed from the Bazant Size-Effect Law a non-dimensional characteristic which indicated whether the behaviour of a given specimen or structure was closer to limit analysis or to linear elastic fracture mechanics. The Bazant brittleness number, β, is defined as follows:

$$\beta = d/\lambda_0 d_a$$

and can be calculated after λ_0 has been determined either experimentally or by finite element analysis. The value of $\beta = 1$ indicates the relative value of d/d_a at the point where the horizontal asymptote for the strength criterion intersects with the inclined line asymptote for the linear elastic fracture mechanics criterion. For $\beta < 1$, the behaviour is closer to plastic limit analysis and for $\beta > 1$ it is closer to LEFM. For $\beta \leq 0{\cdot}1$, the plastic limit analysis may be used as a good approximation and for $\beta \geq 10$, LEFM may be used as an approximation.

Recently, Bache [18] has introduced a characteristic number, B, which is defined as the ratio of elastic energy to fracture energy as follows:

$$B = \frac{\text{elastic energy}}{\text{fracture energy}} = \frac{L^3 f_t^2/E}{L^2 G_F} = \frac{L f_t^2}{E G_F}$$

In the above equation for B, L^3 denotes a volume and f_t^2/E is the strain energy density at failure, while L^2 denotes a crack area and G_F is the fracture energy per unit area. If the stored elastic energy is high in comparison with the energy required for crack propagation then the response of the structure is brittle. A structure is tough where the stored elastic energy is smaller than the energy needed for crack propagation.

A detailed study of the Strain Energy Density in various fracture test specimens has recently been completed by Asghari [19]. One of the objectives of this work was to develop a new brittleness number which could be used for various test specimen geometries and for all notch-depth ratios. In particular three test specimen geometries have been considered—beams in three-point bending, beams in four-point bending and the compact compression test specimen.

Asghari considered the strain energy stored in the uncracked ligament of various fracture test specimens. He assumed that the brittleness number is defined by a length b^* of the ligament length (measured from the tip of the notch). Part of the volume of the ligament (given by $b^* \times$ ligament width $\times$ ligament thickness considered, i.e. notch width) possesses the critical Strain Energy Density $(\mathrm{SED})_{\mathrm{cr}}$. The critical Strain Energy Density is defined as follows:

$$(\mathrm{SED})_{\mathrm{cr}} = \frac{U}{b^* \,.\, w \,.\, t}$$

where U is the total strain energy in the ligament when the applied load becomes the failure load. $(b^* \,.\, w \,.\, t)$ represents that volume of the ligament in which the density of energy is critical—w is the ligament width (also specimen width in most cases) and t is the ligament thickness which corresponds in this study to 4 mm.

If the length of the ligament is l, then the ratio b^*/l is a dimensionless number. Asghari has shown that b can be used as a dimensionless brittleness number which is expressed as follows:

$$b = \frac{b^*}{l} = \frac{U}{(\mathrm{SED})_{\mathrm{cr}} \,.\, w \,.\, t \,.\, l}$$

In the above equation, $(w \,.\, t \,.\, l)$ represents the volume of the ligament and hence $U/(w \,.\, t \,.\, l)$ is the average strain energy density in the ligament, $(\mathrm{SED})_{\mathrm{av}}$. Hence b can be expressed as follows:

$$b = \frac{(\mathrm{SED})_{\mathrm{av}}}{(\mathrm{SED})_{\mathrm{cr}}}$$

Thus the ratio of the two brittleness numbers b_1 and b_2 for two specimens can be written as follows:

$$\frac{b_2}{b_1} = \frac{(\mathrm{SED})_{\mathrm{av2}}}{(\mathrm{SED})_{\mathrm{av1}}}$$

The ratio given above depends upon the failure load in the two specimens. In the case where experimental work has been carried out or reported in the literature, it is possible to determine exactly the ratio of b_2/b_1. Where experimental results are not available the ratio of failure loads for various notch depths can be evaluated from a finite element study. Thus for any given geometry the brittleness number for a range of notch depths can be determined without knowing any failure load values. However, if a comparison is to be made between two test specimen geometries, the relative failure loads must be known.

It is evident that in the equation $b = (\mathrm{SED})_{\mathrm{av}}/(\mathrm{SED})_{\mathrm{cr}}$, when the values of $(\mathrm{SED})_{\mathrm{av}}$ are too high, b tends to zero. When b becomes zero the fracture response corresponds to ductile fracture. It is not clear, however, what is the transition point for a brittle response—this is an area which needs more work.

The initial results for the new brittleness number are shown graphically in Fig. 17. The standard geometry considered was a three-point bend specimen with a notch depth of 50% of the beam depth. It has been assumed that this geometry has a brittleness number of unity. All other brittleness numbers have been determined relative to this standard geometry. The new brittleness number shows

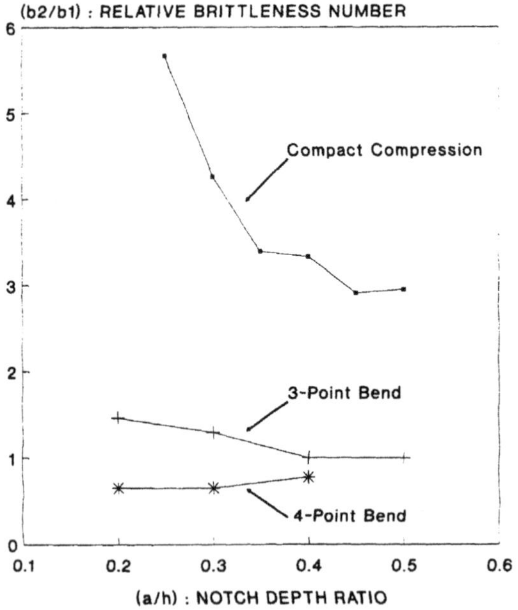

Fig. 17 Brittleness number for three fracture test geometries. a, notch depth; h, height of the specimen.

that the smaller the notch depth in the test specimens, the larger is the brittleness number. Moreover, the brittleness number for the compact compression test specimen is significantly greater than the values for three- and four-point bend specimens. This result is very important as it confirms the view that the compact compression test specimen results in a brittle type of failure and is close to LEFM. This area of research requires further detailed study.

6. CONCLUSIONS

Size effect studies have been carried out on two test specimen geometries. Both geometries are compact test specimens which allow varying sizes to be readily investigated. The first test specimen geometry was developed from earlier work on torsion testing of cores. In this work both solid and circumferentially notched cylinders of varying dimensions have been subjected to torsional loading i.e. Mode III type of loading. The second test specimen geometry used in the study was the compact compression test specimen. The experimental results obtained from the study have been used to investigate the applicability of the Bazant Size-Effect Law to concrete.

In the case of the torsion tests, the approximate Size-Effect Law proposed by Bazant was shown to be applicable to the experimental results obtained. However, the applicability of the results was a little disappointing. In this study various combinations of three out of the four experimental results were considered in turn. These three results were used in conjunction with the Size-Effect Law to predict the fourth result. Thereafter the predicted fourth result was compared with the real experimental values. The comparison between the values obtained from the experimental results and the corresponding results from the application of the Size-Effect Law was not good. The variations between the actual experimental results and the calculated results from the Size-Effect Law were significant.

The experimental results reported some time ago from the compact compression tests have also been used to investigate the applicability of the Bazant Size-Effect Law. In this case good results were obtained for three maximum sizes of coarse aggregate (5 mm, 10 mm and 20 mm)—provided that the notch-depth ratio was 0·25. The results presented here suggest that the compact compression test specimen is an excellent test geometry to study size effects and that in the case of shallow notch depth the type of failure approaches that which corresponds to LEFM.

A new brittleness number is reported in the chapter. The brittleness number is based on strain energy density calculations—in particular, the strain energy density in the uncracked ligament of various fracture test specimens. It is shown that the new brittleness number can be used to compare the relative brittleness of different test specimen geometries. In particular, the study reported here shows that the compact compression test specimen geometry results in brittle fracture and that LEFM can be used for such a test specimen.

ACKNOWLEDGEMENT

The authors wish to thank Dr A. Asghari for permission to use some of his recent studies in the section on Brittleness Numbers.

REFERENCES

[1] Griffith, A. A., The phenomena of rupture and flow in solids. *Royal Soc.* (London), *Phil. Trans. A,* **22** (1920) 163–98.

[2] Weibull, W., A statistical distribution function of wide applicability. *J. Appl. Mech.*, **18** (1951) 273–7.

[3] Bazant, Z. P., Sener, S. & Prat, P. C., Fracture Mechanics Size Effect and Ultimate Load of Beams under Torsion. ACI SP-118 (*Fracture Mechanics: Applications to Concrete,* ed. V. C. Li & Z. P. Bazant). American Concrete Institute, Detroit, 1989, pp. 171–8.

[4] Bazant, Z. P., Size effect in blunt fracture: Concrete, rock, metal. *J. Eng. Mechanics* (ASCE), **110** (1984) 518–35.

[5] Bazant, Z. P., Fracture Energy of Heterogeneous Material and Similitude. In *Proc. SEM/RILEM Int. Conf. on Fracture of Concrete and Rock,* Houston, June 1987, ed. S. P. Shah & S. Swartz. Soc. for Experimental Mechanics, SEM, Bethel, CT, USA, pp. 390–412.

[6] Mindess, S., The effect of specimen size on the fracture energy of concrete. *Cement and Concrete Research,* **14** (1984) 431–6.

[7] Nallathambi, P., Karihaloo, B. L. & Heaton, B. S., Effect of specimen and crack sizes, water/cement ratio and coarse aggregate texture upon fracture toughness of concrete. *Magazine of Concrete Research,* **36**(129) (1984) 227–36.

[8] Bazant, Z. P. & Pfeiffer, P. A., Determination of fracture energy from size effect and brittleness number. *ACI Materials Journal,* **84** (1987) 463–80.

[9] Barr, B., Evans, W. T. & Dowers, R. C., Fracture toughness of polypropylene fibre concrete. *Int. J. Cement Composites and Lightweight Concrete,* **3**(2) (1981) 115–22.

[10] Barr, B. & Sabir, B. B., Fracture toughness testing by means of the compact compression test specimen. *Magazine of Concrete Research,* **37**(131) (1985) 88–94.

[11] Yacoub-Tokatly, Z., Barr, B. & Norris, P., Mode III fracture—a tentative test geometry. In *Fracture of Concrete and Rock: Recent Developments,* ed. S. P. Shah, S. Swartz & B. Barr. Elsevier Applied Science, London, 1989, 596–604.

[12] Barr, B., Hasso, E. B. D. & Weiss, V. J., Effect of specimen and aggregate sizes upon the fracture characteristics of concrete. *Int. J. Cement Composites and Lightweight Concrete,* **8**(2) (1986) 109–19.

[13] Bazant, Z. P. & Prat, P., Measurement of Mode III fracture energy of concrete. *Nuclear Engineering and Design,* **106** (1988) 1–8.

[14] Xu, D. & Reinhardt, H. W., Softening of concrete under torsional loading. In *Fracture of Concrete and Rock: Recent Developments,* ed. S. P. Shah, S. Swartz & B. Barr. Elsevier Applied Science, London, 1989, pp. 39–50.

[15] Hillerborg, A., Modeer, M. & Peterson, P. E., Analysis of crack formation and crack growth by means of fracture mechanics and finite elements. *Cement and Concrete Research,* **6** (1976) 773–82.

[16] Carpinteri, A., Application of fracture mechanics to concrete structures. *J. Struct. Div.* (ASCE), **108** (ST4) (1982) 833–48.

[17] Carpinteri, A., Size effects on strength, toughness and ductility. *J. Eng. Mechanics* (ASCE), **115**(7) (1989) 1375–92.

[18] Bache, H. H., Brittleness/ductility from deformation and ductility points of view. Contribution to Chapter 7 of RILEM Report, *Fracture Mechanics of Concrete Structures; From Theory to Applications,* ed. L. Elfgren. Chapman and Hall, London, 1989, 202–7.

[19] Asghari, A., Strain energy density and the fracture of concrete. PhD thesis, University of Wales College of Cardiff, UK, 1990.

Chapter 5

Scaling in Tensile and Compressive Fracture of Concrete

J. G. M. VAN MIER

Delft University of Technology, Department of Civil Engineering, Stevin Laboratory, P.O. Box 5048, 2600 GA Delft, The Netherlands

ABSTRACT

The heterogeneous nature of cementitious composites introduces an internal length scale which leads to size dependent fracture behaviour. In this chapter a review is given of size and shape dependent fracture behaviour of plain concrete and mortar under compressive and tensile loadings. The approach is rather fundamental, in the sense that physical mechanisms underlying softening are studied. Uniaxial compression tests on prisms with varying height indicate that fracturing in compression is a localised phenomenon, and may be described with a compressive fracture energy concept. Recent results show, however, that boundary condition effects may have a significant effect on the fracture response in compression. In tension the softening can be explained from crack interface bridging near aggregates. It is shown that the post-peak carrying capacity depends on the size of the aggregates in the concrete mix. A crack interface bridge consists of two overlapping cracks with an intact ligament in between. The (flexural) failure of this ligament is a very stable process and may explain the long tail of the softening diagram. Moreover, this type of crack interface bridging has been observed in a variety of materials and scales over nine orders of magnitude. Flexural crack interface bridging seems to be a universal fracture mechanism in brittle heterogeneous materials, and may be responsible for the observed size effects in structures.

1. INTRODUCTION

The salient property of fracture of brittle disordered materials is the size effect. The heterogeneity of the materials introduces an internal length scale, which leads to size dependent fracture behaviour. In fact when considering the fracture of structures at least two different size scales should be considered, viz. the size of the heterogeneity and the size of the flaw(s) which may develop in the material when some external load is applied.

In this chapter a number of scaling effects in the compressive and tensile fracture of concrete will be addressed. Of main interest is the physical interpretation of the size effect. Attention will be given to fracture initiation in concrete specimens subjected to uniaxial tensile or uniaxial compressive load. In considering the fracture of the specimen as a three-dimensional growth process, the nucleation and propagation of a critical flaw must be understood. Basic to the *size effect on strength* is the toughness to flaw nucleation in a specimen. Here we have to consider the statistical variation of material properties, as well as geometry (size and shape) related effects such as strain gradients. When (macro)crack growth is considered, the possible toughening in the material should be clarified (*size effect on localisation*).

Helpful in the discussion is the distinction between different size levels. The fracturing of concrete will be considered at the specimen level (typical dimenson 100 mm) and at the grain level (typical dimension 10 mm). In the remainder of this chapter we will refer to these levels as macro- and microlevel respectively. The macrolevel is the normal level of observation in the laboratory. At this level the material properties for structural engineering are determined. For a thorough understanding of the respective toughening mechanisms, the micromechanics of failure should be studied.

With the introduction of the fictitious crack model by Hillerborg and co-workers [1], the existence of a 'region of discontinuous cracking in front of the continuous (visible) crack' was hypothesised. The existence of such a zone, or at least the elastic equivalent of a crack system for which the correct stress versus displacement ratio and correct fracture energy is known, was essential for explaining the fracture response of various (unreinforced) concrete structures. The above hypothesis implies microcracking ahead of the continuous crack as a possible and sole toughening mechanism. Many researchers have tried to measure the extent of this zone of diffuse microcracking, but the outcomes of the various investigations are extremely contradictory (see the extensive survey by Mindess [2]). Some recent results on the nature of the toughening mechanism in brittle disordered materials will also be presented.

The first part of the chapter deals with experimental observations, both at the macro- and microlevel. The relation between structural changes (crack growth) and load–deformation response is studied. Both in the tensile and compressive case, self similar crack patterns are observed, which suggests that the geometry of the patterns is fractal [3]. In the case of tensile cracking a specific type of flexural crack interface grain bridging is found, which scales in various materials over nine orders of magnitude. In the second part a comparison with various theoretical models is made. For engineering purposes, a continuum description of cracking and localisation is preferred. The validity of recently proposed non-local (e.g. Ref. [4]) and so-called gradient theories (e.g. Ref. [5]) for localisation is discussed in relation to the new experimental findings. On the other hand, micromechanical models may be used for analysing the fracturing of the material in depth. Important in these latter approaches is that the geometry of the fractures as well as the micromechanisms are reproduced to a high degree of accuracy. Numerical micromechanics on the basis of finite element methods (e.g. Ref. [6]), or lattice models (e.g. Ref. [7]) may be helpful in studying the micromechanisms of fracture.

It should be realised that when micromechanics models are employed, the size effect will be an integral part of the solution. The models mimic the microstructure of the composite in detail and various length scales are directly introduced. For obvious reasons this is not the case in continuum-based models, and some additional measures have to be taken to capture the heterogeneity of the material.

2. MACROSCOPIC OBSERVATIONS OF FRACTURE

2.1 Uniaxial Tension

In line with the point of view given in the Introduction, both the size effect on strength and localisation will be discussed. As regards the study of the size effect, specimens should be scaled in all three spatial directions by a constant factor. Normally this is not done, and in fact a shape effect is determined rather than a size effect. The term size effect is then valid only if the problem is truly two-dimensional, and no out-of-plane deformations occur.

As far as the size effect on strength is concerned, the effect of varying the maximum size of the particles in a specimen with geometry of constant size can be studied. In this way, a situation is created where the heterogeneity of the specimen decreases according to the D/d_a ratio, where D is the smallest structural size of the specimen and d_a the maximum size of the aggregate particles. Next to this, the size effect on strength in tension will depend on the curing conditions, and strain (or displacement) gradients directly related to the specimen geometry and loading configuration. Some experimental results will be presented in the following sections.

2.1.1 Non-uniform Fracturing

With the introduction of the fictitious crack model [1], the uniaxial tensile test (using small specimens) was recommended for determining the cohesive stress in the fracture process zone (Fig. 1). Essential was that a '*uniform*' process zone would develop over the specimen's cross-section. The model is based on an analogy with the Dugdale/Barenblatt models for plastic metals, where the plastic crack tip zone is replaced by a zone of diffuse microcracking in front of a macrocracktip. As a result of the relatively small tensile strength of concrete, this process zone would be several orders larger than the plastic zone in metals.

The determination of the cohesive properties of the crack from a uniaxial tensile test is not straightforward. In Ref. [8] it was shown

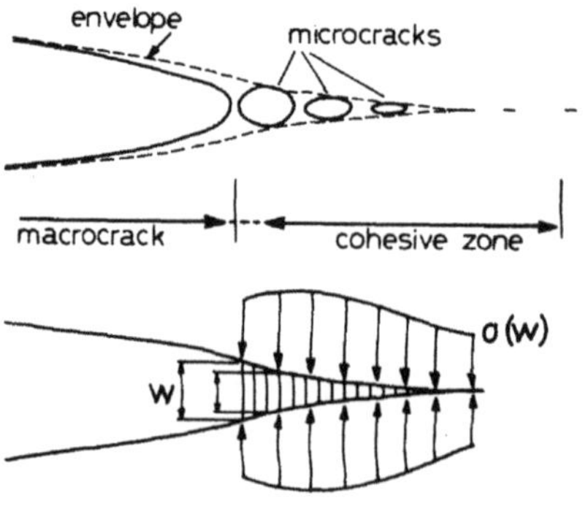

Fig. 1 Microcrack zone in front of a macrocrack and cohesive softening model.

that the fracturing of the tensile specimens is far from uniform because crack nucleation is always from a single (weakest) link in the specimen. From this point the 'crack' will gradually propagate through the specimen's cross-section, which can easily be demonstrated using surface deformation measurements, see Ref. [9]. Bascoul *et al.* [10] and Swartz & Refai [11] have shown that specimens fracture along a curved crack front. Consequently, the surface crack measurements do not reflect the extent of cracking in the specimens interior. All this implies that the fracturing must be considered as a *three-dimensional growth process.*

The non-uniform fracturing in a specimen loaded between non-rotating loading platens leads to a peculiar bump in the descending branch of the load–average crack opening diagram, as shown in Fig. 2. In this figure also the sequence of 'cracking', derived from surface deformation measurements is indicated in the inset. A simple explanation for the occurrence of the bump is the non-uniform fracturing of the specimen. During crack growth from one of the notches, the load eccentricity will increase, thus generating a closing bending moment in the specimen. At the end of the plateau, the crack has reached the other side of the specimen, a process which will usually be accompanied by a dynamic jump in the load–deformation diagram depending on the type of concrete and the specimen size. The problem has gained considerable attention in the past few years, viz. Refs [8], [12]–[17].

When the existence of cohesive softening is accepted, the testing machine stiffness for measuring a stable softening branch can be

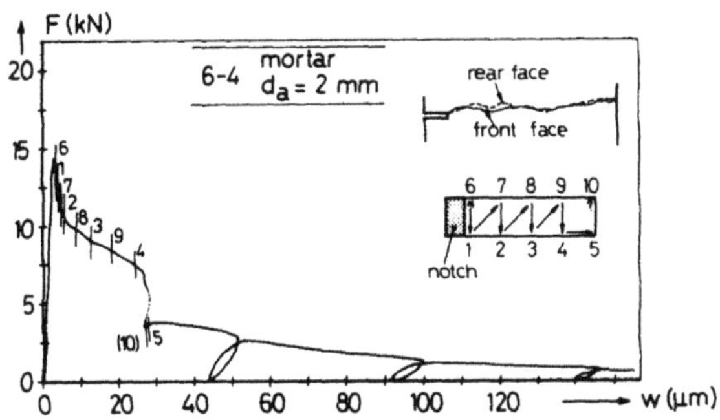

Fig. 2 Load–deformation diagram of a uniaxial tensile test (Single Edge Notched) loaded between fixed end-platens, after Ref. [9].

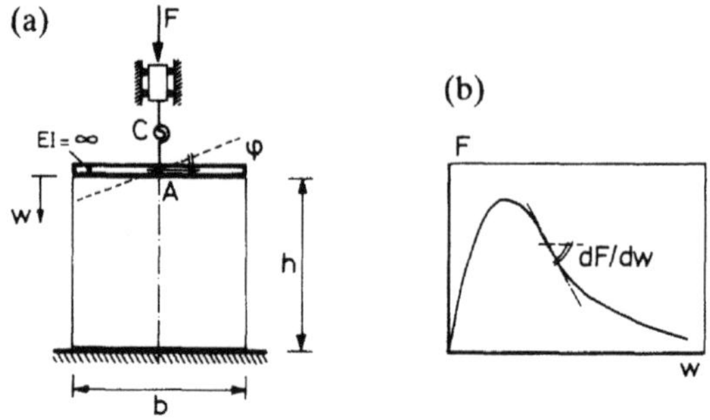

Fig. 3 Specimen/machine geometry (a) and assumed softening law (b) for stability analysis [17].

derived from a simple and straightforward stability analysis. Consider for example the specimen/machine configuration of Fig. 3(a). Assume that the specimen is subdivided into infinitesimal small bar elements in the direction parallel to the loading direction. Each of the bar elements behaves according to the softening law of Fig. 3(b). The load (F_{bar}) carried by a single bar element is equal to

$$F_{bar} = \frac{F_s \, . \, (w_{bar})}{b \, . \, d} \tag{1}$$

where F_s is the total load carried by the specimen; b and d are the width and depth of the specimen. The stability of the system is governed by the moment equilibrium around point A (Fig. 3(a))

$$M_A = M_m + M_s \tag{2}$$

where M_m reflects the rotational stiffness of the testing machine, and M_s the bending moment in the specimen. When a rotation $\Delta\phi$ is given, then

$$M_m = -C \, . \, \Delta\phi \tag{3a}$$

and

$$M_s = -\frac{b^2}{12} \cdot \frac{dF}{dw} \cdot \Delta\phi \tag{3b}$$

Note that as $F = F_s$, the subscript will be left out in the remainder. Equation (2) can now be written as follows

$$M_A = -\left(C + \frac{b^2}{12} \cdot \frac{dF}{dw}\right) . \, \Delta\phi \tag{4}$$

An experiment will become unstable when

$$C+\frac{b^2}{12}\cdot\frac{\mathrm{d}F}{\mathrm{d}w}<0,$$

or differently when

$$\frac{\mathrm{d}F}{\mathrm{d}w}<-\frac{12\,.\,C}{b^2}=\left\{\frac{\mathrm{d}F}{\mathrm{d}w}\right\}_{\mathrm{crit}} \tag{5}$$

Thus, a stable test can be obtained only for a material with softening $\mathrm{d}F/\mathrm{d}w$ when the combination of specimen size and rotational stiffness C of the testing machine fulfils eqn (5). In the analysis it is assumed that the size of the softening zone is equal to the specimen height. This assumption may be correct for compressive loading, yet for tension notched specimens are often used, and generally deformations will localise in a narrow zone. However, the same principle as above may be applied, only a more complex geometry has to be dealt with.

Based on the above described principle, a number of sensitivity studies were carried out by various researchers ([12], [13], [17]), who used various descriptions for the cohesive softening behaviour of the material. A result of the analysis by Hordijk & Reinhardt is given in Fig. 4. Here the effect of rotating versus non-rotating loading platens and the effect of varying the specimen length are shown. In some cases snap-back phenomena were observed. In Section 3.3 this will be elaborated further. Zhou Fanping showed the sensitivity to load eccentricities as well as the dependence of the plateau size (bump) in the descending branch on specimen geometry and size. For example, a load eccentricity of 2·5 mm on a specimen of 50 mm width would lead to a decrease of (apparent) tensile strength which is 25% lower than

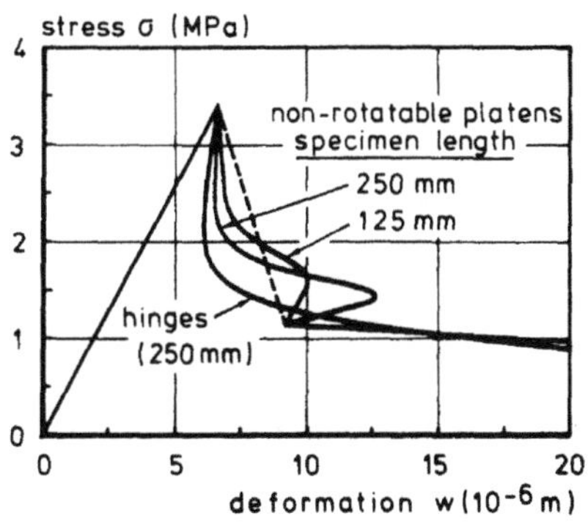

Fig. 4 Influence of boundary condition and specimen length on the softening diagram in uniaxial tension [13].

the true tensile strength. A 10% reduction is already obtained when the load eccentricity is 1 mm.

The major conclusion is that the tensile strength measured in a uniaxial tensile test must be viewed with some scepticism. Furthermore, the shape of the softening curve, measured in a uniaxial tensile test as well, is susceptible to changes in boundary conditions and specimen geometry. Although the use of small specimens is recommended to avoid part of the problem, some doubt about the validity of this approach will always remain. The cohesive softening approach leaves us with the problem that the input softening law cannot be measured directly in an experiment; always an indirect method has to be followed as described in Section 2.3.

2.1.2 Effect of Maximum Aggregate Size

The influence of maximum aggregate size is shown in Fig. 5. In these experiments small prismatic specimens of size 200 mm × 100 mm × 50 mm, single edge notched at half height and made of concrete containing small (2 mm) or large (16 mm) aggregates, were loaded in uniaxial tension. The notch was a 15 mm deep, 5 mm wide sawcut. Control variable in the experiments was the average deformation in the notched region measured with four LVDTs mounted at the

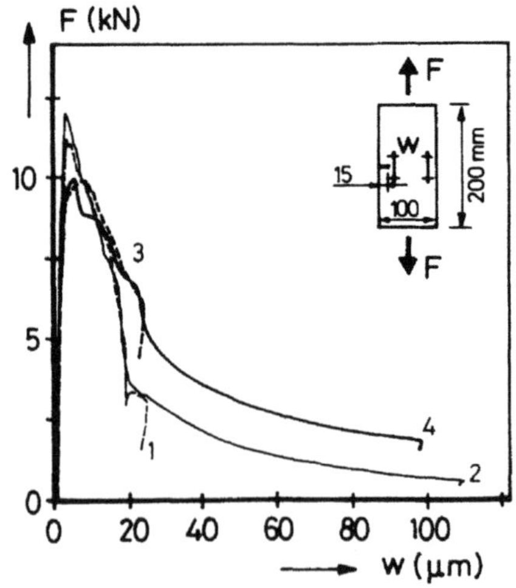

Fig. 5 Influence of maximum aggregate size on load–deformation behaviour in uniaxial tension (see also Table 2).

specimen corners (measuring length 35 mm, see inset of Fig. 5). The complete load–deformation curves are given, both for the 2 mm and 16 mm mix. The details of the experimental technique are given in [18].

Some interesting observations can be made from this figure: the average tensile strength decreases with increasing maximum aggregate size (see also Table 2 in Section 3); the variation in peak stress was considerably larger in the 16 mm experiments; and the slope of the first part of the softening branch was steeper for the small grained material. Note that the 2 mm mortar tests show the same dynamic jump as the result of Fig. 2, indicating that the combination of machine stiffness and specimen size was not optimal yet for measuring the softening response of this material (see Section 2.1.1). Most remarkable however is the rather large difference in load carrying capacity in the tail of the descending branch ($w > 20$–$25\ \mu m$). The experiments confirm previous findings by Petersson [19]. The tests of Fig. 5 were the basis of impregnation tests to study the internal cracking of tensile specimens (see Section 3). Therefore they were unloaded at pre-determined axial crack openings.

Effectively the above results imply that different softening diagrams must be specified when different materials are used. The differences become even larger as soon as fibres are added to the mix, see for example Ref. [20].

2.1.3 Effect of Non-uniform Drying Out

A very serious complication in assessing the size effect is the influence of curing conditions. In Table 1, the strength results of a series of

TABLE 1 Influence of specimen thickness, control mode and curing condition on uniaxial tensile strength (MPa)

Specimen thickness	*Curing condition*	*Control position (Fig. 6)*	
		Middle	*Notch*
50 mm	Oven	2·03 (0·07/2)	2·56 (0·32/2)
	Lab	1·70 (0·19/4)	1·98 (0·18/3)
20 mm	Oven	2·71 (0·28/4)	3·03 (0·05/2)
	Lab	3·06 (0·06/3)	3·13 (0·21/3)

Numbers in brackets denote the standard deviation (in MPa) and number of successful tests.

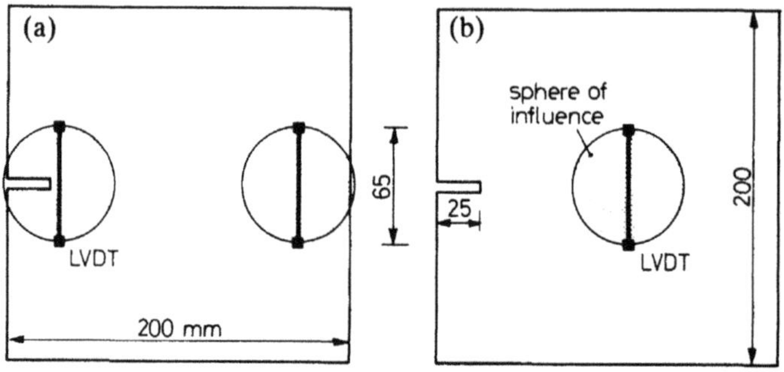

Fig. 6 Notch (a) and middle control (b) and definition of 'sphere of influence' of control LVDTs.

uniaxial tensile experiments on single edge notched specimens of size 200 mm × 200 mm are given. The notch was a 25-mm deep, 5-mm wide sawcut. The specimen geometry is shown in Fig. 6. The specimens were loaded in displacement control, either using the average signal of two centrally placed LVDTs (Fig. 6(b)) as feedback signal in the closed loop test-rig, or by using the average signal measured with four corner LVDTs (Fig. 6(a)). Main variables in the investigation were the specimen thickness (viz. 20 mm or 50 mm) and the curing condition (viz. oven-dried at 60°C, or lab-dried at 17–18°C, 50% RH, both after 14 days of underwater hardening). The age at testing varied between 35 and 42 days.

The results of Table 1 indicate that a pronounced influence of curing condition on peak stress exists. The largest differences are observed for the thicker specimens. The explanation for the observed response is that due to non-uniform drying in the laboratory, stress-gradients will develop through the specimen's cross-section. Tensile stresses will

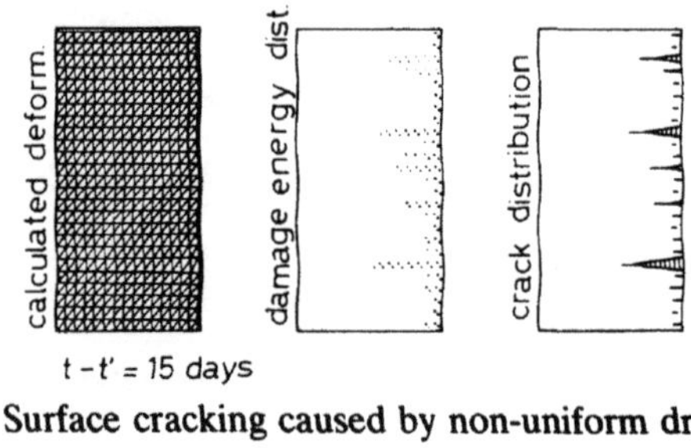

Fig. 7 Surface cracking caused by non-uniform drying [21].

develop near the specimen's surfaces, whereas compressive stresses will appear in the interior, see Fig. 7 [21]. The oven-dried specimens will suffer less from this effect, and indeed in this case the effect may be reversed when the oven-dried specimen is brought into the laboratory environment (50% RH) for testing. The oven-dry specimen will absorb moisture from the air, and likely the moisture gradient will be reversed as compared to the lab-dried specimens: compressive stresses may develop near the specimen's surface, tensile stresses in the interior. The presence of these eigenstresses explains the differences in strength values presented in Table 1. The severity of the effect will depend on the absolute size of the specimen. It should be noted that the effect depends on the moisture gradient, which in turn depends on the drying time [22].

Another effect which may be derived from Table 1, and which is closely related to the moisture gradient effect, is the influence of LVDT position on strength. A higher strength was measured when the feedback signal in the test control was the average signal of the 'notch LVDTs' (Fig. 6(a). Most likely the development of a critical notch is delayed in these experiments, or put differently, the crack growth is more confined. The 'free growth length of the crack' outside the sphere of influence of the LVDTs is reduced by placing them near the notch where the crack will nucleate (Fig. 6).

2.2 Uniaxial Compression

2.2.1 Non-uniform Fracturing

Non-uniform deformations in specimens subjected to uniaxial compression have been observed as well, especially in the softening branch of the complete load–deformation curve, see Fig. 8 after Ref. [23]. In these particular experiments on a 16-mm normal concrete, the compressive load was applied via brush bearing platens (effective rod length 85 mm, 19 × 19 rods of size 5 mm × 5 mm, inter-rod distance 0·2 mm). The brushes are used for eliminating the effect of shear at the specimen-loading platen interface due to differential lateral expansion of the steel loading platens and the concrete specimen. Similarly as in the uniaxial tensile case, fracturing will start from a single initiation point, and grow through the specimen's cross-section. At some moment the deformation distribution is uniform again, and will remain uniform until the end of the experiment. This might point to an 'intact core' in the specimen, or to some kind of bridging action in the

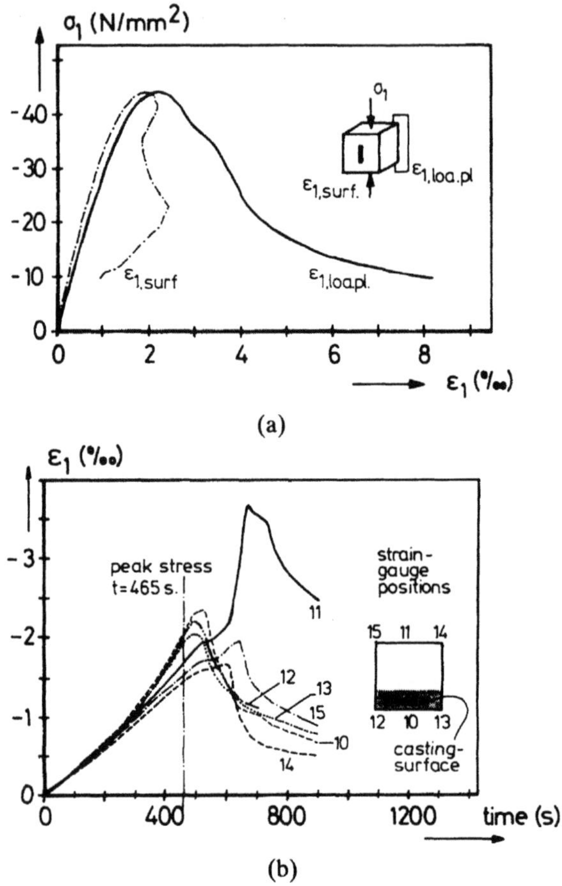

Fig. 8 Non-uniform fracturing in uniaxial compression.

cracks. Note that even by loading the specimens through brushes, cracks will traverse the specimens still at a small inclination [23]. This may either be explained from the observation that the cracks will deviate from the vertical due to aggregate–crack interactions as shown by Zaitsev & Wittmann [24], or from the fact that the boundary shear forces are not completely eliminated by the use of brushes.

2.2.2 Effect of Boundary Shear

When concrete prisms of varying height are loaded between either dry steel loading platens or between flexible platens [25], completely

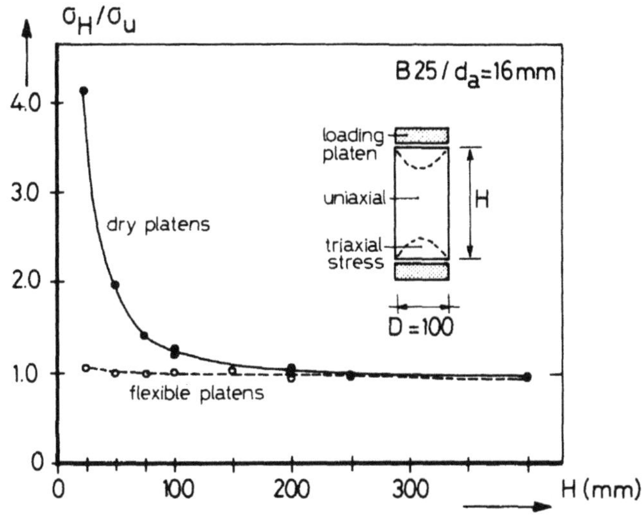

Fig. 9 Effect of prism slenderness and loading system on the uniaxial compressive strength (σ_H). σ_u is the uniaxial compressive strength at $H = 100$ mm for the flexible platens and at $H = 250$ mm for the dry platens [25].

different strength results are obtained, see Fig. 9. In the case of dry steel platens, the 'apparent' compressive strength of the concrete will increase considerably when the prism height is decreased (viz. up to a factor of four when $H/D = 0{\cdot}25$–$0{\cdot}5$). In contrast, when flexible platens are used, the effect disappears, and almost the same strength is measured independent of the prism height. The explanation of the phenomenon is the development of a region of high triaxial stress under the dry platens. The effect of triaxial stress will be more pronounced for the lower prisms, where the lateral expansion is more restricted.

Boundary shear has a significant influence on the post-peak response in compression too [26], [27]. In Fig. 10 the results obtained by Vonk *et al.* [27] are given. These results were obtained in the Eindhoven True Triaxial Machine, from which the results of Fig. 8 were also obtained. The short brushes used by Vonk *et al.* are the same brushes as used in the experiment of Fig. 8. The other loading systems are greased teflon platens, long brushes (rod length extended to 110 mm, rod area 5 mm × 5 mm), and dry steel platens. The results clearly reveal that the slope of the softening curves increases with decreasing boundary shear. Moreover, the inclination of the cracks in the

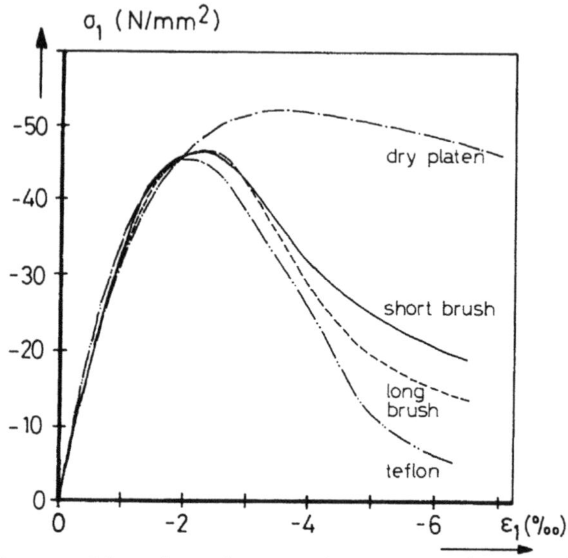

Fig. 10 Influence of boundary shear on the stress–strain curve in uniaxial compression.

specimens changed, depending on the amount of boundary shear. The smallest inclination to the loading axis was found when teflon platens were used. However, when considering the results by Zaitsev & Wittmann [24], it might be questioned whether cracks should run parallel to the loading axis, which is normally considered as proof of a well performed compression experiment.

From this it might be concluded that teflon platens are the 'best' loading system in uniaxial compressive experiments. Yet, recent experiments on prisms with varying shape and size revealed a significant influence of size on strength [28]. In fact when the frictional coefficient of the various loading systems are compared, it is found that brushes perform better (i.e. have the smallest friction coefficient) for small deformations, whereas teflon displays the well known stick-slip behaviour (see Fig. 11). On the other hand as soon as this stick-slip behaviour is overcome, the teflon platens perform better in comparison with the brushes. Evidently the 'best' loading system for the complete regime does not exist.

2.2.3 Effect of Specimen Height

In 1984, van Mier [23] published results on the complete stress–strain and stress–post-peak displacement behaviour measured on prisms of

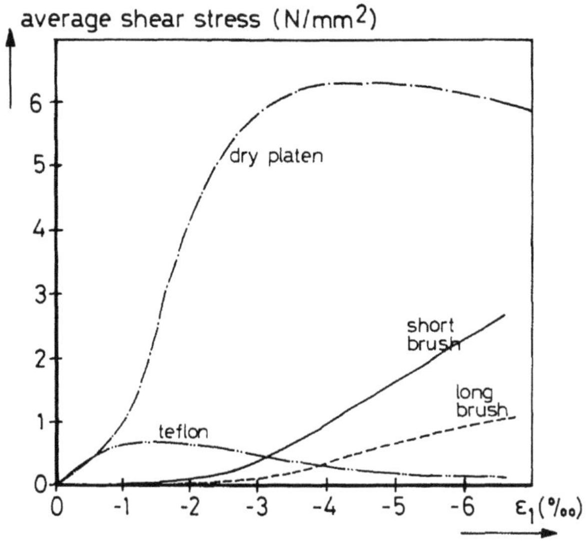

Fig. 11 Calculated shear stress between loading platen and specimen.

constant cross-section ($A = 100\,\text{mm} \times 100\,\text{mm}$) and varying height (viz. 50 mm, 100 mm and 200 mm). The specimens were loaded in uniaxial compression between the short brushes mentioned before. The results are reproduced in Figs. 12(a) and (b). The peak strength was not affected, as would be expected on the basis of the earlier results obtained by Schickert [25], see Fig. 9, and others. The pre-peak stress–strain response was almost identical for the three specimen heights investigated (Fig. 12(a)). In contrast, the post-peak stress–strain curves were completely different, and showed a decreasing slope with decreasing slenderness H/D. However, when the strains were translated to displacements (thus taking into account the specimen height), following

$$\delta = \varepsilon \,.\, H - \varepsilon_{\text{peak}} \,.\, H \tag{6}$$

with $\varepsilon > \varepsilon_{\text{peak}}$, identical post-peak response was measured, indicating that localisation occurs and that some sort of fracture energy approach might be applied in compression too (see Fig. 12(b), [23]). Recently the idea was further elaborated by Hillerborg [29], who used the compressive fracture energy concept for analysing the rotational capacity of reinforced concrete beams.

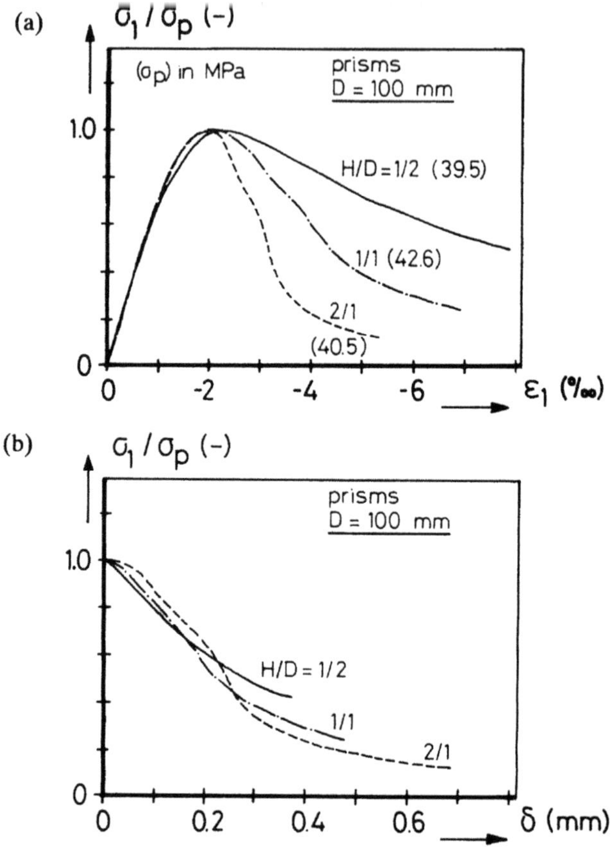

Fig. 12 Influence of specimen height on the stress–deformation response in uniaxial compression [23].

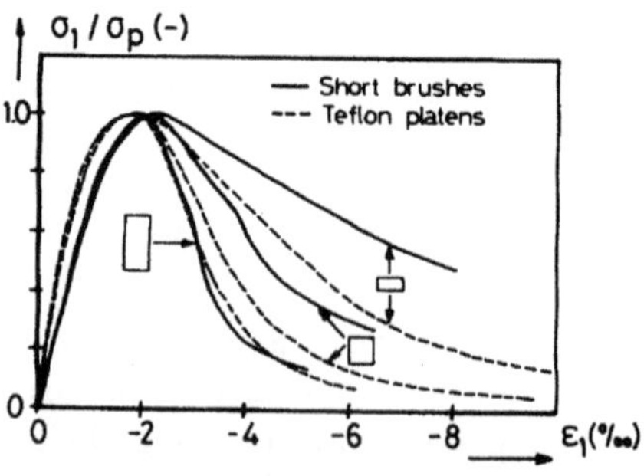

Fig. 13 Influence of specimen height on the stress–strain curve in uniaxial compression, comparison between tests with brushes and teflon.

The compressive size tests were recently repeated by Vonk *et al.* [28], but now teflon platens were used instead of brushes. A comparison between the respective data sets is given in Fig. 13. As expected on the basis of the friction tests of Fig. 10, the softening curves (in terms of stress and strain) are steeper when teflon is used. Furthermore, the unrestricted translation to stress–displacement curves is not possible anymore when the teflon results are considered.

The most likely explanation for the observed differences is that the teflon still has a significant influence on peak strength. This implies that differences may exist in crack nucleation in a specimen loaded between teflon, or in a specimen of identical size but loaded between brushes, leading to different crack (or rather fracture) propagation modes. The results of Fig. 13 suggest that fracture in compression is not a completely surface dominated phenomenon (as is the case for tensile fracture), but rather a mixture between surface and volume dominated mechanisms. It should be mentioned that localisation of deformation in shear bands is observed in triaxial compression (see Ref. [23]), which would suggest that the fracture in compression is a surface dominated phenomenon only. Future research on the intimate relation between boundary shear and size/shape effects in compression should clarify this.

The fact that the compressive strength can be 'regulated' depending on the boundary shear and the size of the specimens makes the determination of the exact (whatever that is in concrete technology) 'size effect' in compression extremely difficult, if not impossible.

2.3 Direct Determination of Macroscopic Fracture Parameters

2.3.1 Inverse Modelling

Specimen and boundary effects should be eliminated when the true fracture properties of a material are to be determined. An important observation is that in heterogeneous materials, fracturing will always start from the weakest discontinuity. For concrete and mortar, which may suffer from eigenstresses due to non-uniform drying out after the manufacturing of specimens, crack nucleation will primarily start at the specimens' surfaces.

The experimental difficulties are tremendous, and most likely the direct measurement of the cohesive stress distribution in the fictitious crack (or in a compressive zone) is impossible. Therefore a procedure of inverse modelling must be followed, as proposed in Ref. [30]. In such an approach the results of a number of non-linear finite element analyses, adopting different computational bilinear softening diagrams, are compared with the outcome of a 'simple' displacement controlled experiment on concrete beams. The bilinear diagram leading to the best fit is then defined as the 'material property'. The computational softening diagram proposed by Wittmann *et al.* [30] is plotted in Fig. 14. The diagram proposed has an extremely steep first slope and a shallow long tail as second branch. The parameters which define the diagram, viz. the tensile strength of the material, the coordinates of the intersection between the two line segments and the maximum crack-opening, depend on the various material parameters such as matrix strength, aggregate–matrix bond strength and aggregate strength, whereas curing conditions also may be of importance as described before.

Using non-linear finite element analysis, it was found that the non-uniform opening observed in uniaxial tensile tests on concrete prisms loaded between non-rotating end-platens can only be modelled if the first part of the computational softening diagram is extremely steep (see Ref. [8], the upper bound for non-uniform opening is indicated in Fig. 14). These results are in agreement with the diagram proposed by Wittmann *et al.* [30]. The bilinear diagram is a rather rough approximation, and more refined fitting procedures may be used to define the computational softening diagram (e.g. Refs [13], [31]). The term *computational* is used here in order to emphasise that the softening diagram cannot be measured directly. It can be regarded as

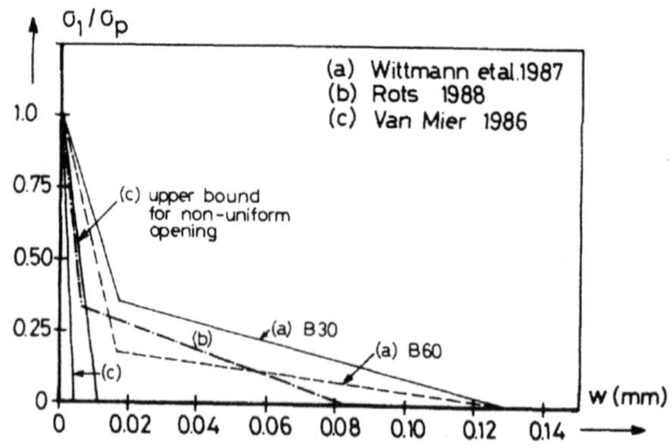

Fig. 14 Computational softening diagram in tension.

the elastic equivalent which is needed in structural analysis; however, the physical mechanisms underlying softening remain completely obscure.

The stability analysis presented in Section 2.1.1 can also be used for determining the slope of the first linear branch of the computational softening diagram (see Ref. 32). The transition from stable to unstable response of specimens of varying size, tested in a machine with known rotational stiffness *C*, will yield the slope of the steep part of the computational softening diagram when eqn (5) is used. Note that eqn (5) must be adjusted for different (notched) geometries.

2.3.2 Physical Interpretation of Softening

As mentioned, the elastic equivalent does not reveal the mechanisms underlying softening. On basis of the bilinear diagram of Fig. 14, Wittmann *et al.* [30] proposed that different physical processes would be responsible for the two different branches of the diagram: large scale debonding of aggregates in the steep part, and frictional pull-out of aggregates in the shallow tail of the diagram. As a matter of fact, recent experimental results, which will be presented in the following section indicate that Wittmann and co-workers were close to the 'truth'.

The growth of 'macrocrack-like' structures at the specimen's surface in uniaxial tensile experiments can be demonstrated using reflection photoelasticity, see Fig. 15. The subsequent stages of 'crack' growth clearly show the discontinuous nature of the fracturing in the descending branch of the load–deformation curve. The deformations indicated in Fig. 15 are the average crack openings measured with two LVDTs (measuring length 65 mm), mounted in the middle of the specimen (size 200 mm × 200 mm × 50 mm), both at the front and back side of the specimen. For all details regarding the test technique, the reader is referred to Refs [9] and [33]. Although the full cross-section of the specimen seems to be fractured at a crack-opening of 31·8 μm, surprisingly, still a considerable load-transfer is possible. The following mechanisms might explain the load-transfer [16]: (a) frictional pull-out of aggregates; (b) the existence of an intact core due to non-uniform drying-out; and (c) flexural resistance of intact ligaments between overlapping crack segments. In fact the latter hypothesis was partly based on the photoelastic result of Fig. 15, which shows the development of overlapping interacting cracks at the specimen's surface. The

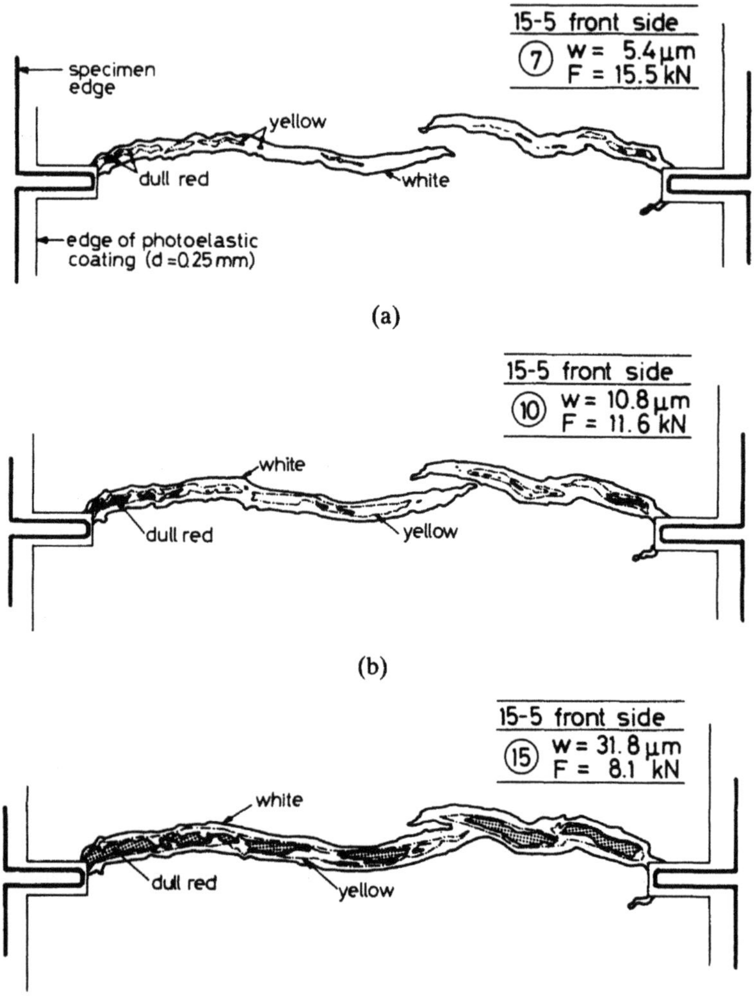

Fig. 15 Photoelastic reflection technique for crack detection in a uniaxial tensile test. Double edge notched plate loaded between fixed end platens.

various phenomena will be discussed in depth in the following sections.

3. STRUCTURAL CHANGES UNDERLYING MODE I FRACTURE

An explanation of softening in terms of structural changes can be found at the microlevel. Recently a number of vacuum impregnation experiments were carried out in order to link the structural changes with the observed load–deformation response in uniaxial tension. In this section the technique and results are presented.

3.1 Vacuum Impregnation Experiments

Single edge notched concrete specimens were loaded in uniaxial tension between non-rotating end-platens. A specimen was loaded up to a prescribed axial deformation at a displacement rate of 0·08 μm/s. The tensile load was applied parallel to the direction of casting. As soon as the prescribed axial deformation was reached, the specimen was unloaded to zero load, and four stiff steel bars were fixed between the glue-platens. Subsequently the specimen with the glue-platens still attached was removed from the test rig and placed in a small steel box for impregnation. The specimens were vacuum impregnated using a low viscosity fluorescing epoxy (110 mpas at 23°C). After impregnation, the specimen was sawn into six slices of approximately 15-mm thickness as indicated in Fig. 16. Following this, the cracks were recorded using ultra-violet photography. The full details of the experimental technique are given in Refs [18] and [34]. Variables in

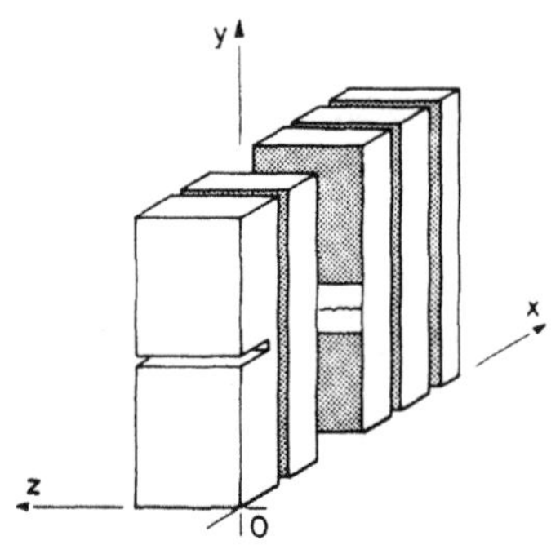

Fig. 16 Slicing of cracked specimens after impregnation.

the investigation were the axial crack-opening and the maximum aggregate size of the concrete (2 mm and 16 mm).

3.2 Relation between *F–w* Response and Crack Growth

In Fig. 5 the *F–w* diagrams measured in the 2 mm and 16 mm impregnation experiments are shown. The unloading displacement is easily identified from these figures: a part of the unloading curve is drawn for each test. The peak loads, displacements and loads at the unloading points of all experiments are summarised in Table 2. The major difference between the *F–w* curves was the residual load, which was approximately three times higher for the coarse grained material.

An example of a crack-graph is shown in Fig. 17. In this figure the cracking over the specimen's depth (*z*-direction, see Fig. 16) is shown at five locations in the specimen. The locations are identified with an *x*-coordinate, which is the distance of the sawcut to the notched face of the specimen as shown in Fig. 16. Besides the information on crack extension over the depth of the specimen, the photographs reveal the complex crack geometry in the *y*-direction, viz. in the direction parallel to the applied tensile loading. Note that originally the five sections of Fig. 17 were parallel to one another. For visualising the

TABLE 2 Results of impregnation experiments

Specimen[a]	*Control length*	d_a *(mm)*	*Age (days)*	F_{max} *(kN)*	σ_{peak} *(MPa)*	w_{unl} *(μm)*	F_{unl} *(kN)*
16A012(1)	35	2	105	11·14	2·64	24·6	2·91
13A005	35	2	104	11·94	2·84	49·5	1·82
14A007(2)	35	2	97	11·99	2·89	109·4	0·60
12A003	35	2	100	10·89	2·60	200·1	0·19
12A004	65	2	107	12·10	2·88	24·7	3·29
15A009	65	2	106	10·70	2·52	49·8	1·65
15A010	65	2	91	10·84	2·55	112·7	0·77
14A008	65	2	99	11·10	2·66	199·4	0·11
41A026	35	16	99	9·49	2·30	8·7	8·14
42A027	35	16	103	11·01	2·59	10·1	8·81
41A025(3)	35	16	100	9·87	2·37	23·9	5·93
44A032	35	16	105	11·18	2·70	47·7	3·39
45A034(4)	35	16	106	9·98	2·36	97·0	1·81

[a] The numbers in brackets after the specimen codes refer to the four *F–w* diagrams of Fig. 5.

Fig. 17 Crack-graph for specimen 16A012, impregnated after loading up to 24·6 μm. $d_a = 2$ mm; $w_{unl} = 24{\cdot}6$ μm; $F_{unl} = 2{\cdot}91$ kN.

extent of cracking, all slices have been projected on the same plane next to each other.

The crack extension can be followed by comparing the results from the subsequent experiments. In Figs 18(a) and (b) the crack fronts at 25 μm and 50 μm average axial crack opening for respectively the 2 mm and 16 mm mixes are shown. The shaded areas reflect the cracked area. Fracturing along curved crack fronts [10], [11], was confirmed. Remarkably, the crack extension is from the *un*-notched side in the 2-mm specimens. The companion specimen (12A004, see Table 2) behaved completely identically. As shown in Fig. 19, crack growth at the surface (indicated by an increase of local axial deformation) was from the notch, which corresponds to previous observations [9]. In Fig. 19, the local deformations at three stages in the descending branch are shown: at $F_{res} = 10$ (just beyond peak), 8 and 2·9 kN (the unloading point at 24·6 μm, see Table 2). In contrast, fracturing (as observed from the impregnation tests as well as from the

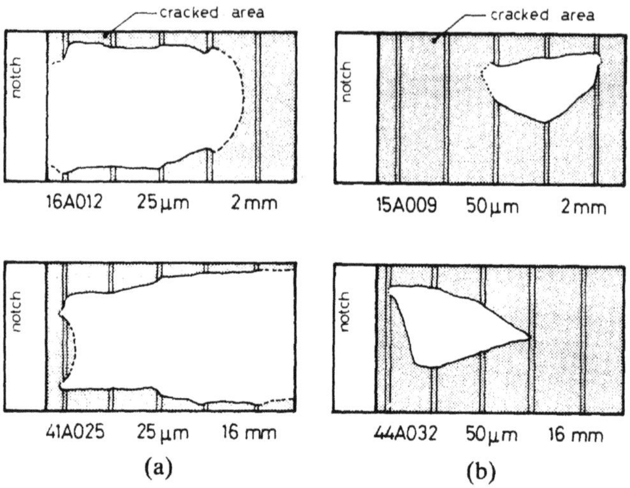

Fig. 18 Crack fronts derived from the impregnation experiments: (a) at 25 μm for the 2 mm and 16 mm mix; (b) *idem.* at 50 μm.

surface deformation measurements) started from the notch in the 16-mm specimens. This was observed in all specimens of this mix loaded up to 10 μm and 25 μm (viz. specimens 41A026, 42A027 and 41A025, see Table 2). This phenomenon will be explained after the presentation of some other (related) results.

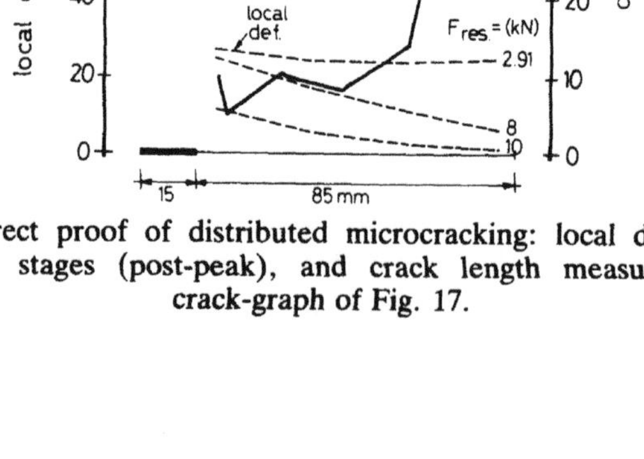

Fig. 19 Indirect proof of distributed microcracking: local deformation at three loading stages (post-peak), and crack length measured from the crack-graph of Fig. 17.

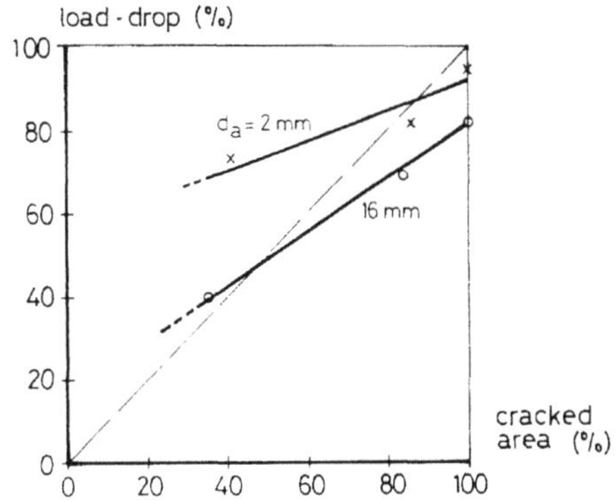

Fig. 20 Load-drop (= peak load minus residual load) versus cracked area for the impregnation experiments.

When the cracked area is compared with the load-drop measured in the $F-w$ diagrams, the relations of Fig. 20 are obtained. Clearly the cracked area is too small to explain the large load-drop, especially for small crack openings (up to 50 μm) in the 2-mm mix. Two explanations are possible, viz. either the 'core' of the specimens is prestressed due to non-uniform drying or, alternatively, some internal microcracking takes place. Note that the development of discontinuous microcracks in the specimen's interior cannot be detected by the impregnation technique, only the continuous surface cracks will be revealed.

3.3 Microcrack Growth and Coalescence

In analysing the response of a cohesive softening zone of the type 'microcrack cloud in front of a stress free crack', in analogy with the Dugdale–Barenblatt model for plastic metals, Ortiz [35] derived the following expression between the cohesive stress σ and the opening displacement w in the plane of the crack (see Fig. 21(a)):

$$w = \frac{\delta U}{\delta \sigma} = -\sigma \frac{8l}{\pi} . \frac{1-\nu^2}{E} . \log \cos \frac{\pi a}{2l} = \frac{\sigma}{k} \tag{7}$$

where l and a are defined in Fig. 21(a), and E and ν are the Young's modulus and the Poisson's ratio of the material which is considered as an isotropic continuum. The stress intensity factor for an array of

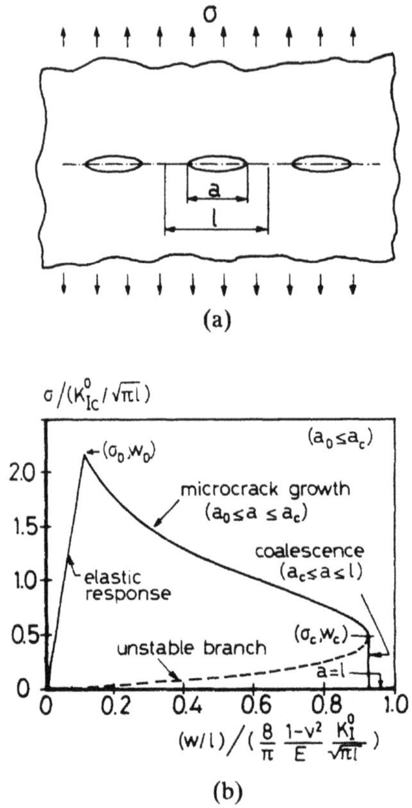

Fig. 21 Effect of microcrack growth and coalescence: (a) regular array of microcracks and (b) dimensionless stress–deformation response [35].

collinear cracks is given by Rice [36]:

$$K_{\rm I} = \sigma\sqrt{\pi l}\,.\,\sqrt{\left(\frac{2}{\pi}\tan\frac{\pi a}{2l}\right)} \tag{8}$$

Microcrack growth will occur only when the crack opening has increased to say $w_0 = k\,.\,\sigma_0$ at which point the critical stress intensity factor $K^0_{\rm Ic}$ of the material is exceeded. Up to this point the σ–w relation is linear as shown in Fig. 21(b). Upon further increasing crack opening, the traction decreases following the equilibrium path $K_{\rm I} = K^0_{\rm Ic}$. A striking feature of the equilibrium path was found by Ortiz [35]: at a microcrack density $a_c/l \approx 0{\cdot}91$, stable microcrack growth cannot be sustained any more. Opening beyond w_c leads to cleavage of the ligament between the cracks, and the cracks coalesce to form a continuous crack. The dynamics of the process were ignored by Ortiz for simplicity, and the load dropped to zero at the critical point as indicated in Fig. 21(b). Similar solutions were obtained by others [37], [38].

When these analytical results are compared with the experiments of the previous section, it might be concluded that in the fine grained material, microcracks seem to coalesce following the above described mechanism at an axial crack opening of approximately 20 μm when the steep part of the F–w diagram is observed in the 2-mm mortar tests (see Fig. 5). On the other hand, it might be argued that the specimen geometry and loading configuration are the cause for the dynamic jump of the crack (see Fig. 4). The stress intensity factor for the problem of a single edge notched plate loaded between non-rotating end platens [39] also leads to an unstable point at relative crack sizes of $a/W = 0{\cdot}95$, where W is the width of the specimen [16]. In this case, the cause for the instability is the cleavage of the ligament which is pulled beyond w_c instantaneously. Van Mier & Schlangen [32] showed that when a deformation gradient exists over the thickness of the specimen, this instability will not occur. At the moment it is not clear how the microcrack coalescence mechanism and the specimen/machine effect interact.

For the coarse grained material, crack growth was found near the notch first (see Fig. 18(a)), but close observation of these results reveals that this concerns debonding cracks near large aggregates (see Fig. 12 in Ref. [34]). The dynamic jump in the test will now be smaller because the closing bending moment (M_s) is smaller due to the increased bridging stress (which causes a smaller load-eccentricity). Note that crack coalescence may be less likely in coarse grained material than in fine grained material. The probability of crack growth in different planes is larger for coarse grained material. Further proof of the development of discontinuous crack branches at small crack openings is shown in Fig. 15. A closer view of these photoelastic results reveals that rather discontinuous fringes are observed for small crack openings, indicating that indeed microcracks develop before the continuous crack emerges. The information from surface cracks may be used when the crack growth sequence is of interest. It should be realised however that the fracture process occurs somewhat earlier at the surface than in the specimen's interior. Therefore, it is impossible to make a direct link between F–w response and surface crack measurements.

3.4 Crack-Interface Grain Bridging

Continuous crack profiles are detected beyond displacements of 25 μm, both in the fine and coarse grained mixes. At 50 μm an internal core was found in the specimens (see Fig. 18(b)). Yet, in view of the findings in the previous section, distributed microcracks may be present in the core. At 100 μm the core has collapsed and the complete cross-section of the specimen is cracked. However, at these crack openings some load can still be carried as indicated in Figs 5 and 20. The mechanism becomes clear upon close observation of the crack geometry: a flexural type of grain bridging has developed in the cracks (Fig. 22). Two overlapping crack tips approach each other, but coalescence seems prohibited due to the presence of the stiff aggregates in the crack path. This type of crack interface grain bridging has been observed in ceramics as well, see Swanson *et al.* [40] and Steinbrech *et al.* [41]. The increased carrying capacity in coarse grained materials can be explained from the larger size of the crack-interface grain bridges as shown in Fig. 22. In fact, from a simple analysis it can be shown that the increase should be a factor of four when the aggregate size is increased fourfold. In the experiments an increase of bridging load by a factor of three has been measured. Note however that in the analysis, flexural failure of the bridges was assumed, whereas failure might be governed by extension of one of the crack branches as depicted in Fig. 23, see Ref. [34].

The existence of crack interface grain bridging can be shown qualitatively by the use of other techniques. In Ref. [42] the so-called double cutting technique was used, and in this way the existence of cohesive forces in the visible (surface) crack was shown. Also acoustic measuring techniques, e.g. Labuz *et al.* [43], or laser holographic measurements may be used (e.g. Castro-Montero *et al.* [44]), yet the true nature of the bridging mechanisms may be revealed by direct observation only.

3.5 Self Similar Crack Patterns

If the results of the impregnation experiments (specifically those of Fig. 23) are compared with the photoelastic 'cracks' of Fig. 15, similar 'bridge patterns' are observed. The experiment of Fig. 15 was carried out on specimens made from the same 2-mm mortar as used in the impregnation experiments. Clearly, bridging also seems to occur at a larger scale, and the large ligament in the middle of the specimen is about 8 mm wide. Theoretical analysis has shown that two approach-

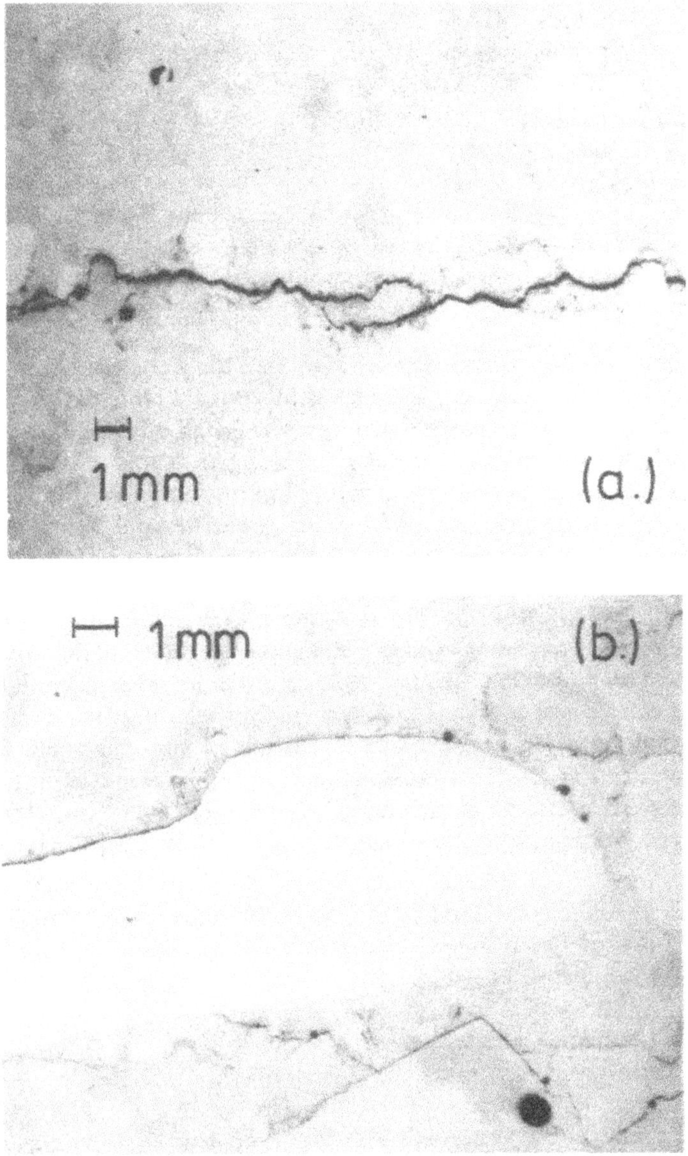

Fig. 22 Crack interface grain bridging at 100 μm average crack opening for (a) 2-mm mortar and (b) 16-mm concrete.

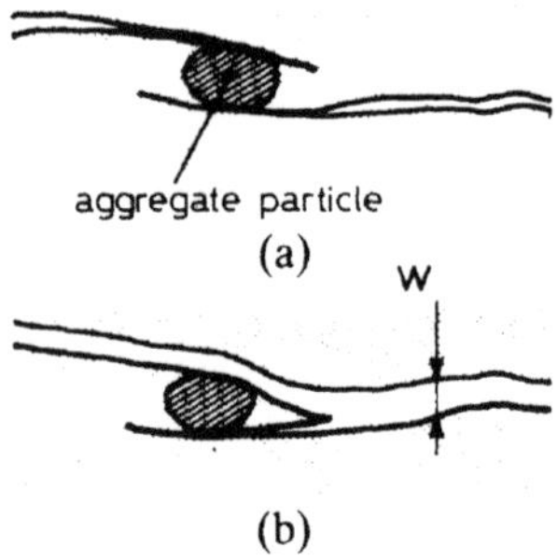

Fig. 23 Interacting overlapping cracks as fundamental stress transfer mechanism in cracks in brittle disordered materials (a), and hypothesised failure mechanism (b).

ing crack tips will avoid each other rather than coalesce [45]. As a matter of fact, as shown by Sempere & Macdonald [46], the cracks will avoid each other more easily when there is a small offset, i.e. when the cracks do not grow in the same plane. This again points to the extreme care which should be addressed in the interpretation of experiments. Figure 15 reveals also the development of a flexural ligament at a smaller scale near the right notch (note the change in fringes from yellow to dull red in these fringes, indicating a local opening of cracks, Fig. 15(b) and (c)). This suggests that the crack patterns are self similar [3], and confirms findings by Sempere & Macdonald [46]. They reported on self-similar shapes over an order of nine magnitudes in glass (with ligament size of 25 μm) and in the Earth's crust (viz. overlapping spreading centres at the ocean floor with a ligament size of 2·5 km), see Fig. 24. This observation is very important, as it may be related to size effects in structures. Clearly the length of a crack is restricted by the size of the structure in which it develops. It might be

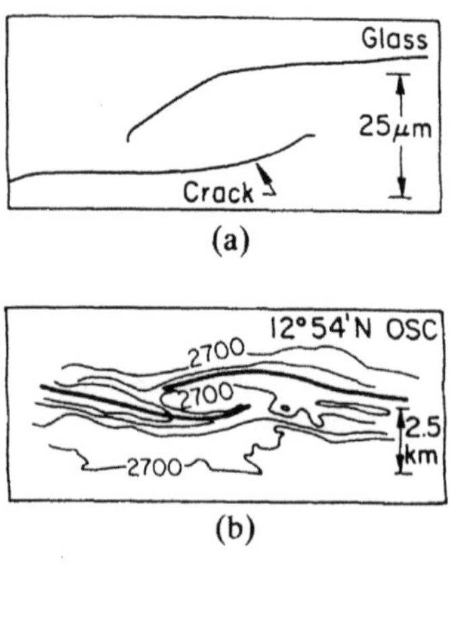

Fig. 24 Interacting overlapping cracks in glass (a) and at the ocean floor (b) [46].

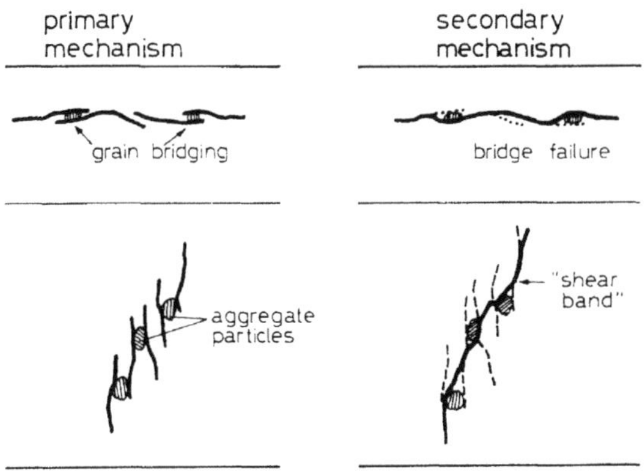

Fig. 25 Primary and secondary crack patterns in tension and compression.

argued that the size of the ligament near crack-overlaps is not only governed by the size of the material heterogeneity, but also by the size of the structure itself. In this way, the capability to stress-redistributions, which seems basic to the size effect in heterogeneous materials, might be affected. Future research should clarify this.

The fact that two approaching cracks avoid each other may lead to the conclusion that the dynamic jump in the $F-w$ diagrams of Figs 2, 4 and 5 is governed by the specimen geometry and loading arrangement only. Crack coalescence as described in Section 3.3 would be of secondary importance only. It would be interesting to determine the conditions under which crack coalescence may occur in heterogeneous materials.

Self-similar crack patterns are observed in compression as well. Shear bands are often formed from en échelon tensile cracks [47]. The en échelon cracks can be found at the specimen level [23], but also at the grain level [48]. The primary and secondary effects in tensile and compressive fracture of brittle disordered materials may be related following Fig. 25.

4. RELATION TO RECENT THEORETICAL DEVELOPMENTS

4.1 Continuum-based Approaches

The experimental results clearly pinpoint the scaling problem in brittle disordered materials. The first length scale introduced in the material is **the heterogeneity of the material** itself, the second length scale is **the size of the cracks in the structure** and now a third length scale is identified, viz. **the size of the crack overlap** which is a direct consequence of the aggregative structure of the material (viz. crack–aggregate interactions). The first two size scales are normally dealt with in fracture mechanics of heterogeneous materials, e.g. Bazant [49]. Crack interactions and crack–aggregate interactions are more difficult to handle.

In the more recent non-local theories the need for an internal length scale was recognised as well [50]. This length scale was determined via a global energy consideration as follows. In large structures, failure is governed by linear elastic fracture mechanics (energy release G_f (J/m^2)), in small structures by the stress–strain curve (energy contained is the area under the curve W_s (J/m^3)). The quotient of these energy quantities yields the internal length scale following

$$l = G_f/W_s \quad \text{(m)} \tag{9}$$

This value is approximately equal to the characteristic length of the non-local continuum [4]. Various values of characteristic length were reported by Bazant. The values are in general related to the maximum aggregate size as $l \approx n \,.\, d_a$, where values of n between 1·88 and 3 are reported. The justification of the non-local models is performed via the analysis of a regular array of microcracks [37].

A different way of handling the heterogeneity of the material in a continuum theory is the use of gradient models (Cosserat continuum, Micropolar continuum) [5], [50]. The introduction of a gradient inevitably leads to an internal length scale in the continuum. In fact this is a rather elegant way of handling the problem, and seems to correspond with the bending in the ligaments between overlapping interacting cracks as described in the previous sections. The characteristic length should then be equal to $l \approx 1 \,.\, d_a$. Du *et al.* [31] also arrived at this 'crack band size' based on laser holography measurements on concrete.

The experimental results described in the previous paragraphs suggest that a *variable characteristic length* should be introduced in the models. Indeed, the width of the localisation zone seems to diminish gradually: fracturing proceeds from microcracking distributed over the complete volume of a tensile specimen to macrocracking which is localised in an infinitesimally narrow zone. Thus, one might argue that during fracturing the characteristic length changes continuously. *In this sense the length scale introduced in the existing higher order continuum models is no more than an average.*

In capturing the scaling in fracture of heterogeneous materials, continuum models have evolved from simple to more complicated. A theory for describing certain phenomena must indeed be as simple as possible, but no more than that. In trying to identify the simplest model possible, one might decide to start the investigation at the other side of the spectrum, viz. by using micromechanics or lattice models.

4.2 Micromechanical Analysis of Fracture

In (numerical) micromechanical models the heterogeneity of the material is directly captured in the schematisation. The advantage of this is that the length scale due to the material itself is directly introduced. The crack-law can be very simple. A disadvantage of the method is that an incredible amount of computer capacity is needed for analysing a simple geometry containing only a few aggregates (see for example the Numerical Concrete developed by Roelfstra *et al.* [6]). The micromechanics models have a wide application. First of all they may be used for obtaining a better understanding of fracture processes. The entanglement between boundary condition influences and specimen size as touched upon in the first part of this chapter might be resolved with the micromechanics models. Based on this, assumptions made in higher order continua may be justified. For example the determination of the characteristic length can be based on micromechanics, as is done already in the analysis of the regular array of microcracks [37], but should be extended to include other crack interactions and crack–aggregate interactions as well. Another use of the micromechanics may be the design of new materials, tailored for specific applications. This is common practice in the field of ceramics [40], [41].

The non-linear element in the 'numerical' micromechanics models is a two-dimensional interface element [6], [14], [51]. The element has a tangential and normal stiffness. Upon fracturing, tensile softening is modelled, normally with a very steep linear branch only; the tangential 'slip' can also be modelled via some sort of softening law. For the combined states of stress a Mohr–Coulomb type of model may be adopted. A major problem is the identification of the various model parameters. The tensile parameters are determined on the basis of macroscopic observations. Note that for the (fine grained) matrix softening behaviour still has to be assumed because with the current computational capabilities in general only a few large aggregates can be modelled. For example, the softening curve for a 2-mm mortar can be used for modelling tensile fracture of the interface. As far as the tangential component is concerned, no macroscopic equivalent is available. Studies on the shear fracture of concrete are currently under way [52], [53], but in view of the path dependency of the tensile/shear fracturing [54], no simple macroscopic model in terms of stress and strain may exist. So far only in a few cases has true sliding been observed [53], but under most conditions secondary tensile cracking prevails. Here we touch upon a rather philosophical problem. The shear fracture of the concrete is, at a lower scale, a tensile problem. The same is true for compressive loading (Fig. 25).

The micromechanics models, which are in principle capable of dealing with the above mentioned path-dependent behaviour, are in need of a number of (microscopic) parameters. In determining these parameters we encounter the same problems as sketched earlier for the determination of the macroscopic properties of the material. The micromechanics models require such an amount of computing time that a parameter study is extremely time consuming. For practical reasons with the current computational resources this is not very good, which implies that handling the problem via a procedure of inverse modelling must be discarded for the moment.

4.3 Lattice Models

When the fundamental laws of fracture are studied, simple models are needed where all input parameters are well controllable, and which will yield a measurable result on real materials. The disadvantage is that, of course, we simplify reality. Recently so-called lattice models have been developed, which serve as an excellent tool for groping after the fundamentals of fracture [7], [55]–[57]. In general a regular lattice of bonds between nodes is constructed, Fig. 26(a). The disorder is brought about either via a probabilistic law for bond breaking or by having a *quenched* disorder with a deterministic breaking law. The quenched disorder may be defined in the threshold displacements (viz. the maximum deformation that a bond can sustain, after this brittle failure occurs), or in the elastic constants. The most common type of lattice of interest for mechanical loading is the central force lattice, which consists of Hookean springs which can freely rotate in the nodes. Another type of model is the beam model, which takes into

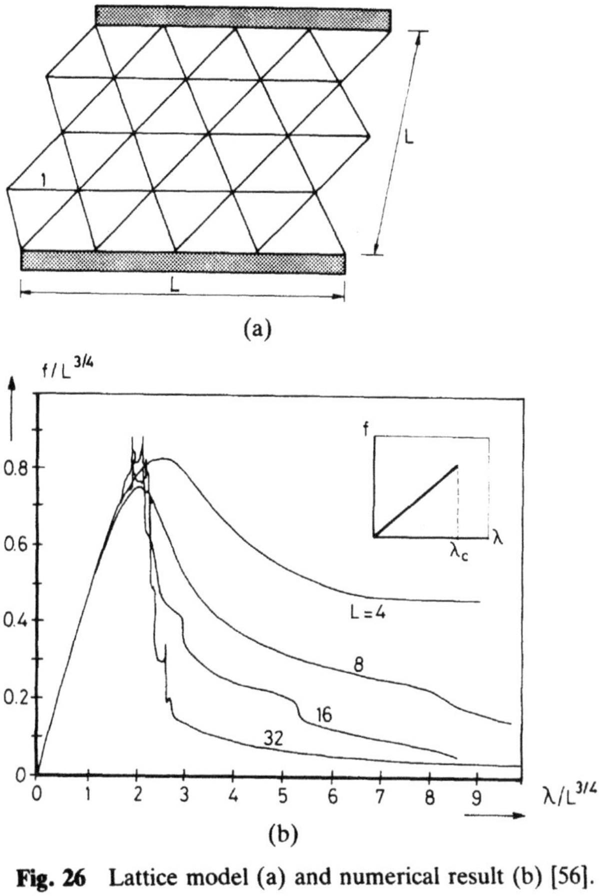

Fig. 26 Lattice model (a) and numerical result (b) [56].

account displacements and rotations. It has been suggested that the beam models can be used to describe solids like concrete [55].

As an example, the results of an analysis with the beam model are shown in Fig. 26(b). The unit size in the lattice is equal to 1, and lattices of size $L = 4$, 8, 16 and 32 have been analysed. The lattice was subjected to uniaxial elongation, and a power law distribution for thresholds $P(\lambda_c) \propto \lambda_c^{-r}$, with exponent $r = 0{\cdot}5$ has been used. The results have been averaged over 50 000 samples for $L = 4$ and less than 10 samples for $L = 64$. In the inset of Fig. 26(b), the breaking law for a single bond is shown.

Three different regimes are identified in the force–elongation curves of Fig. 26(b). The first regime is linear and scales according to

$$f = L^{\alpha}\Phi(\lambda L^{-\beta}) \tag{10}$$

with $\alpha = \beta \approx 0{\cdot}75$, f denotes the external force, and λ is the displacement. This first regime is dominated by disorder.

The second regime is given by the maximum force, and the third regime beyond the maximum is called the catastrophic regime. In this regime the force on the lattice gradually decreases, and a few localised cracks grow together. Strong statistical fluctuations are observed. The local strain distribution is described as being multifractal, which means that when the moments of the distribution are analysed just before the last bond breaks, it is found that each moment has its own exponent [55], [56]. Physically this means that regions of very high strain lie on fractal subsets. The fractal dimensions of these subsets depend on how these strains vary with increasing system size. Because the subsets are fractal, regions of high strain become less frequent when larger system sizes are considered.

When these results are compared with the experimental findings described in Sections 2 and 3, various similarities in behaviour are observed. The regions of high strain found in the computations can be interpreted as the experimentally observed crack interface grain bridging. Furthermore the three regimes of macroscopic response are observed in a similar manner in experiments, e.g. Fig. 12. The scaling of the displacement in Fig. 26 with respect to the lattice size L, is comparable to the introduction of the concept of strain.

Herrmann [56] suggested that the various scaling exponents are independent of the nature of the bonds (e.g. central force of beam type). The models seem very powerful as regards the study of fundamental fracture properties, and fractal dimensions of measured and predicted crack patterns can be compared. Note that the approach is not new, and has been applied by various researchers for analysing concrete fracture, [58]–[60]. The recognition of self similarity and the fractal geometry of the cracks is however new [3], and has given new impetus to this type of modelling.

5. CONCLUSIONS

In this chapter scaling in fracture of concrete under tensile and compressive loading is presented. Three size scales must be distinguished: (1) the size of the heterogeneities in the material; (2) the size of the cracks; and (3) the size of the crack overlap in the cohesive crack. The crack patterns seem to be self similar over a large size scale. In the case of tensile fracture, a flexural type of crack interface grain bridging between overlapping interacting cracks is observed in various brittle disordered materials over a length scale of nine orders of magnitude. Crack interface grain bridging has been identified as a fundamental stress transfer mechanism in cracks in concrete, and the load transfer was shown to depend basically on the size of the maximum aggregates. In compression, the self similarity is observed in en échelon crack patterns, which seem to precede the development of shear bands.

Various theoretical models have been developed which are capable of dealing with the heterogeneity of the material. Higher order continuum models introduce a *constant* length scale, which may be derived using micromechanical considerations. However, it must be realised that this always will be an average value. If the fracture process is studied in more detail, micromechanical models and lattice models are promising tools. The lattice models predict the self similar crack patterns and the fractal geometry of the cracks, which is in agreement with the experiments.

ACKNOWLEDGEMENTS

The author is indebted to Mr G. Timmers and Mr A. Elgersma for their assistance in carrying out the various experiments. Thanks are due to Mr M. B. Nooru-Mohamed, Mr E. Schlangen and Mr R. A. Vonk for the many discussions.

REFERENCES

[1] Hillerborg, A., Modeer, M. & Petersson, P. E., Analysis of crack formation and crack growth in concrete by means of fracture mechanics and finite elements. *Cem. & Conc. Res.*, **6** (1976) 773–82.

[2] Mindess, S., Fracture process zone detection. In *Fracture Mechanics of Concrete*, ed. S. P. Shah & A. Carpinteri. Chapman & Hall, London, New York, 1991, pp. 231–61.

[3] Mandelbrot, B. B., *The Fractal Geometry of Nature*. W. H. Freeman, New York, 1983.

[4] Bažant, Z. P. & Pijaudier-Cabot, G., Measurement of characteristic length of non-local continuum. *J. Eng. Mech.* (ASCE), **115** (1989) 755–67.

[5] Mühlhaus, H.-B. & Vardoulakis, I., The thickness of shear bands in granular materials. *Géotechnique*, **37** (1987) 271–83.

[6] Roelfstra, P. E., Sadouki, H. & Wittmann, F. H., Le béton numérique. *Materials & Structures* (RILEM), **18** (1985) 327–35.

[7] Hansen, A., Roux, S. & Herrmann, H. J., Rupture of central force lattices. *J. Phys. France*, **50** (1989) 733–44.

[8] Van Mier, J. G. M., Fracture of concrete under complex stress. *HERON*, **31**(3) (1986) 1–90.

[9] Van Mier, J. G. M. & Nooru-Mohamed, M. B., Geometrical and structural aspects of concrete fracture. *Eng. Fract. Mech.*, **35** (1990) 617–28.

[10] Bascoul, A., Kharchi, F. & Maso, J. C., Concerning the measurement of the fracture energy of a micro-concrete according to the crack growth in a three point bending test on notched beams. In *Fracture of Concrete and Rock*, ed. S. P. Shah & S. E. Swartz. Springer Verlag, New York, 1989, pp. 396–408.

[11] Swartz, S. E. & Refai, T., Cracked surface revealed by dye and its utility in determining fracture parameters. In *Fracture Toughness and Fracture Energy*, ed. H. Mihashi, H. Takahashi & F. H. Wittmann. Balkema, Rotterdam, 1989, pp. 509–20.

[12] Fanping Zhou, Some aspects of tensile fracture behaviour and structural response of cementitious materials. Report TVBM 1008, Lund Institute of Technology, Lund, Sweden, 1988.

[13] Hordijk, D. A. & Reinhardt, H. W., Macro-structural effects in a uniaxial tensile test on concrete. In *Brittle Matrix Composites 2*, ed. A. M. Brandt & I. H. Marshall. Elsevier Applied Science, London, New York, 1989, pp. 486–95.

[14] Rossi, P., Numerical modelling of the cracking using a non-deterministic approach. In *Fracture Toughness and Fracture Energy*, ed. H. Mihashi, H. Takahashi & F. H. Wittmann. Balkema, Rotterdam, 1989, pp. 383–94.

[15] Rots, J. G., Computational modelling of concrete fracture. PhD thesis, Delft University of Technology, Delft, The Netherlands, 1988.

[16] Van Mier, J. G. M., Mode I behaviour of concrete: influence of the rotational stiffness outside the crack-zone. In *Analysis of Concrete Structures by Fracture Mechanics*, ed. L. Elfgren & S. P. Shah. Chapman & Hall, London, New York, 1990, pp. 19–31.

[17] Vonk, R. A., Uniformity of deformations in compression tests on concrete. Report TUE/BKO-89.17, Eindhoven University of Technology, Eindhoven, The Netherlands, 1989.

[18] Van Mier, J. G. M., Internal crack detection in single edge notched concrete plates subjected to uniform boundary displacement. In *Micromechanics of Failure of Quasi-Brittle Materials*, ed. S. P. Shah, S. E. Swartz & M. L. Wang. Elsevier Applied Science Publishers, London, New York, 1990, pp. 33–42.

[19] Petersson, P. E., Crack growth and development of fracture zones in plain concrete and similar materials. Report TVBM-1006, Division of Building Materials, Lund, Sweden, 1981.
[20] Ward, R. J. & Li, V. C., Simple dependence of structural behavior on material fracture resistance. In *Fracture of Concrete and Rock—Recent Developments,* ed. S. P. Shah, S. E. Swartz & B. Barr. Elsevier Applied Science, London, New York, 1989, pp. 645–60.
[21] Wittmann, F. H. & Roelfstra, P. E., Constitutive equations for transient conditions. In *Proceedings IABSE Colloquium on Computational Mechanics of Concrete Structures—Advances and Applications,* IABSE, Zürich, **54,** 1987, pp. 239–59.
[22] Bonzel, J. & Kadlecek, V., Einfluss der Nachbehandlung und des Feuchtigkeitszustands auf die Zugfestigkeit des Betons. *Beton,* **20** (1970) 303–9; 351–57.
[23] Van Mier, J. G. M., Strain-softening of concrete under multiaxial loading conditions. PhD thesis, Eindhoven University of Technology, Eindhoven, The Netherlands, 1984.
[24] Zaitsev, Y. W. & Wittmann, F. H., Simulation of crack propagation and failure of concrete. *Materials & Structures* (RILEM), **14** (1981) 357–65.
[25] Schickert, G., Schwellenwerte beim Betondruckversuch. *Deutscher Ausschuss für Stahlbeton.* Report No. 312, Berlin, 1980.
[26] Kotsovos, M. D., Effect of testing techniques on the post-ultimate behaviour of concrete in compression. *Materials & Structures* (RILEM), **16** (1983) 3–12.
[27] Vonk, R. A., Rutten, H. S., Van Mier, J. G. M. & Fijneman, H. J., Influence of boundary conditions on softening of concrete loaded in compression. In *Fracture of Concrete and Rock—Recent Development,* ed. S. P. Shah, S. E. Swartz & B. Barr. Elsevier Applied Science, London, New York, 1989, pp. 711–20.
[28] Vonk, R. A., Rutten, H. S., Van Mier, J. G. M. & Fijneman, H. J., Size effect in softening of concrete loaded in compression. In *Proceedings ECF8 Fracture Behaviour and Design of Materials and Structures,* ed. D. Firrao. EMAS Publishers, Warley, UK, 1990, pp. 767–72.
[29] Hillerborg, A., Fracture mechanics concepts applied to moment capacity and rotational capacity of reinforced concrete beams. *Eng. Fract. Mech.,* **35** (1990) 233–40.
[30] Wittmann, F. H., Roelfstra, P. E., Mihashi, H., Huang, Y.-Y., Zhang, X.-H. & Nomura, N., Influence of age of loading, water–cement ratio and rate of loading on fracture energy of concrete. *Materials & Structures* (RILEM), **20** (1987) 103–10.
[31] Du, J. J., Kobayashi, A. S. & Hawkins, N. M., An experimental–numerical analysis of fracture process zone in concrete fracture specimens. *Eng. Fract. Mech.,* **35** (1990) 15–27.
[32] Van Mier, J. G. M. & Schlangen, E., On the stability of softening systems. In *Fracture of Concrete and Rock—Recent Developments,* ed. S. P. Shah, S. E. Swartz & B. Barr. Elsevier Applied Science, London, New York, 1989, pp. 387–96.
[33] Van Mier, J. G. M., Fracture study of concrete specimens subjected to

combined tensile and shear loading. In *Proceedings GAMAC Int. Conf. on Measurement and Testing in Civil Engineering*, ed. J. F. Jullien. Lyon-Villeurbanne, 13–18 Sept. CAST/INSA Publishers, Villeurbanne, 1988. Vol. I, pp. 337–47.

[34] Van Mier, J. G. M., Mode I fracture of concrete: discontinuous crack growth and crack interface grain bridging. *Cem. & Conc. Res.*, **21** (1991) 1–15.

[35] Ortiz, M., Microcrack coalescence and macroscopic crack growth initiation in brittle solids. *Int. J. Solids Structures*, **24** (1988) 231–50.

[36] Rice, J. R., Mathematical analysis in the mechanics of fracture. In *Fracture*, Vol. 2, ed. H. Liebowitz. Academic Press, New York, 1968, pp. 191–311.

[37] Bažant, Z. P., Snapback instability at crack ligament tearing and its implication for fracture micromechanics. *Cem. & Conc. Res.*, **17** (1987) 951–67.

[38] Horii, H., Hasegawa, A. & Nishino, F., Fracture process and bridging zone model and influencing factors in fracture of concrete. In *Fracture of Concrete and Rock*, ed. S. P. Shah & S. E. Swartz. Springer Verlag, New York, 1989, pp. 205–19.

[39] Marchand, N., Parks, D. M. & Pelloux, R. M., K_I-solutions for single edge notch specimens under fixed end displacements. *Int. J. Fract.*, **32** (1986) 53–65.

[40] Swanson, P. L., Fairbanks, C. L., Lawn, B. R., Mai, Y-W. & Hockey, B. J., Crack-interface grain bridging as a fracture resistance mechanism in ceramics: I, experimental study on alumina. *J. Am. Ceram. Soc.*, **70** (1987) 279–89.

[41] Steinbrech, R. W., Dickerson, R. M. & Kleist, G., Characterization of the fracture behavior of ceramics through analysis of crack propagation studies. In *Toughening Mechanism in Quasi-Brittle Materials*, ed. S. P. Shah. NATO-ASI Series, Vol. E-195, Kluwer Academic Publishers, Dordrecht, 1991, pp. 287–311.

[42] Hu, X. & Wittmann, F. H., Experimental method to determine extension of fracture-process zone. *J. Mat. Civil Eng.* (ASCE), **2** (1990) 15–23.

[43] Labuz, J. F., Shah, S. P. & Dowding, C. H., Experimental analysis of crack propagation in granite. *Int. J. Rock Mech. Min. Sci. & Geomech. Abstr.*, **22** (1985) 85–98.

[44] Castro-Montero, A., Miller, R. A. & Shah, S. P., Study of the fracture process in mortar with laser holographic measurements. In *Toughening Mechanism in Quasi-Brittle Materials*, ed. S. P. Shah. Kluwer Academic Publishers, Dordrecht, 1991, 249–65.

[45] Simha, K. R. Y., Fourney, W. L., Barker, D. B. & Dick, R. D., Dynamic photoelastic investigation of two pressurized cracks approaching one another. *Eng. Fract. Mech.*, **23** (1986) 237–49.

[46] Sempere, J.-C. & Macdonald, K. C., Overlapping spreading centers: implications from crack growth simulation by the displacement discontinuity method. *Tectonics*, **5** (1986) 151–63.

[47] Gramberg, J., *A Non-Conventional View of Rock Mechanics and Fracture Mechanics*. Balkema, Rotterdam, 1989.

[48] Stroeven, P., Some aspects of the micromechanics of concrete. PhD thesis, Delft University of Technology, Delft, Netherlands, 1973.
[49] Bazant, Z. P., Mechanics of distributed cracking. *Appl. Mech. Rev.*, **39** (1986) 675–705.
[50] Bazant, Z. P., Recent advances in failure localisation and non-local models. In *Micromechanics of Failure of Quasi-Brittle Materials*, ed. S. P. Shah, S. E. Swartz & M. L. Wang. Elsevier Applied Science Publishers, London, New York, 1990, pp. 12–32.
[51] Stankowski, T., Numerical simulation of progressive failure in particle composites. PhD thesis, University of Colorado, Boulder, USA, 1990.
[52] Hassanzadeh, M. & Hillerborg, A., Concrete properties in mixed mode fracture. In *Fracture Toughness and Fracture Energy*, ed. H. Mihashi, H. Takahashi & F. H. Wittmann. Balkema, Rotterdam, 1989, pp. 565–8.
[53] Van Mier, J. G. M. & Nooru-Mohamed, M. B., Fracture of concrete under tensile and shear-like loadings. In *Fracture Toughness and Fracture Energy*, ed. H. Mihashi, H. Takahashi & F. H. Wittmann. Balkema, Rotterdam, 1989, pp. 549–63.
[54] Nooru-Mohamed, M. B., Schlangen, E. & Van Mier, J. G. M., Fracture of concrete plates subjected to rotating biaxial stress. In *Proceedings ECF8 Fracture Behaviour and Design of Materials and Structures*, ed. D. Firrao. EMAS Publishers, Warley, UK, 1990, pp. 682–7.
[55] Herrmann, H. J., Introduction to modern ideas on fracture patterns. In *Random Fluctuations and Pattern Growth: Experiments and Models*, ed. H. E. Stanley & N. Ostrowsky. Kluwer Academic Publishers, Dordrecht, 1988, pp. 149–60.
[56] Herrmann, H. J., The hunt for universality in fracture. In *Universalities in Condensed Matter*. Proc. in Phys. 32, Les Houches France, 15–24 March. Springer, New York, 1988, pp. 132–35.
[57] Termonia, Y. & Meakin, P., Formation of fractal cracks in a kinetic fracture model. *Nature*, **320** (1986) 429–31.
[58] Bazant, Z. P., Instability, ductility, and size effect in strain-softening concrete. *J. Eng. Mech. Div.* (ASCE), **102** (1976) 331–44.
[59] Burt, N. J. & Dougill, J. W., Progressive failure in a model heterogeneous medium. *J. Eng. Mech. Div.* (ASCE), **103** (1977) 365–76.
[60] Schorn, H. & Rode, U., 3-D modeling of process-zone in concrete by numerical simulation. In *Fracture of Concrete and Rock*, ed. S. P. Shah & S. E. Swartz. Springer Verlag, New York, 1989, pp. 220–8.

Chapter 6

Prediction of Fracture of Concrete and Fiber Reinforced Concrete by the *R*-Curve Approach

CHENGSHENG OUYANG, BARZIN MOBASHER & SURENDRA P. SHAH
NSF Center for Science and Technology of Advanced Cement Based Materials, Northwestern University, 1800 Ridge, Evanston; Illinois 60208, USA

ABSTRACT

An R-curve approach for prediction of fractures of concrete and fiber reinforced concrete is presented. The R-curve is defined as an envelope of fracture energy release rate of specimens with different sizes but the same initial notch length. By assuming that an effective traction-free critical crack is a function of an initial crack length contained in a material, an expression for an R-curve with two parameters can be derived by solving a differential equation. The parameters of the R-curve can be uniquely determined according to K_{IC}^{s} and $CTOD_c$ for positive geometry specimens, and according to K_{IC}^{s} and $\mathrm{d}K_I/\mathrm{d}a = 0$ for negative geometry specimens. Load–CMOD and load–displacement responses can be predicted based on the proposed R-curve by requiring that the crack driving force and crack growth resistance are equal at every equilibrium crack length. Toughening of matrix due to fiber bridging is taken into account by applying closing pressure on crack surfaces. The theoretical predictions are compared with the experimental results of concrete and fiber reinforced concrete containing unidirectional, continuous glass, steel or polypropylene fibers. The predicted responses indicate good agreement with the wide range of experimental results tested by various investigators.

1. INTRODUCTION

Fracture of concrete and fiber reinforced concrete has been studied for several decades. A primary character of fracture of these materials is the existence of stable crack growth prior to the crack reaching its critical length. Such behavior is often referred to as R-curve behavior. Fracture responses of quasi-brittle material, such as concrete and fiber reinforced concrete are indicated in Fig. 1. For brittle materials, there is no stable crack growth. Whenever the strain energy release rate is equal to the fracture toughness of a material, the material fails. This is represented as a horizontal line in Fig. 1. For concrete, however, crack growth is heterogeneous and tortuous accompanied by grain boundary sliding. The existence of a fracture process zone results in stable crack growth prior to the peak load. This increased energy requirement is represented as a rising R-curve in Fig. 1. For fiber reinforced concrete, the presence of fibers introduces additional toughening, which requires more energy dissipation during crack growth. This is indicated by a second rising R-curve in Fig. 1. Note that this enhanced R-curve refers to the matrix in the fiber reinforced cement composite.

Many studies on fracture behavior of concrete have been conducted. Kaplan [1] was first to attempt to apply linear elastic fracture mechanics (LEFM) to concrete. Shah & McGarry [2] realized that due to the existence of a relatively large fracture process zone, LEFM can not be directly applied to concrete. Several models using two or more fracture parameters have been proposed recently for fracture of

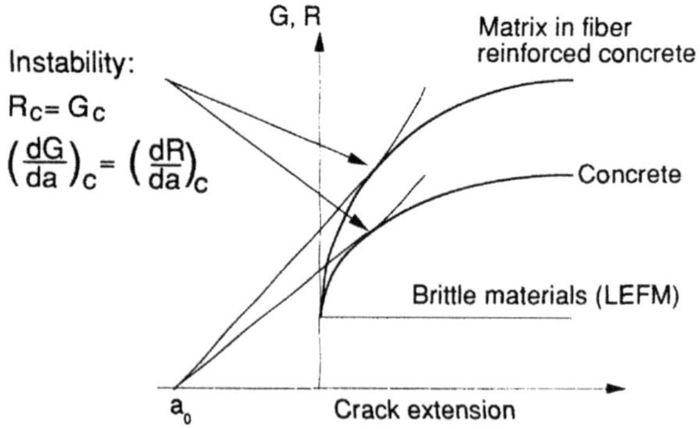

Fig. 1 R-curve behavior of quasi-brittle materials.

concrete. Some of these models can be categorized as models based on cohesive cracks [3], [4], and equivalent-elastic crack models [5], [6]. The cohesive crack models include the fictitious model proposed by Hillerborg *et al.* [3], and the crack band model established by Bazant & Oh [4], where a cohesive force is assumed to bridge the crack surface during crack propagation. The equivalent-elastic crack models are represented by the two-parameter model proposed by Jenq & Shah [5], and the size effect law proposed by Bazant [6], where a crack in concrete is simulated by a traction-free effective crack.

The initial investigation of the matrix cracking behavior in the fiber composites is attributed to Aveston *et al.* [7] who used an energy balance approach in order to formulate the mechanics of multiple fracture, and increase in the matrix fracture strain. Recent developments on the prediction of fracture behavior of fiber composites include work by McCartney [8], and Budiansky *et al.* [9] who provided solutions for the steady state matrix cracking stress using a fracture mechanics energy approach and various fiber–matrix interface conditions. Marshall *et al.* [10] used a stress intensity factor approach in which both short cracks and long cracks approaching a steady state are considered. The effect of fibers across a crack can be modeled by means of a bridging force. This force is a function of position of a fiber along the crack, fiber volume and type, as well as crack opening displacement; equivalently the crack profile can be affected by these bridging forces. Such interdependence of the bridging force on crack opening and crack profile results in a non-linear expression of the bridging force and the crack opening. Solutions for fully or partially bridged cracks in transversely isotropic materials have been provided by Nemat-Nasser & Hori [11]. Ballarini *et al.* [12] applied the closing pressure approach to model fracture of concrete and fiber reinforced concrete. Jenq & Shah [13] have also proposed a fracture mechanics approach based on the two-parameter fracture model [5].

2. *R*-CURVE APPROACH

Fracture of materials can be analyzed by an R-curve, defined as fracture resistance. As shown in Fig. 1, R is the energy rate required for crack propagation in a material, and is an increasing and convex function for quasi-brittle materials. The strain energy release rate due to crack growth is termed G. The following two conditions should be satisfied at a critical crack length, a_c, for crack propagation to occur

$$R_c = G_c \qquad (\partial R/\partial a)_c = (\partial G/\partial a)_c \tag{1}$$

The concept of the R-curve was first introduced using the energy criterion by Irwin in 1954 and 1959 for the study of metals [14]. Krafft *et al.* [15] proposed that the R-curve is a unique function of crack extension for metals. This suggestion has been used by Broek [16] to derive a semi-empirical solution of the R-curve. His solution, however, was only valid for short cracks, i.e. $a \ll b$, where a and b are crack length and width of specimen, respectively.

Wecharatana & Shah [17], [18] used the R-curve to model stable crack growth and the fracture process zone of concrete. Jenq & Shah [19] predicted the R-curve using energy balance and using the two-parameter model, respectively, and indicated that the R-curve should remain constant for cement-based materials after a crack reaches its critical length. Bazant *et al.* [20]–[22] derived geometry-dependent R-curves based on the size effect demonstrated by fracture tests. They also assumed that the R-curve remains constant after the maximum load. Foote *et al.* [23] studied the R-curve in a strain-softening material. Sakai *et al.* [24] and Hsueh & Becher [25] also used the R-curve to study fracture of ceramics.

An R-curve approach for fracture of quasi-brittle materials has been recently proposed by Ouyang *et al.* [26]. An effective traction-free crack is assumed. An R-curve is defined as an envelope of the energy release rates with different specimen size but the same initial notch length, as shown in Fig. 2. By assuming that the critical crack length is the function of an initial crack length a_0, a formulation of the R-curve can be derived by solving a differential equation. The R-curve obtained includes two parameters that can (for example) be uniquely determined from K_{IC}^s and $CTOD_c$, defined by Jenq & Shah [5], for positive geometry specimens, and by K_{IC}^s and the condition of $dK_I/da = 0$ for negative geometry specimens. Load–CMOD curve and load–displacement curve can be computed based on the proposed R-curve. Mobasher *et al.* [27] have extended their R-curve approach to fiber reinforced concrete. A toughening mechanism of fiber on matrix was incorporated into the R-curve by applying closing pressure on crack surfaces, which can reduce stress intensity factor at the crack tip and crack opening displacement at the crack surface. Fracture behavior of concrete and fiber reinforced concrete containing uni-

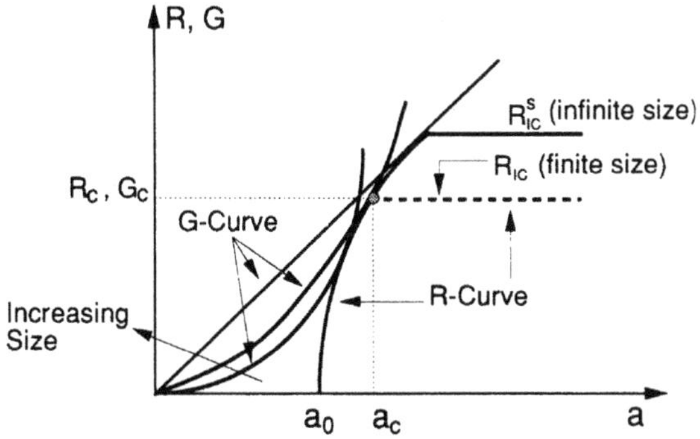

$a_c = a_0 + \Delta a_c = \alpha a_0$

Fig. 2 Definition of *R*-curve.

directional, continuous glass, steel and polypropylene fibers, respectively, can be predicted well by their *R*-curve approach.

The *R*-curve approach developed by Ouyang *et al.* for fractures of concrete and fiber reinforced concrete is presented here.

3. DERIVATION OF THE *R*-CURVE

Based on experimental observations, Krafft *et al.* [15] proposed that the *R*-curve for metals (LEFM) is only a function of crack extension, Δa. This suggestion has been used by Broek [16] to derive a semi-empirical solution for the *R*-curve for a specimen containing a short crack. A similar approach, but which is not restricted to a short crack, is used to derive the *R*-curve for quasi-brittle materials here.

The *R*-curve is defined as an envelope of the energy release rates with different specimen sizes but the same initial notch length, as shown in Fig. 2. Generally, the energy release rate, G, can be expanded as a Taylor's series at a critical point c

$$G = G_c + \sum_{n=1}^{\infty} (1/n!)(\mathrm{d}^n G/\mathrm{d}a^n)_c (a - a_c)^n \tag{2a}$$

where a_c is a critical crack length. Noting $\mathrm{d}G/\mathrm{d}a = \mathrm{d}G/\mathrm{d}\,\Delta a, \ldots,$ $\mathrm{d}^n G/\mathrm{d}\,\Delta a^n$ due to $a = a_0 + \Delta a$, the following equation can be obtained

$$G = G_c + \sum_{n=1}^{\infty} (1/n!)(\mathrm{d}^n G/\mathrm{d}\,\Delta a^n)_c (a - a_c)^n \tag{2b}$$

Using $G = 0$ at $a = 0$, eqn (2b) is reduced to

$$G_c = -\sum_{n=1}^{\infty} (1/n!)(\mathrm{d}^n G/\mathrm{d}\,\Delta a^n)_c(-a_c)^n \tag{3}$$

For materials with pre-critical stable crack propagation, it is assumed that the critical crack length, a_c, is approximately proportional to the initial crack length, a_0. In the coordinate shown in Fig. 2 this relation can be written as

$$a_c = \Delta a_c + a_0 = \alpha a_0 \tag{4}$$

where α is a parameter to describe the pre-critical stable crack growth ($\alpha \geq 1$). Substituting eqn (4) into eqn (3), it is obtained that

$$G_c = -\sum_{n=1}^{\infty} (1/n!)(\mathrm{d}^n G/\mathrm{d}\,\Delta a^n)_c[\alpha/(1-\alpha)]^n(-\Delta a_c)^a \tag{5}$$

Equation (5) must be valid for any critical point c, thus it yields a differential equation with variable coefficients, which governs the critical energy release rate for fracture. Equation (5) can be classified as an Euler equation. By using the following transformation:

$$x = \ln(\Delta a_c) \qquad \text{or} \qquad \Delta a_c = \exp(x) \tag{6}$$

Equation (5) can then be transformed into a linear differential equation with constant coefficients, with its auxiliary equation represented by

$$1 + \sum_{n=1}^{\infty} (1/n!)d(d-1)(d-2)\ldots(d-n+1)[\alpha/(\alpha-1)]^n(-1)^n = 0 \tag{7}$$

The solution of eqn (5) can be written as

$$G_c = \sum_{n=1}^{\infty} \beta_n\,\Delta a_c^{d_n} \tag{8}$$

where β_n are constants to be determined, and d_n the roots of eqn (7). Equation (8) can also be considered a solution of the R-curve due to $G_c = R_c$, i.e.

$$R = \sum_{n=1}^{\infty} \beta_n\,\Delta a^{d_n} \tag{9}$$

where the notations R_c and Δa_c are replaced by R and Δa for convenience.

It can be shown that for $n = 1$ (short crack, where G is a linear function of crack length)

$$d_1 = (\alpha - 1)/\alpha \tag{10}$$

and

$$R = \beta_1(a - a_0)^{d_1} \tag{11}$$

The R-curve hereafter is transformed as a function of crack length a, rather than crack extension Δa. Equations (10) and (11) are the R-curve obtained by Broek [16].

For $n = 2$, G is approximated by a quadratic expression of the crack length. It is obtained that

$$d_{1,2} = 1/2 + (\alpha - 1)/\alpha \pm \{1/4 + (\alpha - 1)/\alpha - [(\alpha - 1)/\alpha]^2\}^{1/2} \tag{12a}$$

$$R = \beta_1(a - a_0)^{d_1} + \beta_2(a - a_0)^{d_2} \tag{12b}$$

Using the boundary condition $(\mathrm{d}R/\mathrm{d}a)_{a=a_c} = 0$, a relationship between β_1 and β_2 is obtained. Equation (12b) can then be rewritten as

$$R = \beta\{1 - (d_2/d_1)[(\alpha a_0 - a_0)/(a - a_0)]^{(d_2 - d_1)}\}(a - a_0)^{d_2} \tag{13a}$$

or

$$R = \beta\psi(a - a_0)^{d_2} \tag{13b}$$

where

$$\psi = 1 - (d_2/d_1)[(\alpha a_0 - a_0)/(a - a_0)]^{(d_2 - d_1)} \tag{14}$$

d_2 is given in eqn (12a), and β is a constant to be determined.

For simplicity, one may only take the second term in eqn (12b) as an R-curve, i.e.

$$R = \beta(a - a_0)^{d_2} \tag{15}$$

Since the value of ψ is very close to unity, the R-curves given by eqns (13b) and (15) are almost the same. But the R-curve given by eqn (13b) ensures a smooth transition to a plateau value at $a = a_c$. The R-curves defined by eqns (13b) and (15) are an increasing and convex function of crack extension, respectively. Only the R-curve given by eqn (15) is used in this chapter.

It is of interest to note that for both $n = 1$ and $n = 2$ the R-curve contains two parameters. A higher order approximation of G ($n > 2$) may be used to derive an R-curve. However the current formulations focus on the R-curve with $n = 2$.

It is noted that the conditions of $G_c = R_c$ and $a_c = \alpha a_0$ rather than eqn (1) are used to derive the critical crack propagation in the present

R-curve. The proposed R-curve does not exactly represent the true mathematical envelope of the fracture energy release rate G. However, if the values of α and β can be determined based on material properties, such as K_{IC}^{s} and $CTOD_c$, eqn (15) can provide a good approximation of the R-curve. Once the two parameters, α and β are determined, the R-curve defined by eqn (15) can be used to compute fracture behavior of quasi-brittle materials.

4. DETERMINATIONS OF α AND β

Because the R-curve depends on both material property and specimen geometry, α and β are material and geometry dependent. As noted by Jenq & Shah [5], Bazant *et al.* [22], and Planas & Elices [28], specimen geometries can be divided into positive and negative geometries. The positive geometry has $dK_I/da > 0$, whereas the value of dK_I/da varies from negative to positive for the negative geometry as shown in Fig. 3. Three types of geometries as shown in Fig. 4 are discussed here. The single edge notched (SEN) specimen subjected to uniaxial tension and the three-point bend (3-PB) specimen are the positive geometry, whereas the center-cracked plate (CCP) subjected to tension is the negative geometry. It is noted that the load in the SEN specimen is represented by σ, whereas the load in 3-PB or CCP specimens is represented by P. But only the load σ, will be mentioned for convenience hereafter. The determinations of α and β for both geometries are discussed here.

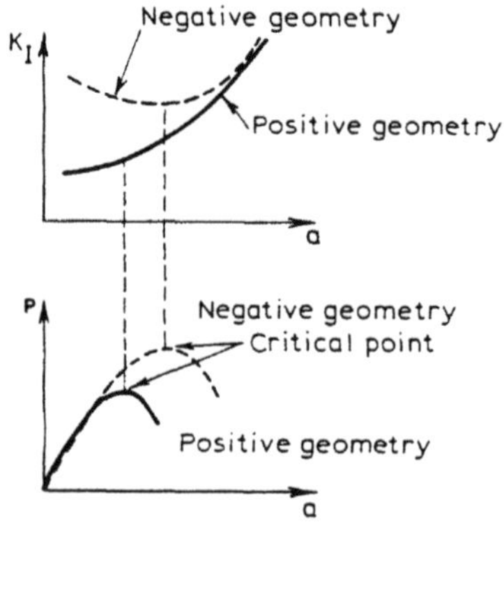

Fig. 3 Positive and negative geometries.

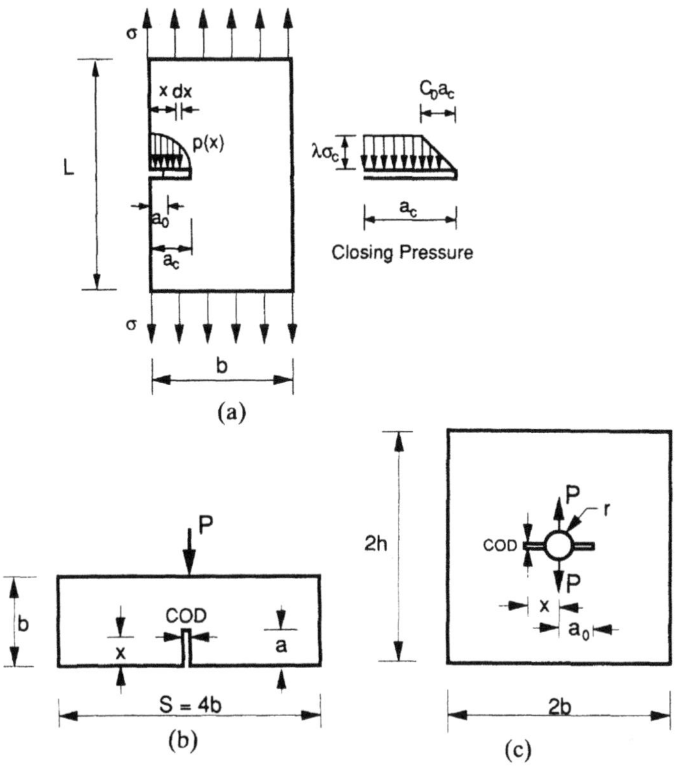

Fig. 4 Geometries of specimens: (a) single edge notched specimen subjected to uniaxial tension; (b) three-point bending beam; and (c) center-cracked plate subjected to tension.

4.1 Concrete

The value of α can be determined based on different conditions for the positive and negative geometries. For the positive geometry, the value of α associated with the critical load σ_c can be determined from

$$\begin{aligned} K_I &= K_{IC}^s \\ CTOD &= CTOD_c \end{aligned} \tag{16}$$

For the negative geometry, the values of α and σ_c can be determined from

$$K_I = K_{IC}^s$$
$$dK_I/da = 0 \qquad (17)$$

The value of β can easily be determined from K_{IC}^s, if α is known, for both positive and negative geometry specimens. The plateau value of R, i.e. the R value at $a_c = \alpha a_0$ (see Fig. 2), can be directly related to K_{IC}^s as,

$$R_{IC}^s = \beta(\alpha a_0 - a_0)^{d_2} = (K_{IC}^s)^2/E'$$

or

$$\beta = (K_{IC}^s)^2/[E'(\alpha a_0 - \alpha_0)^{d_2}] \qquad (18)$$

where E' represents the Young's modulus for the plain stress or the plain strain condition. Once the values of α and β are known, the R-curve is uniquely determined.

K_{IC}^s and $CTOD_c$ are the material properties defined based on the two-parameter fracture model proposed by Jenq & Shah [5]. An effective crack length is assumed for both the two-parameter model and the present R-curve approach. However, the physical meanings of the effective crack length in two models are slightly different from each other. As shown in Fig. 5, in the two-parameter model the effective crack length is a_2 which is determined based on the compliance C_2, whereas in the present R-curve approach the effective crack length is a_1 corresponding to the compliance C_1. When specimen sizes tend to infinity, the relative size of the fracture process zone, where the

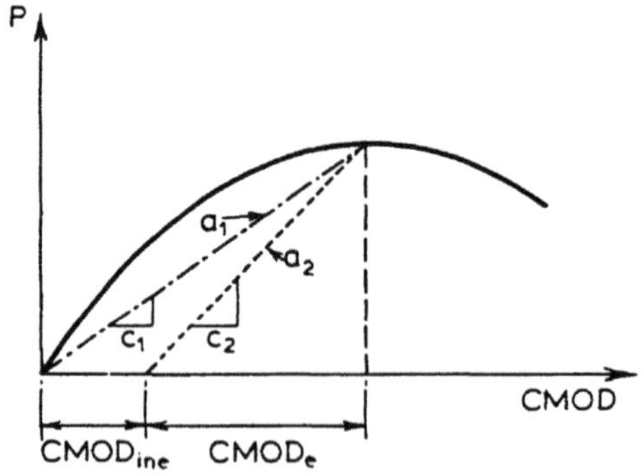

Fig. 5 Compliances between the present R-curve approach and the two-parameter fracture model.

inelastic fracture energy is generated, becomes infinitesimal. As a result, the fracture response in the quasi-brittle material becomes completely elastic and a_1 becomes equal to a_2. Therefore, for the positive geometry, the R-curve obtained based on K_{IC}^s and $CTOD_c$ are valid for an infinite size specimen with an initial crack length a_0. On the other hand, the value of R cannot exceed $R_{IC}^s = (K_{IC}^s)^2/E$, i.e. $R = R_{IC}^s$ when $R > R_{IC}^s$ according to eqn (18). This is due to the assumption that at this point a steady-state crack growth is achieved.

For a finite size specimen with the positive geometry, the initial part of the R-curve is the same as with the infinite size specimen with the same notch. But the plateau value of R is R_{IC} rather than R_{IC}^s, where R_{IC} is size-dependent as shown in Fig. 2. In order to obtain R_{IC}, the effective crack length corresponding to the peak load for the finite size specimen, a_{cf}, should first be determined. By setting $G = R$ and using the LEFM formula, the value of a_{cf} can be determined by maximizing

$$\sigma = (R/G_{\sigma=1})^{1/2} \tag{19}$$

The value of R_{IC} can then be calculated by substituting a_{cf} into eqn (15). Two approximate methods can be used to compute the R-curve after the maximum load. Both of these are approximate since inelastic energy is ignored. One approach is to assume that the value of R remains constant (R_{IC}) after attainment of the maximum load, as was assumed by Bazant *et al* [22]. The other is to assume that the R-curve is defined by eqn (15) until the plateau corresponding to an infinitely large specimen (R_{IC}^s) is reached, i.e. the R-curve is the same as that of the infinite size specimen.

For the negative geometry, since the condition of $dK_I/da = 0$ is size-dependent, the R-curve obtained above is for the finite size specimen whose dimension is used to determine the α value.

Parameter α can be considered as a brittleness index of a material, to describe the pre-critical stable crack growth, for the positive geometry specimens. $\alpha = 1$ represents a perfectly brittle material without stable crack growth, and $\alpha > 1$ represents a quasi-brittle material with stable crack growth. Furthermore, α can be used to predict the effective length of the fracture process zone in an infinite size specimen with positive geometry, such as parameter c_f, introduced by Bazant *et al.* [21], [22]. The definitions of α and c_f directly result in

$$c_f = (\alpha - 1)a_0 \tag{20}$$

The present R-curve formulation allows for the use of various failure criteria, such as the size effect law introduced by Bazant *et al.* [21], [22], to determine the values of α and β.

4.2 Fiber Reinforced Concrete

The present formulation discusses the fiber toughening in the positive geometry. Due to the existence of fibers, both stress intensity factor and crack tip opening displacement consist of contributions from far-field stresses and closing pressure of fibers. The bridging pressure from fibers greatly reduces the stress intensity factor and the crack tip opening displacement. Therefore, crack stability conditions for the matrix can be expressed as,

$$K_{IC}^{s} = K_{I}^{m}(\sigma_c, a_c/b) - \int_{a_c} p(x/a_c)K_{I}^{F}(x/a_c)\,dx \tag{21}$$

$$CTOD_c = CTOD^{m}(\sigma_c, a_c/b, a_0/a_c) - \int_{a_c} p(x/a_c)Q(a_c/b, a_0/a_c, x/a_c)\,dx \tag{22}$$

where K_I^m and $CTOD^m$ are the stress intensity and crack tip opening displacement at the tip of an effective traction free crack in the matrix due to far-field stress; σ_c is the critical far-field stress; a_c is the critical crack length; Q is the Green's function for the crack closure at point a_0 (initial crack mouth) due to a unit force applied at point x along the crack length; $p(x/a_c)$ is the closure force at point x along the crack face, which will be discussed later; a and b are the crack length and the specimen width, respectively; and $K_I^F(x/a_c)$ is the stress intensity due to a unit load applied at point x along the crack surface. K_I^m, $CTOD^m$, K_I^F and Q can be obtained based on LEFM [29], [30].

Equations (21) and (22) present two coupled non-linear integral equations and must be solved for the two unknowns, a_c and σ_c. Once a_c is known, the values of α and β can be calculated. Before eqns (21) and (22) can be applied, however, the formulations of closing pressure, which will be discussed later, should be known.

5. DETERMINATION OF COMPLIANCE

According to LEFM, the energy release rate is

$$G = (\sigma^2 b^2 t/2)(\partial C_m/\partial a) \tag{23}$$

where b and t are the width and the thickness of specimen, and C_m is the loading point compliance of the matrix.

The compliance of the matrix can be obtained by integrating eqn (23) and using $G = R$,

$$C_m = C_m' + \int_{a_0}^{a} 2\beta(a - a_0)/(\sigma^2 b^2 t)\, da \tag{24}$$

where C_m' represents the compliance of the matrix with the initial crack. Knowing the compliance, the load–deflection relationship can be calculated.

6. CLOSING PRESSURE OF FIBERS

The closing pressure, p, should first be formulated before eqns (21) and (22) can be solved. Two types of closing pressure are used to simulate the toughening mechanism of fibers in this chapter. One is the closing pressure obtained from experimental results of fiber pull-out versus slip relationship. The other is a simplified bilinear formulation of closing pressure.

6.1 Closing Pressure Based on Fiber Pullout

Figure 6 represents the average pull out–slip response of a glass fiber strand and steel fibers, respectively. These experiments have been

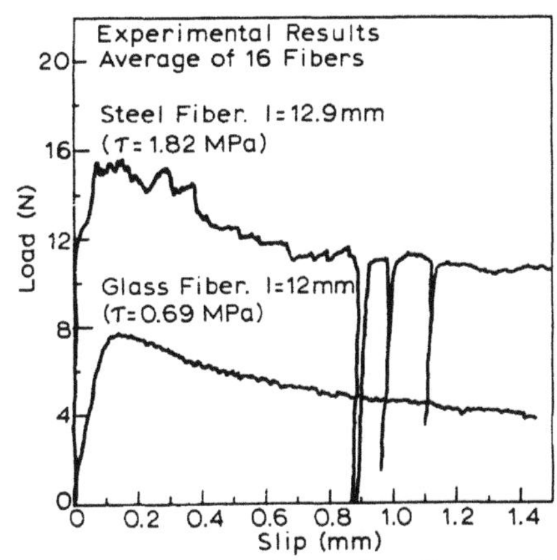

Fig. 6 Closing pressure from experimental pull out–slip relationships of fibers.

recently conducted at Northwestern University where a series of 16 fibers (steel or glass) were pulled out of the matrix socket under closed loop testing conditions [31]. Figure 6 indicates that the initial linear region of the fiber pull-out is followed by a monotonic change in the slope, which characterizes the formation and propagation of a stable debonding process. This debonding process ends at the ultimate load. A stable frictional pull-out governs the post-peak behavior. The interface shear strength evaluated, based on the experimental results, were 1·82 and 0·69 MPa, respectively, for steel fiber and glass fiber using $\tau = P_{max}/(2\pi rl)$, where P_{max} is the maximum pull-out load for a single fiber, r is the fiber radius, and l is the embedment length of the fiber. A single fiber pull-out force was converted to a distributed pressure by dividing the force by the specific fiber spacing. The specific fiber spacing was computed from the fiber volume fraction, and radius:

$$p = (tV_f - \pi r^2/b)P/(\pi r^2) \qquad (25)$$

where P and p represent the single pull-out force and its corresponding distributed pressure (per unit thickness), respectively; r and V_f the fiber radius and volume fraction, and t and b the specimen thickness and width, respectively.

Solutions of eqns (21) and (22) can be achieved by means of an iterative procedure. A marching method was used, where the crack length was incrementally increased. For any given crack length, iterative solutions for the critical stress level were obtained by the following scheme:

(i) A magnitude of the load was assumed, and used to obtain an estimation of the crack opening.
(ii) Based on these crack openings and the pull out–slip response, a closing pressure profile was computed by means of interpolating the experimental results of Fig. 6.
(iii) The closing pressure was substituted in eqns (21) and (22), and these equations were solved independently for updated estimations of the assumed load.
(iv) The procedure was continued until the assumed load, and the solution, met the convergence criterion for the stresses. However, the load magnitudes obtained by solving these two equations independently may not be equal.

(v) Repeat steps (i)–(iv) with updated stress magnitudes until the stresses obtained from eqns (21) and (22) are equal to each other. Satisfaction of both convergence criteria indicates that the solution has been reached.

6.2 Bilinear Formulation of Closing Pressure

A bilinear formulation of closing pressure was used for polypropylene fiber composites, since no pull out–slip relationship was available for this type of fibrillated polypropylene fiber. The formulation is shown in Fig. 4(a). It was assumed that the closing pressure at the crack tip is equal to zero and increases linearly up to a length characterized by C_0a_c, where the magnitude of stress transmitted by the fibers reaches a constant fraction of the load ($\lambda\sigma$, where λ is defined as the stress transmission factor, $0<\lambda<1$),

$$p(x)=\begin{cases}\lambda\sigma & a_0<x<C_0a_c\\ \lambda\sigma[1-(C_0a_c-x)/C_0a_c] & C_0a_c<x<a\end{cases} \tag{26}$$

where $\lambda=0$ represents the case of no closing pressure, (i.e. unreinforced concrete). $\lambda=1$ represents a fully developed crack in which all the force is being transmitted across the crack. There is no stress concentration due to crack length $(1-C_0)a_c$. Note that as λ approaches unity, the length of $(1-C_0)a_c$ does not contribute significantly to the stress intensification since the load is completely transferred through the crack faces. The closing pressure is then similar to the form assumed by Marshall *et al.* [10] to obtain the closed form solutions for matrix cracking strength.

Incorporation of eqn (26) in eqns (21) and (22) results in a set of two equations which are both linear with respect to the critical stress σ_c. Proper substitution of the terms can yield a single non-linear equation for the critical crack length a_c.

7. COMPARISONS WITH EXPERIMENTAL RESULTS

7.1 Concrete and Mortar

Using the present R-curve approach, the fracture response of a specimen can be computed as follows:

(i) Determine α and β based on $K_I = K_{IC}^s$ and $CTOD = CTOD_c$ for the positive geometry specimen and $K_I = K_{IC}^s$ and $dK_I/da = 0$ for the negative geometry specimen by using the LEFM formulations [29].
(ii) Extend crack length step by step, and compute the load, σ, from eqn (19) for each crack extension.
(iii) Compute specimen compliance using eqn (24).
(iv) Calculate the load-point displacement according to the compliance.
(v) Repeat steps (ii)–(iv) until the load decreases considerably after the peak load.

7.1.1 Three-Point Bend (3-PB) Specimens (Positive Geometry)

The 3-PB specimens tested by Jenq & Shah [5], [13] were used. The specimens were made of concrete with a mix proportion by weight of 1:2·6:2·6:0·65 of cement, fine aggregate, coarse aggregate and water. The maximum aggregate size was 19 mm. The dimensions of the beams ($s \times b \times t \times a_0$ as shown in Fig. 4(b)) were 914 mm × 229 mm × 86 mm × 76 mm for the specimen C1L1, 610 mm × 152 mm × 57 mm × 49 mm for the specimen C1M2, and 305 mm × 76 mm × 29 mm × 22 mm for the specimen C1S3, respectively. The values of K_{IC}^s and $CTOD_c$ reported by Jenq & Shah were used to calculate α and β. The Young's modulus, E, is chosen such that the theoretical initial compliance is matched with the experimental one for each specimen. Comparisons between the theoretical and experimental load–displacement curves are shown in Figs 7(a), (b) and (c), where two sets of theoretical results are given. R remains a constant (R_{IC}) after maximum load for the case labeled Theoretical 1, whereas the R-curve is given by eqn (15) until R_{IC}^s is reached for the case labeled Theoretical 2. The stable crack propagation prior to the maximum load in concrete can be reasonably predicted by the model. However, neglecting the inelastic energy in the model causes an underestimation of the post-peak displacement. For all of the subsequent comparison only the results from Theoretical 1 are given for promoting the simplicity of the problem.

7.1.2 Single Edge Notched (SEN) Specimens Subjected to Uniaxial Tension (Positive Geometry)

The formulation discussed was applied to predict the response of mortar subjected to uniaxial tension. The experimental results

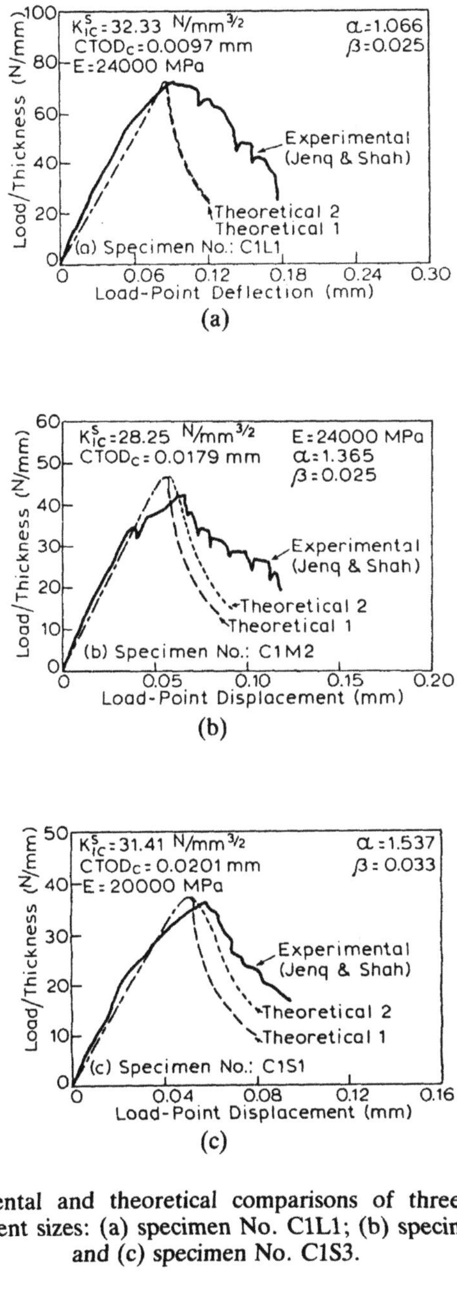

Fig. 7 Experimental and theoretical comparisons of three-point bending beams with different sizes: (a) specimen No. C1L1; (b) specimen No. C1M2; and (c) specimen No. C1S3.

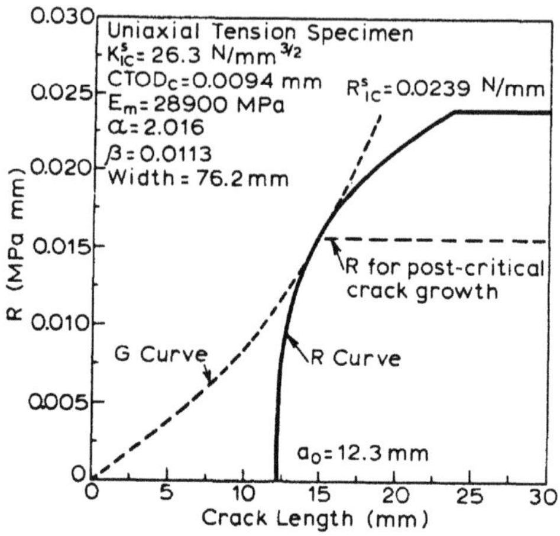

(a)

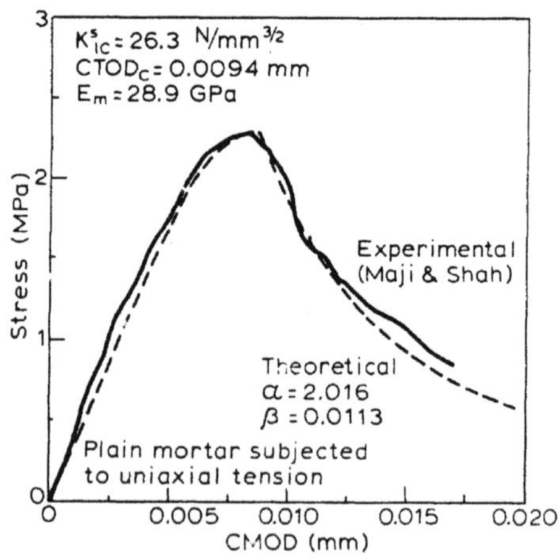

(b)

Fig. 8 Experimental and theoretical comparisons of single notched specimen subjected to uniaxial tension: (a) *R* and *G* curves; (b) stress–CMOD curve; and (c) *R*-curves for specimens with different widths but the same notch.

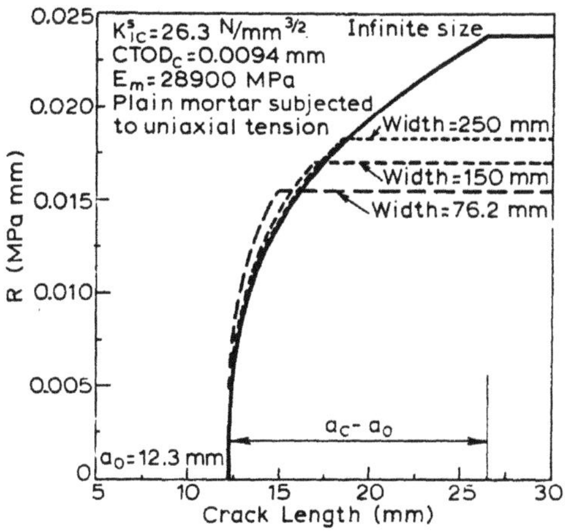

(c)

Fig. 8—(*Continued*)

used were reported by Maji & Shah [32]. The mortar used had a mix proportion of 1:2:0·5 of cement:sand:water by weight. Specimens tested were 304·8 mm × 76·2 mm × 19 mm in size. A single edge notch of 12 mm was used, and the rate of opening of the notch tips was used as the control parameter in loading the specimens. Since the required matrix properties for the mortar mix were not available, the average values of material properties (K^s_{IC}, $CTOD_c$ and E_m) from two mortars, which have the mix proportions of 1:2·6:0·65 and 1:2·6:0·45 (cement:sand:water), respectively, and obtained from a three-point bending specimen by Jenq & Shah [5], were used as an input for theoretical solutions. The R-curve obtained is shown in Fig. 8(a) along with the G curve which was calculated based on equations given by Tada *et al.* [29]. Figure 8(b) indicates a good agreement between the experimental results and the theoretical predictions. The R-curves for the specimens with the same notch size but different widths are shown in Fig. 8(c). Based on the definition of R given in the present theory, these R-curves should be the same. However, some deviation among R-curves is observed, which may be attributed to using up to second order terms in the present theory.

7.1.3 Center-Cracked Plate (CCP) Specimen (Negative Geometry)

The experimental results of mortar reported by Alvarado *et al.* [33] are examined. The mix proportion of the mortar was 1·0:2·6:0·45 of cement, fine aggregate, and water. The dimensions of the specimens $(b \times h \times t \times r \times a_0)$ were 152·4 mm × 152·4 mm × 25·4 mm × 12·7 mm × 22·25 mm. The values of α and β were determined using the K_{IC}^s value given by Alvarado *et al.*, and the condition of $dK_I/da = 0$. The Young's modulus is chosen based on the initial compliance. The theoretical and experimental load–COD curves for both specimens are shown in Fig. 9(a), where COD is defined as the crack opening displacement at a distance of 3·2 mm from the tip of the initial crack. The theoretical R-effective crack extension curves are compared with the experimental R-crack extension curves in Fig. 9(b), where the experimental crack extension is obtained from the com-

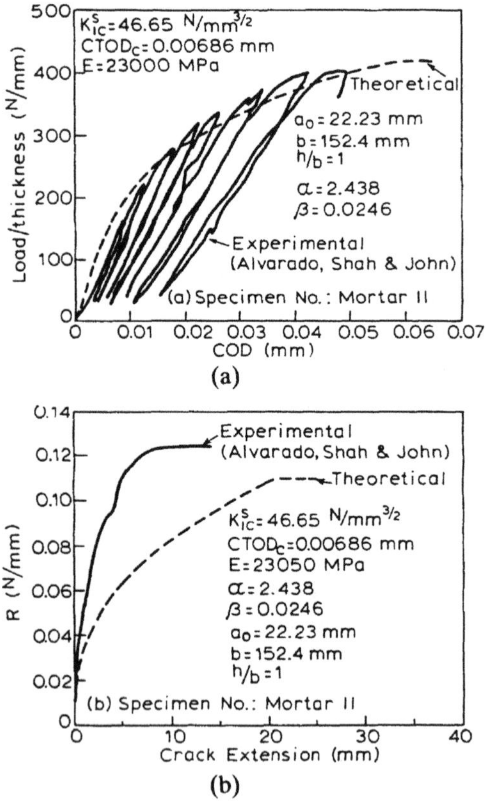

Fig. 9 Experimental and theoretical comparisons of center-cracked plate subjected to tension: (a) load—COD curve; (b) R-crack extension curve.

pliance measurement which is similar with the two-parameter model. The general behavior of R-curves is similar, but as expected the effective crack extension in the present R-curve is larger than the crack extension based on the compliance measurement.

7.2 Fiber Reinforced Concrete

The proposed R-curve approach was applied to single edge notched specimens subjected to uniaxial tension as shown in Fig. 4(a). The specimens were reinforced by unidirectional, continuous steel, glass or polypropylene fibers.

Mobasher, Stang, Castro-Montero & Shah [34]–[36] used an acoustic emission technique and laser holographic interferometry to study crack propagation in this type of specimen. They observed that distributed microcracks first occurred within the cement matrix at an initial loading stage (point A in Fig. 10). These microcracks tended to

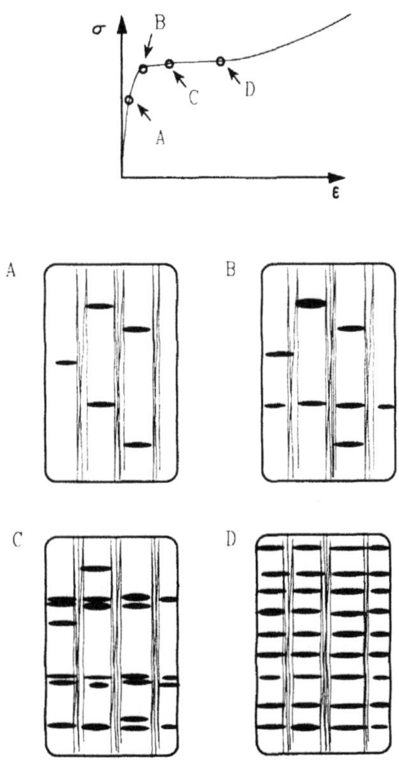

Fig. 10 Crack propagations at different loading stages in a uniaxial tension specimen.

localize and a critical crack propagating through the whole section of the specimen was observed at point B, which is normally referred to as a bend-over-point (BOP) and corresponds to the maximum matrix contribution. After the BOP, several cracks along the specimen length occurred. The researchers reported that the stress and strain at the BOP increases with increasing volume of fiber. For example, for cement-based composites reinforced with about 13% of uniaxial polypropylene fiber, the bend-over-point occurred at 15 MPa which is approximately four times the tensile strength of the unreinforced matrix. Since the present *R*-curve approach is a single crack fracture model, no attempt was made to describe fracture behavior of multiple cracking. The following comparison will only address the stress and strength at the BOP. A single notch of 0·5 mm was assumed in theoretical predictions for the fiber reinforced specimens discussed hereafter.

The predictions from the present *R*-curve approach are also compared with that predicted by the model proposed by Aveston *et al.* [7]. Their model, referred to as the ACK model, predicts that the increase in the matrix fracture strain due to fibers can be obtained by

$$\varepsilon_{mu} = [12\tau\gamma_m E_f V_f^2/(E_c E_m^2 r V_m)]^{1/3} \tag{27}$$

where E and V represent Young's modulus and volume fraction, respectively; subscripts m, f and c represent index for matrix, fiber and composite, respectively; $\gamma_m = 0{\cdot}5(K_{IC}^s)^2/E_m$, which represents the surface fracture energy of the matrix; and τ represents the interfacial shear strength. The ACK model assumed that fracture strain is linearly proportional to the fracture strength.

7.2.1 Steel Fiber Concrete

The experimental results of matrix response in the presence of steel fibers were obtained from the work of Somayaji & Shah [37]. Specimens tested were 305 mm × 25·4 mm × 9·5 mm, using a mortar mix of 1:2:0·5 (cement:sand:water). Continuous steel wires of diameter 1·19 mm were used as the reinforcement. Specimens were tested in tension, and elongations were measured using a LVDT mounted across a gage length of 153 mm.

The procedures to obtain the theoretical results for fiber composites are similar to those for plain concrete except that the second terms of the right hand side (fiber contributions) of eqns (21) and (22) are not zero. The matrix properties previously used in unreinforced concrete

(SEN specimen) were used in the theoretical computation. The experimental bond-slip curve for a steel fiber shown in Fig. 6 was used as the closing pressure in the theoretical formulation. The theoretical bend-over-point is defined as the ultimate strength of the matrix in the presence of the fibers. Figures 11(a) and 11(b) give comparisons of the

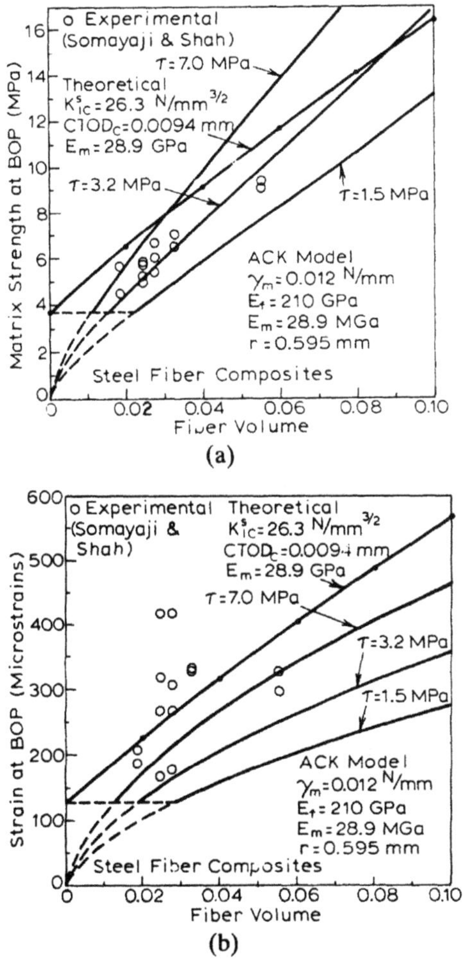

Fig. 11 Experimental and theoretical comparisons for steel fiber reinforced composites: (a) matrix strength at the BOP for different fiber volume fractions; and (b) matrix strain at the BOP for different fiber volume fractions.

theoretically predicted stress and strain at the BOP with the experimental data.

Figures 11(a) and 11(b) also give a comparison between the proposed theory with the ACK model. In order to use the ACK model, an interface shear strength, τ, between fiber and matrix should be known. According to the fiber pull-out results the value of τ for steel fibers is 1·82 MPa. According to the ACK model, the interface shear strength, τ, can be estimated by the following equation:

$$x' = (1 - V_f)\sigma_{mu} r/(2\tau V_f) \quad (28)$$

where x' represents the shear stress transfer length, which is equal to half of crack spacing, σ_{mu} is the matrix crack strength, and r is the radius of the fiber. Substituting the experimental matrix strengths (for zero volume fraction of fiber) and the minimum observed crack spacings reported by Somayaji & Shah [37] into eqn (26), the value of τ was either 3·6 or 6·9 MPa, depending on volume of steel wires. Majumdar [38] also reported a τ value of 5·4 MPa for a steel fiber embedded in a mortar matrix for 28 days. The predictions of the ACK model are shown for $\tau = 1{\cdot}5$, 3·2 and 7·0 MPa, respectively, in Figs 11(a) and 11(b).

According to both models, the ultimate load-carrying capacity of the toughened matrix increases with increasing fiber volume. Note that the reported results are the matrix response only. Furthermore, from a limiting point of view, the present theoretical model approaches the plain matrix response at $V_f = 0$, while the ACK model approaches the matrix strength at the critical volume fraction. Typical critical volume fractions are given by Hannant [39] as 0·3, 0·4, and 0·75% for steel, glass, and fibrillated polypropylene fibers, respectively. The results of the ACK model are shown under the assumption that the unreinforced matrix strength is obtained by the present approach. It is noted that although eqn (26) provides $\tau = 3{\cdot}6$–$6{\cdot}9$ MPa, the ACK model (eqn (27)) fits the BOP data well by using $\tau = 3{\cdot}2$ MPa. There is, however, a significant underestimation for the strain using the same shear magnitude. This phenomenon is perhaps attributed to the assumption that prior to BOP in the ACK model, the matrix remains elastic with a stiffness equal to the E_c. Holographic, optical microscopy, and acoustic emission studies have however confirmed that microcracking does indeed take place prior to the BOP.

Theoretical predictions of the R-curves of an infinite large plate subjected to uniaxial tension are given in Fig. 12(a) for different fiber volume for steel fiber composites. The enhancement in matrix response due to an increase in fiber volume fraction is apparent. An increase in the fiber volume fraction results in both an increase in the plateau value, and an increase in the slope of $R(\partial R/\partial a)$. These changes directly affect the crack instability conditions. The model predicts that as the fiber

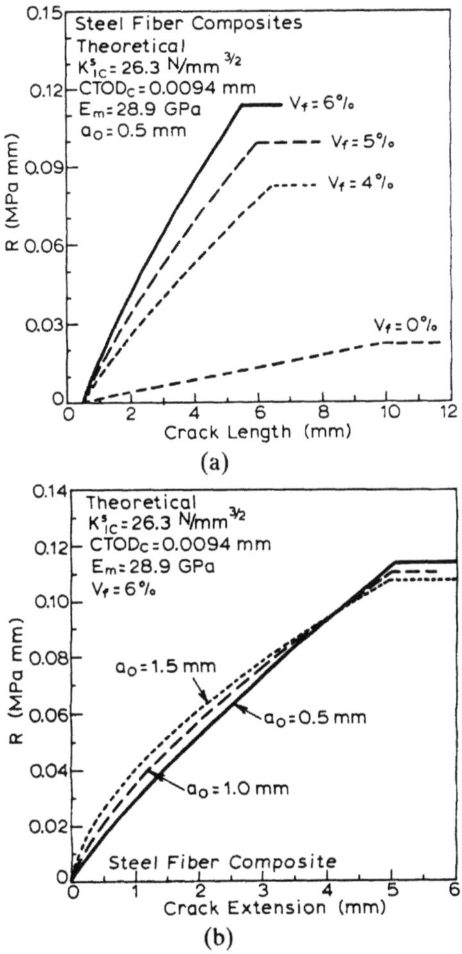

Fig. 12 Influence of fiber volume fraction and initial crack length on R-curve: (a) R-curve for different fiber volume fraction; (b) R-curve for different initial crack lengths.

volume fraction increases the length of stable crack growth decreases. This observation is contrary to the accepted belief that increase in the fiber volume results in an increase in the stable crack growth. It must be noted that the present approach is an equivalent elastic formulation, hence no comparison regarding the crack lengths obtained by various methods can be made. The *R*-curve is not significantly sensitive to various initial crack lengths as shown in Fig. 12(b). Tripling a_0 only causes a slight steepening of slope and a minor decrease of plateau value in the *R*-curve.

7.2.2 Glass Fiber Reinforced Composites

Figure 13(a) represents the theoretical and experimental effects of fiber volume fraction on the BOP of the glass fiber composites. The experimental results were obtained by Mobasher *et al.* [34], Oakley & Proctor [40], and Ali *et al.* [41]. The theoretical results are obtained using the material parameters K_{IC}^{s} and $CTOD_c$ for both mortar and cement paste as reported by Jenq & Shah [5] from the notched-beam tests, because various matrix mix proportions were used by these investigators. The glass fiber pull-out characteristics as shown in Fig. 6 were used as the closing pressure. The theoretical results of the ACK model are also shown. The experimental results of fiber pull-out indicate a shear strength magnitude of $\tau = 0{\cdot}69$ MPa for glass fibers in an unaged form (Fig. 6). This value however is not constant and depends on the curing period, and the age of the composite at the time of the test. Oakley & Proctor [40] reported a value of $\tau = 1{\cdot}1$ MPa for glass fibers. Majumdar [38] has reported values of 2–3 MPa. In order to cover the range of experimental data, the ACK model is shown for $\tau = 0{\cdot}7$ and 2 MPa, respectively. It can be seen that the proposed theoretical model compares well with the experimental data in showing that the matrix strength increases in the presence of fibers. The ACK model fits the data well if $\tau = 0{\cdot}7$ MPa is used for prediction.

The theoretical predictions of matrix strength, σ, are normalized by the strength of the unreinforced matrix, σ_0, and are plotted as a function of fiber volume in Fig. 13(b). Note that the enhancement in the matrix strength is much higher for mortar than cement paste. This may be attributed to the fact that enhancements in the material response can only take place under conditions where pre-critical crack growth exists, causing fibers to become effective. Hence, no enhancement can be gained from the response of a perfectly brittle material,

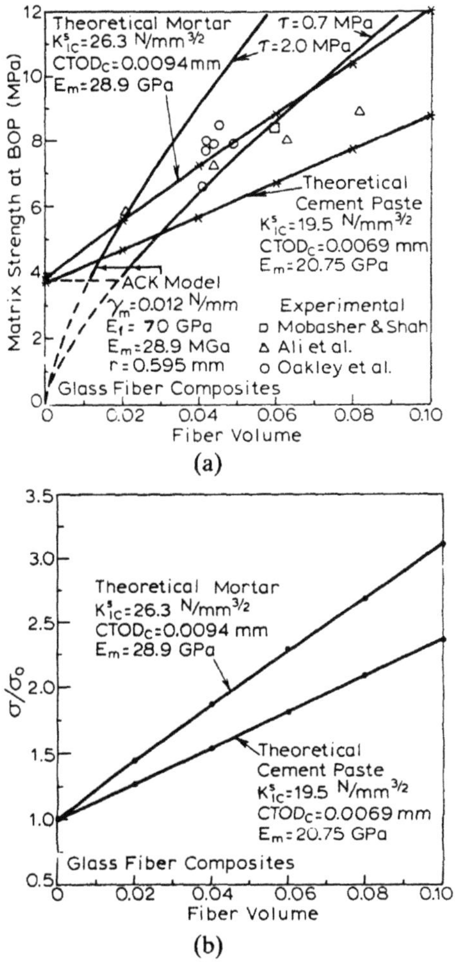

Fig. 13 Experimental and theoretical comparisons for glass fiber reinforced composites: (a) matrix strength at the BOP for different fiber volume fractions; and (b) matrix toughening of mortar and cement paste.

i.e. infinitesimal crack growth prior to onset of instability. The role of fibers is the conversion of the stable cracking process into an increase in load-carrying capacity of the material. Therefore the larger the size of the pre-critical crack growth, the more enhancement can be obtained by reinforcing the material.

7.2.3 Polypropylene Fiber Reinforced Composites

The comparison of theoretical and experimental results for stress at the BOP for polypropylene fiber composites is indicated in Fig. 14(a), where the same matrix properties were used. The experimental results produced by Krenchel & Stang [42], and Mobasher *et al.* [34] are used. Because there is no pull out–slip curve available for polypropylene fiber, the simplified closing pressure given by eqn (28) was used for the theoretical computation. The theoretical predictions are the stress and strain at the BOP for various values of the stress transmission factor when $C_0 = 0{\cdot}1$ was used. It was observed that the present theory using the simplified closing pressure predicts lower values of both stress and strain at the BOP as compared to the data. To enable comparison with the ACK theory, the value of τ is required. Since it is difficult to estimate the radius of polypropylene fibrillated fibers, the values of τ/r are estimated by the crack spacings measured by Mobasher *et al.* [34]. Figure 14(b) indicates the relationship between σ_{mu} and x' according to eqn (26) by using the values of x' and σ_{mu} obtained by Mobasher *et al.* [34], where it is assumed that the majority of the microcrack system has been fully developed at 1% total strain. The slope of the line represents τ/r. One would obtain a value of τ/R in the range of 100 MPa/mm or 200 MPa/mm depending upon whether $V_f = 8\%$ or 12%. These two values of τ/r were used in the ACK model to compute the tensile strength at the BOP, which are given in Fig. 14(a). It is observed that the stress transmission factor λ, which combines the effects of both closing pressure of the fibers, and the traction due to the matrix grain bridging, increases as the fiber volume fraction increases. It is furthermore observed that the ACK model gives good results compared with the experimental data, if the value of $\tau/r =$ 200 MPa/mm is used.

8. CONCLUSIONS

The R-curve approach for fracture of concrete and fiber reinforced concrete is presented in this chapter. By assuming that the critical crack length is a function of an initial crack length a_0, a formulation of the R-curve can be derived by solving a differential equation. The R-curve obtained includes two parameters that can be (for example) uniquely determined from K_{IC}^s and $CTOD_c$ for the positive geometry, and from K_{IC}^s and $dK_I/da = 0$ for the negative geometry. A load–

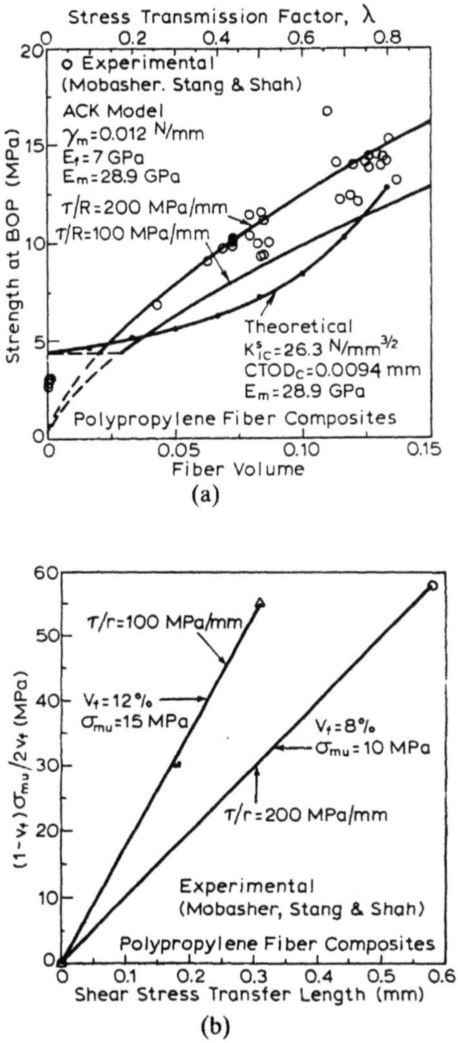

Fig. 14 Experimental and theoretical comparisons of polypropylene fiber reinforced composites: (a) matrix strength at the BOP for different fiber volume fractions; and (b) experimental relation between τ/r and fiber volume fraction.

CMOD curve and a load–displacement curve can be predicted based on the present R-curve. For fiber reinforced composites, toughening of matrix due to fiber bridging is taken into account by applying closing pressure on crack surfaces, which can reduce the stress intensity factor and the crack tip opening displacement. The matrix toughening and fracture response in the fiber reinforced composites can be predicted up to the bend-over-point by incorporating the present R-curve and single fiber pull-out experimental data. The theoretical results show good agreement with experimental results with respect to different materials and different geometries.

ACKNOWLEDGEMENTS

The support of National Science Foundation Center for Science and Technology of Advanced Cement-Based Materials (ACBM) Grant #DMR-8808432, and NSF Grant MSM-8906937 are gratefully appreciated.

REFERENCES

[1] Kaplan, M. E., Crack propagation and fracture of concrete. *Journal of the American Concrete Institute,* **58**(5) (1961) 591–610.

[2] Shah, S. P. & McGarry, F. J., Griffith fracture criterion and concrete. *Journal of Engineering Mechanics Division* (ASCE), **97** (1971) 1663–76.

[3] Hillerborg, A., Modeer, M. & Petersson, P. E., Analysis of crack formation and crack growth in concrete by means of fracture mechanics and finite elements. *Cement and Concrete Research,* **6** (1976) 773–82.

[4] Bazant, Z. P. & Oh, B. H., Crack band theory for fracture of concrete. *Materials and Structures,* **16** (1983) 155–77.

[5] Jenq, Y. S. & Shah, S. P., Two parameter fracture model for concrete. *Journal of Engineering Mechanics* (ASCE), **111** (1985) 1227–41.

[6] Bazant, Z. P., Size effect in blunt fracture: concrete, rock and metal. *Journal of Engineering Mechanics* (ASCE), **110** (1984) 518–35.

[7] Aveston, J., Cooper, G. A. & Kelly, A., Single and multiple fracture. *The Properties of Fiber Composites, National Physical Laboratory,* Guildford, Surrey: IPC Science and Technology Press Ltd, Surrey, UK, 1971.

[8] McCartney, L. N., Mechanics of matrix cracking in brittle-matrix fibre reinforced composites. *Proceedings of Royal Society of London,* **A409** (1987) 329–50.

[9] Budiansky, B., Hutchinson, J. W. & Evans, A. G., Matrix fracture in

fiber-reinforced ceramics. *Journal of Mech. Phys. Solids*, **34** (1986) 167–89.

[10] Marshall, D. B., Cox, B. N. & Evans, A. G., The mechanics of matrix cracking in brittle-matrix fiber composites. *Acta Metal.*, **33** (1985) 2013–23.

[11] Nemat-Nasser, S. & Hori, M., Toughening by partial or full bridging of cracks in ceramics and fiber reinforced composites. *Mechanics of Materials*, **6** (1986) 245–69.

[12] Ballarini, R., Shah, S. P. & Keer, L. M., Crack growth in cement-based composites. *Engineering Fracture Mechanics*, **20** (1984) 433–45.

[13] Jenq, Y. S. & Shah, S. P., Crack propagation in fiber-reinforced concrete. *Journal of Structural Engineering* (ASCE), **112** (1986) 19–34.

[14] McCabe, D. E. (Ed), *Fracture Toughness Evaluation by R-Curve Method, ASTM-STP 527.* American Society for Testing and Materials, Philadelphia, Penn., 1973.

[15] Krafft, J. M., Sullivan, A. M. & Boyle, R. W., Effect of dimensions on fast fracture instability of notched sheets. *In Proceedings of Crack Propagation Symposium*, College of Aeronautics, Cranfield, UK, Vol. I, 1961, pp. 8–28.

[16] Broek, D., *Elementary Engineering Fracture Mechanics*, 4th Edn. Martinus Nijhoff, Hingham, Mass., 1987, pp. 185–200.

[17] Wecharatana, M. & Shah, S. P., Slow crack growth in cement composites. *Journal of Structural Engineering* (ASCE), **108** (1982) 1400–1413.

[18] Wecharatana, M. & Shah, S. P., Predictions of nonlinear fracture process zone in concrete. *Journal of Engineering Mechanics* (ASCE), **109** (1983) 1231–46.

[19] Jenq, Y. S. & Shah, S. P., A fracture toughness criterion for concrete. *Engineering Fracture Mechanics*, **21** (1985) 1055–69.

[20] Bazant, Z. P., Kim, J. K. & Pfeiffer, P. A., Nonlinear fracture properties from size effect tests. *Journal of Structural Engineering* (ASCE), **112** (1986) 289–307.

[21] Bazant, Z. P. & Kazemi, M. T., Determination of fracture energy, process zone length and brittleness number from size effect, with application to rock and concrete. *International Journal of Fracture*, **44** (1990) 111–31.

[22] Bazant, Z. P., Gettu, R. & Kazemi, M. T., Identification of nonlinear fracture properties from size effect tests and structural analysis based on geometry-dependent *R*-curve. *International Journal of Rock Mechanics, Mining Science & Geometry Abstract*, **28** (1991) 43–51.

[23] Foote, M. L., Mai, Y. W. & Cotterell, B., Crack growth resistance curves in strain-softening materials. *Journal of Mechanics, Physics and Solid*, **34** (1986) 593–607.

[24] Sakai, M., Yoshimura, J., Goto, Y. & Inagaki, M., *R*-curve behavior of a polycrystalline graphite: microcracking and grain bridging in the wake region. *Journal of American Ceramics Society*, **71** (1986) 609–616.

[25] Hsueh, C. H. & Becher, P. F., Evaluation of bridging stress from *R*-curve behavior for nontransforming ceramics. *Journal of American Ceramics Society*, **71** (1988) 234–7.

[26] Ouyang, C., Mobasher, B. & Shah, S. P., An *R*-curve approach for

fracture of quasi-brittle materials. *Engineering Fracture Mechanics*, **37** (1990) 901–13.

[27] Mobasher, B., Ouyang, C. & Shah, S. P., Modelling of fiber toughening in cementitious materials using an R-curve approach. *International Journal of Fracture* (1991) (in press).

[28] Planas, J. & Elices, M., Conceptual and experimental problems in the determination of the fracture energy of concrete. In *Fracture Toughness and Fracture Energy-Test Methods for Concrete and Rock*, eds H. Mihashi, H. Takahashi & F. H. Wittmann, A. A. Bolkema Publishers, Brookfield, Vermont, 1989, pp. 165–81.

[29] Tada, H., Paris, P. C. & Irwin, G. R., *The Stress Analysis of Cracks Handbook*, 2nd Edn. Paris Production Inc., St. Louis, Missouri, 1985.

[30] Mobasher, B., Reinforcing mechanism of fibers in cementitious composites. PhD thesis, Department of Civil Engineering, Northwestern University, Evanston, IL., (1989).

[31] Li, Z., Mobasher, B. & Shah, S. P., Characterization of interfacial properties in fibre-reinforced cementitious composite. *Journal of the American Ceramic Society*, **74** (1991) 2156–3164.

[32] Maji, A. K. & Shah, S. P., Process zone and acoustic emission measurement in concrete. *Experimental Mechanics*, **28** (1988) 27–33.

[33] Alvarado, M. A., Shah, S. P. & John, R., Mode I fracture in concrete using center-cracked plate specimens. *Journal of Engineering Mechanics* (ASCE), **115** (1989) 366–83.

[34] Mobasher, B., Stang, H. & Shah, S. P., Microcrack in fiber reinforced concrete. *Cement and Concrete Research*, **20** (1990) 665–76.

[35] Stang, H., Mobasher, B. & Shah, S. P., Quantitative damage characterization in polypropylene fiber reinforced concrete. *Cement and Concrete Research*, **20** (1990) 540–58.

[36] Mobasher, B., Castro-Montero, A. & Shah, S. P., A study of fracture in fiber reinforced cement-based composites using laser holographic interferometry. *Experimental Mechanics*, **30** (1990) 286–94.

[37] Somayaji, S. & Shah, S. P., Bond stress versus slip relationship and cracking response of tension members. *Journal of American Concrete Institute*, **78** (1981) 217–25.

[38] Majumdar, A. J., Properties of fiber cement composites. *RILEM Symposium on Fiber Reinforced Cement and Concrete*, ed. A. Neville. The Construction Press, Lancaster, UK, 1975, pp. 279–313.

[39] Hannant, D. J., *Fibre Cements and Fiber Concretes*. John Wiley & Sons, New York, 1978, p. 219.

[40] Oakley, D. R. & Proctor, B. A., Tensile stress–strain behavior of glass fiber reinforced cement composites. *RILEM Symposium on Fiber Reinforced Cement & Concrete*, Ed. A. Neville. The Construction Press, Lancaster, UK, 1975, pp. 347–59.

[41] Ali, M. A., Majumdar, A. J. & Singh, B., Properties of glass-fiber cement—the effect of fiber length and content. *Journal of Materials Science*, **10** (1975) 1732–40.

[42] Krenchel, H. & Stang, H., Stable microcracking in cementitious materials. *Proceedings of 2nd International Symposium on Brittle Matrix Composites*, Cedzyna, Poland, Sept. 1988.

Chapter 7

Size Effect in Concrete Structures: an R-Curve Approach

M. ELICES & J. PLANAS

Departamento de Ciencia de Materiales Universidad Politécnica de Madrid, E.T.S. Ing. Caminos, Ciudad Universitaria, 28040-Madrid, Spain

ABSTRACT

The maximum load size effect is analysed with the help of the R–Δa curve approach. To this end the concept of equivalent elastic crack is reviewed and it is shown that the R–Δa curve approach is a special case of this equivalence. The analysis includes positive, negative and uncracked geometries. The concepts of nominal *stress intensity factor and of* intrinsic *size are generalized to get consistent results for the asymptotic behaviour up to first order for any geometry and to avoid negativeness of the objective size. Practical examples of these three geometries are included, where the simple predictions based on the equivalent elastic crack are compared with the numerical computations of more involved cohesive cracks.*

1. INTRODUCTION

The change of a structural property when the size of a structure increases is known as a *size effect* related to this property. It is well known that structures become more brittle as their size increases but, nevertheless, the classical procedure is to scale up a successful design keeping the working stresses the same. Hence, a knowledge of the size effect on the maximum load a structure can withstand is relevant and this contribution summarizes some of the recent trends related to the *maximum load size effect* for plain concrete structures.

Neither limit analysis nor linear elastic fracture mechanics (LEFM) are able to explain experimental size effect results for ordinary sizes, as has been stated in all recent symposia devoted to this subject. In this chapter the problem is approached from two directions: one is based on the cohesive crack model and the other relies on R–Δa curves that may be considered a special case of the cohesive models valid only under certain conditions, as will be shown later.

Furthermore, most of the published work on size effects deals with cracked geometries belonging to a class known as *positive* (or *normal*) geometries [1]. This contribution takes into account *negative* geometries and *uncracked* geometries. Despite the practical significance of uncracked samples, relevant work on this problem is scarce.

The plan of this chapter is as follows:

(a) The basic concepts related to the size effect are introduced; a powerful approach based on the cohesive crack model is summarized and positive and negative geometries introduced as well as the concepts of nominal stress intensity factor and intrinsic size. To gain insight, the R–Δa curve approach is considered because it allows a simple analytic and graphic treatment.

(b) The size effect law for *positive geometries* is reviewed from both points of view and it is shown that both approaches converge for large sizes. The *negative geometries* are analysed and it is shown that the nominal stress intensity factor and intrinsic size may be redefined to eliminate negative values. Finally, the *uncracked geometries* are investigated showing the existence of two limiting behaviours, the brittle limit for large sizes, that may be represented by an R-curve model, and the plastic limit for small sizes, which requires the use of a cohesive formulation.

2. BASIC CONCEPTS

2.1 Linear Elastic Fracture Mechanics Approach

For linear perfectly brittle materials, the evolution of strength with size is described by the equations of Linear Elastic Fracture Mechanics

(LEFM). When the specimen is symmetric with respect to the crack plane and subjected to proportional symmetric loading, promoting crack growth in pure opening mode (Mode I), the basic LEFM loading parameter is the stress intensity factor (SIF) K_I which may always be written, for a crack of length a, as

$$K_I(a) = \sigma_N \sqrt{D}\, S(a/D) \tag{1}$$

where the dependence of the SIF on a has been made explicit, D is a characteristic dimension of the structure that we call in short *size*, $S(a/D)$ is the geometrical shape factor and σ_N, the *nominal* stress, is the load divided by a conventional cross-section.

The SIF for the initial crack length a_0 is uniquely related to the nominal stress and it appears convenient to define a *nominal SIF* as

$$K_{IN} = \sigma_N \sqrt{D}\, S(a_0/D) \tag{2}$$

Since for the geometrically similar structures the a_0/D ratio is constant, the maximum nominal SIF may be used as a measure of the strength instead of the maximum nominal stress.

For *positive geometries*, defined as those in which the stress intensity factor is an increasing function of crack length under constant load ($S'(a/D) > 0$), the peak load is reached at crack initiation, i.e. when $K_I(a_0) = K_{IN} = K_{IC}$. Hence $K_{INmax} = K_{IC} = constant$ for these kind of geometries.

In consequence, when LEFM applies, the prediction for the maximum load as size increases is the well known result

$$\sigma_{Nmax} = \frac{K_{IC}}{\sqrt{D}\, S(a_0/D)} \tag{3}$$

showing a decrease of the strength with the square root of the size.

The specific fracture energy G_F is defined as:

$$G_F = \frac{K_{IC}^2}{E'} \tag{4}$$

where E' is the equivalent modulus of elasticity for plane strain or generalized plane stress. This magnitude will be useful when comparing this model against other models.

The size effect equation of LEFM may then be written in the form:

$$\frac{E' G_F}{[K_{INmax}]^2} = 1 \tag{5}$$

2.2 Cohesive Crack Approach

Cohesive crack models were reviewed by the authors in Refs [2] and [3]. Based on the cohesive crack model set up by Hillerborg and co-workers for concrete [4], [5], the authors developed a general expression for the size effect equation using only dimensional analysis and imposing LEFM behaviour for large sizes [6], [7]. Positive geometries were implicitly assumed. This expression takes the form of a power series expansion and may be written in terms of the maximum nominal SIF as:

$$\frac{E'G_F}{[K_{\text{INmax}}]^2} = 1 + C_1\frac{l_{ch}}{D} + C_2\left(\frac{l_{ch}}{D}\right)^2 + \ldots \tag{6}$$

where C_i are dimensionless coefficients that depend on geometrical ratios and on the shape of the softening curve and

$$l_{ch} = \frac{E'G_F}{f_t^2} \tag{7}$$

where f_t is the tensile strength.

For very large sizes $l_{ch}/D \rightarrow 0$, and eqn (6) converges to the LEFM equation (5). Equation (6) shows that brittleness is a *relative* concept, relating a material size (l_{ch}) to a structural size (D). In a given engineering context one may unambiguously use the term 'brittle material' if this material possesses a l_{ch} much less than any practical size in this particular engineering context. However this denomination may be misleading in another engineering field in which much smaller sizes are routinely used.

The first order deviation from LEFM in eqn (6) is given by the coefficient C_1. The structure of this coefficient was determined by the authors using a special asymptotic analysis [8], [9]. C_1 is seen to be expressible as the product of a geometrical factor times a material constant. The resulting size effect expression is

$$\frac{E'G_F}{[K_{\text{INmax}}]^2} = 1 + \Delta a_{C\infty}\frac{2S_0'}{S_0 D} + o\left(\frac{l_{ch}}{D}\right) \qquad \text{only for} \qquad S_0' > 0 \tag{8}$$

where the $\Delta a_{C\infty}$ is the *critical effective crack extension for infinite size*, which may be obtained from the softening curve of the postulated cohesive model as described in the quoted references, and S_0 and S_0' are the values of the shape factor and of its first derivative for the initial crack length. $o(x)$ stands for a function vanishing faster than its argument. The limitation to positive geometries (i.e. when the SIF increases with crack length, or $S_0' > 0$) is essential because otherwise the structure of the size effect equation is controlled by higher order terms and the form (8) is no longer applicable.

This result is the same as that obtained by Bazant [10], based on energetic considerations and on the crack band concept, which may be written in terms of the nominal SIF as

$$\frac{E' G_F}{[K_{\mathrm{INmax}}]^2} = 1 + \frac{\lambda_0}{D} \tag{9}$$

where λ_0 is a parameter having dimensions of length that depends both on material and specimen geometry.

From eqn (8) it may be noticed that two cracked samples of the same material are asymptotically equivalent when the factor $(2S_0'/S_0)D^{-1}$ takes the same value for both structures. Consequently, an *intrinsic size* D_i was defined [1] as

$$\frac{1}{D_i} = \frac{2S_0'}{S_0 D} \tag{10}$$

With this definition, eqn (8) takes the simpler form

$$\frac{E' G_F}{[K_{\mathrm{INmax}}]^2} = 1 + \frac{\Delta a_{C\infty}}{D_i} + o\left(\frac{l_{ch}}{D_i}\right) \tag{11}$$

A general size-effect relation for the maximum load and positive geometries is sketched in Fig. 1 for cohesive cracks together with the asymptotic expression (11) which, with the variables used, takes the form of a straight line. Two other extreme cases are also depicted: when LEFM applies, K_{INmax} is a constant and a horizontal line is found. When limit analysis is appropriate, then σ_{Nmax} is constant and a straight line through the origin is the result. For small sizes, cohesive crack models tend to limit analysis, or strength-of-materials models. This is shown by an asymptote parallel to the limit analysis straight line. For large sizes, cohesive crack models tend towards LEFM and should converge to $E'G_F(K_{\mathrm{INmax}})^{-2} = 1$ and must do so linearly as shown in the figure. This representation is not the usual size effect plot (σ_{Nmax} versus D) but it is helpful in performing asymptotic analysis because large sizes are kept in focus near the origin.

For negative geometries the analysis is not so simple and a redefinition of the intrinsic size is needed. The same happens with uncracked samples. Both problems will be dealt with later.

2.3 R–Δa Curve Approach

The analysis of the size effect is made more simple when using the R–Δa curve approach because a more intuitive graphical treatment is possible, and because some interesting results may be analytically obtained, while cohesive models require numerical procedures. On the other hand, one has to rely on models based on elastic crack equivalences with their limitations. Specifically, one has to be aware that R–Δa curves are geometry and size dependent although in practice, for standard geometries and usual sizes, this dependence is not strong for the purposes of the analyses to follow.

To review the R–Δa curve approach we start with a summary of the elastic crack equivalence models and its relation to cohesive cracks. Then the R–Δa equivalence is reviewed and, finally, its connection with maximum load size effect is introduced.

2.3.1 The Equivalent Elastic Crack [11]

The simplest approach to nonlinear fracture problems is to use an equivalent elastic crack, with its tip somewhere in the nonlinear zone surrounding the true (non-stress transferring) crack. The location of the effective or equivalent crack tip has to be determined according to some criterion. There are many possible criteria, so that many

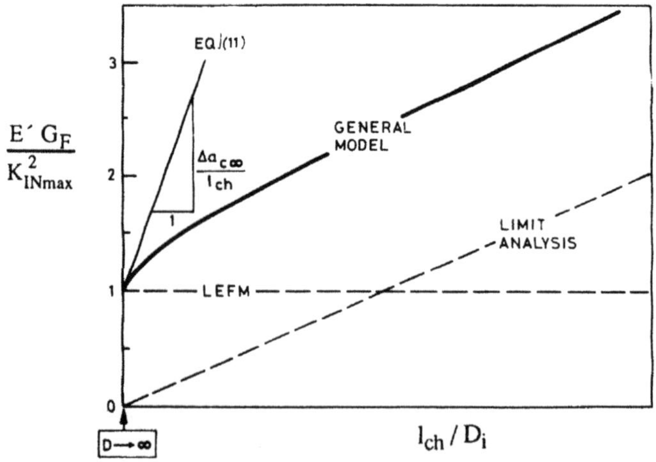

Fig. 1 A size-effect plot for the maximum load and positive geometries.

different equivalences between the actual specimen and the effective elastic one may be defined.

Let us consider a particular equivalence—the force–displacement equivalence—to fix the idea. Two geometrically identical cracked samples, as shown in Fig. 2, are loaded under displacement control u. One sample is made with a cohesive material, as defined above, and the other is made with a linear elastic material. The measured response of the two samples—i.e. the loads P and P_{eq}—for every displacement u will be different, but we can force the responses to match each other, $P = P_{eq}$, by choosing a suitable *equivalent crack length* a_{eq} and a suitable *equivalent crack growth resistance* R_{eq} at each deformation level. In doing so, we force both samples to exhibit the same P–u behaviour but, in general, the equivalence ends here; stress or displacement fields, or relevant parameters like CMOD or CTOD, are not the same. Moreover the R–Δa curve obtained is not a material property, it depends on the geometry and specimen size.

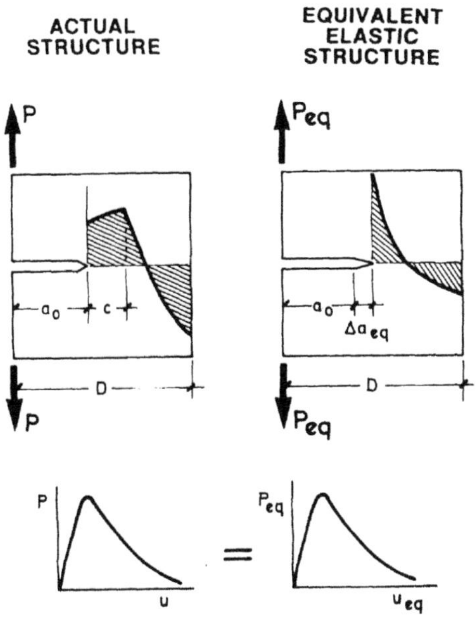

Fig. 2 Elastic crack equivalence: in a P–u equivalence the load and crack length are determined to obtain the same P–u curve as for the actual structure.

The load–displacement equivalence is one of the many possible equivalences. For other equivalences two conditions are imposed to hold simultaneously for the actual and the equivalent specimen. At each loading step of the actual specimen, the equivalence conditions allow the unique determination of the equivalent load P_{eq} and the equivalent crack length a_{eq}, from which we may compute the SIF for the equivalent crack, which will always take the form:

$$K_{Ieq}(P_{eq}, a_{eq}) = \frac{P_{eq}}{B\sqrt{D}} S\left(\frac{a_{eq}}{D}\right) \tag{12}$$

where B is the specimen thickness.

A point of the equivalent R–Δa curve may be obtained at each step by imposing that in equilibrium, the driving specific energy

$$J_{eq} = \frac{[K_{Ieq}(P_{eq}, a_{eq})]^2}{E'} \tag{13}$$

which may be computed from eqn (12), equals the specific resisting energy $R_{eq}(\Delta a_{eq})$.

The equivalence concept and some examples were presented in [11]. Of these, the following equivalence due to Bazant is of special concern for the work to follow.

2.3.2 Bazant Size-Independent R–Δa Equivalence [12]–[14]
Bazant introduced a method for computing a size-independent—but geometry dependent—R–Δa curve from the knowledge of the maximum load size effect curve: variation of peak load P_{peak} with size D for geometrically similar pre-cracked specimens.

Two conditions are imposed in this approach:

(a) The first condition is that the peak load must be the same on the actual and on the equivalent elastic specimen for every size D; $P_{max} = P_{eq,max}$.
(b) The second condition is that the R–Δa curve must be 'size independent'.

The only input is the size effect curve, known from experiments or by computation, i.e.

$$P_{max} = P_{max}(D) \tag{14}$$

The unknowns are the equivalent resistance and the equivalent crack extension at maximum load for every size. When the size is eliminated, the $R-\Delta a$ curve appears. The result, derived and used by Bazant in many papers, is that the R-curve is the envelope of the following uniparametric family of functions, with parameter D:

$$R^{\mathrm{BG}} = \frac{1}{E'}\left[\frac{P_{\max}(D)}{B\sqrt{D}} S\left(\frac{a_0 + \Delta a^{\mathrm{BG}}}{D}\right)\right]^2 \tag{15}$$

where superscript BG stands for Bazant and General, because no special size effect curve is postulated here. The size effect may come from a cohesive model, from Bazant's size effect law, or from any other imaginable model. The result depends, of course, on the maximum load size effect and also on a_0/D and other hidden geometrical parameters inside the shape function S. It is size-independent (remember that it is the envelope of eqn (15)), but geometry-dependent.

Further support for the use of $R-\Delta a$ curves as a useful tool to make rough or qualitative estimates is provided by a recent analysis made by the authors of the performance of R–CTOD models [15]. It was shown that the use of a geometry and size independent R–CTOD curve, directly obtained from a cohesive model, gives a good approximation of the size effect curve for usual laboratory geometries (less than 5% error in maximum load for notched beams of depth above 8 cm, for a concrete with $l_{\mathrm{ch}} = 0{\cdot}3$ m). It was also shown that an R–CTOD curve model was equivalent to an $R-\Delta a$ curve model for large sizes by proving that under these circumstances there is a unique relationship between Δa and CTOD, which indicates that at least for large sizes the use of size- and geometry-independent $R-\Delta a$ curves is reasonable. For small sizes, the estimates should be considered as purely qualitative if no further analysis is available to prove the contrary.

2.3.3 Analysis of Size Effect using R–Δa Curves

According to the preceding paragraph there is a one-to-one correspondence between a size effect curve and the corresponding $R-\Delta a$ curve. If one makes the assumption that this $R-\Delta a$ curve is universal—i.e. geometry-independent—it can be used to compute a size effect relation for a new geometry. When only the trend of the size effect is investigated, only the trend of the $R-\Delta a$ curve is needed, so that the geometric independence is not a strong restriction as far as the analysis of trends is concerned. The procedure to follow to extract the size effect curve from the postulated $R-\Delta a$ curve is as follows:

The crack is in equilibrium under a given load when the driving specific energy J (eqn (13)) equals the specific resisting energy $R(\Delta a)$. The equilibrium is unstable when the rate of J equals that of $R(\Delta a)$ and the second order rate of J exceeds that of R. The point of unstable equilibrium corresponds to the peak load. Computing this maximum load for different sizes one gets the maximum load size effect relation. This procedure will allow us to perform a graphical analysis of the behaviour of positive and negative geometries and to redefine the concept of intrinsic size for negative and uncracked geometries.

For convenience, the specific resistance can be written to give explicitly the dependence on the material parameters G_F and $\Delta a_{C\infty}$ identified now as the infinite crack growth resistance $G_F = R_{max} = R(\infty)$ and as the crack extension for which this limiting value is reached $R(\Delta a_{C\infty}) = G_F$. We then write:

$$R = G_F R^*(\Delta a/\Delta a_{C\infty}) = G_F R^*(D\,\Delta\alpha/\Delta a_{C\infty}) \tag{16}$$

where $R^*(u)$ is a dimensionless material function satisfying $R^*(u) = 1$ for $u \geq 1$ and $\Delta\alpha$ is the reduced increment of crack length defined as $\Delta\alpha = \Delta a/D$.

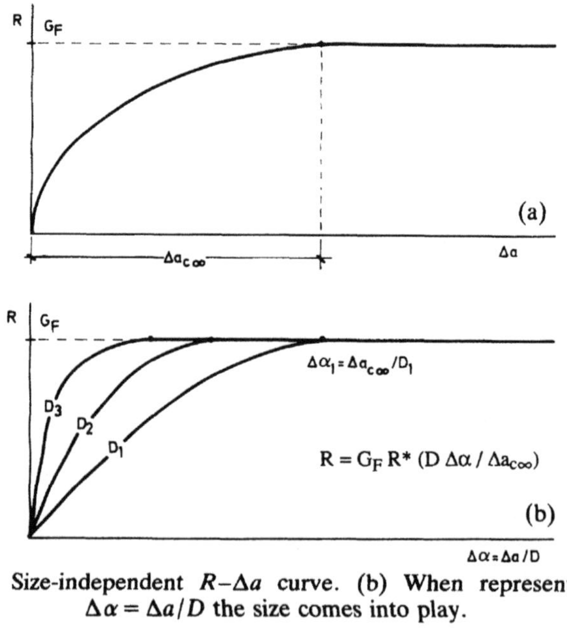

Fig. 3 (a) Size-independent R–Δa curve. (b) When represented versus $\Delta\alpha = \Delta a/D$ the size comes into play.

Using the variable $\Delta\alpha$, size independence is lost but this new representation will help us to better analyse the size effect. A family of R–$\Delta\alpha$ curves, for different sizes D, is sketched in Fig. 3. All the R-curves are generated from a unique curve by horizontal affinity, with affinity factor $\Delta a_{C\infty}/D$, inversely proportional to the size according to eqn (16). As the size increases they tend towards a step function. Notice that these curves attain the plateau at a crack extension given by $\Delta a_{C\infty}/D$, as previously defined.

In the same way, the specific driving force can be written as

$$J = \frac{K_I^2(a_0 + \Delta a)}{E'} = \frac{K_{IN}^2}{E'} \frac{S^2[(a_0 + \Delta a)/D]}{S_0^2} = \frac{K_{IN}^2}{E'} \cdot \frac{S^2(\alpha)}{S_0^2} \tag{17}$$

where the nominal stress intensity factor (eqn (2)) is used and the reduced crack length is $\alpha = \alpha_0 + \Delta\alpha$ (where $\alpha_0 = a_0/D$).

A family of J–α curves, for different values of the nominal SIF, are sketched in Figs 4(a) and 4(b). The first is for positive geometries and

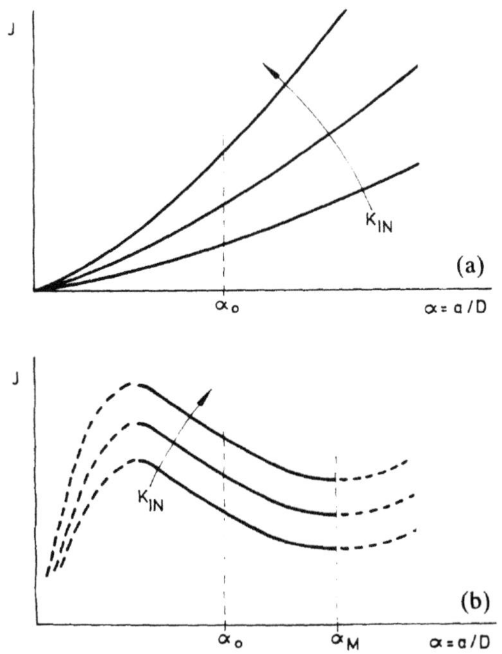

Fig. 4 (a) Driving J curves for positive geometries. (b) Driving J curves for negative geometries.

the second for negative ones. All the J curves are generated from a unique curve, scaled by the factor K_{IN}^2 according to eqn (17). With this new variable these curves are now size-independent. Notice that negative geometries exhibit a minimum, where $S'(\alpha_M) = 0$, located at the same abscissa α_M. We will come back to this point during the analysis of negative geometries.

3. POSITIVE GEOMETRIES

The maximum load size effect for cohesive materials and positive geometries can be conveniently analysed using the variables K_{INmax} (maximum nominal stress intensity factor) and D_i (intrinsic size) or, more precisely:

$$\frac{E'G_F}{K_{INmax}^2} \quad \text{and} \quad \frac{1}{D_i} \tag{18}$$

as described in Section 2.2. A general expression for the size effect, in the form of a power series expansion, was given in eqn (6) and the trend sketched in Fig. 1. This representation has the advantage of bringing the infinite size into focus—near the origin. Also the use of the intrinsic size D_i, instead of D, has the advantage of gathering together the results from different geometries [7], [16].

For large sizes—when $(l_{ch}/D_i)^2$ is negligible relative to l_{ch}/D_i—the size effect for maximum load takes the simple form

$$\frac{E'G_F}{[K_{INmax}]^2} = 1 + \frac{\Delta a_{C\infty}}{D_i} \tag{19}$$

already introduced in eqn (11) for cohesive materials. However this form is much more general and appears for very different models, such as Bazant's and all the R–Δa models. The use of the variables defined in eqn (18) has, then, the additional advantage that the large size behaviour is linear. Bazant's size effect plot is also a straight line in this kind of plot.

As an example, the size dependence of the maximum load was analysed for notched beams made with three different cohesive materials and tested in three-point bending. The beams had a span-to-depth ratio of four and an initial notch-to-depth ratio of 0·5 [3]. The results are sketched in Fig. 5. It is interesting that the

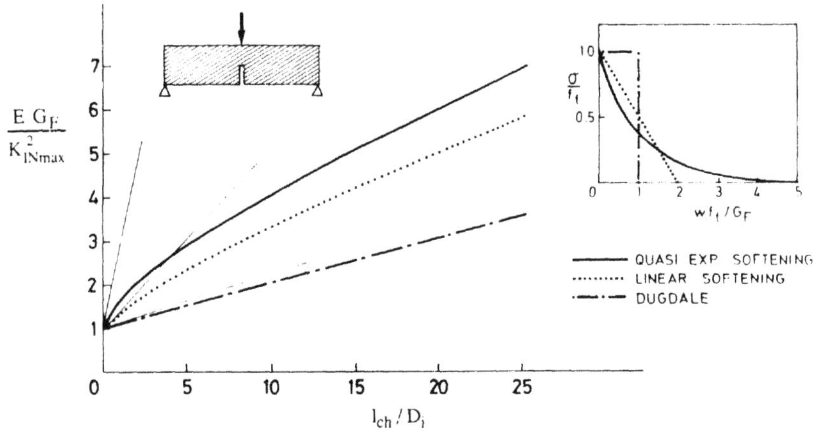

Fig. 5 Size effect curves for three different cohesive materials.

Dugdale model gives a result very close to Bazant's law and that the deviation from a straight line is larger the longer the tail of the softening curve.

The evolution of the maximum load with specimen size can be tracked using the R–Δa curve equivalence. The procedure, explained at the end of the last section, is depicted in Figs 6(a) and 6(b).

Figure 6(a) shows the evolution of the load (for a specific size D) until instability occurs (point 3). The load is measured indirectly by the nominal stress intensity factor K_{IN} and the ordinates at the tip of the initial crack (segments AB, AC, . . . , AD) are a measure of the applied load. The maximum load is related to AD, i.e. $K_{INmax}^2/E' =$ AD.

Figure 6(b) shows the evolution of the maximum load for different sizes ($D_1 < D_2 < D_3$), the maximum SIF increasing with increasing size, in accordance with the size effect curve depicted in Fig. 1 (notice that in Fig. 1 inverse values are plotted). The maximum attainable load corresponds to point E, where AE = G_F. This happens for infinite size ($D \to \infty$) or for linear elastic brittle materials where the R–Δa curve is a step function. This representation clearly shows that peak loads for cohesive materials are always less than the corresponding ones for elastic brittle materials.

For every size D, the evolution of the load with the equivalent crack extension is obtained by putting $J = R$, with J given by eqn (15) and R by eqn (14).

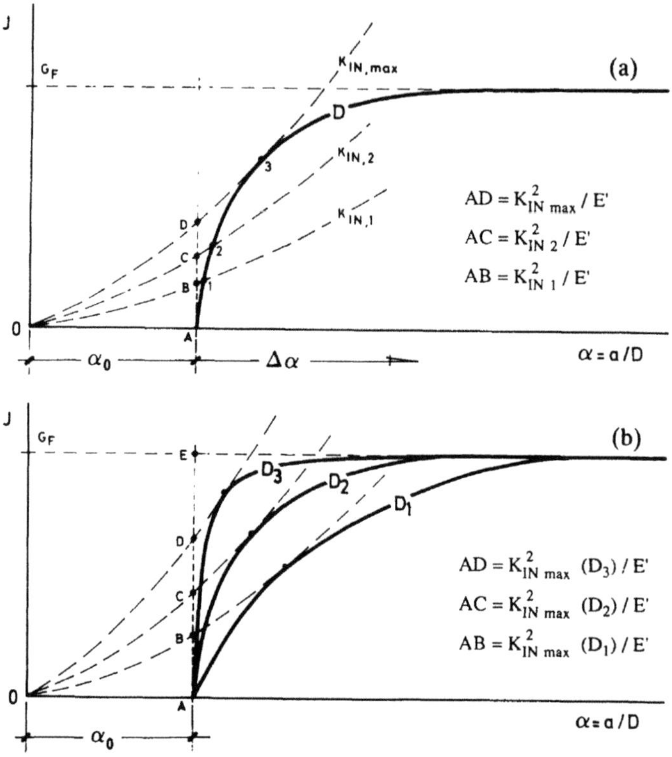

Fig. 6 Positive geometries: (a) crack and load evolution for a given size; (b) evolution of maximum load with specimen size.

For very large sizes, one may obtain expressions with different degrees of approximation by expanding J in powers of $\Delta a/D \ll 1$.

The first approximation (zeroth order) corresponds to the case when $\Delta a/D$ is negligible relative to $\alpha_0 = a_0/D$. In this situation, $J = R$ is simply written as:

$$\frac{K_{IN}^2}{E'G_F} = R^*\left(\frac{\Delta a}{\Delta a_{C\infty}}\right) \tag{20}$$

Obviously, then, the maximum of K_{IN} coincides with the maximum of R^* which by hypothesis is reached for $\Delta a = \Delta a_{C\infty}$ and takes the value $R^*_{max} = 1$, hence

$$\frac{E' G_F}{K^2_{INmax}} = 1 \quad \text{and} \quad \Delta a_C = \Delta a_{C\infty} \quad \text{for} \quad D = \infty \tag{21}$$

The next degree of approximation (first order) is obtained by retaining the first order term in $\Delta a / D$; in this case the condition $J = R$ reduces to

$$\frac{K^2_{IN}}{E'} \left(1 + \frac{2S'_0}{S_0} \frac{\Delta a}{D}\right) = G_F R^* \left(\frac{\Delta a}{\Delta a_{C\infty}}\right) \tag{22}$$

Hence, in a J–Δa plot, the driving J is a straight line of very small slope. The maximum load is obtained by imposing the tangency condition. Two general cases may be considered:

Case 1. The R-curve reaches the plateau at a cusp point as in Fig. 7(a). In this case the tangency occurs at the cusp point for $\Delta a = \Delta a_{C\infty}$ and $R^* = 1$, so that after substitution in eqn (22) the first order approximation of the size effect curve is easily obtained as

$$\frac{E' G_F}{K^2_{INmax}} = 1 + \frac{2S'_0}{S_0} \frac{\Delta a_{C\infty}}{D} \tag{23}$$

Case 2. The R-curve reaches the plateau smoothly as in Fig. 7(b). In this case the size effect curve is the same, to first order in $\Delta a_{C\infty}/D$. To prove this assertion we may assume, with great generality, that $R^*(\Delta a)$ may be expanded in a Taylor Series on the left of $\Delta a = \Delta a_{C\infty}$, and we may take two terms for $\delta = \Delta a_{C\infty} - \Delta a$ small

$$R^*\left(1 - \frac{\delta}{\Delta a_{C\infty}}\right) \approx 1 - \frac{R^{*''}}{2} \left(\frac{\delta}{\Delta a_{C\infty}}\right)^2 \tag{24}$$

The condition for J and R to be tangent to each other is, then

$$\frac{K^2_{IN}}{E'} \frac{2S'_0}{S_0} \frac{1}{D} = G_F R^{*''} \frac{\delta}{\Delta a^2_{C\infty}} \tag{25}$$

or

$$\frac{\delta}{\Delta a_{C\infty}} = \frac{K^2_{IN}}{E' G_F} \frac{2S'_0}{S_0} \frac{\Delta a_{C\infty}}{D} \frac{1}{R^{*''}} \tag{26}$$

Hence, the deviation δ is first order in $\Delta a_{C\infty}/D$ so that when substituted back in eqn (22), it gives rise to quadratic terms in $\Delta a_{C\infty}/D$ which, in consistency with the initial approach, must be deleted.

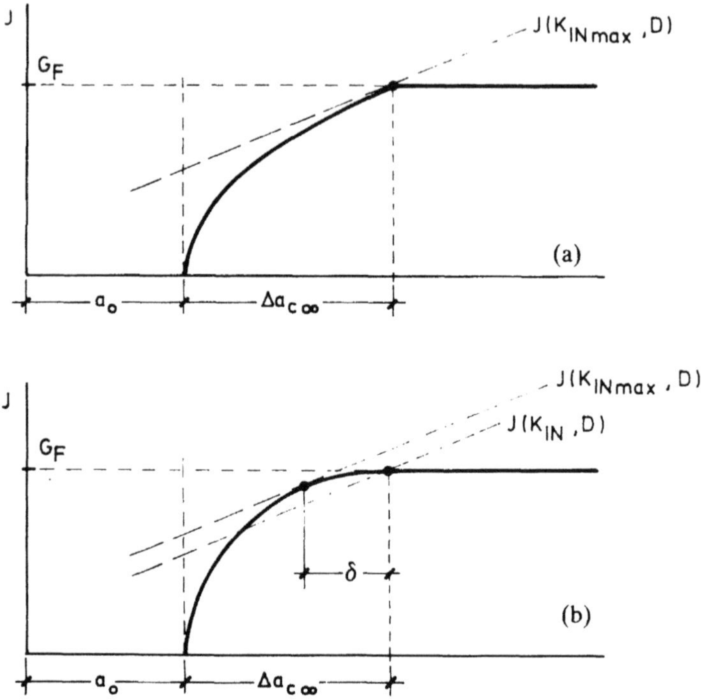

Fig. 7 Peak load situation for large sizes ($\Delta a/D \ll 1$); (a) R-curve with a cusp point; (b) R-curve with smooth maximum.

In this way, the first order result happens to coincide again with the expression (23) which has, in turn, the same shape as the asymptotic expression (8) for the cohesive cracks. It has been proven, then, that use of adequately chosen R–Δa curves leads to size effect curves possessing the same general and asymptotic trends as the cohesive models, while being easier to handle analytically.

4. NEGATIVE GEOMETRIES

The previous sections served to present some essential results concerning the size effect behaviour for positive geometries. The first result is that Bazant's work proves that there is a one-to-one correspondence between the size effect curve and the size-independent

R–Δa curve. The second fact is that the R-curve model, however simplified, can capture the *general trend* of the size effect curve for positive geometries, and may serve, then, to introduce the material property $\Delta a_{C\infty}$ and the definition of the intrinsic size D_i (as done by Bazant with somewhat different terminology). This avoids the use of cohesive models, which require special asymptotic analyses, when only looking for qualitative features.

In the present section we use first the R–Δa model to reach some general conclusions for the size effect in the case of negative geometries ($S_0'<0$) and then verify that these trends are also followed by the cohesive model.

The introduction of the intrinsic size was not just a matter of formal simplification of eqn (8). Indeed, it fills a need for the objectification of the definition of structural size. When the specimens considered are all geometrically similar (homothetical) we have a single degree of freedom and the selection of D does not matter. However, when we consider more degrees of freedom the arbitrariness of the selection of the size makes it difficult to compare the results and the difficulty increases if we consider different specimens and loadings. This is avoided when D_i is used as a measure of size, because it is independent of the selection of D. Indeed, it may easily be shown that eqn (10) is equivalent to the definition

$$\frac{1}{D_i} = \left.\frac{\partial \ln K^2}{\partial a}\right|_{a=a_0} \tag{27}$$

wherein the partial derivative with respect to crack length a, the load and all remaining dimensions of the specimen remain constant.

Apart from this desirable objectivity, computational results of size effect for different usual laboratory geometries and material models showed that when this variable was used, all the size effect curves came to fit in a relatively narrow band, which increases the interest of this definition [7], [16].

At the time of the introduction of this definition, the authors pointed out that this intrinsic size could become negative, which was a serious drawback for a definition of 'size'. But it was emphasized that since this definition was intended to be meaningful only for positive geometries where eqn (8) applies and the intrinsic size (eqn (10)) is also positive, this fact was not very troubling. However there are geometries where $S_0'<0$ and eqns (8) and (11) do not apply, and it is essential to understand the differences between these geometries and to find, if possible, a unified treatment. These non-positive geometries lead to asymptotic size effect equations which differ from those corresponding to the *normal* positive case, so they are termed *anomalous*.

In the whole family of non-positive geometries there are two sub-families that bear very different properties and are physically easy to distinguish. The first case corresponds to those structures with positively vanishing intrinsic size $D_i = 0^+$ whatever the actual size; physically they correspond to specimens with no initial crack. The second is represented by the geometries where the SIF initially decreases with crack length, usually going through a minimum for a fixed a/D relation; their intrinsic sizes as defined so far are negative. However, one of the results of the discussion to follow is that a more efficient definition of the intrinsic size will attribute them, surprisingly enough, the size value $+\infty$, whatever the actual size.

The case corresponding to apparently negative intrinsic size is shown in Fig. 8. Again the variable $\alpha = a/D$ is used; this provides J-curves depending only on one parameter K_{IN} (and not on size D). On the contrary $R(\alpha)$-curves are now size dependent, as already noted. The driving J-curve in this case has negative slope at the initial crack length (α_0) and goes through a minimum for a certain value α_M of the relative crack length.

The evolution of the maximum load with specimen size can be analysed using the R–$\Delta\alpha$ curve equivalence, following the same steps as in the positive geometries.

Figure 8(a) shows the evolution of the load (for a specific size D) until instability occurs at point 3 and K_{IN} reaches K_{INmax}.

Figure 8(b) shows the evolution of the maximum load for different sizes ($D_1 < D_2 < D_3 < D_4$). The maximum SIF increases with increasing size until size D_3 where J is tangent to the maximum value of the R-curve ($R = G_F$). At this moment the peak load reaches its maximum possible value and remains stationary for larger sizes, as is seen in the figure for $D_4 > D_3$.

This fact suggests a redefinition of the nominal SIF taking into account the fact that the initial crack length no longer plays a key role. Now the relevant parameter is the crack length where the minimum in the J-curve occurs. To distinguish the new definition, the underlined symbol $\underline{K}_{IN}$ is used in the following, its definition being a direct modification of eqn (2),

$$\underline{K}_{IN} = \sigma_N \sqrt{D}\, S(\alpha_M) \tag{28}$$

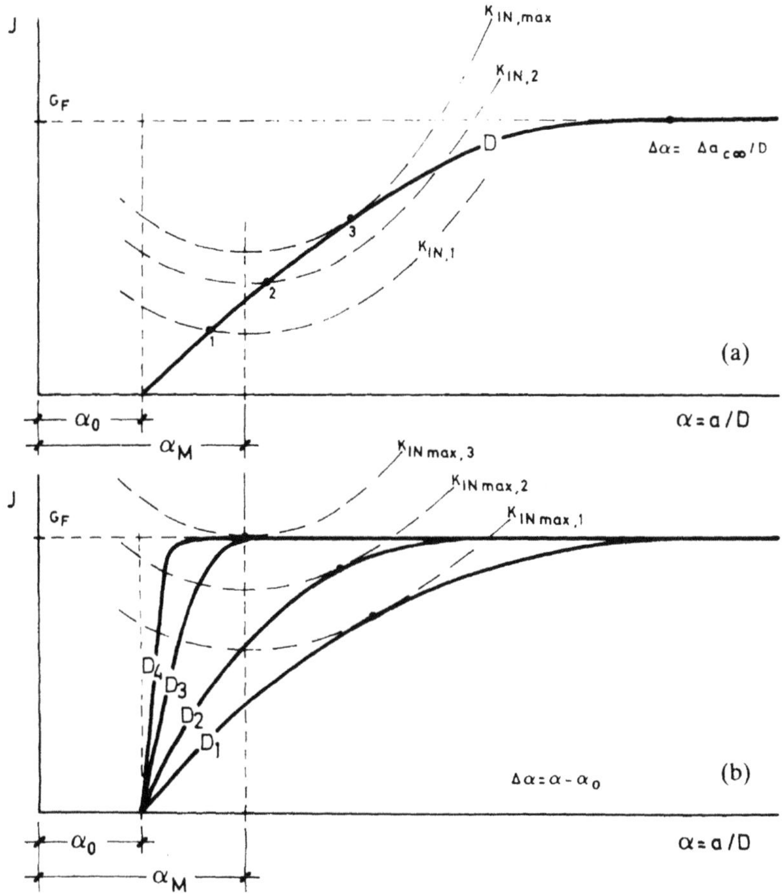

Fig. 8 Negative geometries: (a) crack and load-evolution for a given size; (b) evolution of maximum load with specimen size.

Since for large sizes the initial geometry is 'erased' along the loading process up to the peak load (for sizes larger than D_3 in Fig. 8(b)), it appears logical to define the intrinsic size using the final dominant geometry rather than the initial one, so that the new underlined intrinsic size $\underline{D}_i$ results from a direct modification of eqn (10)

$$\frac{1}{\underline{D}_i}=\frac{2S'(\alpha_M^+)}{S(\alpha_M)}\frac{1}{D}=+0 \qquad \text{or} \qquad \underline{D}_i=+\infty \tag{29}$$

where the second equality comes from the condition $S'(\alpha_M)=0$

associated with the condition of $S(\alpha)$ being minimum at $\alpha = \alpha_M$. This logical redefinition removes the negativeness of the intrinsic size for these geometries, which was an unsatisfactory feature of the previous definition. It does not, however, remove the degeneracy since different geometries and sizes fall into this category. In the size effect plot of Fig. 1 these different cases would all give points on the vertical axis because the intrinsic size is ∞. Hence, this first order intrinsic size cannot be used to extract all the information regarding size effect and, until more information is available, a plot as shown in Fig. 9 must be used, where the size D is an arbitrarily chosen structural dimension and the underlined nominal SIF is defined by eqn (28).

Let us analyse the shape of the maximum load size effect when represented against the coordinates $E'G_F/\underline{K}^2_{\text{INmax}}$ and l_{ch}/D, as shown in Fig. 9. For large sizes the underlined nominal SIF remains constant as already stated. More specifically, one gets:

$$\frac{E'G_F}{[\underline{K}_{\text{INmax}}]^2} = 1 \qquad \text{for} \qquad \frac{l_{ch}}{D} \leq \frac{l_{ch}(\alpha_M - \alpha_0)}{\Delta a_{C\infty}} := \frac{l_{ch}}{D_M} \tag{30}$$

where D_M has been defined as the limit of the horizontal portion.

For smaller sizes the underlined nominal SIF starts to decrease. Now let us explore the structure of the size effect near the tangency point $(1, l_{ch}/D_M)$ attempting to extract some additional information. A

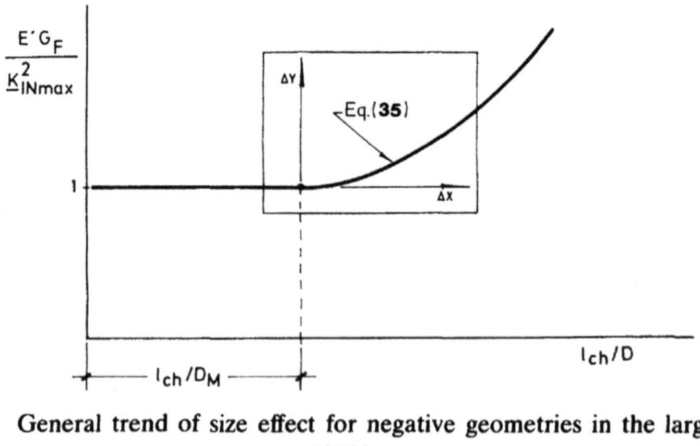

Fig. 9 General trend of size effect for negative geometries in the large size range.

zoom of this region is obtained when the coordinates

$$\Delta Y = \frac{E'G_F}{[\underline{K}_{\text{INmax}}]^2} - 1 \qquad \text{and} \qquad \Delta X = \frac{l_{\text{ch}}}{D} - \frac{l_{\text{ch}}}{D_M} \tag{31}$$

are used, as sketched in Fig. 9.

To obtain an expression for the size effect valid near the origin of this new set of coordinates, both expressions—J and R—can be expanded in powers of $\Delta\alpha_M$ up to $\Delta\alpha_M^2$. The procedure is best followed with the help of Fig. 10.

The driving J may be approximated by

$$J = \frac{\underline{K}_{\text{IN}}^2}{E}\left[\frac{S(\alpha_M + \Delta\alpha_M)}{S_M}\right]^2 \approx \frac{\underline{K}_{\text{IN}}^2}{E}\left(1 + \frac{S''_M}{S_M}\Delta\alpha_M^2\right) \tag{32}$$

and the resisting R becomes

$$R = G_F R^*\left[\frac{D}{\Delta a_{C\infty}}(\alpha_M - \alpha_0 + \Delta\alpha_M)\right]$$

$$\approx 1 - \frac{|R^{*\prime\prime}|}{2}\left[1 - \frac{D}{\Delta a_{C\infty}}(\alpha_M - \alpha_0 + \Delta\alpha_M)\right]^2 \tag{33}$$

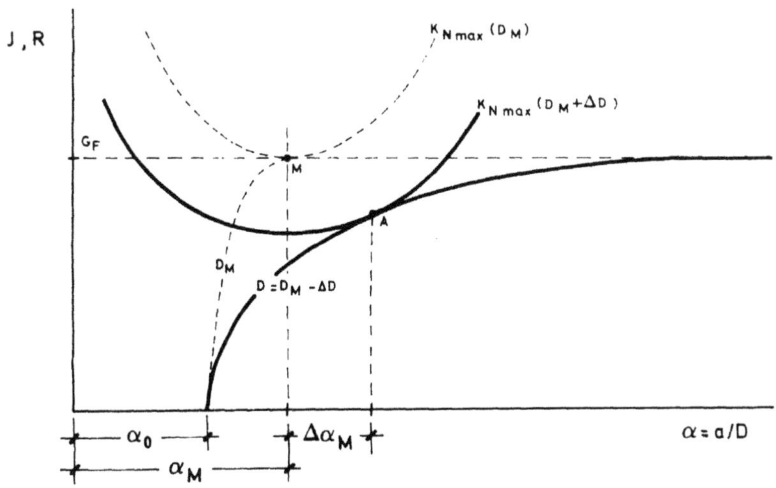

Fig. 10 Peak load J–R situation for negative geometries and sizes close to D_M.

where $S''(\alpha_M)$ is the second derivative of the shape function at the stationary point and $|R^{*''}|$ stands for the absolute value of the second derivative of the function $R^*(u)$ at the point where it attains the plateau, $u = 1$.

The maximum load for size D can be inferred from the two equations that reflect the behaviour of J and R at point A, i.e.

$$J = R$$
$$\frac{\partial J}{\partial \alpha} = \frac{\partial R}{\partial \alpha} \tag{34}$$

Using expressions (32) and (33) valid for small values around the stationary point M and taking into account the coordinates ΔX and ΔY, one arrives at

$$\Delta Y = \left(\frac{\Delta a_{C\infty}}{l_{ch}}\right)^2 \frac{|R^{*''}|\, S''(\alpha_M)}{|R^{*''}|\, S(\alpha_M) + 2S''(\alpha_M)(\alpha_M - \alpha_0)^2} \Delta X^2 \tag{35}$$

This equation can be grouped in three factors; the first factor is a material property, also appearing in the normal size effect eqn (6). The last one is the increment of the inverse of the size, and contains all the relevant information about size. The intermediate factor is a mixed function of geometrical properties (curvature of the shape function, and distance from initial crack tip to the minimum point) and of material properties (curvature of R–Δa curve) and there is no general way to uncouple them, except for particular cases (such as an R–Δa curve attaining the plateau in a cusp point with infinite curvature). Hence, it seems that for this class of geometries there is not a physically or mathematically useful definition of higher order intrinsic size (which must obviously be material-independent).

These results should be qualitatively capable of extrapolation to cohesive materials, as the following paragraphs will show.

The convergence to the LEFM result eqn (30) for very large sizes in cohesive materials may be justified as follows: upon loading, first a cohesive zone grows in front of the initial crack up to a critical size, which occurs for $K_{IN}^2 = E'G_F$ if the size of the structure is large enough. At this point the cohesive zone starts to translate self similarly, or closely so, under increasing load. The maximum happens when, roughly speaking, the cohesive zone goes through the minimum of the J-curve. At this point $\underline{K}_{\mathrm{IN}}^2 = E'G_{\mathrm{F}}$ and the previous result (eqn (30)) is recovered.

Now, how long may one expect the plateau to be? This may be roughly answered by saying that the necessary condition for the previous description of the process to hold is that the critical (fully developed) fracture zone may be achieved before reaching the maximum load. This requires that the distance between the initial crack tip and the location of the minimum be larger than the critical cohesive zone size. A measure of the critical size of the cohesive zone is the critical size extension for infinite size $\Delta a_{\mathrm{C}\infty}$ which may be obtained from the softening curve following the procedures indicated in [3]. The previous condition may be then written as $D(\alpha_{\mathrm{M}} - \alpha_0) \approx\ <\Delta a_{\mathrm{C}\infty}$, or,

$$\frac{l_{\mathrm{ch}}}{D} \leq \frac{l_{\mathrm{ch}}(\alpha_{\mathrm{M}} - \alpha_0)}{\Delta a_{\mathrm{C}\infty}} \tag{36}$$

which coincides exactly with the condition obtained in eqns (30) from R–Δa curves. The essential difference is that in a cohesive model $\Delta a_{\mathrm{C}\infty}$ is determined from the shape of the softening curve. Using the values previously found for the quotient $\Delta a_{\mathrm{C}\infty}/l_{\mathrm{ch}}$ [1], [3], [9] for the softening curves displayed in Fig. 5, the conditions are

$$\frac{l_{\mathrm{ch}}}{D} \leq 7{\cdot}64(\alpha_{\mathrm{M}} - \alpha_0) \quad \text{for rectangular softening (Dugdale)}$$

$$\frac{l_{\mathrm{ch}}}{D} \leq 2{\cdot}39(\alpha_{\mathrm{M}} - \alpha_0) \quad \text{for linear softening} \tag{37}$$

$$\frac{l_{\mathrm{ch}}}{D} \leq 0{\cdot}40(\alpha_{\mathrm{M}} - \alpha_0) \quad \text{for quasi-exponential softening}$$

The first step towards evaluating the accuracy of the above approximation has been carried out for linear and quasi-exponential softening materials in a single edge notched strip subjected to opening point loads at the crack mouth, for an initial notch/depth ratio $a_0/D = 0{\cdot}025$. The SIF is taken from the Tada handbook ([17], page 2.25) and the location of the minimum and the revised nominal SIF

$\underset{\sim}{K}_{IN}$ were determined as

$$a_M/D = 0{\cdot}179\,44$$
$$\underset{\sim}{K}_{IN} = 4{\cdot}787\frac{P}{\sqrt{D}} \tag{38}$$

where P is the load per unit thickness.

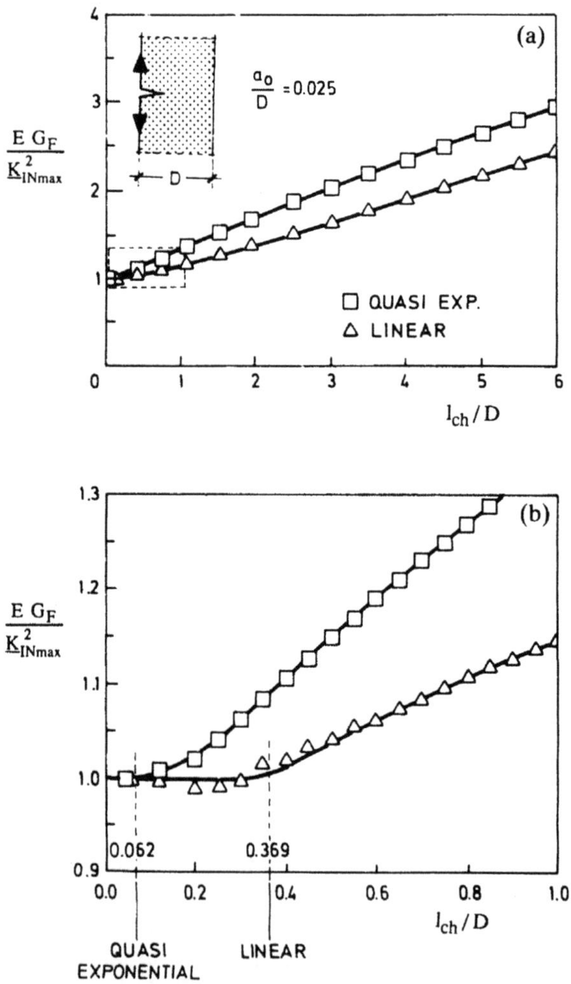

Fig. 11 (a) Computed size effect curves for a mouth-loaded SEN specimen of a cohesive material with linear and quasi-exponential softening. (b) Zoom in the large size range.

According to this, the plateau in the size effect curve has to occur in the following approximate intervals:

$$\frac{l_{ch}}{D} \approx < 0{\cdot}369 \qquad \text{for linear softening}$$
$$\frac{l_{ch}}{D} \approx < 0{\cdot}062 \qquad \text{for quasi-exponential softening} \tag{39}$$

The first approximation to the size effect curve has been determined numerically using an influence method as described in [3]. Only 40 equal elements were used on the central section, so that appreciable numerical error may be present for this estimate, which should not be pushed too far in the long size range. Despite these limitations, the results displayed in Fig. 11 show a good trend for the linear softening case, and, although in the very limit of resolution, the extent of the plateau seems to agree also with the estimate (39) for the quasi-exponential softening.

We may conclude that R–Δa models may be used to obtain the general trend of the size effect in the case of negative geometries, too.

5. UNCRACKED GEOMETRIES

Uncracked specimens also exhibit a maximum load size effect, this is a well known experimental fact that has also been numerically reproduced (see Ref. [18], for example). For such geometries, the equivalent elastic crack is of no use for predicting a size effect, simply because the specimen has no crack. Nevertheless, for very large sizes the R–Δa curve approach can still be used to predict the *brittle limit* behaviour, as will be shown.

Consider an uncracked specimen in which, under elastic conditions, the maximum tensile stress occurs at some point at the outer boundary. This is the case of unnotched bend beams, for example. Let us define the load level by this (elastically computed) maximum tensile stress, and call it $\underline{\sigma}_N$. For a beam of depth D and thickness B subjected to a bending moment M, this underlined nominal stress is known to be $6M/(BD^2)$.

We may ask what the expressions are for the nominal SIF and for the intrinsic size of such a specimen when a *small crack* normal to the surface at the location of the maximum stress is present, more specifically, when the initial relative crack depth a_0/D is vanishingly small. The SIF for such limiting condition is well known to be (crack at the edge of a half-space):

$$K_I(a) = \sigma_N \sqrt{D}\, 1{\cdot}12\sqrt{\pi a/D} \tag{40}$$

from which one easily obtains from eqns (2) and (10)

$$K_{IN} = \sigma_N 1{\cdot}12\sqrt{\pi a_0} \tag{41}$$

$$D_i = a_0 \tag{42}$$

Hence, when the initial crack tends to zero, both the nominal SIF and the intrinsic size vanish. This means that in the size effect plot of Fig. 1, all these geometries coalesce into the (+∞, +∞) corner which implies a singular behaviour. It is clear that this plot is no longer useful for uncracked geometries and has to be modified if we want to retain some of its advantages. This is done according to the following reasoning.

Figure 12 shows a particular case where the onset of unstable crack

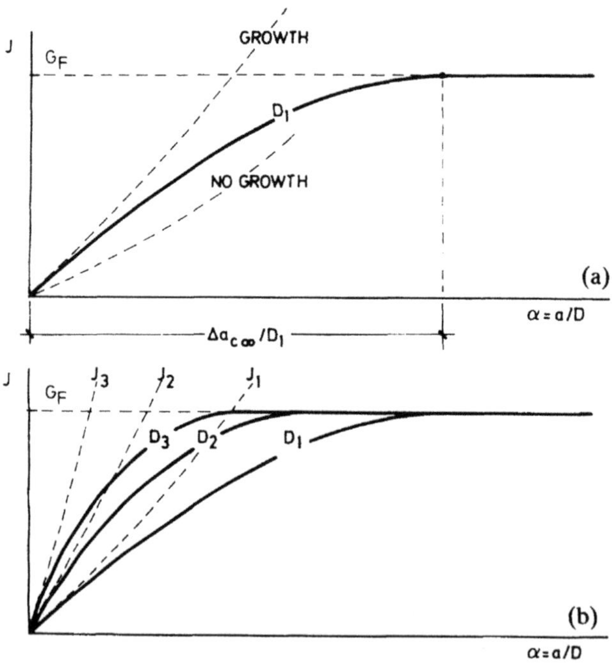

Fig. 12 Uncracked geometries: (a) limiting crack growth situation for a given size; (b) evolution of stability condition with specimen size.

growth (at fixed load) may be obtained. In this case the load may be increased without crack growth up to the point where the *driving-J* curve and the *resisting-R* curve become mutually tangent at the origin. If, moreover, the curvature of the J-curve is larger than that of the R-curve—as depicted—the crack growth will be unstable and initiation will coincide with the peak load.

For very *large sizes* the expression (40) for the SIF is applicable, and the J-curve is a straight line. For a reasonably shaped R-curve (convex curve, negative curvature), the instability condition is then met when the J- and R-curves are tangent at the origin, and maximum load coincides with crack initiation. This is the so-called *brittle limit*.

The expression for the slopes of the $R(\alpha)$ curve—from eqn (16)—and the $J(a)$ curve—from eqns (17) and (40)—are

$$R'(\alpha)|_{\alpha=0} = G_F \frac{R^{*\prime}(0)}{\Delta a_{C\infty}} D \tag{43}$$

$$J'(\alpha)|_{\alpha=0} = \frac{\sigma_N^2 1{\cdot}12^2 \pi}{E'} D \tag{44}$$

Equating both slopes and remembering the definition of eqn (7) of l_{ch}, one arrives at the following expression for the peak load at the brittle limit

$$\left(\frac{f_t}{\sigma_{Nmax}}\right)^2 = \frac{1{\cdot}12^2 \pi\ \Delta a_{C\infty}}{R^{*\prime}(0) l_{ch}} \tag{45}$$

As expected, this result for the maximum load does not depend on D because it was derived for very large sizes. It depends on $R^{*\prime}(0)$ and on material properties.

If we identify this brittle limit result with the classical result of brittle fracture in the frame of the strength of materials—where the peak load is reached as soon as the tensile strength, f_t, is attained at some point inside the body—then $\sigma_{Nmax} = f_t$, and both members of eqn (45) must be equal to unity. Hence, the initial slope of the R-curve must be

$$R^{*\prime}(0) = \frac{3{\cdot}95\ \Delta a_{C\infty}}{l_{ch}} \tag{46}$$

Let us go back to the initial representation (Fig. 1) of the maximum load size effect. It is clear that the nominal SIF, as defined in eqn (2), is no longer usable but we can use a similar representation where the four following interesting features are still achieved:

1. large sizes are in focus near the origin;
2. horizontal straight lines correspond to LEFM results (for pre-cracked specimens);
3. straight lines through the origin correspond to limit analysis results;
4. Bazant's size effect law retains a linear form.

Such representation is obtained by noting that the result for the brittle limit, i.e.

$$\left(\frac{f_t}{\underline{\sigma}_{Nmax}}\right)^2 = 1 \tag{47}$$

can be written, multiplying both members by the inverse of the dimensionless size, as

$$\frac{E'G_F}{[\underline{\sigma}_{Nmax}]^2 D} = \frac{l_{ch}}{D} \tag{48}$$

which, in a plot with coordinates $(E'G_F/(\underline{\sigma}^2_{Nmax}D) - l_{ch}/D)$ is a straight line through the origin with slope unity, as shown in Fig. 13. This plot has the additional advantage of being affine to that in Fig. 1, i.e. corresponding to a scale change on both axes. The vertical axis

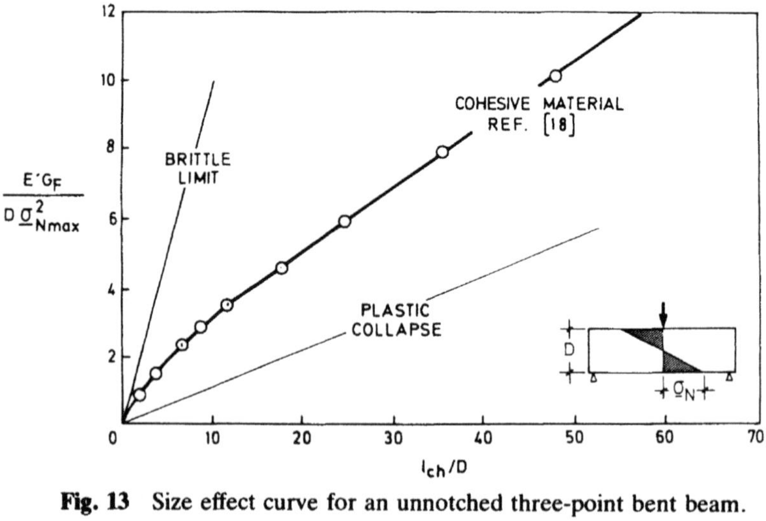

Fig. 13 Size effect curve for an unnotched three-point bent beam.

appearing in Fig. 13 is obtained from that in Fig. 1 by multiplying by S_0^2, with the precaution of making the product first and then taking the limit as a_0 tends to 0. Similarly, the horizontal axis in Fig. 13 is obtained by factoring that in Fig. 1 by $S_0^2/(2S_0')$, with analogous precaution.

A comparison with the size effect behaviour computed for a cohesive material is also plotted in Fig. 13, corresponding to a three-point bent beam. These results were obtained numerically by Gustafsson [18] using a cohesive crack with bilinear softening to simulate concrete behaviour. The convergence to the result of eqn (48) is obvious, but is reached for very large sizes as shown in the figure.

To compare the predictions of the $R{-}\Delta a$ approach, as in previous examples, the maximum load size effect was computed for cohesive materials with linear and quasi-exponential softening. Such results are shown in Fig. 14 and, again, the *brittle-limit* predictions from the effective elastic crack are only useful for large sizes (for $l_{ch}/D < 0{\cdot}2$), as expected.

The other extreme behaviour—the plastic collapse, for very small sizes—depicted in Fig. 1 also appears in the diagrams of Figs 13 and 14. Such behaviour corresponds to small sizes and is represented by an asymptote ($\sigma_{Nmax} = 3f_t$ in the figure). The Gustafsson results tend

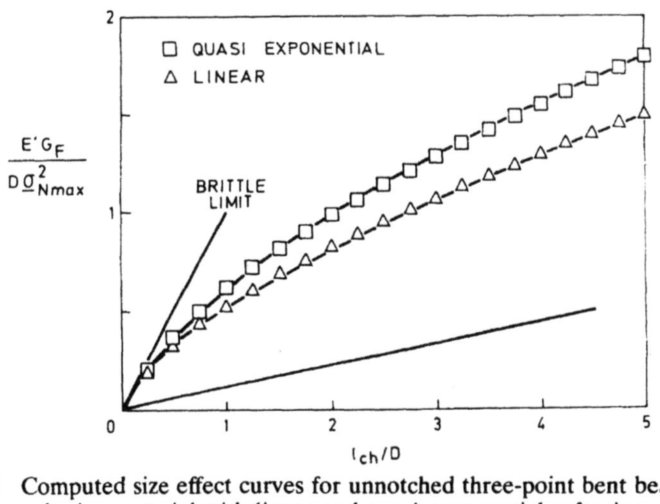

Fig. 14 Computed size effect curves for unnotched three-point bent beams of a cohesive material with linear and quasi-exponential softening.

towards this asymptote for unrealistic small sizes. For a concrete with characteristic size l_{ch} around 0·3 m, most practical situations are far from both asymptotic behaviours—the brittle limit and plastic collapse—but for other materials this may not be the case. The same comments apply to the results shown in Fig. 14.

6. SUMMARY AND CONCLUSIONS

The following conclusions may serve as a summary of the previous analyses:

1. It is possible to generalize the notions of *nominal SIF* and of *intrinsic size* to get consistent results for the asymptotic behaviour up to first order for any geometry and to avoid negativeness of the objective size. This is done by defining these values for a crack length $\underline{\alpha}$ corresponding to the minimum of the stress intensity factor to be found as the crack grows

$$\underline{\alpha}: \qquad S(\underline{\alpha}) = \min\{S(\alpha); \alpha \geq \alpha_0\} \tag{49}$$

which takes the value α_0 for positive geometries and α_M for negative geometries. With this convention, eqns (2) and (28) and (10) and (29) may be summarized as

$$K_{IN} := \sigma_N \sqrt{D}\, S(\underline{\alpha}) \qquad \frac{1}{D_i} := \frac{2S'(\underline{\alpha})}{S(\underline{\alpha})D} \tag{50}$$

2. Despite this unification, the uncracked and negative geometries are singular, or 'first order degenerate', in the sense that the magnitudes defined in eqn (50) are ill-defined, and somewhat special plots have to be used to remove the degeneracy.
3. The size effect of negative geometries has to be represented using an arbitrarily chosen size, as in Fig. 9. The essential feature is that the curve displays initial horizontal tangent and then starts to grow quadratically.
4. The size effect of uncracked specimens may be conveniently represented in a plot like the one in Fig. 13 where two limiting behaviours are found. For very large sizes the brittle limit appears as a straight line through the origin. For small sizes the plastic limit appears as an asymptote parallel to the 'true' plastic size effect line.

ACKNOWLEDGEMENTS

The authors gratefully acknowledge financial support for this research provided by CICYT, Spain, under grants PB86-0494 and CE89-0012.

REFERENCES

[1] Planas, J. & Elices, M., Size effect in concrete structures: mathematical approximations and experimental validation. In *Cracking and Damage, Strain Localization and Size Effect,* ed. J. Mazars & Z. P. Bazant. Elsevier Applied Science, London, 1989, pp. 462–76.

[2] Elices, M. & Planas, J., Material models. In *Fracture Mechanics of Concrete Structures: From Theory to Applications,* ed. L. Elfgren, Chapman and Hall, London, 1989, Ch. 3, pp. 16–66.

[3] Planas, J. & Elices, M., A nonlinear analysis of a cohesive crack, (Accepted for publication in *Int. J. of Fracture*).

[4] Hillerborg, A., Modeer, M. & Petersson, P. E., Analysis of crack formation and crack growth in concrete by means of fracture mechanics and finite elements. *Cement Concr. Res.*, **6** (1976) 773–82.

[5] Hillerborg, A., The theoretical basis of a method to determine the fracture energy G_F of concrete. *Materials and Structures*, **18** (1985) 291–6.

[6] Planas, J. & Elices, M., Towards a measure of G_F: an analysis of experimental results. In *Fracture Toughness and Fracture Energy of Concrete,* ed. F. H. Wittmann. Elsevier Science, Amsterdam 1986, pp. 381–90.

[7] Elices, M. & Planas, J., Size effect and experimental validation of fracture models. In *Analysis of Concrete Structures by Fracture Mechanics,* ed. L. Elfgren. Chapman and Hall, London, 1990, pp. 99–127.

[8] Planas, J. & Elices, M., Un nuevo método de análisis del comportamiento asintótico de la propagación de una fisura cohesiva en Modo I. *Anales de Mecánica de la Fractura,* **3** (1986) 219–27.

[9] Planas, J. & Elices, M., A New Method of Asymptotic Analysis of the Development of a Cohesive Crack in Mode I Loading. Report 87-02, Departamento de Ciencia de Materiales, E.T.S. de Ingenieros de Caminos, Ciudad Universitaria, 28040-Madrid, Spain, 1987.

[10] Bazant, Z. P., Size effect in blunt fracture: concrete, rock, metals. *J. Eng. Mech.* (ASCE), **110** (1984) 518–38.

[11] Planas, J. & Elices, M., Asymptotic Analysis of Cohesive Cracks and its Relation with Effective Elastic Cracks. In *Toughening Mechanisms in Quasi-Brittle Materials,* ed. S. P. Shah, Kluwer, Dordrecht, 1990, 189–202.

[12] Bazant, Z. P., Kim, J. K. & Pfeiffer, P. A., Nonlinear fracture properties from size effect tests. *J. Structural Engineering,* (ASCE), **112**(2) (1986) 289–307.

[13] Bazant, Z. P. & Pfeiffer, P. A., Determination of fracture energy from size effect and brittleness number. *ACI Materials Journal*, **84** (1987) 463–80.

[14] Bazant, Z. P., Gettu, R. & Kazemi, M. T., Identification of Nonlinear Fracture Properties from Size Effect Tests and Structural Analysis based on Geometry-Dependent *R*-Curves. Report 89-3/498p, Center for Advanced Cement-Based Materials, The Technological Institute, Northwestern University, Evanston, Illinois, USA, 1989.

[15] Planas, J., Elices, M. & Toribio, J., Approximation of cohesive models by *R*–CTOD Curves. In *Fracture of Concrete and Rock*, ed. S. P. Shah, S. E. Swartz & B. Barr. Elsevier Science, London, 1989, pp. 203–12.

[16] Llorca, J., Planas, J. & Elices, M., On the use of maximum load to validate or disprove models of concrete fracture behaviour. In *Fracture of Concrete and Rock*, ed. S. P. Shah, S. E. Swartz & B. Barr. Elsevier Science, London, 1989, pp. 357–68.

[17] Tada, H., Paris, P. & Irwin, G., *The Stress Analysis of Cracks Handbook*. Del Research Corp., St. Louis, Missouri, 1985.

[18] Gustafsson, J., Fracture Mechanics Studies of Non-Yielding Materials like Concrete: Modelling of Tensile Fracture and Applied Strength Analyses. Report TVBM-1007, Division of Building Materials, Lund Institute of Technology, Lund, Sweden, 1985.

Chapter 8

Modelling Crack Toughness Curves in Fibre-Reinforced Cement Composites

YIU-WING MAI
Centre for Advanced Materials Technology,
Department of Mechanical Engineering,
University of Sydney,
Sydney, NSW 2006,
Australia

ABSTRACT

Fracture in fibre-reinforced cement composites can be usefully described by the crack-toughness (T) curve methodology based on linear elastic fracture mechanics analysis using the K-superposition principle. With the simplified assumptions of a power law strain-softening characteristics of the damage zone (i.e. fibre bridging zone and/or matrix fracture process zone) and a linearized crack profile the stable crack growth in fibre–cement composites can be readily modelled with T-curves and the associated load–deflection diagrams obtained. The relationship between the T-curve and the damage zone over which bridging stresses are transferred in accordance with the strain-softening law is emphasized. It is also shown that whilst specimen geometry and size have some significant effects on the T-curve, the matrix fracture process zone effect is quite small considering the usually large scatter of the experimental data. Using the crack-toughness curve concept the Weibull strength distribution of short fibre cement composites is also studied and it is shown that the Weibull modulus is considerably increased due to the T-curve effect.

1. INTRODUCTION

The application of fracture mechanics analysis to the design of structures and selection of materials for metals, polymers, ceramics and fibre composites is well-established [1]. However, its application to fibre-reinforced concrete structures and cementitious materials has a very recent history and many international conferences and workshops were held in the last few years to enable a rational understanding and appreciation of the applicability and limitations of this new knowledge to cement-based quasi-brittle materials [2], [3]. This contribution discusses the application of fracture mechanics principles to describe the stable crack growth behaviour of short-fibre cement composites, in which the fibres pull out rather than break, in terms of the crack-toughness (T)-curve concept. Only the 'opening' Mode I crack growth process is considered although the analysis can be extended to 'shearing' Mode II and mixed-Modes I/II fracture propagation.

Conventional linear elastic fracture mechanics using a single-parameter fracture criterion such as the critical potential energy release rate G_c and the critical stress intensity factor K_c are unsuitable to describe the final failure event in fibre-reinforced cement composites. This is because, unlike ideally brittle materials which have very small damage or fracture process zones, fibre cements usually have a small matrix fracture process zone (FPZ) at the crack tip region where microcracking activities take place and a large fibre bridging zone (FBZ) at the crack wake where fibre debond and fibre pull-out occur, Fig. 1. Failure of the structure does not eventuate until some amount of stable crack growth (defined here as the growth of the FBZ) has happened. Because this stable crack growth depends on the specimen geometry, size and initial notch depth, a single-parameter K_c and G_c does not apply. Instead, the failure behaviour has to be characterized in terms of the crack-toughness (T)-curve usually plotted in the form of stress intensity factor (K_R) or potential energy release rate (G_R) versus crack growth (Δa). The single most important physical mechanism giving rise to the increasing crack-resistance of short-fibre cement composites is the progressive development of the fibre bridging zone at the crack wake which effectively shields the crack tip from the applied load. More detailed discussions on crack shielding and crack-toughness are given in Section 2 and interested readers may refer to a review given earlier by Mai & Lawn [4].

If the FBZ is treated in the classical sense as a pseudo-plastic zone

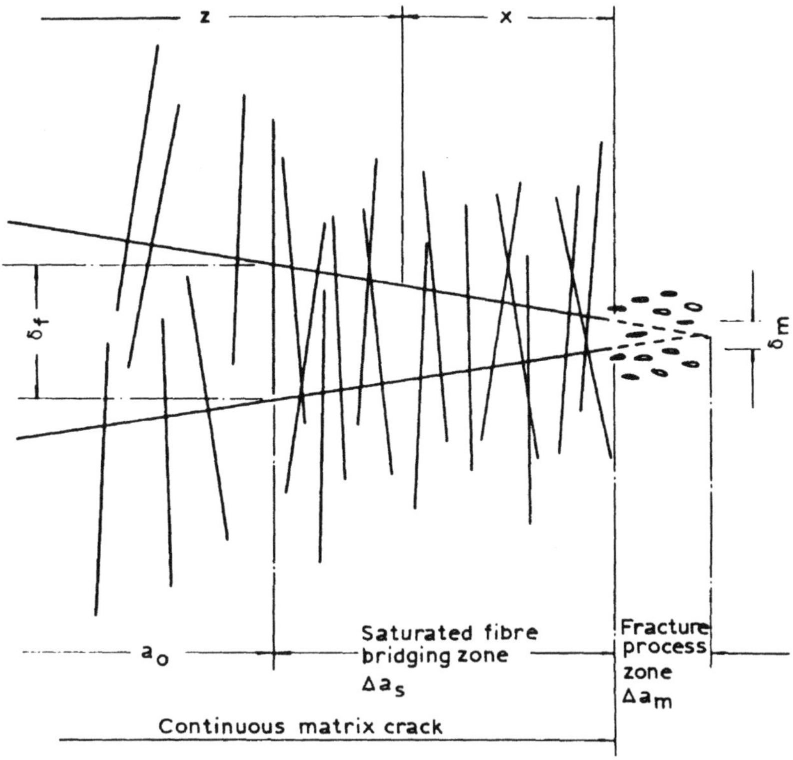

Fig. 1 Fibre bridging and matrix fracture process zones at the notch tip of a fibre-reinforced cement composite.

then it is not surprising that many early investigators have suggested that linear elastic fracture mechanics concepts cannot apply to fibre cements because the FBZ for fibres with large aspect ratios may be in the order of a hundreds of millimetres [5], [6]. Hence the determination of a *valid* fracture toughness value would require prohibitively large size specimens. While this size requirement is still true if we are interested in a single-parameter fracture toughness measurement it does not mean that linear elastic fracture mechanics analysis cannot apply to the modelling of the development of the FBZ and/or real crack growth after the saturated FBZ is developed for small size samples and structures. Provided that the crack closure stress (σ)–crack face separation (δ) relationship is known, *a priori*, either calculated theoretically or determined experimentally, for the FBZ we can evaluate the crack-toughness (T)-curve using the K-superposition principle with increasing amount of stable crack growth for any given specimen geometry and size. Details of the determination of crack bridging stresses and theoretical T-curves are discussed in Sections 3 and 4. It has often been said that the size of the matrix fracture process zone (FPZ) may not be small particularly for concrete and mortar matrices. The measurement of the FPZ and its effect on the crack-toughness curve are given in Section 5. Finally, an illustration of the applicability of the T-curve analysis for the prediction of tensile strength of short-fibre cement composites is given in Section 6.

2. CRACK SHIELDING AND CRACK TOUGHNESS

In the classical Griffith fracture analysis, equilibrium fracture propagation occurs when the potential energy release rate G_a is exactly dissipated by the intrinsic fracture energy absorption at the crack tip R_0; or when the applied stress intensity factor K_a equals the intrinsic toughness T_0, i.e.

$$G_a = R_0 \qquad \text{and} \qquad K_a = T_0 \tag{1}$$

For ideally brittle materials $R_0 = 2\gamma$ where γ is the surface energy; R_0 and T_0 are related by the usual relationship: $T_0^2 = ER_0 = 2E\gamma$. Unstable fracture occurs if

$$\frac{\partial G_a}{\partial(\Delta a)} \geq \frac{\mathrm{d}R_0}{\mathrm{d}(\Delta a)} \qquad \text{and} \qquad \frac{\partial K_a}{\partial(\Delta a)} \geq \frac{\mathrm{d}T_0}{\mathrm{d}(\Delta a)} \tag{2a}$$

$$G_a = G_c = R_0 \qquad \text{and} \qquad K_a = K_c = T_0 \tag{2b}$$

The RHS expressions in the inequalities of eqn (2a) are only dependent on the material properties but the LHS expressions depend on specimen size, geometry and initial notch depth. Since R_0 and T_0 are intrinsic material properties independent of crack growth then the critical fracture event can be described by either the critical value of G_a (i.e. G_c) or K_a (i.e. K_c) alone.

If R and T are functions of crack size due to crack shielding effects then K_c and G_c vary with specimen size and geometry and they do not have a single value at failure any more. In this case, we have

$$R(\Delta a) = R_0 + \sum_{i=1} R_i(\Delta a) \tag{3a}$$

$$T(\Delta a) = T_0 + \sum_{i=1} T_i(\Delta a) \tag{3b}$$

where $R_i(\Delta a)$ and $T_i(\Delta a)$ are the fracture resistance or toughness increase due to the ith crack shielding process. We may conveniently call these the crack-resistance (R) and crack-toughness (T) curves with units of J/m^2 and $Nm^{-3/2}$ respectively. At equilibrium fracture $G_a(\Delta a) = G_R(\Delta a) = R(\Delta a)$ and $K_a(\Delta a) = K_R(\Delta a) = T(\Delta a)$. $G_R(\Delta a)$ and $K_R(\Delta a)$ can be obtained from experiments using the appropriate expressions for $G_a(\Delta a)$ and $K_a(\Delta a)$ from stress intensity factor handbooks [7], [8]. These procedures are already quite well established [4]. To theoretically compute G_R and K_R curves we need to know how to evaluate the terms $\sum R_i(\Delta a)$ and $\sum T_i(\Delta a)$. Because stress intensity factors are additive we take the T expression in eqn (3b) and consider that for every T_i there is an associated K_i so that applying the equilibrium crack growth condition we obtain

$$K_a(\Delta a) - \sum K_i(\Delta a) = K_e(\Delta a) = T_0 \tag{4}$$

This means that the effective stress intensity factor K_e at the crack tip is not due to the applied K_a alone but is subtracted from this by the contributions due to the crack shielding $\sum K_i$. It is generally possible to express the crack shielding terms K_i as an integral over tractions $\sigma(x, \delta)$ at the crack-interface (such as fibre bridging stresses in short-fibre cement composites) by

$$K_i = \int g(a, x)\sigma(x, \delta)\,\mathrm{d}x \tag{5}$$

where x is a distance coordinate measured from the origin of the crack plane and $g(a, x)$ is an appropriate Green's function. The failure condition is given by

$$\frac{\partial K_e}{\partial(\Delta a)} = \frac{\partial K_a}{\partial(\Delta a)} - \partial\left[\sum_{i=1} K_i\right] \Big/ \partial(\Delta a) \geq 0 \tag{6a}$$

or, if we choose to associate crack shielding terms with the toughness term T rather than the stress intensity factor term K, then this condition becomes

$$\frac{\partial K_a}{\partial(\Delta a)} \geq \frac{dT}{d(\Delta a)} \tag{6b}$$

and

$$K_a = K_R = K_c \tag{6c}$$

evaluated at $\Delta a = \Delta a_c$ at failure. Of course Δa_c is not a unique quantity and this makes the critical fracture toughness K_c also not an invariant material property (although it is invariably larger than the intrinsic toughness T_0).

In an earlier review [9], the author discussed the various crack shielding mechanisms associated with a range of engineering materials. Generally, they can be classified into two categories: those mechanisms that take place ahead of the crack tip and those that occur in the crack wake. Relevant to this chapter are the cementitious matrices such as cement pastes, mortars and concretes in which microcracks and aggregate debonding and pull-out occur at the crack tip and fibre-reinforced cement composites in which there is an additional fibre bridging zone at the crack wake. Whatever the mechanisms, whether ahead or behind the crack tip, they all contribute to the crack shielding $\sum K_i$ terms. In modelling stable crack growth in cementitious matrices it is common to replace the shielding zone with a fictitious extension of the same length to the initial notch depth and over which closure stresses σ act in opposition to the applied opening stresses σ_a. Real physical crack growth only happens after the shielding zone of closure stresses has reached its saturated length. Stable crack growth in fibre cements is modelled in a slightly different way depending on the size of the matrix fracture process or shielding zone. This is discussed in Sections 4 and 5 later. Equation (5) suggests that the shielding terms K_i can be estimated rather easily if the closure stress (σ)–crack face separation (δ) relationship is known.

3. DETERMINATION OF CRACK-INTERFACE BRIDGING STRESSES

The functional relationship between the closure stress (σ)–crack face opening (δ) relationship for the fibre bridging zone (FBZ) at the crack wake can in principle be determined from a simple tensile test in which *stable* crack growth across the width of the specimen is obtained.

Experimentally, this may be difficult, even though the fibres tend to stabilize the crack growth (see Section 6 later), and often a very stiff testing machine and a short 'gauge-length' sample are required. Figure 2 shows a schematic tensile stress–displacement curve for a fibre cement composite. Region I represents the elastic response of the specimen and in region II the non-linear curve is caused by the *uniform* dispersed cracking within the gauge section until at σ_m the ultimate strength is reached and the cracking becomes localized on a future failure plane. In region III a continuous matrix crack is established and in region IV the fibres are being pulled out. Usually, $\delta_m \ll \delta_f$ and $\sigma_f \approx \sigma_m$. (See Fig. 1). Hence the closure stress (σ)–crack face separation (δ) relationship is given by the descending portion of the stress–displacement curve in Fig. 2 after the maximum stress σ_m.

In some simple cases the σ–δ relationship can be directly worked out from theory such as in fibre pull-out, but more often than not, indirect evaluations using other measurements are required. In the following sections we discuss two independent methods to determine

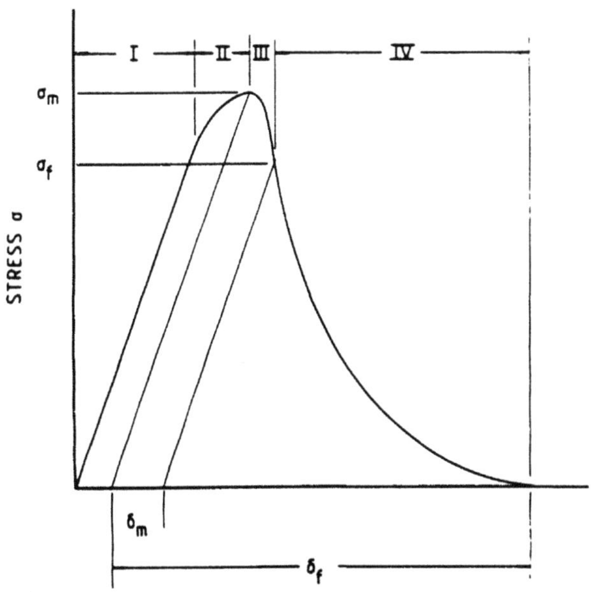

Fig. 2 Schematic tensile stress–displacement curve for a fibre-cement composite.

the σ–δ functional relationship for the FBZ of a fibre–cement composite.

3.1 Single-Fibre Pull-out Test

We consider a simplified model in which the mechanical shear bond strength τ is *constant* along the fibre–matrix interface. Defining L/d as the fibre aspect ratio, then the maximum fibre pull-out force $F_m = \pi dL\tau/2$ and the pull-out force when the crack face has separated an amount δ is given by

$$F(\delta) = F_m(1 - 2\delta/L) = \pi d\tau L(0{\cdot}5 - \delta/L) \quad (7)$$

Since the number of fibres per unit area of bridged crack surface is $\beta V_f/(\pi d^2/4)$, where β is the fibre orientation efficiency factor and V_f is the fibre volume fraction, the average closure stress–displacement relationship is linear as expected. Thus

$$\sigma = [\beta V_f \tau L/d][1 - 2\delta/L] \quad (8)$$

Figure 3 shows the representative σ–δ relationships for a range of short-fibre cement composites taken from Li & Ward [10]. Quite clearly, this is different to the straight line relationship of eqn (8).

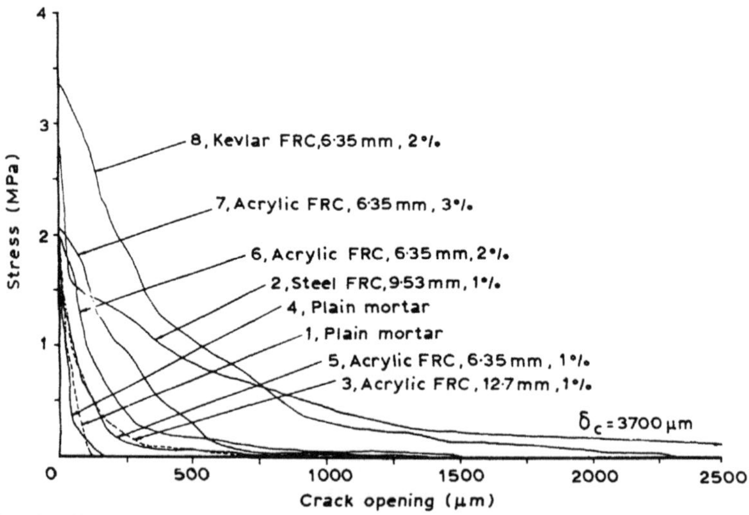

Fig. 3 Closure stress-crack opening curves for various short-fibre cement composites.

Ballarini *et al.* [11] have also shown that for a steel-fibre reinforced concrete the σ–δ function is non-linear.

A simple explanation of this discrepancy is that the single-fibre pull-out model fails to consider that when the crack-face separation is δ there is already a proportion $2\delta/L$ of the embedded fibres pulled out completely leaving only $(1-2\delta/L)$ of the fibres yet to be pulled out. Thus, the correct σ–δ relationship should be

$$\sigma = [\beta V_f \tau L/d][1-2\delta/L]^2 \tag{9}$$

which is no longer linear but quadratic.

There are several other problems associated with the simple fibre pullout model described above. For example it does not consider the Poisson effects of the fibre and matrix which lead to a *non-constant friction* at the debonded fibre–matrix interface. The assumption of a constant friction stress is hardly justified because many short-fibre cement composites contain steel and polymeric fibres whose Poisson contractions during fibre pullout must have reduced the clamping stresses on the fibres. This factor alone can produce a non-linear σ–δ relationship. Gao *et al.* [12] have recently produced a vigorous mathematical analysis on the single fibre pull-out and debond problem by including these additional factors. Their analytical results can apply to continuous fibre composites in which fibre debond proceeds with increasing applied stress *and* to short-fibre composites in which the fibres are pulled out after complete debonding. The latter is relevant to the present discussion and for this problem the fibre pull-out force as a function of the embedded fibre length l is given by

$$F(l) = \bar{F}[1-\exp(-\lambda l)] \tag{10}$$

where

$$\bar{F} = \left(\frac{4q_0}{\pi d^2}\right)\left[\frac{\alpha v_f + \gamma v_m}{\alpha v_f k}\right] \tag{11}$$

$$\lambda = 4\mu/dk \tag{12}$$

$$k = (\alpha v_f + \gamma v_m)/[(1-v_f)\alpha + 1 + v_m + 2\gamma] \tag{13}$$

$$\alpha = E_m/E_f \quad \text{and} \quad \gamma = d^2/(D^2-d^2) \tag{14}$$

q_0 is the initial clamping stress due to fabrication of the composite, μ is the frictional coefficient at the debonded interface, v and E are the Poisson's ratio and Young's modulus, D is the matrix cylinder diameter in the fibre pull-out sample, and f and m are the subscripts for fibre and matrix, respectively. Without going through the complicated mathematics, if we consider the fibre to be rigid and ignore all the elastic stretch of the fibre during pull-out, then the fibre pull-out force as a function of the fibre-end displacement Δ ($\approx\delta$) is given by

$$F(\delta) = \bar{F}[1 - \exp(-\lambda(l - \delta))] \tag{15}$$

which of course is not a straight line but a non-linear relationship. Since the maximum fibre embedded length is $L/2$ and taking into consideration the proportions of pulled out and yet to pull out fibres at a crack-face separation δ we obtain the average closure stress as

$$\sigma(\delta) = \left(\frac{4\beta V_f}{\pi d^2}\right)\bar{F}\left(1 - \frac{2\delta}{L}\right)\left\{1 - \exp\left[-\frac{\lambda L}{2}\left(1 - \frac{2\delta}{L}\right)\right]\right\} \tag{16}$$

which is not very different to eqn (9). However, the advantage of the more exact analysis given by Gao *et al.* is that the single-fibre pull-out data allow us to determine the fibre–matrix interfacial properties such as q_0 and μ and hence $\bar{F}$ to be used in eqn (16) to evaluate the σ–δ relationship.

3.2 Compliance Method

It has been proven in earlier work by Mai & Hakeem [13] that the compliances of a given specimen geometry containing a saw-cut notch and a crack of the *same* length but bridged by unbroken fibres, are different. Figure 4 shows that the compliances of a natural crack (i.e. with a fibre bridging zone) are always smaller than those of a saw-cut notch for a double-cantilever-beam (DCB) cellulose-fibre cement composite tested both dry and wet. The bridging fibres tend to close up the crack face and hence reduce the compliance. Following this observation and using an assumed linearized crack profile within the saturated bridging zone Δa_s as suggested by Foote *et al.* [14].

$$\delta/\delta_f = x/\Delta a_s \tag{17}$$

Hu & Wittmann [15] developed a multi-cutting technique to determine the fibre bridging stresses $\sigma(x)$ using the following compliance expression

$$\frac{\sigma(x)}{\sigma_m} = -\frac{C^2(a)}{C_,(a)} \cdot \frac{C_,^*(x)}{C^{*2}(x)} \tag{18}$$

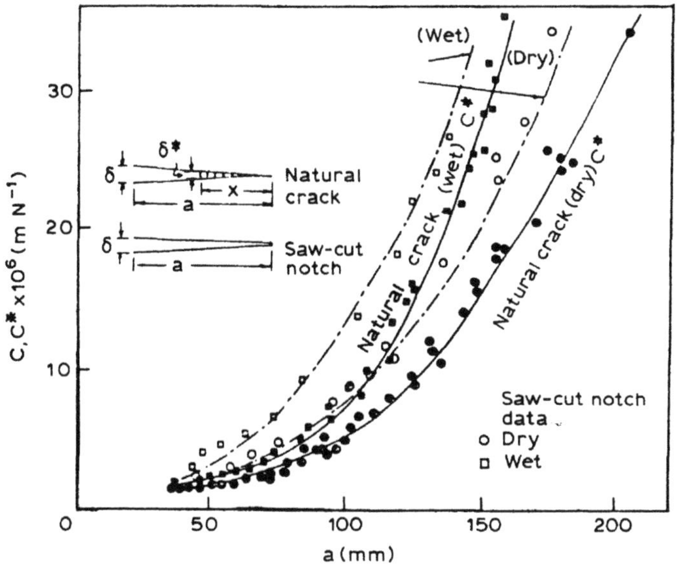

Fig. 4 Effect of fibre bridging at crack-interface on compliance measurements. C and C^* refer to measurements without and with a FBZ.

where $a(=a_0 + \Delta a)$ and x are defined in Fig. 1. C is the theoretical compliance without the bridging zone and C^* is the compliance after a saw-cut is made into the bridging zone. The lower subscript (,) denotes the first derivatives. This is a very useful method of measuring $\sigma(x)$ except that it is experimentally cumbersome. Recently, Hu & Mai [16] have extended this method which does not involve saw-cutting the fibre bridging zone by simply using the unloading compliances of a partially grown FBZ. This is because an unsaturated bridging zone is the *same* as a fully developed bridging zone but partially removed by saw-cutting. This argument leads to the same expression as in eqn (18) for the bridging stresses. By assuming that the σ–δ relationship follows a power law equation

$$\frac{\sigma(x)}{\sigma_m} = \left[1 - \frac{\delta(x)}{\delta_f}\right]^n \qquad n > 0 \tag{19}$$

Hu & Mai further obtained

$$\Phi(\Delta a) = \frac{C(a)}{C,(a)}\left[\frac{C(a)}{C^*(\Delta a)} - 1\right] \tag{20a}$$

$$= \begin{cases} \Delta a_s/(n+1) & \Delta a \geq \Delta a_s \\ (\Delta a_s/n+1)\left\{1 - \left[1 - \dfrac{\Delta a}{\Delta a_s}\right]^{n+1}\right\} & \Delta a < \Delta a_s \end{cases} \tag{20b}$$

Figure 5 shows the $\Phi(\Delta a)$ curves constructed using eqn (20a) above for the wet and dry cellulose-fibre cement mortar composites in the DCB geometry studied by Mai & Hakeem [13]. After $\Delta a_s \simeq 50$ mm, Φ reaches a plateau value of 21 mm and 11 mm respectively for the wet and dry samples. Hence from eqn (20b) we can obtain $n \simeq 1{\cdot}40$ and 3·50 for the wet and dry conditions and the non-dimensionalized fibre bridging stresses can be determined from eqn (19) and shown in Fig. 6. It is quite clear that the bridging stresses vanish at a faster rate with distance from the crack tip in the dry cement composites. This is also consistent with the observations of a less stable fracture propagation in the dry specimens with not only pulled out fibres but a small proportion of broken fibres.

Equations (9) and (16) may be recast in the dimensionless forms of

$$\frac{\sigma(x)}{\sigma_m} = \left[1 - \frac{\delta(x)}{\delta_f}\right]^2 \tag{21}$$

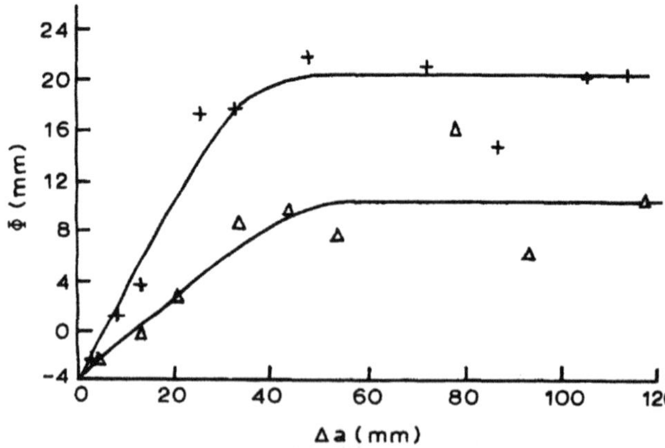

Fig. 5 Φ–Δa curves for wet (+) and dry (Δ) cellulose-fibre cement mortar composites.

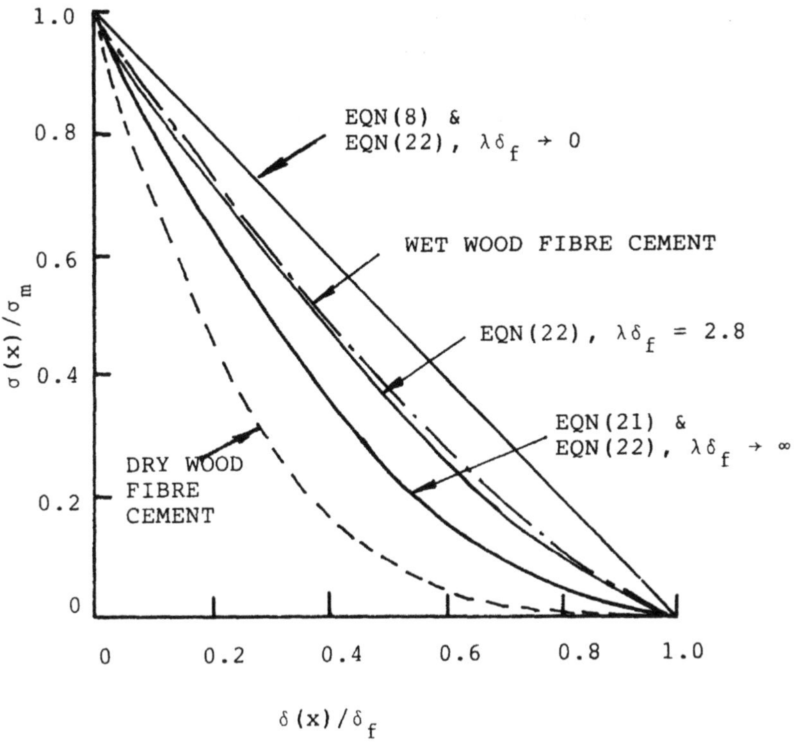

Fig. 6 Normalized closure stress-crack face separation relationship for wet and dry cellulose-cement composites.

and

$$\frac{\sigma(x)}{\sigma_m}=\frac{[1-\delta(x)/\delta_f]\{1-\exp[-\lambda\delta_f(1-\delta(x)/\delta_f)]\}}{[1-\exp(-\lambda\delta_f)]} \tag{22}$$

respectively. It is interesting to note that for large $\lambda\delta_f$, eqn (22) is reduced to the linear form of eqn (8) and for small $\lambda\delta_f$ the limiting solution is the same as eqn (21). These limiting solutions are given in Fig. 6. Note that the wet composite data are bounded by these two curves and in fact they fit eqn (22) very well by an adjusting parameter $\lambda\delta_f = 2{\cdot}8$. The dry composite data lie outside the bounding solutions and this confirms that an additional fibre failure mechanism has occurred in addition to the assumed fibre pull-out mechanism.

4. THEORETICAL MODELLING OF CRACK-TOUGHNESS CURVES

4.1 *K*-Superposition Principle and Assumptions

Various fracture models have been proposed to describe the stable crack growth in fibre-reinforced cement composites; but they almost all follow the fictitious crack model of Hillerborg and co-workers [17], [18] originally developed for cementitious matrices. These models include, for example, Wecharatana & Shah [19], Ballarini *et al.* [11], Visalvanich & Naaman [5], etc. Because the closure stress in the FBZ is dependent on the crack face separation distance, the problem becomes non-linear and an iterative numerical solution is required [18], [19]. Recently, Jenq & Shah [20] and Mobasher *et al.* [21] have extended the two-parameter fracture model proposed initially for concrete to fibre–cement composites. It is shown that such a model can be used to predict the crack toughness curves of fibre cements and the stable crack growth prior to failure instability [21].

Parallel to these fracture models a simple but rigorous *K*-superposition method [6], [14], [22], [23] has been developed for short-fibre cement-based composites. It is postulated that crack growth occurs when the sum of the applied stress intensity factor K_a and the crack shielding stress intensity factor due to the closure stresses $-K_r$ is equal to the intrinsic toughness of the cement matrix material T_0 (more often called K_{Ic} elsewhere), i.e.

$$K_a - K_r = T_0 \tag{23}$$

which is in fact eqn (4) when there is only one crack shielding term. Hence the crack-toughness curve is given by

$$K_R(\Delta a) = T_0 + K_r(\Delta a) \tag{24}$$

where K_r may be readily evaluated by eqn (5) if $\sigma(x, \delta)$ is known. The complications encountered here are that K_r depends on σ which in turn depends on δ. But δ is the sum of the displacements due to K_a and $-K_r$. Iterations are therefore required to determine σ that is consistent with δ within the FBZ. An exact method of analysis for DCB and notched-bend (NB) specimens in which the stress field is non-uniform has been given by Foote *et al.* [14], [23], [24]. Once $\sigma(x)$ is known within the fibre bridging zone Δa, $K_r(\Delta a)$ can be determined and the crack-toughness $K_R(\Delta a)$ calculated.

The method of analysis can be much simplified if we make two additional assumptions already discussed in the previous section. The first is a linearized crack profile for the FBZ given by eqn (17) and the second is a power law strain-softening σ–δ relationship for the FBZ in accordance with eqn (19). Figure 7 provides some justification to the assumed linear crack profile for the NB geometry. One iterative solution is still required to determine the saturated fibre bridging zone Δa_s whilst satisfying the crack growth criterion of eqn (24). For crack growth Δa less than Δa_s, $\sigma(x)$ can be determined from eqns (17) and (19), i.e.

$$\frac{\sigma(x)}{\sigma_m} = \left[1 - \frac{x}{\Delta a_s}\right]^n \tag{25}$$

and K_r calculated from eqn (5). No further iterations are necessary. The accuracy of this approximation method relative to the exact iterative solution can be judged from the K_R-curves for a DCB cellulose-fibre cement composite shown in Fig. 8.

4.2 Specimen Geometry and Size Effects

The fact that the FBZ in many fibre-reinforced cement composites is of the orders of tens to hundreds of millimetres, it can hardly be expected that the crack-toughness curve is invariant with specimen size and geometry. A unique toughness curve can only be obtained in

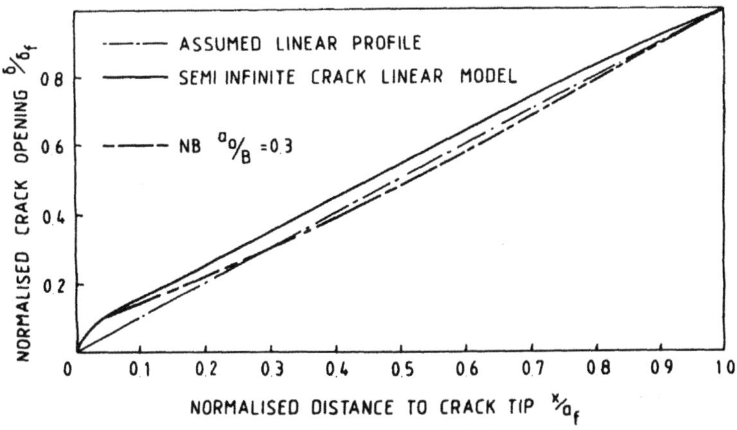

Fig. 7 Linearized crack profile for semi-infinite and NB specimens.

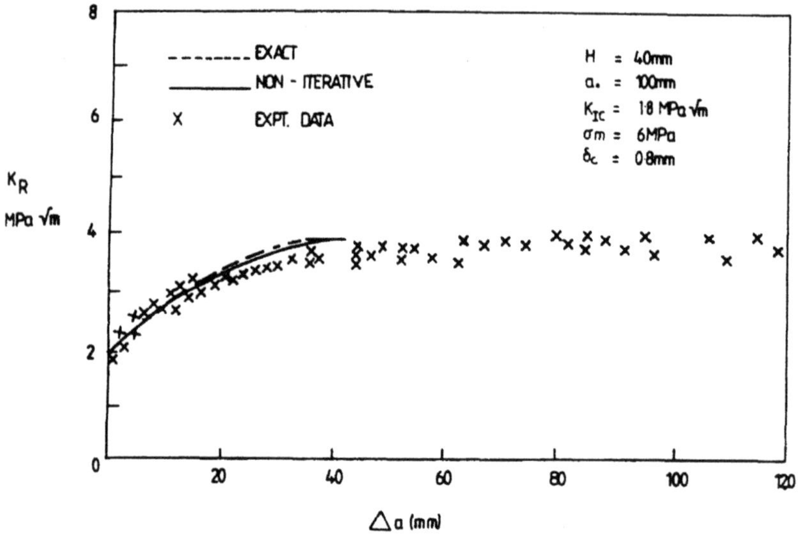

Fig. 8 Comparison of 'exact' and approximate K_R-curves for DCB geometry.

semi-infinite specimens where the FBZ is small relative to the initial notch depth, unbroken ligament and overall specimen size [25].

For the two most common specimen geometries of DCB and NB used for the determination of crack-toughness curves in fibre-cement composites, Figs 9 and 10 compare the effects of size on the

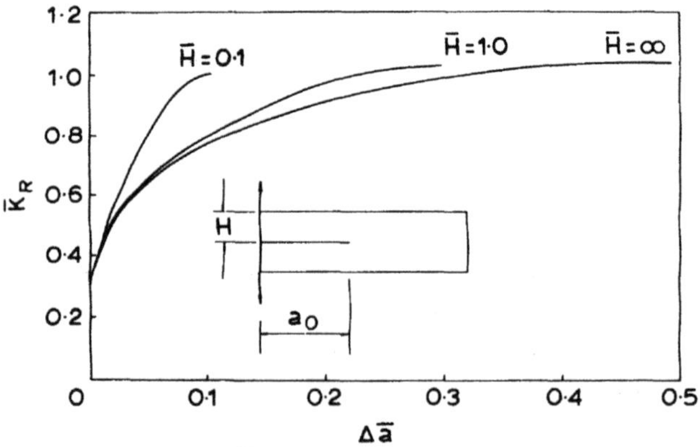

Fig. 9 Non-dimensional $\bar{K}_R$-curves for DCB geometry with different $\bar{H}$.

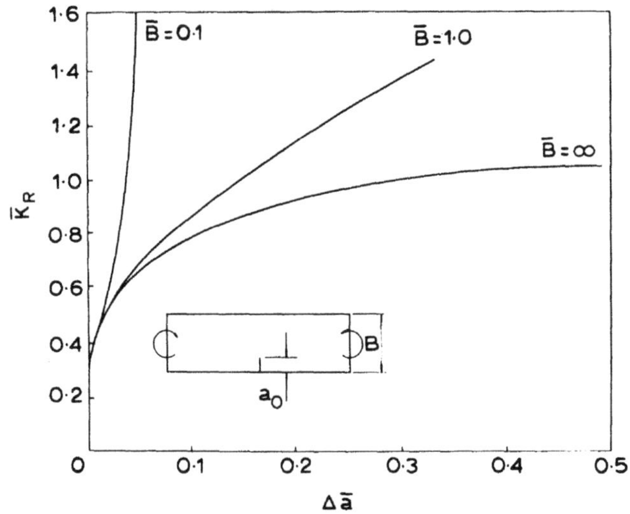

Fig. 10 Non-dimensional $\bar{K}_R$-curves for NB geometry with different $\bar{B}$.

non-dimensionalized K_R-curves. Here $\bar{K}_R = K_R/K_\infty$ (where K_∞ is the plateau value of the crack-toughness curve), $\Delta\bar{a} = \Delta a/(K_\infty/\sigma_m)^2$, $\bar{H} = H/(K_\infty/\sigma_m)^2$ (where H is the DCB height) and $\bar{B} = B/(K_\infty/\sigma_m)^2$ (where B is the NB depth). In the DCB geometry $a_0/H = 3$ and in the NB geometry $a_0/B = 0{\cdot}3$. We have taken $\bar{T}_0$ $(=T_0/K_\infty)$ or $\bar{K}_{Ic}$ as 0·3 in these calculations for the cellulose-fibre cements. K-solutions for the DCB and NB geometries are taken from Foote & Buchwald [26] and Tada *et al.* [7] respectively. The conclusions from these theoretical results may be summarized as follows:

(a) For the DCB geometry the crack-toughness curves are dependent on the beam height H and the length of the saturated FBZ increases with H.
(b) All the plateau values of the K_R-curves for the DCB geometry tend to the same theoretical limit K_∞.
(c) When crack growth is normalized to the saturated FBZ, $\Delta a/\Delta a_s$, the crack-toughness data all fall on the *same* curve independent of the initial notch depth to beam height ratio (a_0/H), Fig. 11, in the DCB geometry.
(d) In the NB geometry the crack-toughness curves are also dependent on the beam depth B and the saturated FBZ length also increases with B.

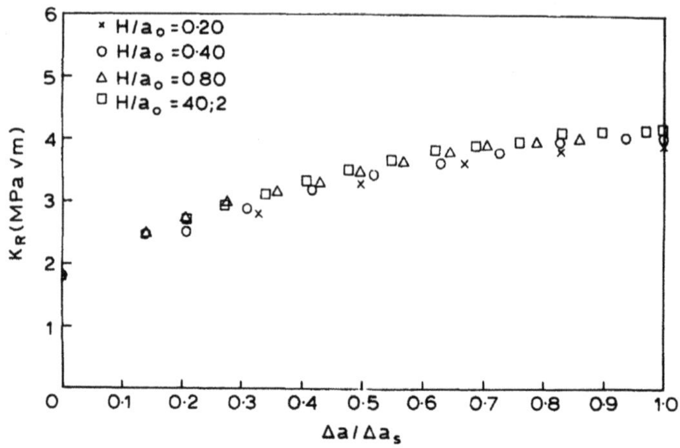

Fig. 11 Normalized plot of K_R versus $\Delta a/\Delta a_s$ for DCB geometry with different H/a_0 ratios.

(e) Except for $\bar{B}\rightarrow\infty$, the crack-toughness curves for small NB depths do not reach a limiting plateau value and indeed the toughness is much larger than K_∞. This effect has been proven experimentally on asbestos–cement composites with different NB sizes [27].
(f) For the NB geometry with a given B, the initial notch depth a_0 also produces a significant effect on the crack-toughness curve, Fig. 12.
(g) When H and B both approach infinite sizes the crack-toughness curves are identical as expected.

4.3 Comparison with Experimental Results

Experimental crack-toughness curves for an asbestos/cellulose–cement mortar composite have been obtained by Mai *et al.* [27] for NB specimens of different beam depths and initial notch depth/beam depth ratios. These are shown in Figs 13 and 14 respectively. The composite and fibre properties are given in Table 1. The bond strengths τ were not measured directly but simply selected to give reasonable agreement with the experimental tensile strength σ_m of this fibre–cement composite material. The fibres were not randomly aligned and an efficiency factor β was estimated to be 0·31. Because the closure stress (σ)–crack face separation (δ) relationship was not

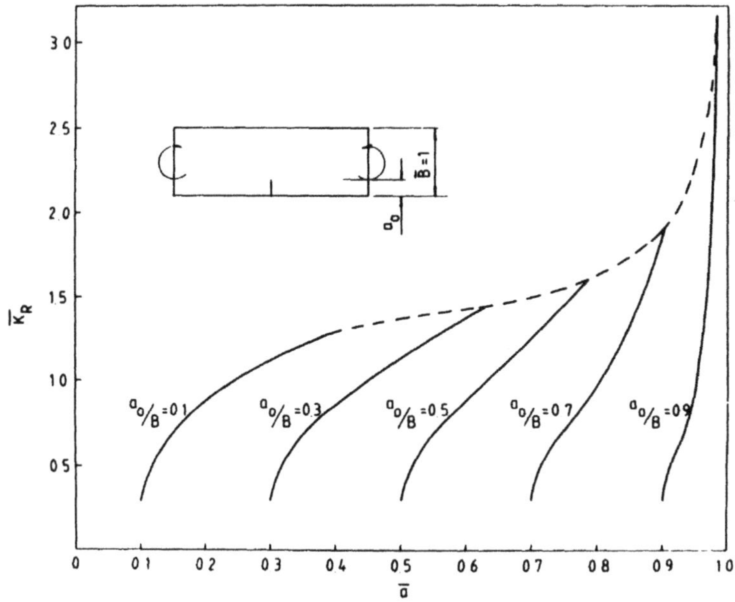

Fig. 12 Effect of a_0/B on non-dimensionalized $\bar{K}_R$-curves for NB geometry for $\bar{B} = 1{\cdot}0$.

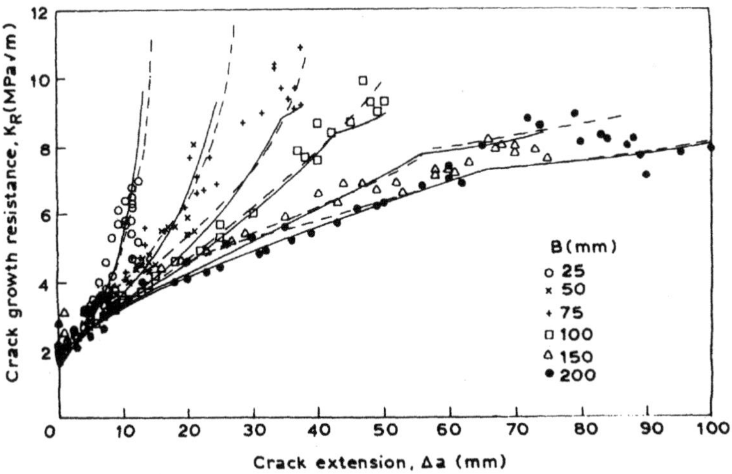

Fig. 13 Experimental K_R-curves for asbestos/cellulose fibre cement composites. NB geometry with $a_0/B = 0{\cdot}3$ and varying B. Theory based on $\sigma_m = 10\,\text{MPa}$, $K_\infty = 5\,\text{MPa}\sqrt{\text{m}}$, $T_0 = 1{\cdot}9\,\text{MPa}\sqrt{\text{m}}$. —— without matrix FPZ; - - - - with matrix FPZ.

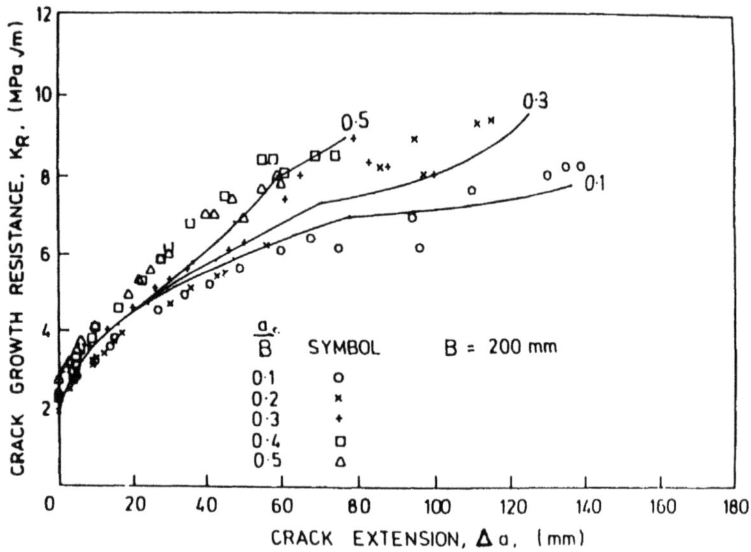

Fig. 14 Experimental K_R-curves for asbestos/cellulose fibre cements with $B = 200$ mm and varying a_0/B.

determined for this hybrid composite material we cannot compute $K_R(\Delta a)$ using eqn (24). On a more practical level, however, we can model the stable crack growth by determining the most appropriate fracture parameters (T_0, K_∞, E, σ_m) or (T_0, σ_m, δ_f, E) to give the best fit to the experimental K_R-curve of a given geometry and size of specimen. In this way, values of T_0 (or K_{Ic}) and K_∞ given in Table 1 are chosen to fit the data of the largest NB specimen with $B = 200$ mm.

TABLE 1 Properties of an asbestos/cellulose-reinforced mortar composite

(a) *Composite properties*		
Young's modulus (E)		6 GPa
Intrinsic toughness (T_0 or K_{IC})		1·9 MPa$\sqrt{m}$
Plateau toughness (K_∞)		5·0 MPa$\sqrt{m}$
(b) *Fibre composites*		
	Asbestos	*Cellulose*
Aspect ratio	80	135
Fibre length	2·0 mm	3·5 mm
Volume fraction	0·08	0·07
Bond strength (τ)	≈2·0 MPa	0·88 MPa

Theoretical K_R-curves for smaller NB specimens and different a_0/B ratios were determined using these estimated parameters and $n = 1$ with the approximation method outlined in Section 4.1. Clearly from Figs 13 and 14, the theoretical curves agree well with the experimental data and in particular they also replicate the trend of the K_R-curves to exceed K_∞ for small B.

It is also possible to predict the load–crack mouth deflection curves for the three-point bending tests conducted on the NB specimens of different beam depths. For any given Δa there is a corresponding δ (less than δ_f) at the receding edge of the FBZ which can be related via fracture mechanics analysis to the crack mouth displacement Δ at the original machined notch. Thus P–Δ curves can be estimated with crack growth until final failure occurs. Comparisons of theoretical and experimental P–Δ curves are given elsewhere.

5. CRACK-TOUGHNESS CURVE WITH LARGE MATRIX FPZ

5.1 Matrix Fracture Process Zone in Fibre–Cement Composites

In the theoretical model presented in Section 4 we have not included the effect of the matrix fracture process zone size (FPZ) because we assume it is small and can be sufficiently represented by the intrinsic toughness term T_0 (or K_{Ic}). If this matrix FPZ is not small then it has to be incorporated in the stable crack growth analysis as discussed in the next section. There is some indication from previous work that in asbestos [28] and cellulose [29] fibre cement mortars the matrix FPZ can be as large as 28 mm. In concrete-matrix short-fibre composites this size can be even bigger.

Whilst there is a large range of methods to measure the matrix FPZ the simple 'section and bend' technique developed by Foote *et al.* [28] is particularly suitable for fibre–cement composites whose FPZ sizes are reasonably large. Essentially, what is needed is to grow a stable crack in a specimen such as the compact tension geometry, Fig. 15, and then narrow strips approximately 4 mm in width are sectioned in the ligament region. Strips remotely located in the unstressed region are also sectioned to represent the undamaged material. The bending stiffness of each strip is measured in pure bending with the compressive surface facing the crack growth direction and the fracture location is also noted. If the failure sites lie along the prolongation of the initial

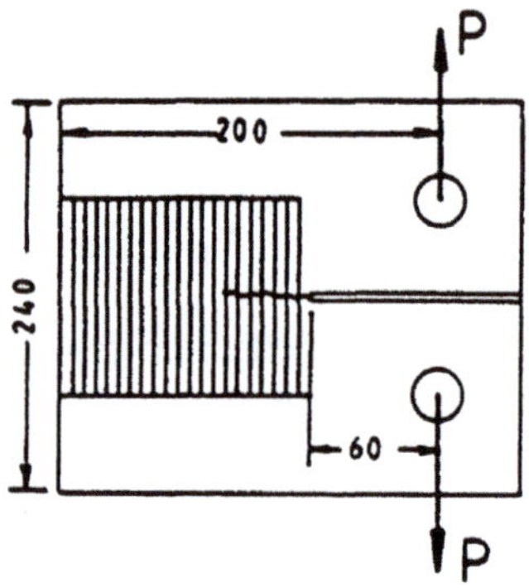

Fig. 15 Compact tension specimen used for matrix FPZ evaluation experiments. (All dimensions in mm.)

notch these strips belong to the FBZ. But if the failure sites are randomly distributed the strips belong to the matrix FPZ. Coupled with the normalized stiffness plot as a function of crack length we can easily determine the extent of the matrix FPZ, Fig. 16. When this technique is used for a cellulose-fibre cement mortar composite we estimate the matrix FPZ to be about 28–40 mm as shown in Fig. 16.

5.2 Modelling Crack Toughness Curve with a Matrix Fracture Process Zone

The easiest method of including the matrix FPZ in the stable crack growth analysis in fibre–cement composites is to consider this as a fictitious extension of the continuous matrix crack with a constant stress σ_m acting on it. Because of this finite stress at the fictitious crack extension, the sum of the stress intensity factors at its tip when crack growth occurs must be equal to zero. Hence

$$K_a - K_r - K_m = 0 \tag{26}$$

where K_m is the stress intensity factor caused by the closure stresses σ_m in the matrix FPZ. Note that all K_a, K_r and K_m are calculated at the end of the fictitious crack tip. An additional crack growth condition is that the crack opening displacement at the tip of the continuous matrix crack must be equal to δ_m. (See Fig. 1). Hence

$$\delta(a) = \delta_m. \tag{27}$$

We can assume a linear σ–δ relationship and a straight crack face profile in the FPZ to model crack growth as before. Δa_s and Δa_m (the matrix FPZ) are obtained by iteration to satisfy both eqns (26) and (27) above. Unfortunately, for $\Delta a < a_s$, we still have to obtain the

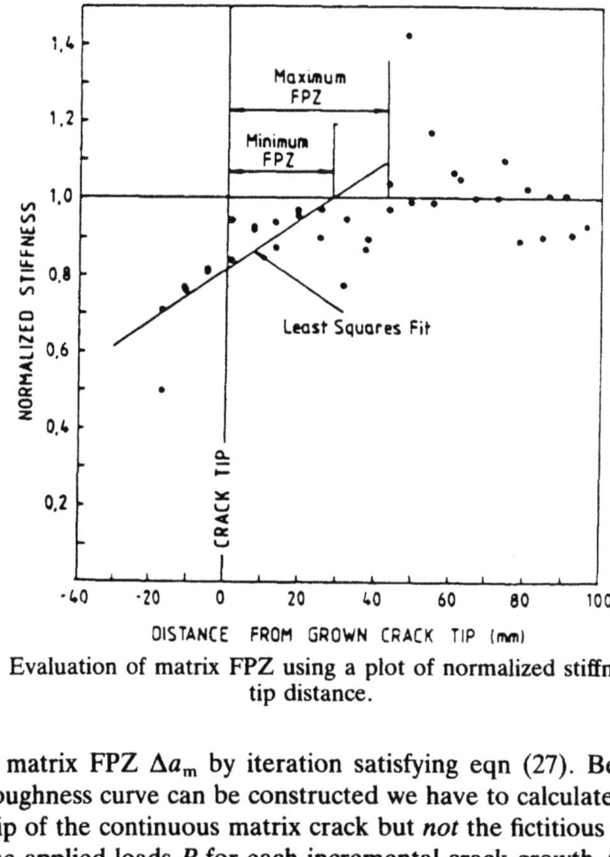

Fig. 16 Evaluation of matrix FPZ using a plot of normalized stiffness–crack tip distance.

current matrix FPZ Δa_m by iteration satisfying eqn (27). Before the crack-toughness curve can be constructed we have to calculate $K_R(\Delta a)$ at the tip of the continuous matrix crack but *not* the fictitious crack tip using the applied loads P for each incremental crack growth Δa.

Theoretical K_R curves calculated with the matrix FPZ for the cellulose-cement mortar composite are also shown in Fig. 13 for the three-point NB geometry with varying beam depths. For this composite material at least, the effect of the matrix FPZ is not important and its inclusion in the analytical model is not necessary. It must, however, be cautioned that this statement may not be correct if the matrix FPZ is very large, such as in steel fibre-reinforced concrete composites.

6. TENSILE STRENGTH OF SHORT-FIBRE CEMENT COMPOSITES WITH CRACK-TOUGHNESS CURVE CHARACTERISTICS

In the analytical analysis of stable crack growth in fibre-reinforced cement composites given in the earlier sections, only a single crack situation is considered. We now consider a typical tensile specimen containing many inherent defects which are bridged by the short fibres. Each defect will display a crack-toughness curve characteristic upon loading. Extension and arrest of these growing cracks depend on whether eqn (23) is satisfied or not. When K_e is less than T_0 the crack will arrest, and when K_e is bigger than T_0 the crack will extend to the next nearest fibres in its future path. Because of the presence of many growing cracks a statistical fracture mechanics analysis is required to solve this problem. Such a solution has been provided by Hu *et al.* [30], [31] and Hu [32]. In the following only the main results of the tensile strength are given. Details of the analysis are given in the original references.

Let us assume that in a rectangular plate under uniform tension there is a random distribution of aligned fibres of density ρ_f and aspect ratio L/d. The matrix cracks are considered as equivalent Griffith cracks lying normal to the applied stress whose sizes vary in accordance with a Pareto distribution function, i.e.

$$q(a) = (\rho_m m/2a_0)(a_0/a)^{(m+2)/2} \tag{28}$$

for $a > a_0$ and ρ_m is the matrix crack density, m is the Weibull modulus of the matrix material and a_0 is the reference crack size. The Weibull strength distribution equation is

$$\begin{aligned} F(\sigma) &= 1 - \exp\left\{-V \int_{a(\sigma)}^{\infty} q(a)\,\mathrm{d}a\right\} \\ &= 1 - \exp\{-V\rho_m(\sigma/\sigma_0)^m\} \end{aligned} \tag{29}$$

where V is the volume of material under tensile stress and

$$\sigma_0 = T_0(2/\pi a_0)^{1/2} \tag{30}$$

If these flaws are bridged by a high density ρ_f of unbroken fibres the bridging stress across the crack is given by

$$\sigma_b = \pi\tau d\rho_f L^2/4d \tag{31}$$

which is the same for each flaw. (Here, we do not consider the reduction in σ_b as the fibres are being pulled out to simplify the analysis). The Weibull strength distribution in this case becomes

$$F(\sigma) = 1 - \exp\{-V\rho_m[(\sigma - \sigma_b)/\sigma_0]^m\} \tag{32}$$

and it is essentially still the weakest link theory because there is little variation of the fibre bridging stress across the matrix flaws.

If, however, the density of the fibres is low to medium where the number of bridging fibres in each flaw is not constant, Hu *et al.* [30], [31] have shown that a considerable amount of stable crack growth can occur before failure. For simplicity, the theoretical equations are omitted here. Instead the tensile strength results obtained from computer simulations [31] are shown in Figs 17–19 for $\rho_f = 0$, 0·1 and 0·2 mm^{-2} respectively. The specimen size is 100 mm^2, $\rho_m =$ 0·003 mm^{-2}, $m = 8$, $a_0 = 2$ mm, $T_0 = 0{\cdot}6$ MPa$\sqrt{\text{m}}$, $L = 5$ mm, $d =$ 0·1 mm and $\tau = 4$ MPa. Both the first cracking and final fracture strength distributions are shown in those figures. There are some differences between the first and final crack strengths for the unreinforced mortar matrix material, Fig. 19, due to some edge effects in the computer simulation study. The Weibull modulus of the failure strength is 8·74 in good agreement with the assumed value of 8. As the fibre density increases there are also improvements in the Weibull moduli of both the first cracking and final failure strengths. Hence, $m = 16{\cdot}7$ and 14·8 for the final failure strength distribution for $\rho_f = 0{\cdot}1$ and 0·2 mm^{-2} respectively. Physically, this means that there is much less scatter in the tensile strength of the short-fibre-reinforced cement mortars due to the crack-toughness curve effect thus promoting stable

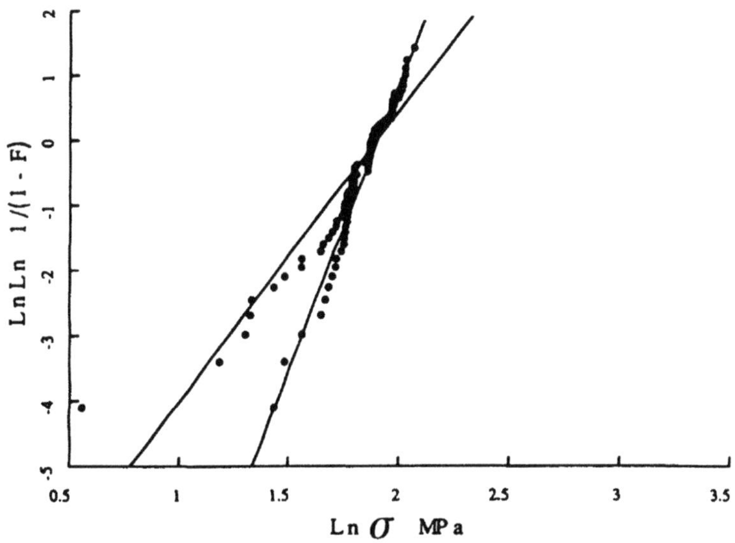

Fig. 17 Weibull strength distribution of unreinforced cement mortar.

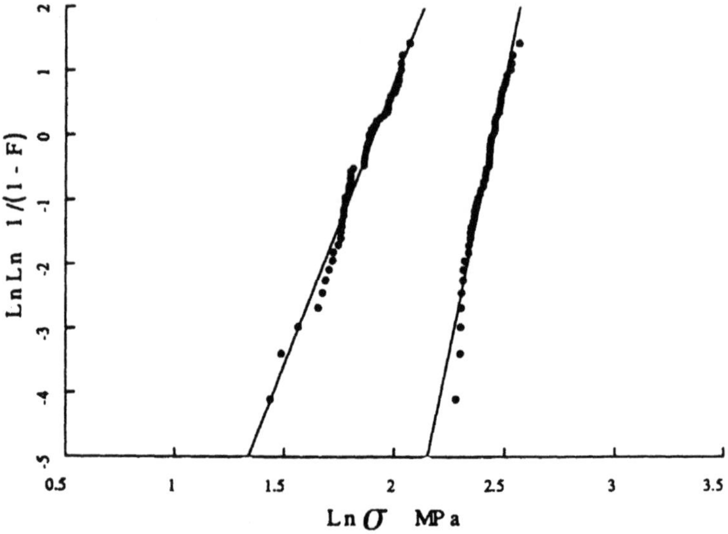

Fig. 18 Weibull strength distributions of first crack and final failure for a fibre cement composite with $\rho_f = 0{\cdot}1\ \mathrm{mm}^{-2}$.

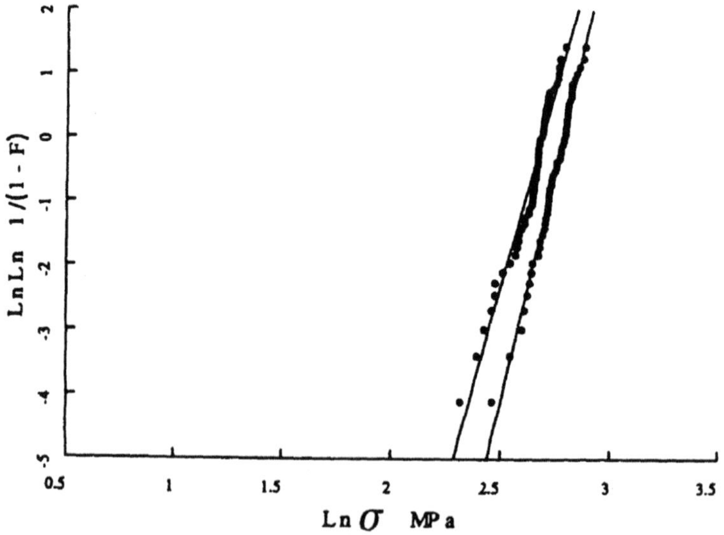

Fig. 19 Weibull strength distributions of first crack and final failure for a fibre cement composite with $\rho_f = 0{\cdot}2\ \mathrm{mm}^{-2}$.

crack growth prior to failure instability. The distance between the first cracking and failure strength lines in Figs 18 and 19 is an indication of the occurrence of stable crack growth. As ρ_f increases the amount of stable crack growth first increases and then decreases until at very large ρ_f there is no stable crack growth like the case for $\rho_f = 0$. However, the tensile strength σ_m will always increase with ρ_f.

7. CLOSURE

The application of fracture mechanics analysis to predict the stable crack growth of fibre-reinforced cement composites is demonstrated in these notes. By assuming a linearized crack profile and a simple power law for the strain-softening relationship of the FBZ in conjunction with a K-superposition principle we have shown that the crack-toughness curve characteristics for two specimen geometries of DCB and NB with different sizes are different. These predicted K_R–Δa curves are in agreement with the experimental data obtained for a cellulose-fibre-reinforced mortar composite. Notably, for short NB specimens the crack-toughness can rise above the limiting value K_∞. It is also shown how the matrix FPZ can be included in the stable crack growth analysis. For the cellulose-fibre cement composite concerned this inclusion is not significant enough to cause any substantial changes in the predicted K_R-curves.

Using the crack-toughness curve concept and a statistical fracture mechanics analysis, the Weibull tensile strength of short-fibre cement composites, in which fibres pull out rather than break, is shown to have a narrower distribution than the unreinforced matrix material. That is, the tensile strength is not only higher but is more consistent because of the larger Weibull modulus in the fibre-reinforced material.

ACKNOWLEDGEMENTS

The author wishes to acknowledge the continuing financial support provided by the Australian Research Council. The contributions of two colleagues, B. Cotterell and X.-Z. Hu, to many aspects of this work in the form of original data and discussions are appreciated. In particular X.-Z. Hu developed the Φ-curve concept to measure crack-interface bridging stresses.

REFERENCES

[1] Atkins, A. G. & Mai, Y.-W., *Elastic and Plastic Fracture.* Ellis Horwood/John Wiley, Chichester, UK, 1985.

[2] Shah, S. P. (Ed.), *Application of Fracture Mechanics to Cementitious Composites.* Martinus Nijhoff Publishers, Dordrecht, 1985.

[3] Shah, S. P. (Ed.), *Mechanisms in Quasi-Brittle Materials.* NATO ASI Series, Series E: Applied Sciences, Vol. 195, Kluwer Academic Publishers, Dordrecht, The Netherlands, 1991.

[4] Mai, Y.-W. & Lawn, B. R., Crack stability and toughness in brittle materials. *Ann. Rev. Mater. Sci.*, **16** (1986) 415–39.

[5] Visalvanich, K. & Naaman, A. E., Fracture model for fibre-reinforced concrete. *J. Amer. Concr. Inst.*, **80** (1983) 128–38.

[6] Cotterell, B. & Mai, Y.-W., Modelling crack growth in fibre-reinforced cementitious materials. *Mater. Forum,* **11** (1985) 341–51.

[7] Tada, H., Paris, P. C. & Irwin, G. R., *Stress Intensity Factor Handbook.* Del Research Corporation, Hellertown, PA., 1973.

[8] Sih, G. C., *Handbook of Stress Intensity Factors.* Institute of Solid and Fracture Mechanics, Lehigh University, Bethlehem, USA, 1973.

[9] Mai, Y.-W., Fracture resistance and fracture mechanisms of engineering materials. *Mater. Forum,* **11** (1988) 232–67.

[10] Li, V. C. & Ward, R., A novel testing technique for post-peak tensile behaviour of cementitious materials. In *Proc. Int. Workshop on Fracture Toughness and Fracture Energy: Test Methods for Concrete and Rock,* ed. H. Mihashi, Balkema, Rotterdam, 1988, pp. 139–56.

[11] Ballarini, R., Shah, S. P. & Keer, L. M., Crack growth in cement-based composites. *Eng. Fract. Mech.,* **20** (1984) 433–45.

[12] Gao, Y.-C., Mai, Y.-W. & Cotterell, B., Fracture of fibre-reinforced composites. *ZAMP,* **39** (1988) 550–73.

[13] Mai, Y.-W. & Hakeem, M. I., Slow crack growth in cellulose fibre cements. *J. Mater. Sci.,* **19** (1984) 501–8.

[14] Foote, R. M. L., Mai, Y.-W. & Cotterell, B., Crack growth resistance curves in strain-softening materials. *J. Mech. Phys. Solids,* **34** (1986) 593–607.

[15] Hu, X.-Z. & Wittmann, F. H., Fracture process zone and K_r-curve of hardened cement paste and mortar. In *Fracture of Concrete and Rock,* ed. S. P. Shah, S. E. Swartz & B. Barr. Elsevier Applied Science, London, 1989, pp. 307–16.

[16] Hu, X.-Z. & Mai, Y.-W., A general method for determination of crack-interface bridging stresses. *J. Mater. Sci.*, (in press).

[17] Hillerborg, A., Modeer, M. & Petersson, P. E., Analysis of crack formation and crack growth in concrete by means of fracture mechanics and finite elements. *Cement & Concrete Research,* **6** (1976) 773–82.

[18] Hillerborg, A., Analysis of one single crack. In *Fracture Mechanics of Concrete,* ed. F. H. Wittmann. Elsevier, Amsterdam, 1983, pp. 223–49.

[19] Wecharatana, M. & Shah, S. P., A model for predicting fracture resistance of fibre reinforced concrete. *Cement & Concrete Research,* **13** (1983) 819–23.

[20] Jenq, Y. & Shah, S. P., Crack propagation in fibre-reinforced concrete. *J. Struct. Eng. Div.* (ASCE) **112,** (1986) 19–24.

[21] Mobasher, B., Ouyang, C. & Shah, S. P., A *R*-curve approach to predict toughening of cement-based matrices due to fibre reinforcement. (To be published).

[22] Cotterell, B. & Mai, Y.-W., The effect of a fracture process zone on a model for crack growth in fibre-reinforced cementitious composites. *Adv. Cement Research,* **1** (1988) 75–83.

[23] Foote, R. M. L., Cotterell, B. & Mai, Y.-W., Analytical modelling of crack growth resistance curves in DCB fibre-reinforced cement specimens. In *Fracture Toughness and Fracture Energy of Concrete,* ed. F. H. Wittmann. Elsevier, Amsterdam, 1986, pp. 535–44.

[24] Cotterell, B., Mai, Y.-W. & Foote, R. M. L., Bounding solutions for crack growth resistance curve in fibre-reinforced cement composites. In *Engineering Applications of New Composites,* ed. S. A. Paipetis & G. C. Papanicolaou. Omega Scientific, UK, 1988, pp. 186–96.

[25] Foote, R. M. L., Cotterell, B. & Mai, Y.-W, Crack growth resistance curves for cement composites. In *Adv. Cement Matrix Composites,* ed. D. M. Roy. Materials Research Society, Penn., 1980, pp. 135–44.

[26] Foote, R. M. L. & Buckwald, V. T., An exact solution for the stress intensity factor for a double cantilever beam. *Int. J. Fract.,* **29** (1985) 125–34.

[27] Mai, Y.-W., Foote, R. M. L. & Cotterell, B., Size effects and scaling laws of fracture in asbestos cements. *Int. J. Cement Composites,* **2** (1980) 23–34.

[28] Foote, R. M. L., Mai, Y.-W. & Cotterell, B., Process zone size and slow crack growth measurements in fibre cements. Paper presented at ACI Symposium on Fibre Reinforced Concrete, SP-105, ed. S. P. Shah & G. B. Barton. Amer. Concrete Inst., Detroit, 1987, pp. 55–70.

[29] Lenain, J. C. & Bunsell, A. R., The resistance to crack growth of asbestos cement. *J. Mater. Sci.,* **14** (1979) 321–32.

[30] Hu, X.-Z., Mai, Y.-W. & Cotterell, B., Computer simulation of fracture behaviour of short-fibre reinforced cement. In *Proceedings 3rd Int. Symp. on Developments in Fibre Reinforced Cement and Concrete,* ed. R. N. Swamy *et al.* RILEM Technical Committee 49-TFR, Sheffield, 1988, Vol. 2, Paper 6.4.

[31] Hu, X.-Z., Mai, Y.-W. & Cotterell, B., Tensile strength of short-fibre reinforced brittle materials with pre-existing defects. *Phil. Mag.*, (in press).

[32] Hu, X.-Z., Statistical fracture of brittle materials. PhD thesis, The University of Sydney, Australia, 1988.

Chapter 9

Fracture Mechanics Evaluation of Anchorage Bearing Capacity in Concrete

PIETRO BOCCA, ALBERTO CARPINTERI & SILVIO VALENTE
Politecnico di Torino, Department of Structural Engineering, 10129 Torino, Italy

ABSTRACT

This chapter illustrates the mechanical behaviour observed during pull-out tests performed on over 60 steel anchor bolts embedded at variable depths (28–55 mm) in slabs of concrete and mortar. The test results are compared to those obtained from a numerical simulation of the growth of an axial-symmetrical crack in a solid characterized by elasto-softening behaviour (cohesive crack model). Maximum loads are compared to those predicted by the most common design formulas and 'size-effect' is found to play a decisive role.

1. INTRODUCTION

The calculation of the bearing capacity of steel anchor bolts embedded in concrete has taken on ever greater importance with the rapid growth of anchorage applications in the field of structural engineering.

The methods employed so far are generally based on simplified formulations or on the theory of plasticity. ACI Standards [1], for instance, propose formulas for the determination of the bearing capacity, of both bolt and hook type anchorages, based on concrete tensile strength and on the assumption that failure occurs along a truncated cone shaped surface.

However, a comparison with actual test results has shown that the values provided by such formulas are in poor agreement with experimental data, the latter being greatly affected by the size and shape of the specimens.

The adoption of correction factors, such as those introduced to take into account size-effect, while enabling the theoretical formulas to be improved, did not solve the problem. Likewise, the application of the theory of plasticity yielded no significant improvement. The criteria of concrete plasticity, employed by several authors [2]–[5] in either the two- or three-dimensional field, in fact, have shown poor agreement with the results obtained experimentally, again on account of the decisive role played by anchorage size and shape.

A totally new way to address and overcome these difficulties now comes from Fracture Mechanics, as borne out by the works of various authors [6]–[10]. The choice of an energy model to gain an understanding of the interaction taking place between the anchor bolt and the concrete makes it possible to take into account concrete geometry and offers the advantage of yielding expressions for the determination of anchorage bearing capacity of unrestricted validity: these expressions are linked to fracture energy (G_F), the elastic modulus, the tensile strength of the material and bolt embedment depth.

In this investigation, the problem is addressed by means of the cohesive crack model, as described in some detail later on, in Section 4. Over other methods, the cohesive crack model offers the advantage of being suitable for problems in which the crack path is unknown *a priori*. Thus, the application of this model, based on the finite elements method, makes it possible to predict not only the bearing capacity of the steel bolt, but also the evolution of the crack path.

The results obtained from tests performed at the laboratory of the Dipartimento di Ingegneria Strutturale of the Politecnico di Torino, Italy, are also presented: the mechanical behaviour was observed in pull-out tests performed on over 60 steel anchor bolts of limited length embedded to varying depths in slabs of different materials (concrete, mortar). For each bolt, the pull-out force, the complete pull-out force versus bolt displacement diagram and the crack path were determined. This was done to provide a sizeable series of experimental results and thereby fill a gap in this field, hoping the results can also be used by other authors. However, the ultimate aim of this work is to interpret the experimental data through a general theoretical formulation based on the cohesive crack model, and hence on fracture mechanics, which might solve this type of structural problem in a systematic way.

TABLE 1 Bolt sizes

Bolt	V_i *[mm]*	d_i *[mm]*
a	28	34
b	35	34
c	40	40
d	44	34
e	52	45
f	55	34

2. THE TESTING PROGRAMME

2.1 Specimen Characteristics

Sixty steel bolts possessing the geometrical characteristics listed in Table 1 were placed, before casting, in 12 slabs of concrete and mortar, sized 1 m × 1 m × 0·15 m (Fig. 1). Six different types of bolt were used, of different lengths V_i and with different head diameters. After 30 days of ageing in a controlled environment ($T = 20°C$; $RH = 65\%$), five bolts of the same kind were pulled out from each slab, according to the arrangement shown in Fig. 1.

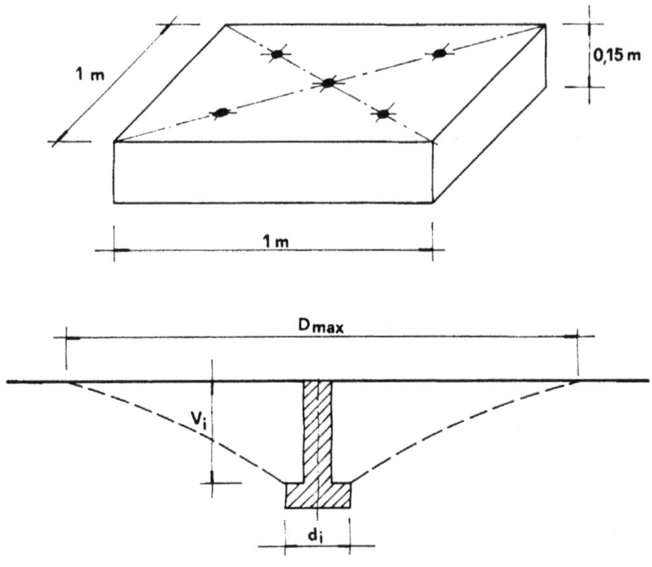

Fig. 1 Layout of the slabs and steel bolts.

TABLE 2 Material properties

	Compression strength		Young's modulus		Fracture energy		Tensile strength	
	n^a	$R[MPa]$	n	$E[MPa]$	n	$G_F[N/m]$	n	$\sigma_u[MPa]$
Concrete	6	51·6	3	27500	3	119	3	2·67
Mortar	6	27·6	3	17200	3	39	3	1·03

[a] n = number of tests.

Six slabs were manufactured with concrete having the following composition:

Portland cement type 425	450 kg/m^3
Sand with 0–8 mm diameter	910 kg/m^3
Aggregate with 4–12 mm diameter	910 kg/m^3
Plasticizer	1 kg/m^3

Six slabs were manufactured with mortar having the following composition:

Portland cement type 425	400 kg/m^3
Sand with 0–8 mm diameter	1700 kg/m^3

Table 2 gives the mechanical characteristics of concrete and mortar. The specimens used were manufactured at the same time as the slabs and aged for 30 days in identical environmental conditions.

Compressive strength was determined from crushing tests on 16-cm sized cubes, the secant elastic modulus from tests on 16 cm × 16 cm × 50 cm prisms. Three-point tests on notched prisms (10 cm × 10 cm × 84 cm in size) were performed according to RILEM Recommendations [11] in order to determine fracture energy and tensile strength values. Load versus displacement diagrams were produced by controlling notch opening by means of a Hottinger Baldwin DD1 displacement transducer at a speed of 2×10^{-6} m/s.

2.2 Testing Procedure

The bolts were pulled out by means of an MTS with 250 kN maximum load. The bolt displacement, η, was calculated as the arithmetic mean of the η_1, η_2 values measured by a pair of inductive displacement transducers (T_1, T_2) arranged as shown in Fig. 2. The load versus displacement curves were plotted at a constant bolt displacement speed of $d\eta/dt = \dot{\eta} = 5 \times 10^{-6}$ m/s.

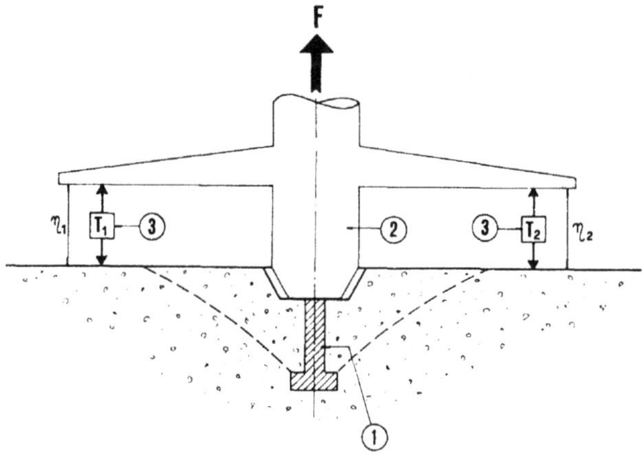

Fig. 2 Pull-out apparatus: (1) bolt; (2) MTS; (3) inductive displacement transducer.

When pulling out the bolts, a reaction ring of over 50 cm in diameter was rested against the slab surface. As observed in earlier tests [10], under this arrangement, the state of stress applied by the ring pushing against the slab surface does not interfere with the state of stress produced by the pull-out force, *F*.

3. TEST RESULTS

Figures 3(a)–(f) show the pull-out force versus displacement curves measured for the bolts embedded in concrete. For each pair of cartesian axes a bundle of curves relating to bolts embedded to the same depth is shown. Figures 4(a)–(f) show the corresponding curves as measured on mortar specimens.

Figures 5 and 6 illustrate the mean curves as obtained for varying embedment depths. In this case, size does not substantially alter the material behaviour. It should be noted, instead, that a more brittle behaviour was observed overall in the case of bolts pulled out from mortar compared with those pulled out from concrete.

Tables 3 and 4 give the values of maximum pull-out force (third column), maximum elastic displacement (fifth column) and total fracture energy (seventh column). The latter, relating to the area under the F versus η curve, is plotted in Fig. 7: as expected, it

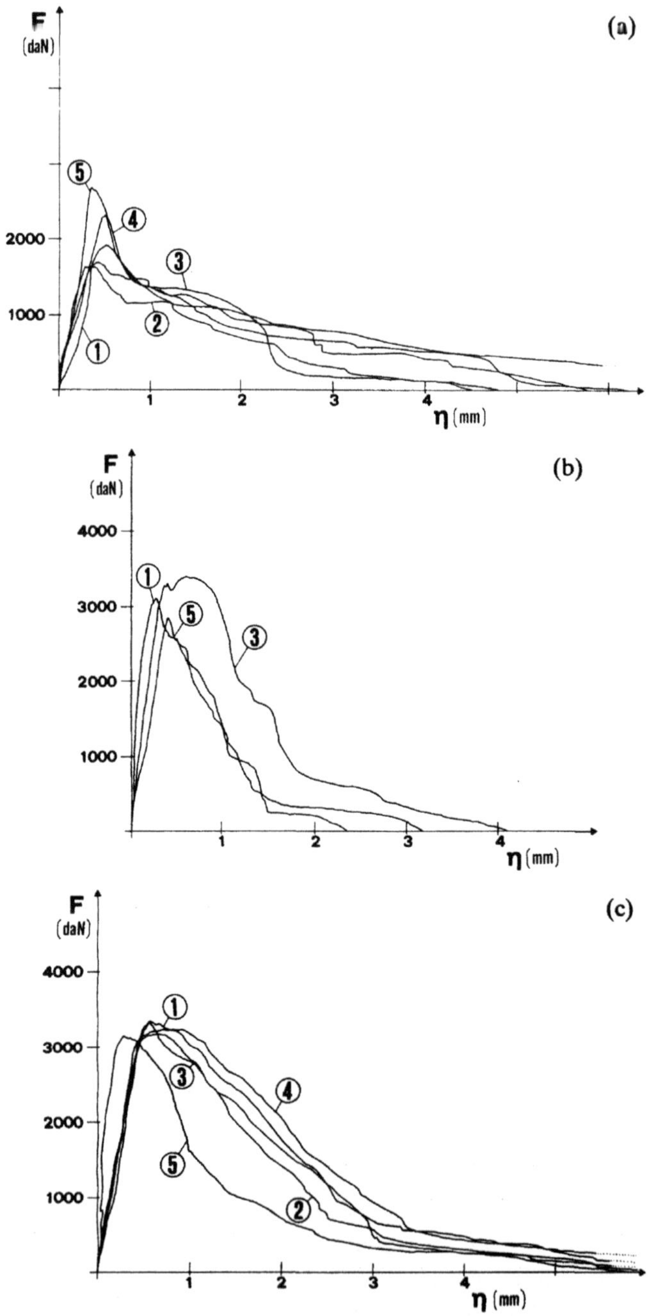

Fig. 3 Force versus displacement diagrams for six (a–f) embedment depths: concrete slabs.

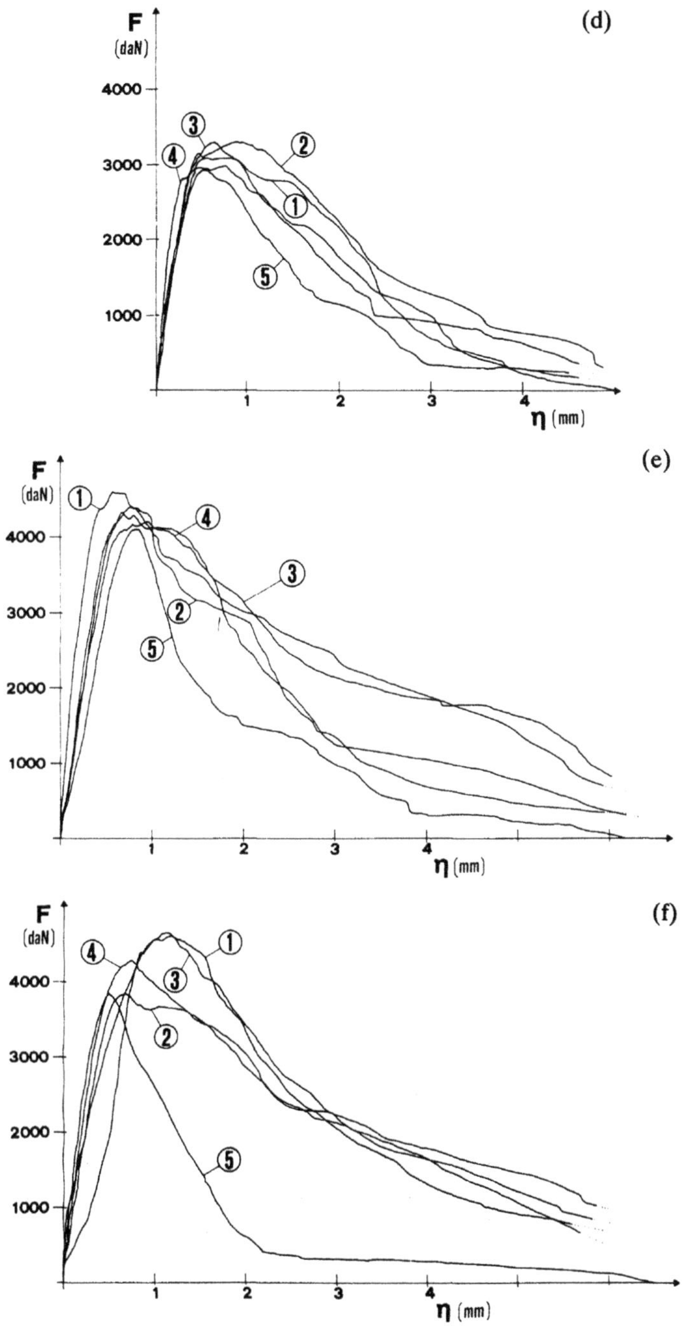

Fig. 3—(*continued*)

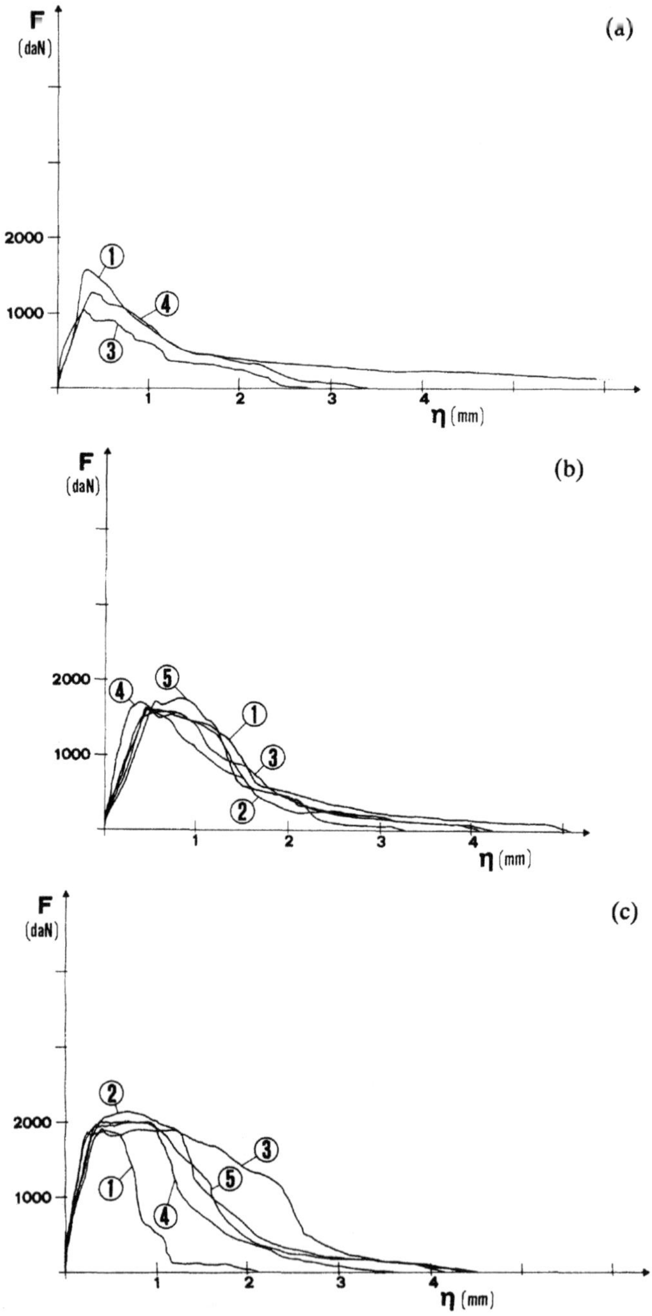

Fig. 4 Force versus displacement diagrams for six (a–f) embedment depths: mortar slabs.

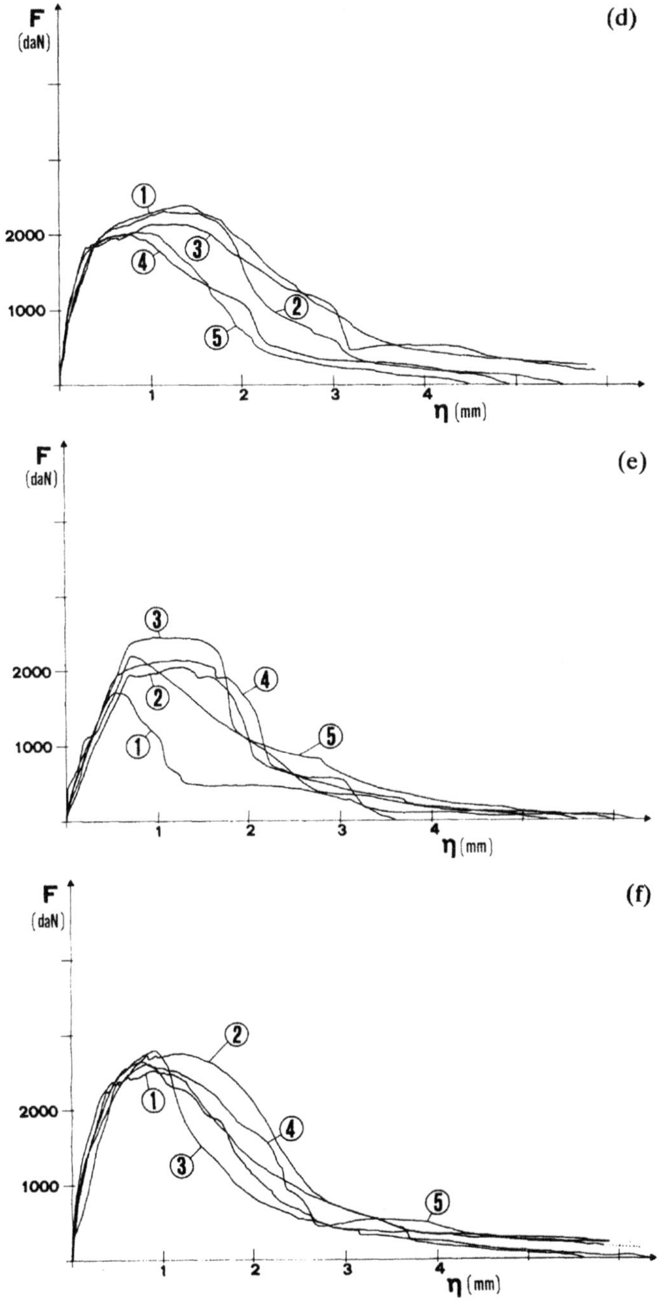

Fig. 4—(*continued*)

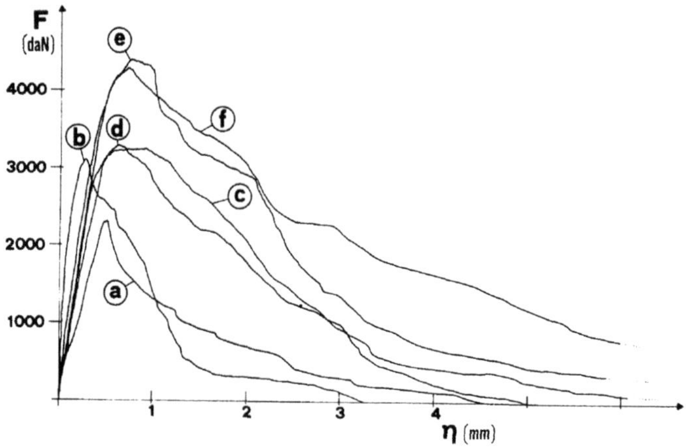

Fig. 5 Mean values of force versus displacement diagrams for six (a–f) embedment depths: concrete slabs.

increases with increasing bolt embedment depth. The difference in the results obtained for the two materials is quite apparent.

The failure surfaces of concrete (top) and mortar (bottom) for type (a) and (b) specimens can be observed in Fig. 8, see also Table 5.

In the following photographs (Figs 9 and 10), embedment depth is progressively increased. It can be seen that the initial cracking angle,

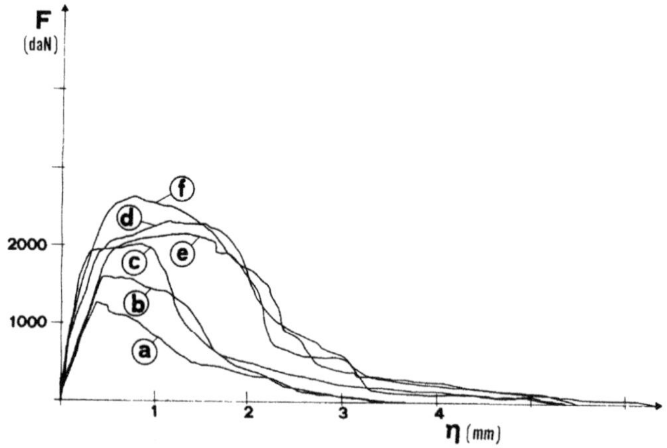

Fig. 6 Mean values of force versus displacement diagrams for six (a–f) embedment depths: mortar slabs.

TABLE 3 Experimental results—concrete slabs

Bolt	*V (mm)*	F_{max} *(daN)*	F_{av} *(daN)*	η_{el} *(mm)*	η_{av} *(mm)*	W_f *(daN ·mm)*	$W_{f,av}$ *(daN ·mm)*
a_1		1 700		0·42		4 058	
a_2		1 625		0·34		4 383	
a_3	28	1 937	2 049	0·56	0·43	3 384	3 989
a_4		2 312		0·52		3 123	
a_5		2 675		0·34		4 998	
b_1		3 075		0·26		2 964	
b_2		—		—		—	
b_3	35	3 375	3 100	0·60	0·41	4 876	3 385
b_4		—		—		—	
b_5		2 850		0·38		2 316	
c_1		3 325		0·58		7 365	
c_2		3 300		0·55		5 910	
c_3	40	3 175	3 230	0·64	0·57	6 317	6 377
c_4		3 225		0·80		7 696	
c_5		3 125		0·27		4 600	
d_1		3 325		0·55		7 154	
d_2		3 300		0·86		8 835	
d_3	44	3 275	3 150	0·62	0·62	6 603	6 948
d_4		2 950		0·54		7 025	
d_5		2 900		0·52		5 127	
e_1		4 575		0·58		15 008	
e_2		4 375		0·74		10 343	
e_3	52	4 175	4 300	0·95	0·71	14 673	11 697
e_4		4 300		0·48		11 322	
e_5		4 075		0·80		7 140	
f_1		4 625		1·14		13 071	
f_2		3 800		0·62		12 647	
f_3	55	4 575	4 215	1·16	0·82	15 116	12 051
f_4		4 250		0·72		14 043	
f_5		3 825		0·48		5 383	

as measured with respect to the horizontal, increases with increasing embedment depth.

Cracking paths, as determined on two orthogonal diameters, are plotted in non-dimensional form in Figs 11 and 12 (as the mean over three identical specimens). For concrete (Fig. 11), it can be seen that the initial cracking angle and the mean angle increase with increasing embedment depth (from small (a) to large (f)). In the case of mortar (Fig. 12), the initial angle increases, but the mean angle does not.

TABLE 4 Experimental results—mortar slabs

Bolt	V *(mm)*	F_{max} *(daN)*	F_{av} *(daN)*	η_{el} *(mm)*	η_{av} *(mm)*	W_f *(daN ·mm)*	$W_{f,av}$ *(daN ·mm)*
a_1		1 562		0·32		2 627	
—		—		—		—	
a_3	28	1 025	1 284	0·30	0·31	1 221	1 848
a_4		1 265		0·32		1 697	
—		—		—		—	
b_1		1 575		0·44		2 915	
b_2		1 525		0·64		2 350	
b_3	35	1 518	1 603	0·76	0·61	2 374	2 463
b_4		1 662		0·40		2 165	
b_5		1 737		0·82		2 511	
c_1		1 875		0·44		1 451	
c_2		2 125		0·60		3 531	
c_3	40	1 900	1 985	0·40	0·57	4 308	3 073
c_4		2 000		1·00		2 880	
c_5		2 025		0·40		3 196	
d_1		2 375		1·32		6 719	
d_2		2 300		1·16		5 298	
d_3	44	2 125	2 165	1·34	1·04	6 304	5 194
d_4		2 000		0·58		4 002	
d_5		2 025		0·78		3 647	
—		—		—		—	
e_2		2 025		1·24		4 254	
e_3	52	2 437	2 203	1·34	1·15	4 637	4 538
e_4		2 125		1·28		4 524	
e_5		2 225		0·74		4 740	
f_1		2 500		0·90		5 608	
f_2		2 725		1·20		6 854	
f_3	55	2 775	2 645	0·90	0·91	5 211	5 836
f_4		2 625		0·74		5 718	
f_5		2 600		0·80		5 793	

4. THE COHESIVE CRACK MODEL

The test results were compared to those obtained from numerical simulations and based on the cohesive crack model. The behaviour of concrete at collapse, in fact, is characterized by the localization of a non-linear zone within a very narrow band, whilst the rest of the material retains its linear behaviour. The cohesive model describes this band as a fictitious crack (or process zone) where the material, albeit

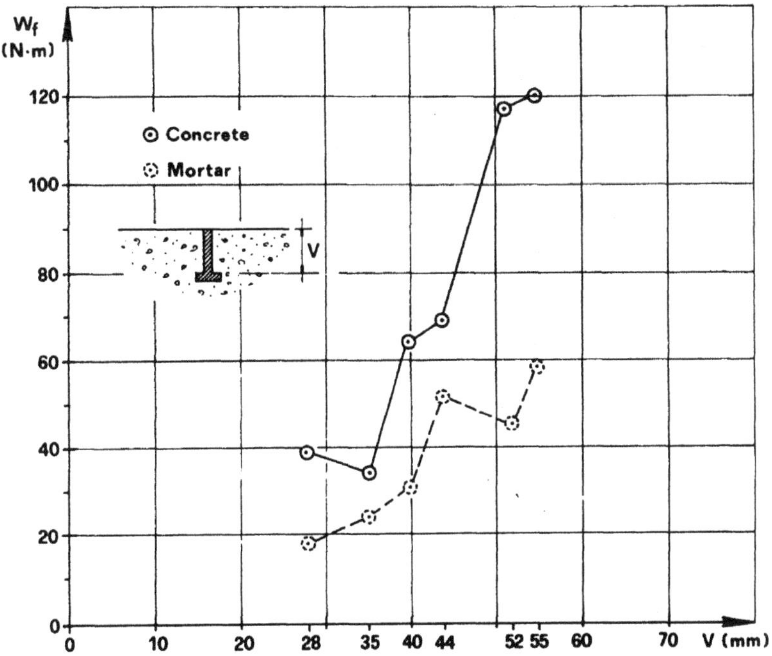

Fig. 7 Total failure energy as a function of embedment depth.

damaged, can still transfer stresses. The stresses involved are described as decreasing (strain-softening) functions of the relative displacements of the opposing crack surfaces (displacement discontinuity). The process zone originates perpendicularly to the principal tensile stress when the latter reaches its ultimate value, σ_u. The point at which this condition occurs is called *fictitious crack tip*: it represents the boundary between the integral and the damaged material. The point at which the displacement discontinuity reaches the critical value w_c, beyond which no stresses are transferred, is called *real crack tip*: this point divides the stress-free crack from the process zone (Fig. 13). According to the foregoing assumptions, there are no stress singularities in this model.

The cohesive model was first proposed by Barenblatt [12] and, based on independent studies, by Dugdale [13] as a tool for the analysis of collinear propagation (Mode I) problems. Later on, Hillerborg *et al.* [14] applied it to concrete for the interpretation of

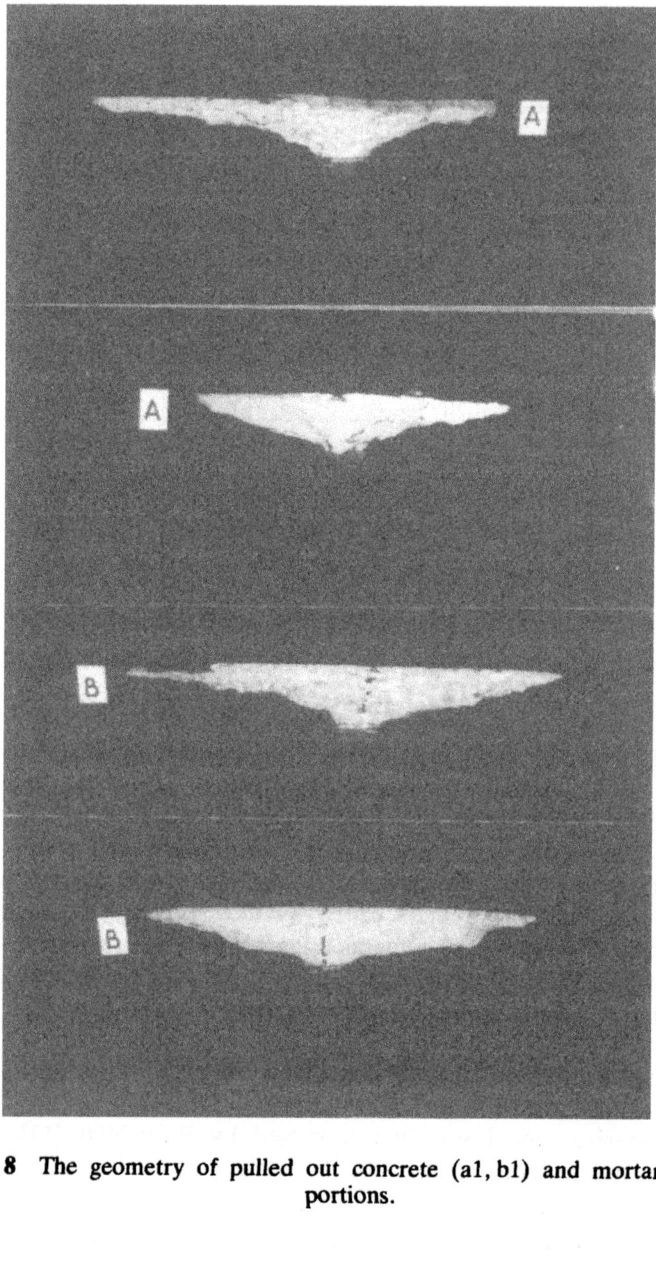

(a1)

(a2)

(b1)

(b2)

Fig. 8 The geometry of pulled out concrete (a1, b1) and mortar (a2, b2) portions.

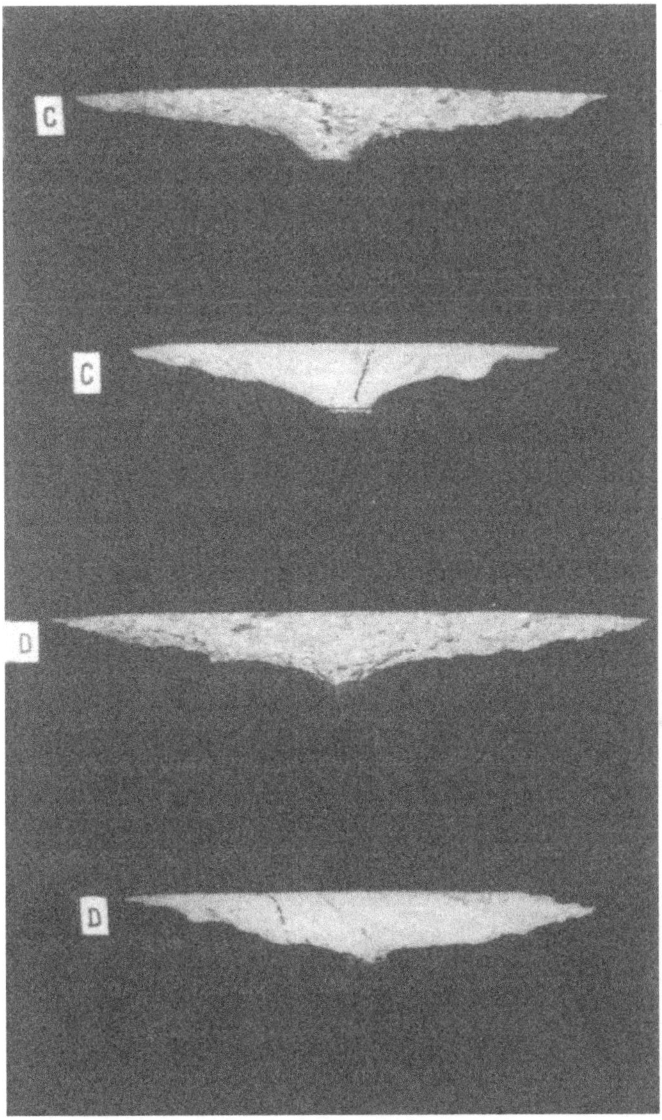

(c1)

(c2)

(d1)

(d2)

Fig. 9 The geometry of pulled out concrete (c1, d1) and mortar (c2, d2) portions.

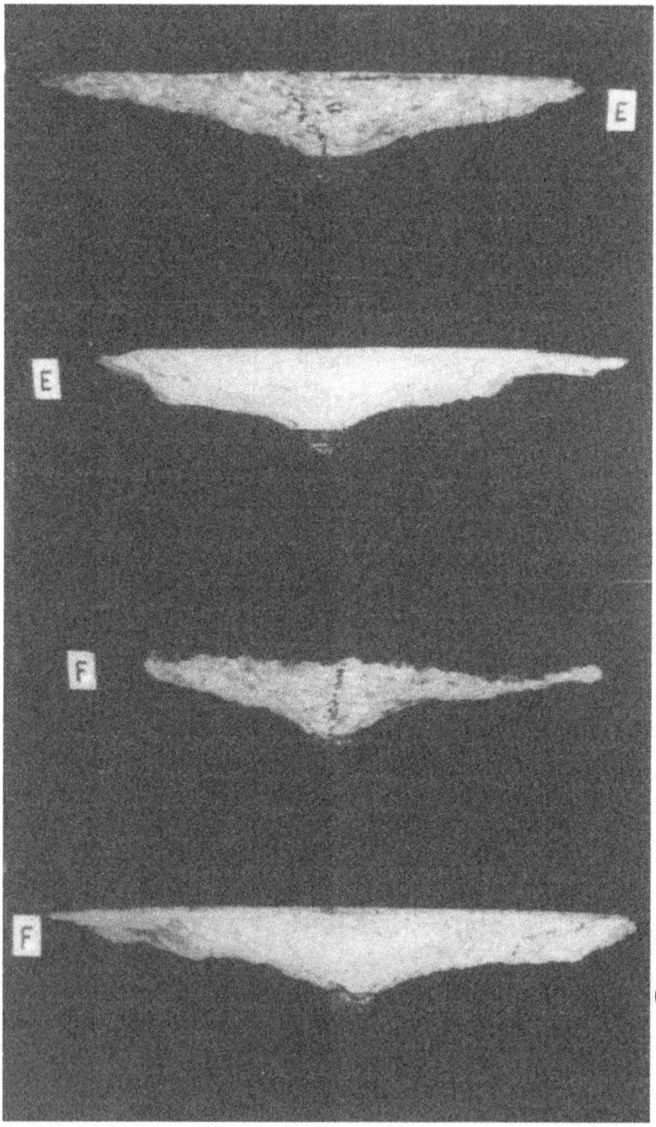

(e1)

(e2)

(f1)

(f2)

Fig. 10 The geometry of pulled out concrete (e1, f1) and mortar (e2, f2) portions.

TABLE 5 Cone-shaped failure surfaces

Bolt	*Concrete*		*Mortar*	
	D_{max} $(m \cdot 10^{-2})$	A_{lig} $(m^2 \cdot 10^{-4})$	D_{max} $(m \cdot 10^{-2})$	A_{lig} $(m^2 \cdot 10^{-4})$
a	31·3	776	28·0	622
b	38·0	1 148	36·5	1 060
c	43·0	1 469	45·0	1 607
d	48·0	1 835	47·0	1 760
e	49·5	1 958	55·5	2 452
f	55·5	2 463	57·0	2 595

three-point bending tests. In the same field, Carpinteri [15], [16] has explained the phenomenon of ductile–brittle transition, by defining the brittleness number, s_E, which depends on the properties of the material and the structural dimensions (size-effect). Recently (see Bocca, Carpinteri and Valente [17]–[20]), the cohesive model has been employed to cope with problems in which the crack path is unknown *a priori* (Mixed Mode). These applications, based on the finite elements method, concern the simulation of the four-point shear test and the pull-out test (axial-symmetrical case). During the latter test, it has been assumed that the front of the fictitious crack retains its

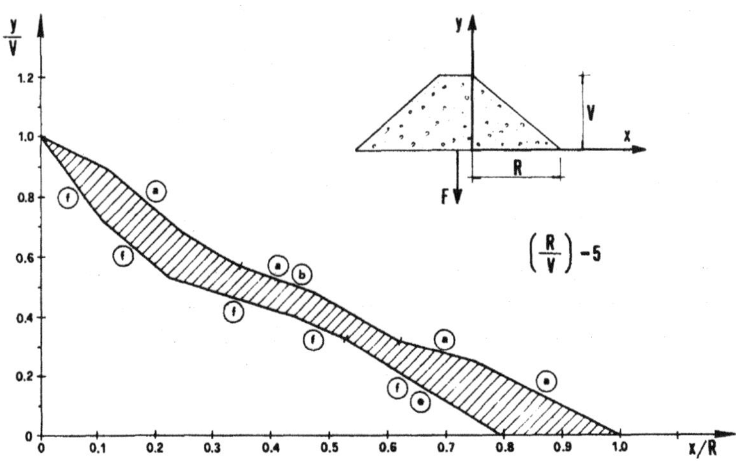

Fig. 11 Cone-shaped fracture surfaces for six (a–f) embedment depths: concrete slabs.

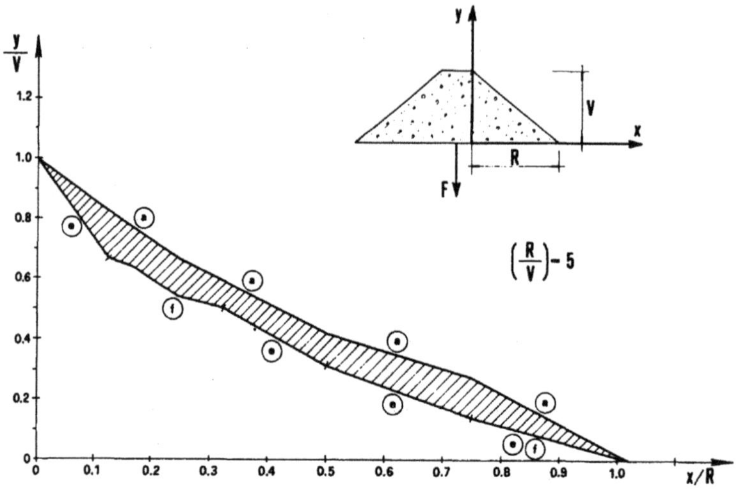

Fig. 12 Cone-shaped fracture surfaces for six (a–f) embedment depths: mortar slabs.

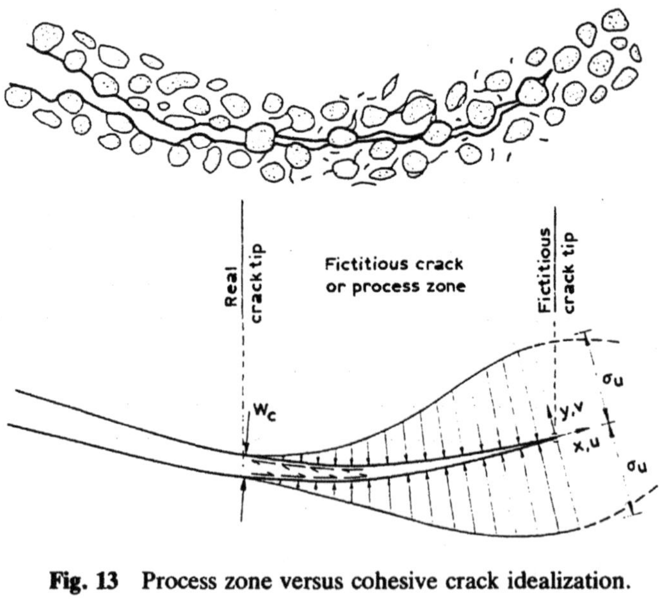

Fig. 13 Process zone versus cohesive crack idealization.

circular form and is not ovalized as it gets larger. This assumption is substantiated by the experimental evidence that the portions of material (concrete or mortar) pulled out together with the metal bolt are approximately axial-symmetrical.

5. THE CONSTITUTIVE LAW OF THE FICTITIOUS CRACK

The cohesive stresses (σ_c, τ_c) transferred through the process zone (fictitious crack) are decreasing functions of the displacement discontinuities (w_n, w_t). As already observed in earlier testing programmes (Bocca *et al.* [10]), the amount of fracture energy dissipated during pull-out tests is greater than that which would be strictly necessary to cause the spread of a crack in Mode I conditions. Hence, the crack has been characterized by a constitutive law designed to take τ_c into account, i.e. friction dissipation.

The correlation between cohesive stresses and the crack opening path was observed experimentally by Hillerborg [21] and Hassanzadeh [22] through tests on a concrete having the same characteristics as the concrete used in this investigation (maximum aggregate diameter 8 mm, water–cement ratio 0·5, 28 days of ageing in a moist environment). In their works, after comparing different displacement paths, some linear and some parabolic, Hillerborg and Hassanzadeh came to the conclusion that the most likely path is parabolic, governed, that is, by an equation of the form (Fig. 3)

$$w_n = \beta w_t^{1/2} \tag{1}$$

where w_n is a monotonic increasing function during the irreversible process of crack growth. In this connection, if the crack path is sufficiently regular, the authors come to the conclusion that its influence can be neglected. From Fig. 14 it can be seen that initially it is: $dw_t/dw_n = 0$. This is in agreement with the cohesive model assumption under which the tip of the fictitious crack propagates orthogonally to the principal tensile stress.

Based on the above considerations, Hillerborg's [21] test results have been approximated with piecewise linear functions: the diagrams obtained in this manner (Figs. 15 and 16) have been taken as the constitutive laws of the fictitious crack.

Stresses σ_c and τ_c relating to values of β not explicitly shown in Figs. 15 and 16 are determined by linear interpolation. It is important to

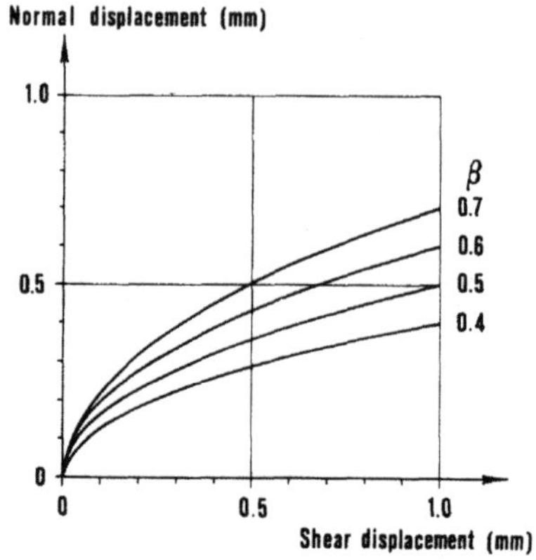

Fig. 14 Parabolic crack displacement paths.

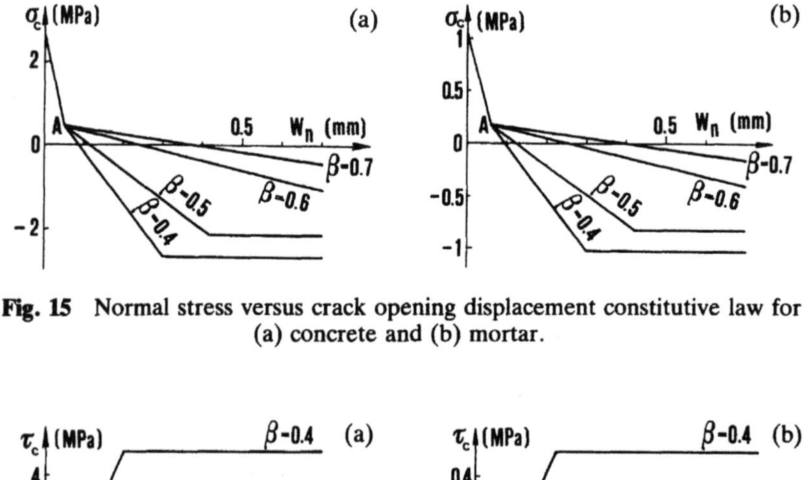

Fig. 15 Normal stress versus crack opening displacement constitutive law for (a) concrete and (b) mortar.

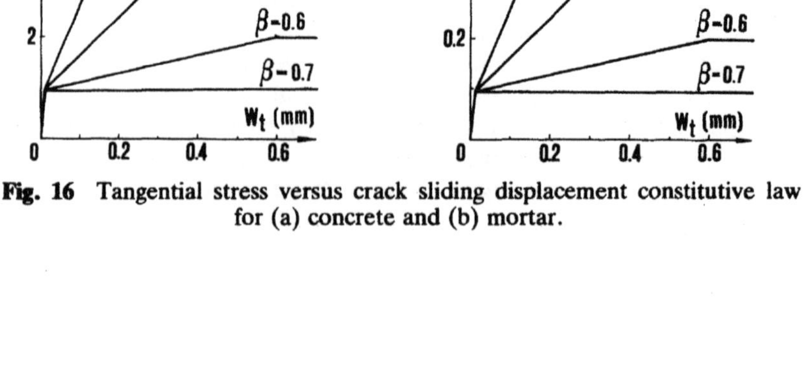

Fig. 16 Tangential stress versus crack sliding displacement constitutive law for (a) concrete and (b) mortar.

note that Hillerborg's [21] experimental results were obtained by means of a testing set-up designed to impose strict compliance with the load path denoted by eqn (1). In a numerical simulation, no such conditions can be imposed *a priori* and all one can do is to check the crack opening path *a posteriori* to make sure it is not too far a deviation from a parabolic curve. In other words, during a numerical simulation, dw_n/dw_t may happen to be slightly different from $0{\cdot}5\beta w_t^{-1/2}$ as is obtained from eqn (1) by derivation.

Hence, the incremental constitutive law can be written in this form:

$$\sigma_c = \sigma_c(\beta, w_n) \qquad \tau_c = \tau_c(\beta, w_t) \qquad \beta = w_n w_t^{-1/2} \tag{2}$$

$$\begin{Bmatrix} d\sigma_c \\ d\tau_c \end{Bmatrix} = \begin{bmatrix} \dfrac{\partial \sigma_c}{\partial w_n} & \dfrac{\partial \sigma_c}{\partial w_t} \\ \dfrac{\partial \tau_c}{\partial w_n} & \dfrac{\partial \tau_c}{\partial w_t} \end{bmatrix} \begin{Bmatrix} dw_n \\ dw_t \end{Bmatrix} \tag{3}$$

Equations (2) are of the piecewise linear type, as assumed in Figs 15 and 16. Hence, they display some cuspidal points, where the prime derivatives depend on the sign of speed. At these points, the uniqueness of the prime derivatives requires

$$\dot{w}_n > 0 \tag{4a}$$

$$\dot{w}_t > 0 \tag{4b}$$

where the dot denotes derivation with respect to time. Expressions (4) are imposed *a priori* in the works of Hillerborg [21] and Hassanzadeh [22], while they are verified *a posteriori* in the numerical simulations illustrated in this chapter. On the other hand, instances of violation of eqn (4b) have been observed at the initial stage of crack growth, i.e. for small values of w_n and w_t. However, this phenomenon is deemed to be negligible, as, during this initial stage, σ_c does not depend on β, i.e. it does not depend on w_t, whilst τ_c is small.

Now, assuming conditions (4) to have been verified, the constitutive law can be expressed in holonomic terms

$$\mathbf{P}_c^{(i+1)} = \mathbf{P}_0^{(i)} + \mathbf{L}^{(i)}\mathbf{w}^{(i+1)} \tag{5}$$

where $\mathbf{L}$ is the 2×2 matrix contained in eqn (3) and

$$\mathbf{P}_c = [\sigma_c, \tau_c]^T \qquad \mathbf{w} = [w_n, w_t]^T \qquad \mathbf{P}_0^{(i)} = \mathbf{P}_c^{(i)} - \mathbf{L}^{(i)}\mathbf{w}^{(i)}$$

Since this is a non-linear evolutionary phenomenon, the index (i) denotes the step to which the quantities refer.

6. SOLUTION FOR A SINGLE CRACK GROWTH STEP

By subdividing the domain into a finite number of elements, arranged so that the (real and fictitious) crack is at the interface between two elements, considering the constitutive law (5) and making use of the Principle of Virtual Work (see Bocca, Carpinteri and Valente [17]–[20]), we get the following system of linear equations

$$(\mathbf{K}-\mathbf{C}^{(i)})\mathbf{u}^{(i+1)}=\mathbf{F}_0^{(i)}+\lambda\mathbf{F}_1 \tag{6}$$

where $\mathbf{K}$ = stiffness matrix $(n\times n)$

$\mathbf{C}^{(i)}$ = non-symmetrical strain-softening matrix, assembled from $\mathbf{L}$

$\mathbf{u}^{(i+1)}$ = vector of the n unknown nodal displacements

$\mathbf{F}_0^{(i)}$ = load vector, assembled from $\mathbf{P}_0^{(i)}$

$\mathbf{F}_1$ = external load vector

λ = external load multiplier

If the matrix $(\mathbf{K}-\mathbf{C}^{(i)})$ is non-singular, it is useful to solve the linear system of eqns (6) separately for the two right hand side vectors:

$$(\mathbf{K}-\mathbf{C}^{(i)})\mathbf{u}_0^{(i+1)}=\mathbf{F}_0^{(i)} \qquad (\mathbf{K}-\mathbf{C}^{(i)})\mathbf{u}_1^{(i+1)}=\mathbf{F}_1 \tag{7}$$

Denoting by $\boldsymbol{\sigma}_0^{(i+1)}$ and $\boldsymbol{\sigma}_1^{(i+1)}$ the stresses at the tip of the fictitious crack (centroid of the element dashed in Fig. 17), relating to $\mathbf{u}_0^{(i+1)}$ and $\mathbf{u}_1^{(i+1)}$, respectively, we can write (axial-symmetrical problem):

$$\boldsymbol{\sigma}=[\sigma_x,\sigma_y,\sigma_z,\tau]=\boldsymbol{\sigma}_0^{(i+1)}+\lambda\boldsymbol{\sigma}_1^{(i+1)} \tag{8}$$

The load multiplier λ may therefore be obtained by imposing that the main tensile stress, at the tip of the fictitious crack, should reach

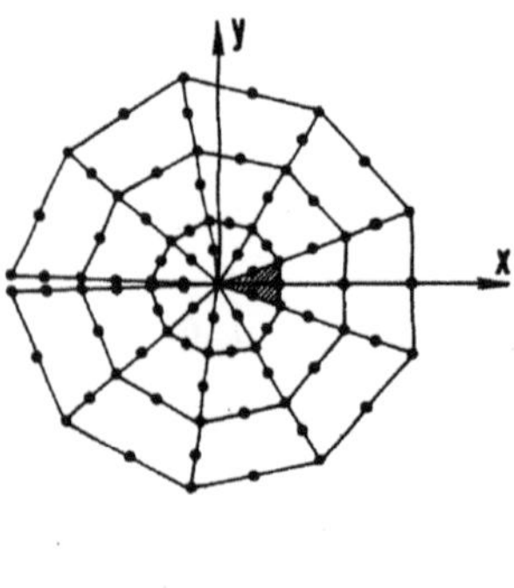

Fig. 17 The finite element rosette at the fictitious crack tip.

the concrete tensile strength, σ_u. From Mohr's circle we therefore get:

$$\sigma_u^3 - (\sigma_x + \sigma_y + \sigma_z)\sigma_u^2 + (\sigma_x\sigma_y + \sigma_x\sigma_z + \sigma_y\sigma_z - \tau^2)\sigma_u + \sigma_x\sigma_y\sigma_z - \sigma_z\tau^2 = 0 \quad (9)$$

By substituting eqn (8) into eqn (9) we obtain a cubic equation in λ and we can determine its value. The vector of nodal displacements then turns out to be

$$\mathbf{u}^{(i+1)} = \mathbf{u}_0^{(i+1)} + \lambda \mathbf{u}_1^{(i+1)} \quad (10)$$

Since the coefficients and known terms of the linear system of eqns (6) depend on $\mathbf{u}$, it proves necessary to assemble matrix $\mathbf{C}^{(i+1)}$ and vector $\mathbf{F}_0^{(i+1)}$, to increment index (i) by a unit and to solve the system of eqns (6) and (9) again. This sequence of equilibrium iterations comes to an end when the Euclidean norm of the vector $(\mathbf{u}^{(i+1)} - \mathbf{u}^{(i)})$ drops below a predetermined tolerance value. On the other hand, before the solution found in this manner can be deemed acceptable, it is necessary to verify that conditions (4) have been met and that, at all points of the domain, the main tensile stress does not exceed σ_u.

If the solution is found to be acceptable, by substituting λ into eqn (8) the state of stress at the fictitious crack tip is determined. If we denote the corresponding principal stresses by $\sigma_3 \leq \sigma_2 \leq \sigma_1$, and take the tensile stresses to be positive, either of the following situations may occur

$$\sigma_3 \leq \sigma_2 < \sigma_1 = \sigma_u \quad (11a)$$

$$\sigma_3 \leq \sigma_2 = \sigma_1 = \sigma_u \quad (11b)$$

In the case of eqn (11a), the crack is assumed to propagate orthogonally to stress σ_1. In the case of eqn (11b), one of Mohr's three circles has degenerated into a point and hence the above mentioned propagation criterion cannot be applied.

7. CONTROLLING THE NUMERICAL PROCEDURE THROUGH THE LENGTH OF THE FICTITIOUS CRACK

Based on the considerations made in the previous section, it can be seen that the external load, represented by the multiplier λ, is not the control variable, but rather an unknown which is determined, at each crack growth step, by substituting eqn (8) into eqn (9).

Similarly, the displacements $\mathbf{u}$ are not control variables, as they have to be determined by substituting the current value of λ into eqn (10).

According to a method called 'Fictitious Crack Length Control Scheme' (see Bocca, Carpinteri and Valente [17]–[20]), the length of the fictitious crack serves as the control variable, since it is a surely monotonic increasing function during the irreversible cracking process. This does not apply to the load since, once it has reached its maximum value, it decreases. Likewise, it may happen that the load-point displacement decreases with increasing crack opening. The latter phenomenon is known as catastrophic collapse or snap back.

In the initial stage of the cracking process the matrix $(\mathbf{K} - \mathbf{C})$ is positive definite and hence all its eigenvalues are positive. This property derives from the fact that the matrix $\mathbf{K}$, representing elastic stiffness, is positive definite, whereas matrix $\mathbf{C}$ initially is negligible. Under these conditions, the system of linear eqns (7) has one and only one solution. Since the cubic equation in λ which is obtained by substituting eqn (8) into eqn (9) offers only one physically acceptable solution, it can be concluded that system (6) admits one and only one solution.

It is important to note that this uniqueness of the solution holds under the assumptions illustrated in the foregoing sections. If the assumption of the crack propagating orthogonally to the main tensile stress (eqn (9)), for instance, is replaced with the assumption of the crack propagating according to Mohr–Coulomb's failure criterion, we obtain a different crack path and a different maximum load. Numerical tests, however, have yielded negligible differences. Similarly, if we carry out a three-dimensional analysis and assume that the front of the fictitious crack is ovalized, we shall probably obtain slightly different solutions with respect to the ones illustrated in Fig. 18.

With increasing fictitious crack dimensions, the smallest eigenvalue of $(\mathbf{K} - \mathbf{C})$ decreases, until it vanishes. Before this last condition is verified, in all the geometrical cases subjected to numerical analysis, condition (11b) has been seen to occur. This condition corresponds to the degeneration of one of Mohr's three circles and hence entails the impossibility of determining the direction of the next crack growth step.

Figure 19 provides a qualitative illustration of the phenomenon from the mechanical point of view: a tensile stress distribution, which helps to balance the external load F, acts on the surface generated by the revolution of segment AB; the stress distribution acting on the surface

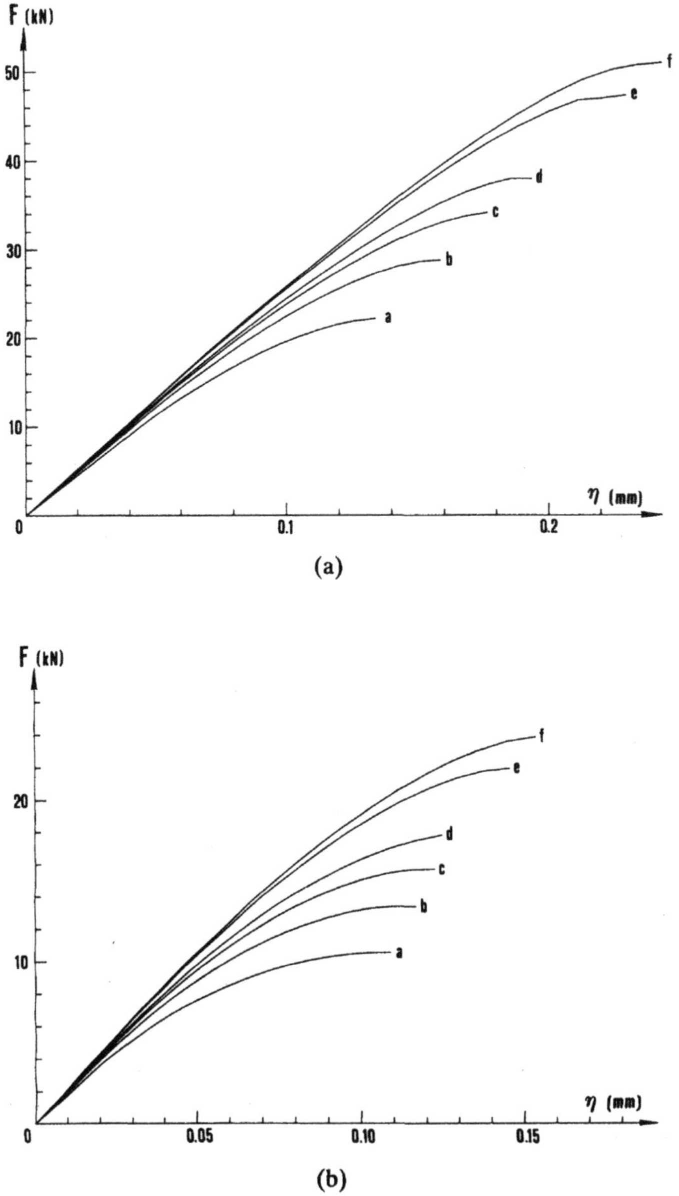

Fig. 18 Load versus displacement diagrams obtained numerically for different embedment depths in the case of (a) concrete and (b) mortar.

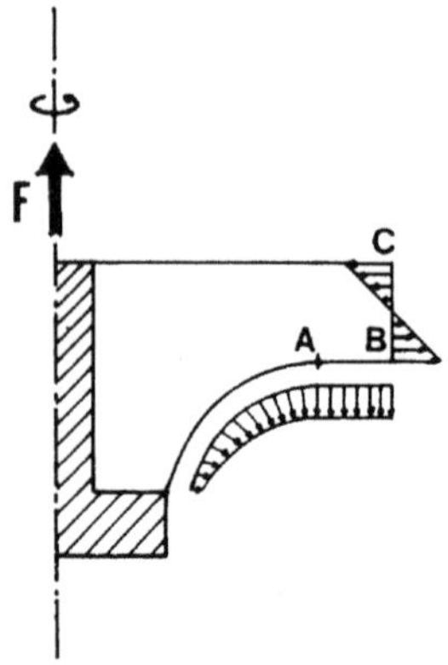

Fig. 19 Stresses acting on the portion of material pulled out together with the steel bolt.

generated by the revolution of segment BC is of a flexural type. If, on the two surfaces defined in this manner, τ vanishes and the tensile stresses become equal, then one of Mohr's circles degenerates into a point. Under these conditions, then, the same tensile stress acts on all the elementary areas passing through B and orthogonal to the plane shown in Fig. 19, and hence the direction of growth of the crack turns out to be indeterminate.

In order to go on with the numerical simulation, from this point on, the crack path has been imposed *a priori,* by adopting a curve based on the experimental results.

Proceeding with the analysis in this manner, the smallest eigenvalue of the $(\mathbf{K}-\mathbf{C})$ matrix is found to decrease monotonically until it vanishes. As this gives rise to the danger of the uniqueness of the solution being lost, all numerical simulations were halted at this stage.

8. RESULTS OF THE NUMERICAL SIMULATION

From the numerical analysis it is possible to draw a first conclusion about the bolt: maximum load is virtually unaffected by head size, provided that the latter is sufficient to prevent local concrete crushing phenomena.

The analysis developed so far refers to both the materials being considered, i.e. concrete and mortar. As illustrated in Table 2, concrete is characterized by a greater tensile strength than mortar. Similarly, from Fig. 16, it can be seen that concrete displays greater shear strength, thanks to aggregate interlocking. It can also be noted, moreover, that the curves shown in Fig. 16(b) can be derived from those in Fig. 16(a) by multiplication by a constant factor. As for the behaviour in normal stress conditions, the curves in Fig. 15(b) can be obtained from those in Fig. 15(a) through multiplication by a constant factor and by moving the cuspidal point, denoted by letter A in Fig. 15, slightly to the right.

The diagrams shown in Fig. 18, illustrate the pulling force F, calculated numerically, as a function of the displacement of its application point η. These curves, in good agreement with the experimental ones (Figs 3, 4, 5 and 6), are characterized by a monotonic decrease in tangential stiffness until the maximum load is reached. From an examination of the curves shown in Fig. 18, it can be concluded that the numerical analysis has been carried on far enough to determine, with good approximation, the maximum load. Figure 20 gives an example of the meshes used in numerical simulations for Case f, relating to concrete, and shows the numerically predicted crack path which is in good agreement with the path determined experimentally (Fig. 10). The corresponding displacement fields, enlarged 100 times, are depicted in Fig. 21.

9. COMPARISON BETWEEN EXPERIMENTAL AND THEORETICAL RESULTS

Figure 22 shows the pull-out force versus embedment depth curves in the case of concrete. The test results are marked out by triangles. Four curves have been plotted, according to as many different formulas: ACI 349, ACI as modified on the basis of Bazant's size effect law [23], Eligehausen–Sawade's formula [9], fictitious crack model. Figure 23 shows the corresponding curves for mortar.

In order to clarify the comparison between experimental and theoretical results, a non-dimensional diagram, in logarithmic form, has been prepared for the concrete slabs (Fig. 24). The pull-out force has been divided by the value predicted according to ACI 349 and plotted on the Y axis. Thus, the ACI curve becomes a horizontal straight line.

For small embedment depths, the ACI formula underestimates the bearing capacity of the bolt, and therefore all the experimental results are seen to lie above the reference line. On the contrary, for large

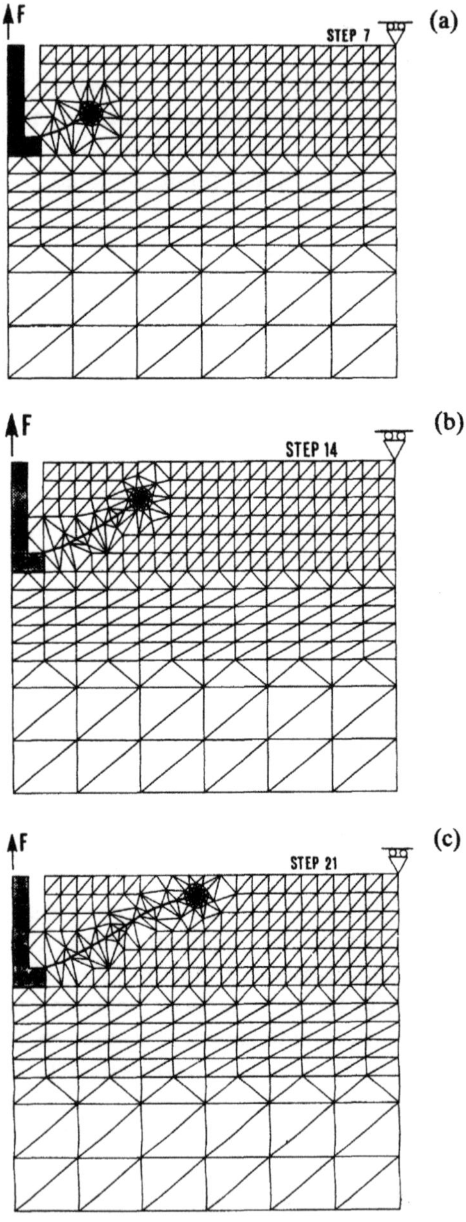

Fig. 20 Subsequent finite element meshes for Case f, related to concrete.

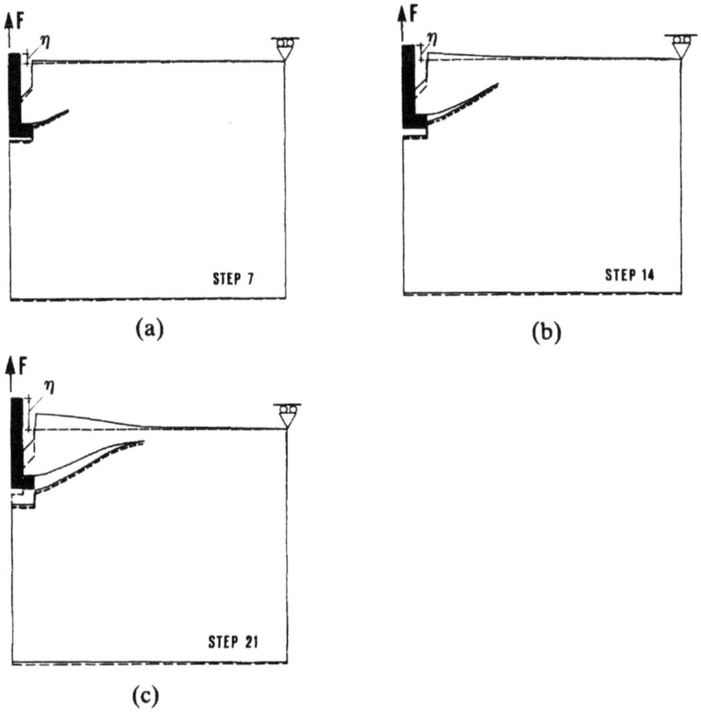

Fig. 21 Displacement fields, enlarged 100 times, for Case f, related to concrete.

embedment depths, the formula proposed by ACI 349 does not take into account size-effect, which is clearly reflected in the experimental results.

In order to get around this problem, Bazant proposed the so called 'size-effect law' [23] which was used to calculate the second curve, marked by the wording 'ACI modified'. This curve is a better approximation of the test results.

The third curve represents the failure loads predicted from the formula proposed by Eligehausen & Sawade [9], which was derived from Linear Elastic Fracture Mechanics.

The fourth and last curve shows the failure loads predicted according to the cohesive crack model.

Figure 25 shows the corresponding logarithmic diagram for mortar slabs. In this case too, the test results clearly reflect a strong

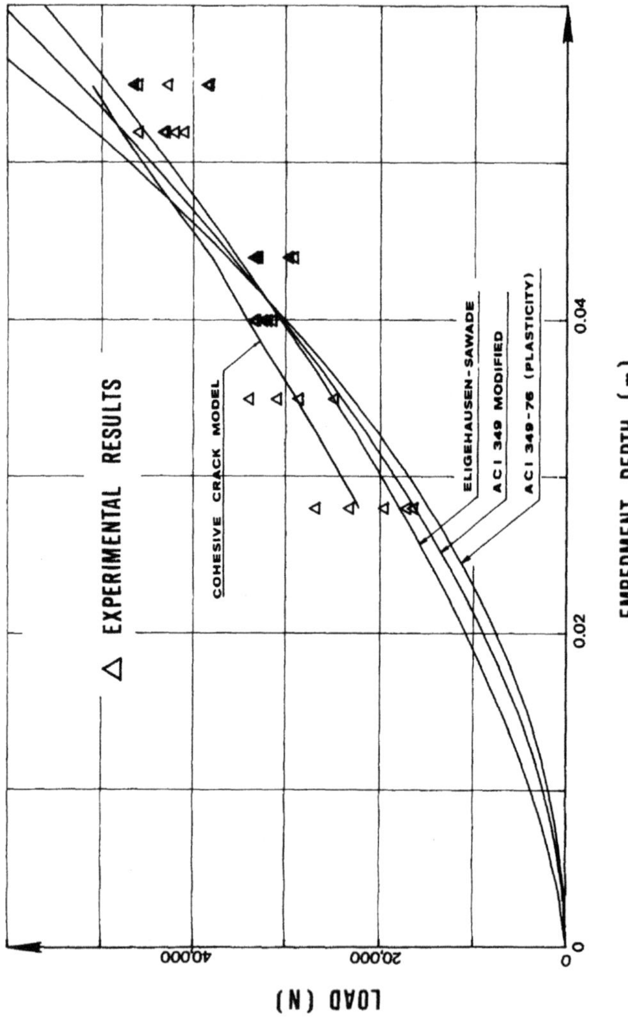

Fig. 22 Maximum load versus embedment depth: concrete slabs.

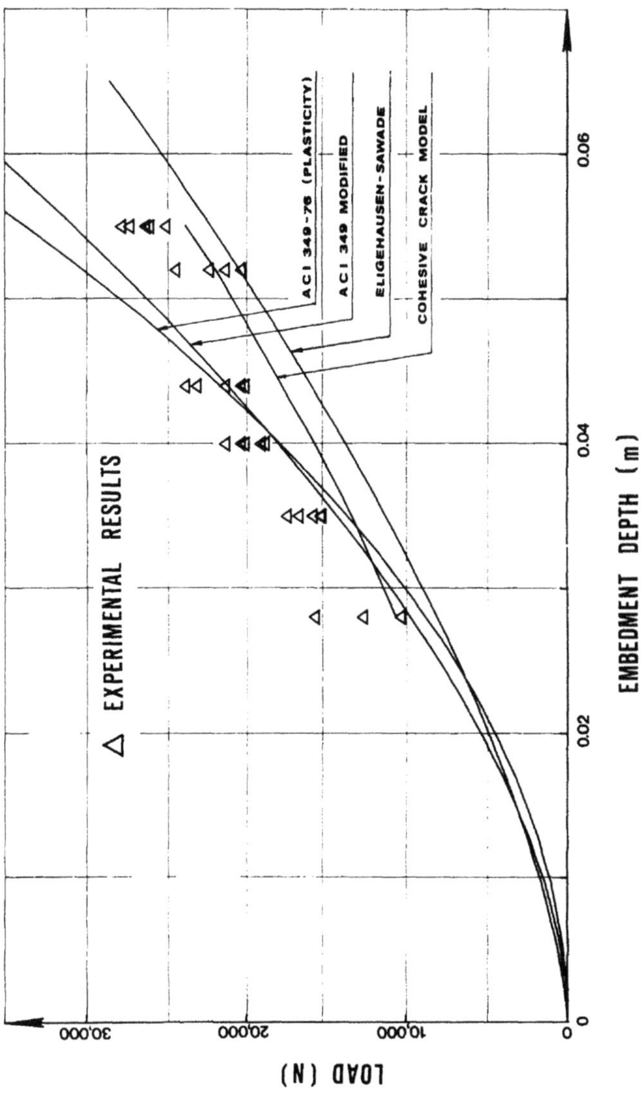

Fig. 23 Maximum load versus embedment depth: mortar slabs.

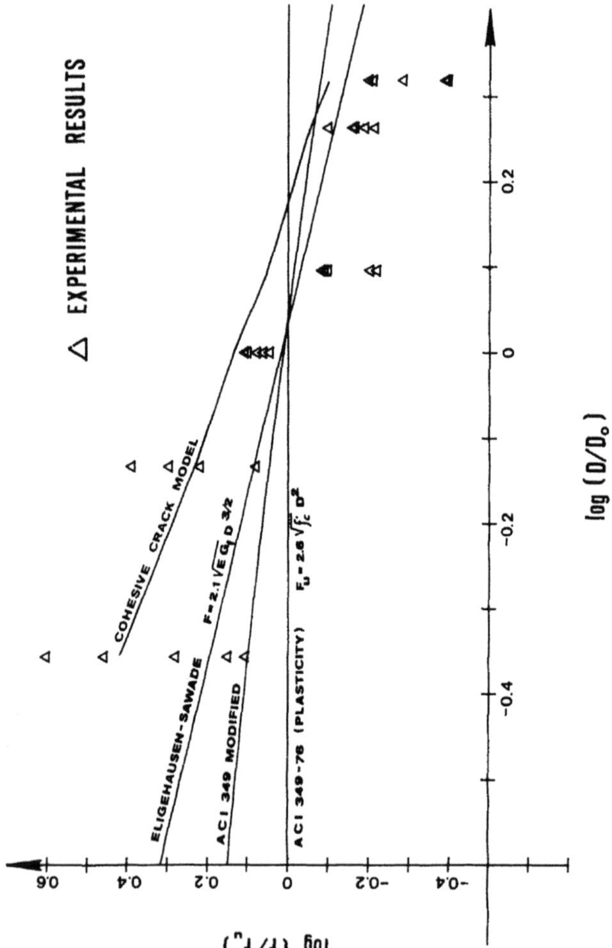

Fig. 24 Non-dimensional maximum load versus embedment depth diagram in bi-logarithmic scale: concrete slabs.

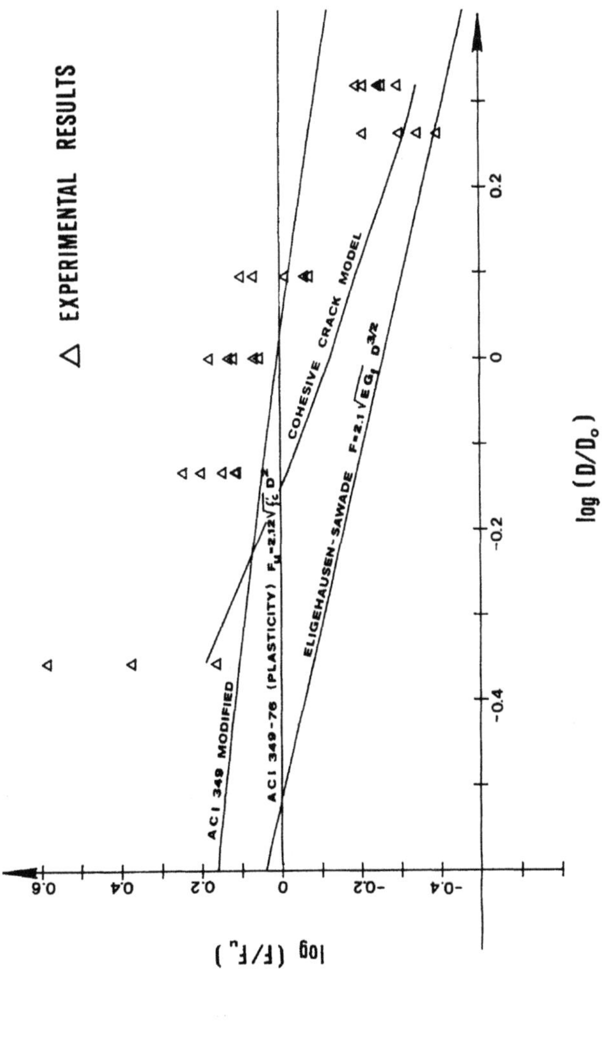

Fig. 25 Non-dimensional maximum load versus embedment depth diagram in bi-logarithmic scale: mortar slabs.

'size-effect' which is neglected by the ACI formula and is taken into account by the other three formulas.

10. CONCLUSIONS

This extensive testing programme on short anchor bolts has confirmed the theoretical predictions expressed by linear and non-linear fracture mechanics, with special reference to size-effect.

Compared to the traditional formulas, whether based on the elasticity or the plasticity theory, fracture mechanics provides a more realistic approach to the problem being considered, opening up new and wider scope for research into this as well as other problems of a structural nature.

ACKNOWLEDGEMENTS

This study was carried out with the financial support of the Ministry of University and Scientific Research (M.U.R.S.T.) and the National Research Council (C.N.R.).

REFERENCES

[1] ACI-349-76, Code Requirements for Nuclear Safety Related Concrete Structures. *ACI Journal*, **75** (1978) 329–47.

[2] Kierkegaard-Hansen, P., Lok-strength. *Nordisk Betong* (Stockholm, Sweden) **3** (1975) 19–28.

[3] Jensen, B. C. & Braestrup, H. W., Lok-Tests determine the compressive strength of concrete. *Nordisk Betong* (Stockholm, Sweden) **20**(2) (1976) 9–11.

[4] Ottosen, N. S., Non-linear finite element analysis of a pull-out test. *Journal of the Structural Division*, **107**(4) (1981) 591–603.

[5] Peier, R., Model for pull-out strength of anchors in concrete. *Journal of the Structural Division*, **109**(5) (1983) 1155–73.

[6] Miller, G. & Keer, L., Approximate analytical model of anchor pull-out tests. *Transactions of the ASME*, **49** (1982) 768–72.

[7] Krenchel, H. & Shah, S. P., Fracture analysis of pull-out testing. *Materials & Structures*, **18**(108) (1985) 439–46.

[8] Ballarini, R., Shah, S. & Keer, L., Failure characteristics of short anchor bolts embedded in a brittle material. In *Proceedings of the Royal Society (London)* **A404** (1986) 35–54.

[9] Eligehausen, R. & Sawade, G., Fracture mechanics of concrete structures. *RILEM Report*, ed. L. Elfgren. Chapman and Hall, 1989, pp. 281–99.

[10] Bocca, P., Carpinteri, A. & Valente, S., Evaluation of concrete fracture energy through a pull-out testing procedure. In *Proceedings of the International Conference on Recent Developments on the Fracture of Concrete and Rock*. ed. S. P. Shah, S. E. Swartz & B. Barr. Elsevier Applied Science, 1989, pp. 347–56.

[11] RILEM Recommendations, determination of the fracture energy of mortar and concrete by means of three point bend tests on notched beams. *Materials & Structures*, **18** (1985) 287–90.

[12] Barenblatt, G. I., The formation of equilibrium cracks during brittle fracture: general ideas and hypotheses. Axially-symmetric cracks. *Journal of Applied Mathematics and Mechanics*, **23** (1959) 622–36.

[13] Dugdale, D. S., Yielding of steel sheets containing slits. *Journal of the Mechanics and Physics of Solids*, **8** (1960) 100–104.

[14] Hillerborg, A., Modeer, M. & Petersson, P. E., Analysis of crack formation and crack growth in concrete by means of fracture mechanics and finite elements. *Cement and Concrete Research*, **6** (1976) 773–82.

[15] Carpinteri, A., Interpretation of the Griffith Instability as a bifurcation of the global equilibrium. In *Proceedings of NATO Advanced Research Workshop on Application of Fracture Mechanics to Cementitious Composites*, ed. S. P. Shah. Martinus Nijhoff, 1985, pp. 284–316.

[16] Carpinteri, A., Softening and snap-back instability in cohesive solids. *International Journal for Numerical Methods in Engineering*, **28** (1989) 1521–37.

[17] Carpinteri, A. & Valente, S., Size-scale transition from ductile to brittle failure: a dimensional analysis approach. In *Proceedings of the CNRS-NSF Workshop on Strain Localization and Size Effects Due to Cracking and Damage*, ed. J. Mazars & Z. P. Bazant. Elsevier Applied Science, 1988, pp. 477–90.

[18] Carpinteri, A., Valente, S. & Bocca, P., Mixed mode cohesive crack propagation. In *Proceedings of the Seventh International Conference on Fracture (ICF-7)*, ed. K. Salama, K. Ravi-Chandar, D. M. R. Taplin & P. Rama Rao. Pergamon Press, 1989, pp. 2243–57.

[19] Bocca, P., Carpinteri, A. & Valente, S., Size effects in the mixed mode crack propagation: softening and snap-back analysis. *Engineering Fracture Mechanics*, **35** (1990) 159–70.

[20] Bocca, P., Carpinteri, A. & Valente, S., Mixed mode fracture of concrete. *International Journal of Solids and Structures*, **27** (1991) 1139–53.

[21] Hillerborg, A., Mixed mode fracture in concrete. In *Proceedings of the Seventh International Conference on Fracture (ICF-7)*, ed. K. Salama, K. Ravi-Chandar, D. M. R. Taplin & P. Rama Rao. Pergamon Press, 1989, pp. 2259–68.

[22] Hassanzadeh, M., Determination of fracture zone properties in mixed Mode I and II. *Engineering Fracture Mechanics*, **35** (1990) 845–53.

[23] Bazant, Z. P., Size effect in blunt fracture: concrete, rock, metals. *Journal of Engineering Mechanics* (ASCE), **110** (1984) 518–35.

Chapter 10

Anchor Bolts Modelled with Fracture Mechanics

LENNART ELFGREN & ULF OHLSSON
Luleå University of Technology, Luleå, S-95187, Sweden

ABSTRACT

Size effects and design criteria for anchor bolts can be studied by non-linear fracture mechanics. In the last few years many such studies have been carried out. In this chapter theoretical results are analysed and compared to some experimental values.

1. INTRODUCTION

Anchor bolts or headed studs, see Fig. 1, embedded in concrete or grouted in concrete or rock are important structural details. They are used to transfer local loads into structures e.g. from cantilevers to beams and walls, from beams to columns and from columns to slabs and foundations. They are also used as roof bolts in tunnels and as tie backs in rocks.

Various design procedures have been developed over the years based on simplified assumptions and on empirical test results, see e.g. Klinger & Mendonca [1] and Rehm *et al.* [2]. Contributions to the development of fracture mechanics methods for anchor bolts have among others been given by Ottosen [3], Elfgren *et al.* [4], [5], Peier [6], de Borst [7], Ballarini *et al.* [8], Hellier *et al.* [9], Rots [10], Eligehausen & Sawade [11], Ohlsson [12], Eligehausen & Ozbolt [13], Cervenka *et al.* [14], Ballarini & Shah [15] and Ohlsson & Elfgren [16].

One general way of writing the *empirical capacity* F_{max} has been

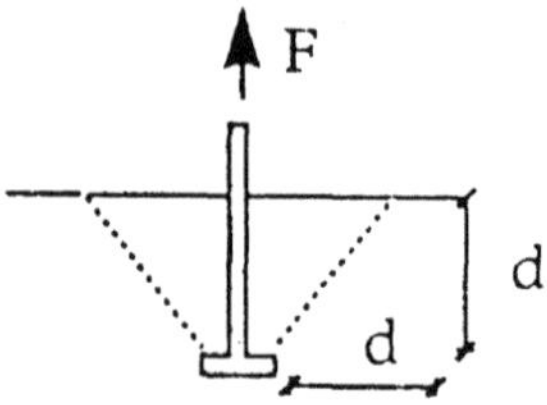

Fig. 1 Anchor bolt embedded in concrete.

proposed by Eligehausen & Sawade [11]:

$$F_{max} = a_1 f_c^{a_2} d^{a_3} \tag{1}$$

where a_1, a_2 and a_3 are empirical constants
f_c = concrete compressive strength (N/mm^2)
d = embedded depth (mm)

The factor a_1 is used to ensure dimensional correctness of the formulae and to calibrate measured loads to predicted values. In Eligehausen & Sawade [11], $a_1 = 15{\cdot}5$ is proposed.

The expression $f_c^{a_2}$ represents an approximate value of the tensile strength of the concrete derived from the compressive strength. Values of $a_2 = 0{\cdot}5$ and $0{\cdot}66$ are used.

The influence of the embedment depth d is given by $a_3 = 1{\cdot}5$ to $1{\cdot}54$ by Eligehausen & Sawade [11]. This means that the failure load does not increase in proportion to the surface of the failure cone which would instead give $a_3 = 2$. However, the value $a_3 = 2$ has been used extensively due to its straightforwardness and its simplicity, e.g. in the US code ACI 349-76 [17]. This can lead to unsafe bolts as the influence of the embedment depth is overestimated in this way especially for large embedment depths.

With recommended numerical values inserted we get

$$F_{max} \approx 15 f_c^{0{\cdot}5} d^{1{\cdot}5} \tag{2}$$

One basic way to model the failure with *fracture mechanics* is to study the *elastic energy* and the *fracture energy* as proposed by Bache, see e.g. Elfgren [18], [19] and Di Tommaso [20].

The elastic energy is the sum of the strain energy $0{\cdot}5\sigma\varepsilon = 0{\cdot}5\sigma^2/E$ over the volume of the anchor bolt. If we approximate the volume with $V = \pi \cdot d^2 \cdot d$, see Fig. 1, we may write

$$W_{el} \sim \frac{1}{2}\frac{\sigma^2}{E}\pi d^3 \sim \frac{\sigma^2 d^3}{E} \tag{3}$$

The fracture energy needed to create a crack is proportional to the crack area.

$$W_{cr} \sim G_F \pi d^2 \sim G_F d^2 \tag{4}$$

Here G_F denotes the fracture energy [Nm/m^2] of the concrete. A *brittleness number* B can now be defined as the ratio of elastic to fracture energy at failure

$$B = \frac{\text{Elastic energy}}{\text{Fracture energy}} \sim \frac{\sigma^2 d^3}{EG_F} = \frac{\sigma^2 d}{EG_F} \tag{5}$$

or with $\sigma = f_t$ at failure

$$B = \frac{f_t^2 d}{EG_F} \tag{6}$$

(Here EG_F/f_t^2 is often called the characteristic length l_{ch} and B can then be written $B = d/l_{ch}$, see e.g. Hillerborg [21]).

Fracture mechanics approaches based on linear analysis and stress intensity factors K_I and K_{II} have been presented by Ballarini *et al.* [8] and by Ballarini & Shah [15]. Methods based on non-linear analysis have been presented by Hellier *et al.* [9] and Rots [10]. An energy approach presented by Eligehausen & Sawade [11] leads to the following formula

$$F_{max} = 2{\cdot}1(E_c G_F)^{0{\cdot}5} d^{1{\cdot}5} \tag{7}$$

where E_C = modulus of elasticity of the concrete
G_F = fracture energy of the concrete
d = embedment depth

In order to compare the various methods, RILEM TC 90-FMA [22] has initiated a Round Robin Analysis of Anchor Bolts. Preliminary results have among others been presented by Cervenka *et al.* [14] and Eligehausen & Ozbolt [13].

2. PLANE STRESS SPECIMEN

2.1 Experimental work

Test specimens according to Fig. 2 were cast. The test configuration was proposed by Arne Hillerborg. A similar test set up was earlier used by Shah & Ballarini, see Ballarini *et al.* [8]. The specimens were

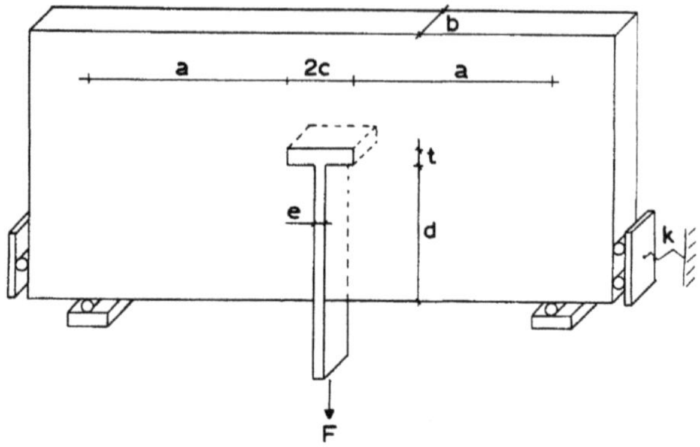

Fig. 2 Test specimens. In series 1, $b = 50$ mm, $2c = 100$ mm, $t = 25$ mm, $e = 15$ mm. The ratio a/d was $300/200 = 1{\cdot}5$, $150/100 = 1{\cdot}5$ and $200/100 = 2{\cdot}0$. In series 2, $b = 50$ mm, $2c = 15$ mm and 45 mm, $t = 5$ mm and 15 mm, $e = 5$ mm and 15 mm. The ratios a/d were $100/50 = 2{\cdot}0$, $150/150 = 1{\cdot}0$, and $300/150 = 2{\cdot}0$. The stiffness k was obtained by the tension in two 12-mm reinforcement bars.

cast in steel moulds and were made with two different geometries, here denoted series 1 and series 2. The geometry in series 2 is according to the RILEM, TC 90-FMA (1990) Round Robin Analysis of Anchor Bolts [22]. Test series 1 has earlier been presented in Ohlsson [12] and in Ohlsson & Elfgren [16] and is only summarized here. Information about the tests is also given in Ghasemlou & Johansen [23] and Ordqvist & Soutokorva [24]. The concrete was made with Portland cement and natural aggregates, see Table 1.

Cubes (150 mm) were cast in order to determine the splitting strength f_{cspl} and the compressive strength f_{cc}. The modulus of elasticity, E, was determined on drilled concrete cores, diameter 70 mm, length 150 mm. The fracture energy, G_{FE}, was determined

TABLE 1 Concrete mixes

Concrete	*Cement* (kg/m³)	*Water* (kg/m³)	*Aggregate 0–8 mm* (kg/m³)	*Aggregate 8–12 mm* (kg/m³)	*Micro silica* (kg/m³)	*Super plasticizer* (kg/m³)
C40	237	177	1 064	934	13	
C45	374	173	933	859		
C80	450	156	770	1 149	70	10

TABLE 2 Material properties

Concrete	f_{cc} *(MPa)*	f_{cspl} *(MPa)*	E_c *(GPa)*	G_{FE} *(N/m)*
C40	44 ± 1(12)	4·1 ± 0·5(6)	—	151 ± 23(5)
C45	55 ± 2(3)	3·4 ± 0·2(3)	38·2 ± 0·9(3)	91 ± 7(3)
C80	85 ± 7(3)	5·9 ± 0·3(6)	40·4 ± 0·9(4)	121 ± 1(3)

(Numbers in parenthesis denote number of tests; ± values denote standard deviations.)

with the RILEM three-point bending test on notched beams (840 mm × 100 mm × 100 mm), see Hillerborg [21]).

The tests were performed in a servohydraulic testing machine. The C45 concrete specimens in series 1 were loaded by pulling the anchor rod with a constant velocity. The other specimens were tested in displacement control with the anchor head displacement (the displacement of the anchor head relative to the supports) as the feedback signal to the servohydraulic system.

The failure generally consists of three main cracks, see Fig. 3. Besides the pullout cone, a bending crack is formed. The bending crack can be avoided if the specimen is very high compared to the embedment depth, but the specimen would then be very difficult to handle.

Figure 4 shows examples of load–displacement curves. The first part of the curves is linear elastic with no crack growth. At a certain load level, cracking starts. The specimen weakens. The load–displacement diagram has a peak and an unloading part. At the end of this second part the crack system is visible. As the deformation increases the stiffness of the specimen also increases. The stiffening is caused by shear stresses in the existing tensile cracks. Some of the specimens with small embedment depths do not show this stiffening effect.

Table 3 shows the peak loads obtained.

2.2 Finite Element Analysis

The pull-out of the series 1 specimens was analysed with the finite element method. The analysis has focused on the influence of different embedment depths and material properties.

The pull-out tests were modelled using the finite element program ABAQUS. Two embedment depths, 100 mm and 200 mm, were studied.

(a)

(b)

Fig. 3 Anchor bolts No. 14 (a) and (b) No. 16 after testing.

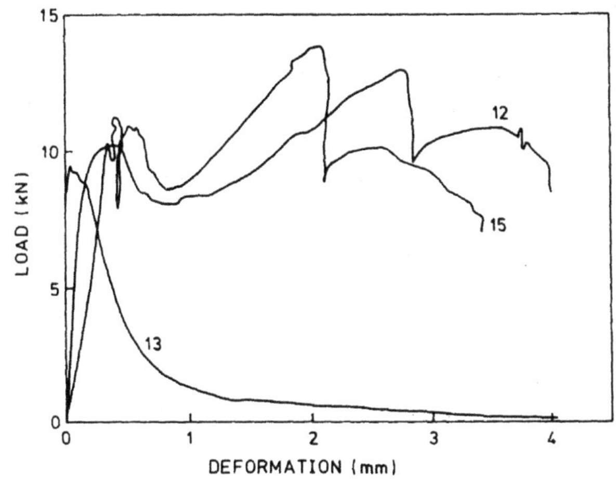

Fig. 4 Load–displacement curves for tests Nos 12, 13 and 15.

A discrete cracking model was used. The crack path was modelled as a straight line from the upper edge of the anchor head to the support on the top surface, see Fig. 5. The crack path is marked as a solid line in the figure.

Four-noded plane stress elements, type CPS4 with four integration points were used. The fracture zone was modelled with spring elements, type SPRING2. A load–displacement relation in the 1-direction (crack opening direction) was chosen to give a linear

TABLE 3 Peak loads

Test no.	*Concrete type*	*d (mm)*	*a (mm)*	F_{max} *(kN)*	*Test no.*	*Concrete type*	*d (mm)*	*a (mm)*	F_{max} *(kN)*
Series 1					Series 2				
1	C45	200	300	16·0	12	C40	50	100	12·9
2	C45	200	300	20·1	13	C40	50	100	9·4
3	C45	100	150	15·2	14	C40	50	100	11·4
4	C45	100	150	15·4	15	C40	50	100	13·8
5	C80	200	300	27·5	16	C40	150	150	51·3
6	C80	200	300	31·8	17	C40	150	150	33·5
7	C80	200	300	25·8	18	C40	150	300	17·0
8	C80	100	150	38·1	19	C40	150	300	16·5
9	C80	100	150	23·6					
10	C80	100	200	24·8					
11	C80	100	200	21·3					

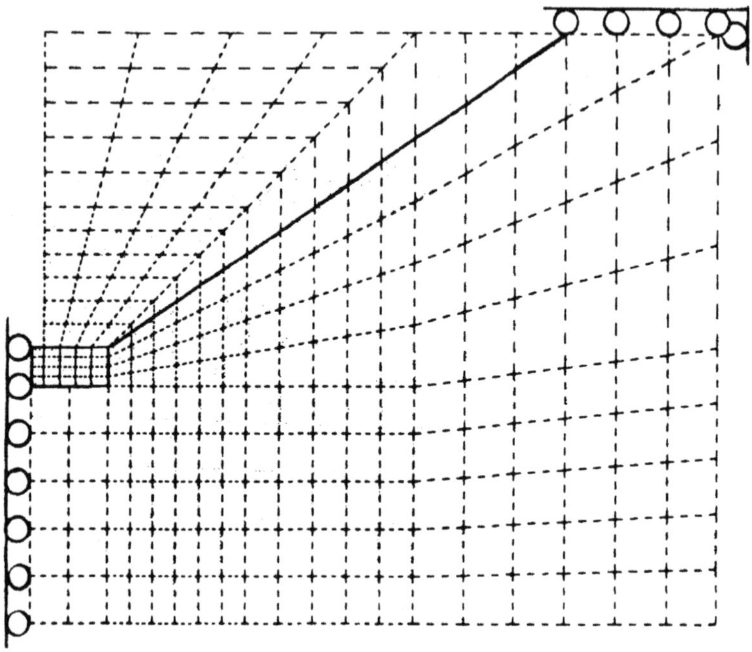

Fig. 5 Finite element model. Embedment depth $d = 100$ mm and $d = 200$ mm.

stress–crackwidth relation. In the 2-direction (crack sliding direction) the stiffness was zero. Bond between steel and concrete was modelled only on the horizontal parts of the anchor head which transferred compressive stresses to the concrete.

The material properties were varied according to Table 4.

Load–displacement curves showing the influence of the parameters

TABLE 4 Material properties in the analysis

Series	G_F *(N/m)*	f_{ct} *(MPa)*	E_c *(GPa)*	*d* *(mm)*
A	50,100,200,400	3·0	30	200
B	100	0·75,1·5,3·0,6·0	30	200
C	100	3·0	15,30,60,120	200
D	50,100,200,400	3·0	30	100
E	100	0·75,1·5,3·0,6·0	30	100

G_F, f_t, E_c and d are shown in Fig. 6. A high fracture energy gives a higher peak load and a more ductile behaviour after peak load. The peak load increases with a higher tensile strength but the brittleness increases too. A low tensile strength gives a very ductile behaviour. A high modulus of elasticity gives a higher peak load and a stiffer initial behaviour.

The influence of different embedment depths, $d = 100$ mm and $d = 200$ mm, has also been studied. The peak load is higher for the larger embedment depth. The ratio between the peak loads is ≈1·3. The studied pull-out test with two different embedment depths is not a true size effect problem. The size of the anchor bolt is the same for both the embedment depths. The smaller specimen will be stiffer compared to the larger specimen. This stiffness increase will increase the peak load.

2.3 Brittleness Number

The brittleness number is the ratio between elastic energy and fracture energy in a structure, as shown in Section 1.

The fracture energy is the amount of energy that the structure can absorb in the fracture zone. When the elastic energy in a structure is large compared to the fracture energy, the structure will have a brittle behaviour, compare Gustafsson & Hillerborg [25], Bazant & Pfeiffer [26] and Di Tommaso [20].

The elastic energy in a specimen is correlated with the stiffness of the specimen. An effective modulus of elasticity, E_{eff}, proportional to the stiffness of the specimen is therefore introduced. E_{eff} is determined from linear elastic finite element calculations. A finite element calculation with $E_{\text{steel}} = E_{\text{concrete}}$ is made. The stiffness F/d obtained is denoted K_{ref}. In the second calculation E_{steel} is raised to its real value. The stiffness obtained is denoted K_i. The effective modulus of elasticity is then calculated as

$$E_{\text{eff}} = E_{\text{concrete}} \cdot \frac{K_i}{K_{\text{ref}}} \tag{8}$$

E_{eff} is therefore a function of E_{concrete}, E_{steel} and the geometry of the specimen. The brittleness number can now be written $B = f_t^2 d/(E_{\text{eff}} G_F)$. Investigating the influence of the brittleness number, B, on the peak load, F_{max}, we plot a dimensionless diagram with the brittleness number, B, versus a non-dimensional load $F_{\text{max}}/d\,.\,b\,.\,f_t$, see Fig. 7.

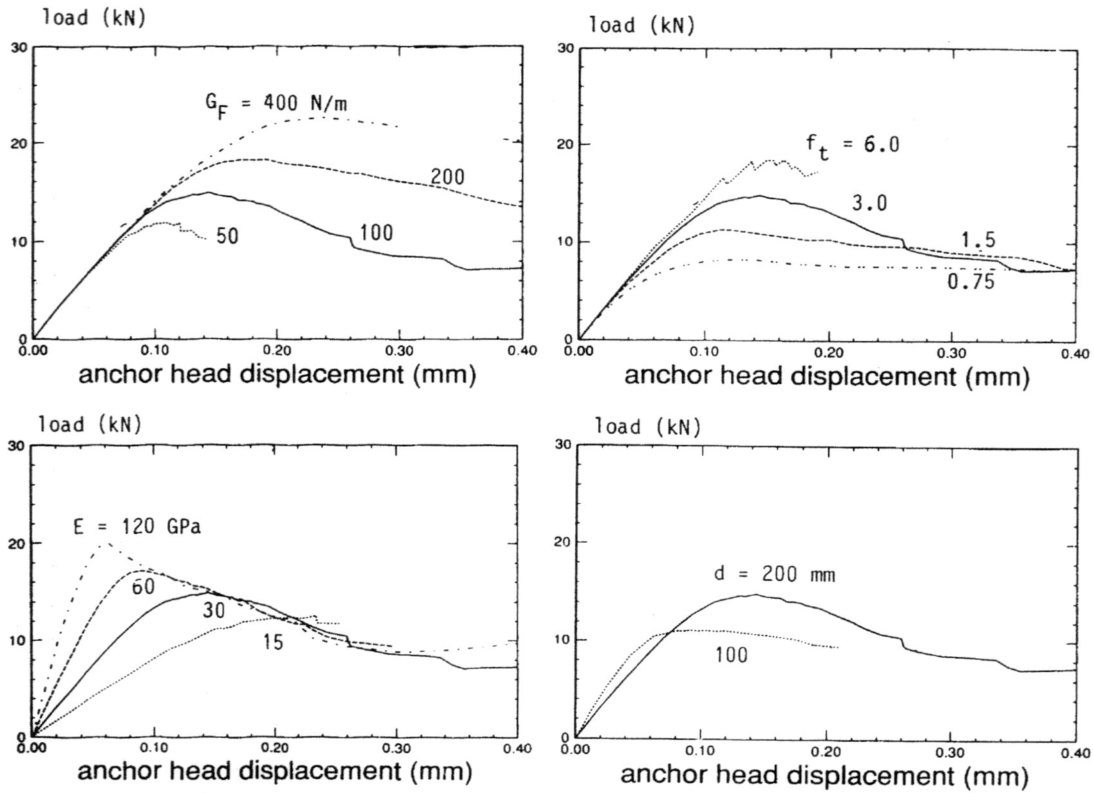

Fig. 6 Load–displacement curves from Finite Element Analysis. Material properties according to Table 4.

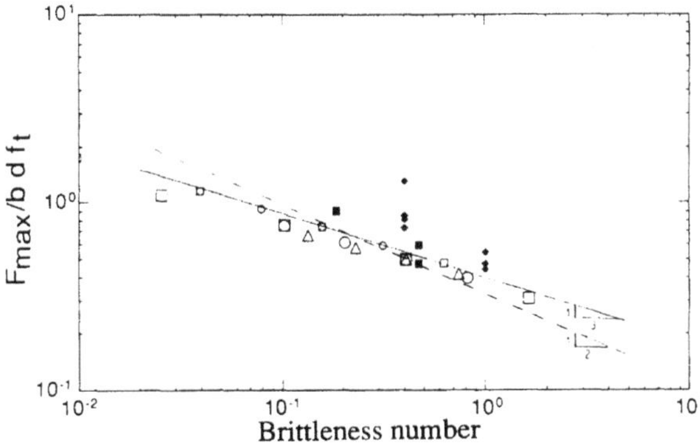

Fig. 7 Results from finite element calculations and laboratory tests. Open symbols denote results from finite element calculations and filled symbols denote results from laboratory tests.

Figure 7 shows a correlation between the brittleness number and the peak load in the finite element calculations. The laboratory tests also show this correlation, but the scatter is very large. The finite element calculations seems to predict a lower limit for the bearing capacity. A curve between the plotted results from the finite element calculations will have a slope of $k \approx -1/3$. This means that

$$\frac{F_{max}}{b\,.\,d\,.\,f_t} \sim \left(\frac{1}{B}\right)^{1/3} \quad \text{or} \quad F_{max} \sim b(E_{eff}G_F f_t\,.\,d^2)^{1/3} \qquad (9)$$

Thus, according to the finite element calculations, E_{eff}, G_F, and f_t have equal influence on the bearing capacity of the anchor bolt.

The slope of the B versus $F_{max}/b\,.\,d\,.\,f_t$ curve may be a starting point when considering design formulae

$k=0$	will give $F_{max} \sim b\,.\,d\,.\,f_t$ (no size effect)
$k=-1/4$	will give $F_{max} \sim b\,.\,d^{3/4}\,.\,f_t^{1/2}\,.\,E^{1/4}\,.\,G_F^{1/4}$
$k=-1/3$	will give $F_{max} \sim b\,.\,d^{2/3}\,.\,f_t^{1/3}\,.\,E^{1/3}\,.\,G_F^{1/3}$
$k=-1/2$	will give $F_{max} \sim b\,.\,d^{1/2}\,.\,E^{1/2}\,.\,G_F^{1/2}$ (linear fracture mechanics)

3. AXISYMMETRIC SPECIMEN

3.1 Experimental Work

Test specimens according to Fig. 8 were cast. The geometry of the tests are approximately according to the RILEM, TC 90-FMA (1990)

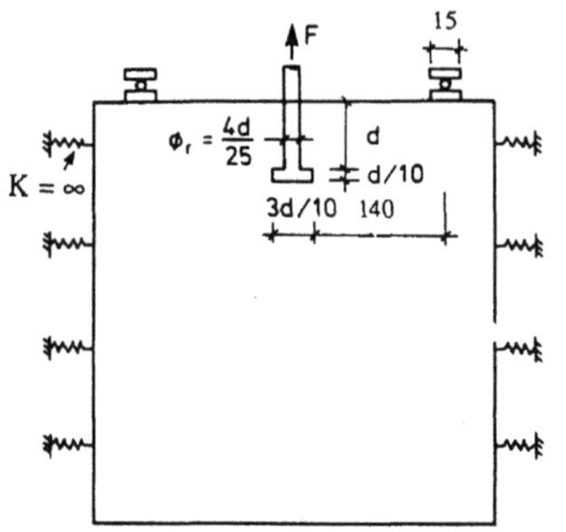

Fig. 8 Test specimen.

Round Robin Analysis of Anchor Bolts [22]. The concrete was from the same batch as the series 2 plane stress specimen (C40).

The test set up is shown in Fig. 9. A steel tube ($\varnothing_{outer} = 325$ mm, $\varnothing_{inner} = 295$ mm, $h = 100$ mm) was placed as a support on top of the concrete. A stiff steel beam was placed over the steel tube, anchoring the test specimen to the floor. The anchor displacement was measured on the anchor at the same level as the concrete surface.

The tests were performed in displacement control with a servohydraulic testing machine with the anchor displacement as the feedback signal to the servohydraulic system.

3.2 Results

Load–displacement curves from the tests are shown in Fig. 10.

Figure 11 shows a tested specimen. In tests 1 and 3, a pull-out cone was formed. In test 2 radial cracking occurred. The peak loads are shown in Table 5 together with some other test results.

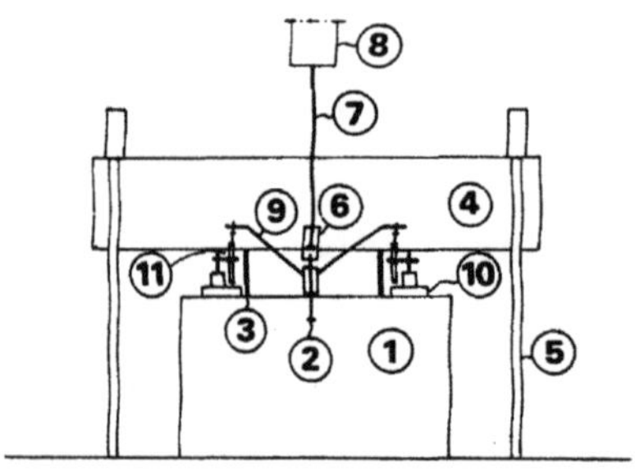

Fig. 9 Test set up. 1 = test specimen, 2 = anchor bolt to be tested, 3 = steel tube support, 4 = steel beam, 5 = steel bar supporting the steel beam, 6 = hinge, 7 = steel bar, 8 = servohydraulic actuator, 9 = steel bar, 10 = steel plate, 11 = displacement gauge (LVDT).

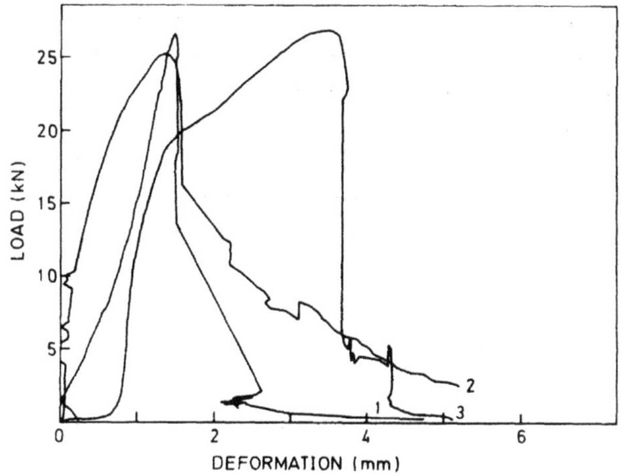

Fig. 10 Load–displacement curves from axisymmetric tests.

In Fig. 12 the test results are shown in a diagram with brittleness number versus strength. Analytical results by Eligehausen & Ozbolt [13] indicate a slope changing from $k = -1/2$ to $k = -1/3$ for decreasing brittleness numbers. A slope of $k = -1/2$ (corresponding to linear fracture mechanics analysis) will give $F_{max} \sim d^{3/2}E^{1/2}G_F^{1/2}$ compare with eqn (7).

This is the theoretical limit value for specimens with high brittleness numbers. For smaller brittleness numbers the slope gradually changes

TABLE 5 Test results

Test	*d* *(mm)*	f_t *(MPa)*	*E* *(GPa)*	G_F *(N/m)*	F_u *(kN)*
Barr [27]	50	1·5	20	100	22·5
Barr [27]	50	1·5	20	100	29·75
Barr [27]	150	1·5	20	100	159
Eligehausen &	120	1·8	23·5	70	97·2
Sawade [11]	250	1·8	23·5	70	290
Eligehausen & Sawade [11]	500	1·8	23·5	70	885
Ohlsson [12] 1	50	4·1	30	150	26·6
2	50	4·1	30	150	25·2
3	50	4·1	30	150	26·9

(a)

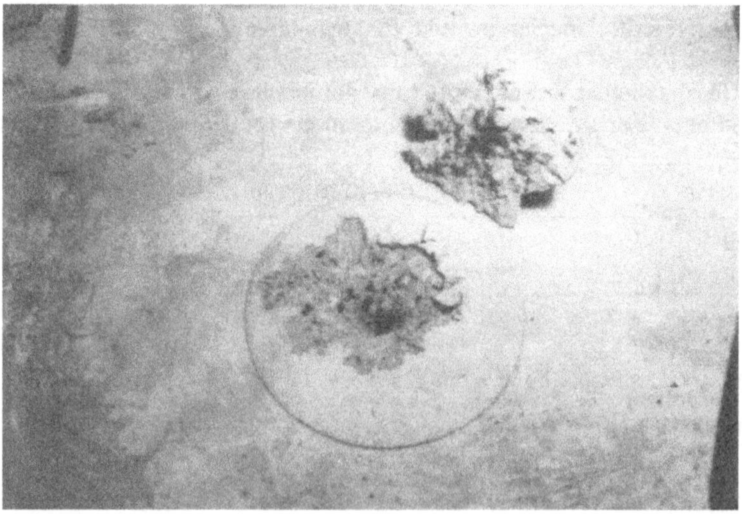

(b)

Fig. 11 Test specimen No. 1 after failure.

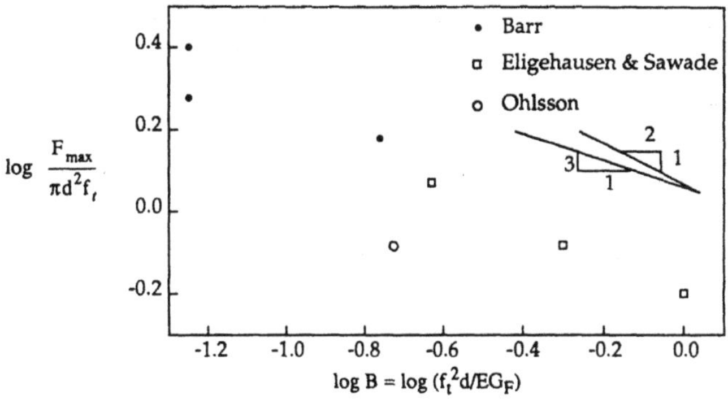

Fig. 12 Failure load F_{max} as function of brittleness number for axisymmetric specimens.

to reach the limit $k=0$ for specimens with very small brittleness numbers. This general trend is well known and has been demonstrated many times, see e.g. Bazant & Pfeiffer [26] and Di Tomasso [20]. However, what is not yet generally accepted, is to base design formulae on this knowledge. This would be a logical next step which can give formulae of the following form:

Very small B	$k=0$	$F_{max} \sim d^2 . f_t$ (no size effect)
Medium B	$k=-1/3$	$F_{max} \sim d^{5/3} . f_t^{1/3} . E^{1/3} . G_F^{1/3}$
Very large B	$k=-1/2$	$F_{max} \sim d^{3/2} . E^{1/2} . G_F^{1/2}$ (LEFM)

More results are needed before any definitive conclusions can be reached. However, it is already now obvious that fracture mechanics models are able to describe size effects in a reasonably accurate way.

REFERENCES

[1] Klinger, R. E. & Mendonca, J. A., Tensile capacity of short anchor bolts and welded studs: a literature review. *ACI Journal* (Detroit), **79**(4) (1982) 270–9.

[2] Rehm, G., Eligehausen, R. & Mallee, R., Befestigungstechnik (Fastening technology. In German). In *Beton Kalender 1988*. Ernst & Sohn, Berlin, 1988, pp. 569–663.

[3] Ottosen, N. S., Non-linear finite element analysis of pull-out tests. *Journal of the Structural Division*, (ASCE), **107**(ST4) (1981) 591–603.

[4] Elfgren, L., Broms, C. E., Johansson, H. & Rehnström, A., Anchor bolts in reinforced concrete foundations. Short time tests. Research Report TULEA 1980:36, Division of Structural Engineering, Luleå University of Technology, 1980, 117 pp.

[5] Elfgren, L., Broms, C. E., Cederwall, K. & Gylltoft, K., Fatigue of anchor bolts in reinforced concrete foundations. *IABSE Colloquium Fatigue of steel and concrete structures*, Lausanne, March 1982. Proceedings, IABSE Report, Vol. 37, Zürich, pp. 105–117.

[6] Peier, W. H., Model of pull-out strength of anchors in concrete. *Journal of Structural Engineering*, (ASCE), **109**(5) (1983) 1155–73.

[7] de Borst, R., Non-linear analysis of frictional materials. Proefschrift ter verkrijging van de graad van doctor in te technische wetenshappen aan de Technische Hogeschool Delft. Delft, 1986, 140 pp.

[8] Ballarini, R., Shah, S. P. & Keer, L. M., Failure characteristics of short anchor bolts embedded in a brittle material. *Proceedings of the Royal Society* (*London*). **A404** (1986) 35–54.

[9] Hellier, A. K., Sansalone, M., Ingraffea, A. R., Carino, N. J. & Stone, W. C., Finite element analysis of the pull-out test using a nonlinear discrete cracking approach. *Cement, Concrete and Aggregates*, (ASTM), **9**(1) (1987) 20–29.

[10] Rots, J., Computational modelling of concrete fracture. Dissertation, Delft University of Technology, Department of Civil Engineering, Delft, 1988, 132 pp.

[11] Eligehausen, R. & Sawade, G., Analysis of anchorage behaviour (Chapter 13.1). A fracture mechanics based description of the pull-out behaviour of headed studs embedded in concrete (Chapter 13.2). In *Fracture Mechanics of Concrete Structures. From theory to applications*, ed. L. Elfgren. Chapman & Hall, London, 1989, pp. 263–80, 281–99.

[12] Ohlsson, U., Fracture Mechanics Studies of Concrete Structures. Licentiate thesis 1990:07L. Division of Structural Engineering, Luleå University of Technology, 1990.

[13] Eligehausen, R. & Ozbolt, J., Size effect in anchorage behaviour. *ECF 8 Fracture Behaviour and Design of Materials and Structures*, Vol. II, ed. D. Firrao. EMAS, Warley, UK, 1990, pp. 721–7.

[14] Cervenka, V., Pukl, R. & Eligehausen, R., FEM simulation of concrete failure. *ECF 8 Fracture Behaviour and Design of Materials and Structures*, Vol. II, ed. D. Firrao. EMAS, Warley, UK, 1990, pp. 728–33.

[15] Ballarini, R. & Shah, S. P., Fracture mechanics based analysis of pull-out tests and anchor bolts. *Analysis of Concrete Structures by Fracture Mechanics, Proceedings of the RILEM workshop dedicated to Professor Arne Hillerborg*, ed. L. Elfgren & S. P. Shah. Chapman & Hall, London, 1991, pp. 245–80.

[16] Ohlsson, U. & Elfgren, L., Anchor bolts in concrete structures. Two-dimensional modelling. *Analysis of Concrete Structures by Fracture Mechanics. Proceedings of the RILEM Workshop dedicated to Professor Arne Hillerborg*, ed. L. Elfgren & S. P. Shah. Chapman & Hall, London, 1991, pp. 281–301.

[17] ACI 349-76 (1978) Steel embedments. Proposed addition to code requirements for nuclear safety related concrete structures. *ACI Journal* (Detroit), **75**(8) (1978) 329–47.

[18] Elfgren, L. (Ed.), *Fracture Mechanics of Concrete Structures. From theory to applications, A RILEM Report by TC 90-FMA*. Chapman & Hall, London, 410 pp.

[19] Elfgren, L., Applications of fracture mechanics to concrete structures. *Fracture Toughness and Fracture Energy. Test methods for concrete and rock*, ed. H. Mihashi, H. Takahasi & F. Wittman. Balkema, Rotterdam, 1989, pp. 575–90.

[20] Di Tommaso, A., Size effect and brittleness. In *Fracture Mechanics of Concrete Structures. From theory to Applications*, ed. L. Elfgren. Chapman & Hall, London, 1989, Chapter 7, pp. 191–207.

[21] Hillerborg, A., The theoretical base of a method to determine the fracture energy G_F of concrete. *Materials and Structures*, **18**(106), (1985) 291–6.

[22] RILEM RC 90-FMA (1990) Round-Robin Analysis of Anchor Bolts—Invitation, *Materials and Structures*, **23**(133) (1990) 78. Revised invitation in *RILEM News* (Cachan Cedex), No. 1991.1, March 1991.

[23] Ghasemlou, F. & Johansen, R., Hakförankringar i betong (Hook anchors in concrete. In Swedish). Diploma work 1990:26E. Division of Structural Engineering, Luleå University of Technology, 1990, 71 pp.

[24] Ordqvist, C. & Soutokorva, M., Hakförankringar i höghållfast betong (Hook anchors in high strength concrete. In Swedish). Diploma work 1990:40E. Division of Structural Engineering, Luleå University of Technology, 1990, 59 pp.

[25] Gustafsson, P. J. & Hillerborg, A., Sensitivity in shear strength of longitudinally reinforced concrete beams to fracture energy of concrete. *ACI Structural Journal* (Detroit), **85**(3) (1988) 286–94.

[26] Bazant, Z. P. & Pfeiffer, P. A. Determination of fracture energy from size effect and brittleness number. *ACI Materials Journal*, (Detroit), **84**(6) (1987) 463–80.

[27] Barr, B. I. (1990) Contribution to Round-Robin Analysis of Anchor Bolts (Pers. Comm.)

Chapter 11

Simulation of Bond and Anchorage: Usefulness of Softening Fracture Mechanics

JAN G. ROTS

TNO Building and Construction Research, Delft University of Technology, P.O. Box 49, 2600 AA Delft, The Netherlands

ABSTRACT

This chapter presents two applications of softening fracture mechanics to bond and anchorage in reinforced concrete. Smeared and discrete crack models including tensile softening are used to simulate the fracture behavior in a tension-pull specimen and in an anchorage structure. Both problems involve axi-symmetry and it is demonstrated that in such cases transverse secondary cracking, transverse primary cracking as well as longitudinal splitting cracking can be simulated. The bond computations support the verification and validation of traction-slip rules. The anchorage computations provide a clear demonstration of the elastic-softening theory. The outcome correctly reveals that the elastic energy stored in the structure is released for creating fracture surface, as accompanied by dangerous brittle load–displacement behavior.

1. INTRODUCTION

Roughly speaking, computational models for reinforced concrete can be applied in three different ways (see e.g. Blaauwendraad [1]):

- to support experimental research;
- to develop, verify or validate design rules;
- to incidentally analyse particular structures.

In this chapter two applications from fracture mechanics will be worked out. First, a computational resolution of the bond-slip problem in reinforced concrete will be presented. This application belongs to both the first and the second category. On the one hand, it will be demonstrated that the fracture computations provide additional insight regarding internal cracking patterns and non-uniqueness of bond-slip laws, issues on which no consensus has been reached in the experimental world. On the other hand, it will be shown that the computations support the verification of engineering design rules like bond stress-slip laws.

Secondly, a direct application of the third class will be presented. It involves the study of an anchorage for mechanical components in the nuclear power industry, whereby safety requirements asked for precise nonlinear analysis of the pull-out fracture.

A common feature of both applications is the presence of axisymmetry, leading to transverse as well as longitudinal cracks. Guidelines for quantifying the fracture parameters in such cases will be presented, especially with regard to the band width of the smeared longitudinal crack.

2. THREE-LEVEL APPROACH TO BOND-SLIP

Some ten years ago the numerical modeling of bond-slip received attention and various types of bond-slip elements have been proposed (e.g. Schäfer [2], Dörr [3], de Groot *et al.* [4]). At that time the available fracture models were less sophisticated, and the analysis was sometimes frustrated by shortcomings of the crack model while the bond-slip formulations were adequate [4]. To date, the reverse seems to hold true. Crack models have undergone rapid developments, but most existing simulators adopt the coarse and unsafe assumption of perfect bond. Consequently, there is a need to rejuvenate bond-slip research with the achievements of fracture mechanics and elastic-softening models.

The broad scope of the subject calls for a distinction into three approaches of decreasing degree of precision:

- *Resolution of bond-slip*
  This strategy zooms at the micro-behavior in the vicinity of the reinforcing bar (rebar), where secondary transverse and lon-

gitudinal cracks are crucial mechanisms. The method aims at explaining the fundamentals of traction-slip behavior.

- *Bond-slip interface analysis*

  This approach lumps traction-slip behavior into an interface, with a view to predicting the spacing and width of localized primary cracks in reinforced concrete members.

- *Tension-stiffening*

  When the concrete is densely reinforced, distributed fracture occurs and even the above approach that zooms at primary cracks becomes too delicate. Tension-stiffening then accounts for the bond characteristics in an indirect manner.

In this chapter attention is confined to the impact of fracture mechanics on the first level of approach. For a fracture mechanics view towards the second and third level of approach, the reader is referred to e.g. Rots [5].

3. DETAILED MODELING OF THE BOND-SLIP PROBLEM

3.1 Bond Mechanisms

Confining attention to deformed rebars, the bond-slip, i.e. the tangential relative displacement between the rebar and the concrete (measured some distance away from the rebar), is controlled by four mechanisms [6]–[8]:

- elastic deformation;
- conical transverse cracking behind the ribs of the rebar (these secondary cracks arise internally at either side of an externally visible primary crack);
- longitudinal cracking in response to tensile ring-stresses;
- 'crushing' in compressive cones radiating out from the ribs.

Although experimental data are not decisive [9], [10], the crushing is likely to occur only in cases of lateral confinement, which is typical of anchorage bond and short-embedment pull-out tests. Such cases have been studied [11] and induce significant compression in front of the ribs. This chapter considers the long-embedment flexural bond, where the compressive stresses remain relatively small: from the output we did not observe stresses larger than $25\,\text{N/mm}^2$, while a manual calculation that incorporates the rib surface characteristics provides additional justification for this assumption. For this reason, the crushing has been ignored and the analysis renders tensile cracking and elasticity.

3.2 Geometrical Modeling

We consider a tension-pull configuration, i.e. a center-plaçed rebar protruding from a concrete cylinder or block. The specimen is pulled at both ends and represents the conditions in the tension zone of a reinforced concrete member, between two primary flexural cracks. Figure 1 details the idealization which assumes axi-symmetry as well as mirror-symmetry. The dimensions correspond to a portion of the specimens tested by Dragosavić & Groeneveld [12] and also come close to the popular 6 × 6 inch specimens with a 1 inch diameter rebar [9], [13], [14]. The rebar is composed of four-node rectangles and the concrete of three-node triangles in a cross-diagonal pattern which minimizes directional bias.

Adhesion and fraction along the steel–concrete interface have been neglected in favor of mechanical interlock provided by the ribs [6]. The mechanical interlock was modeled by rigidly attaching the steel to the concrete at rib locations, while only partially attaching it between the ribs. The overlapping nodes at rib locations were tied to each other in the radial direction as well as in the axial direction, while the overlapping nodes at locations in between the ribs were tied to each

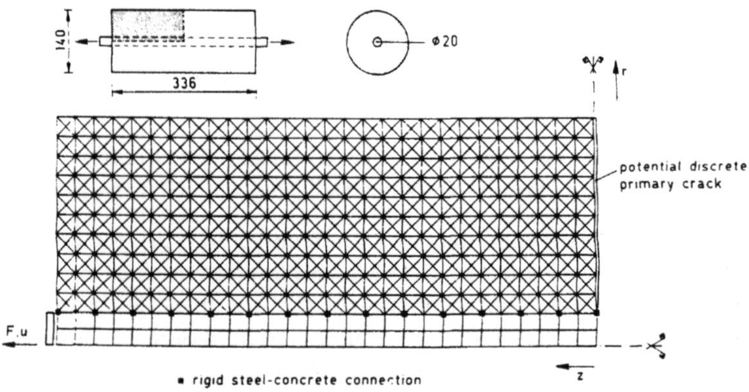

Fig. 1 Computational set-up to investigate bond-slip related cracks in a tension-pull configuration. The configuration is typical for a tension zone in bending.

other only in the radial direction. For a rib spacing of 12 mm and an element size of 6 mm this corresponds to an interchangeable tying scheme.

3.3 Material Modeling

An elastic-softening law has been assumed for the concrete. The elastic-softening parameters were taken as Young's modulus $E = 25\,000\ \mathrm{N/mm^2}$, Poisson's ratio $\nu = 0{\cdot}2$, tensile strength $f_{ct} = 3{\cdot}0\ \mathrm{N/mm^2}$, fracture energy $G_f = 75\ \mathrm{J/m^2}$, a nonlinear shape of the softening curve according to Reinhardt *et al.* [15] and a secant unloading/reloading model.

An axi-symmetric version of the smeared crack concept was adopted whereby the direction of longitudinal cracks was naturally fixed because of axi-symmetry, while transverse cracks were allowed to rotate so as to maintain coaxiality and to avoid overstiff response [5]. The transverse cracks were assumed to localize within a band width $h = 6$ mm which was estimated with a view to the particular finite element configuration [5]. For the longitudinal cracks the definition of a band width was less evident, as will be discussed later.

Along the axis of mirror-symmetry discrete interface elements were incorporated with a view to capturing a probable primary crack. In order to activate this crack, the discrete elements were assigned a material imperfection. Although experimental researchers generally add a geometrical imperfection in the form of a starter notch [3], [12], the use of a material imperfection turned out to better suit the present study.

The investigation can be seen as an extension of the only existing computational study in this field by Ingraffea *et al.* [16] which ignored longitudinal cracking and which was terminated at the onset of primary cracking.

4. TRANSVERSE CRACKS IN TENSION-PULL SPECIMEN

4.1 Transverse Secondary Cracking

The combined smeared/discrete fixed/rotating crack analysis progressed as shown in Figs 2 and 3, presenting the incremental deformations and crack patterns at typical load stages. Initially, transverse cracks form at the location where the steel exits the concrete (Fig. 2(a)). These cracks, discovered by Goto [17], nucleate

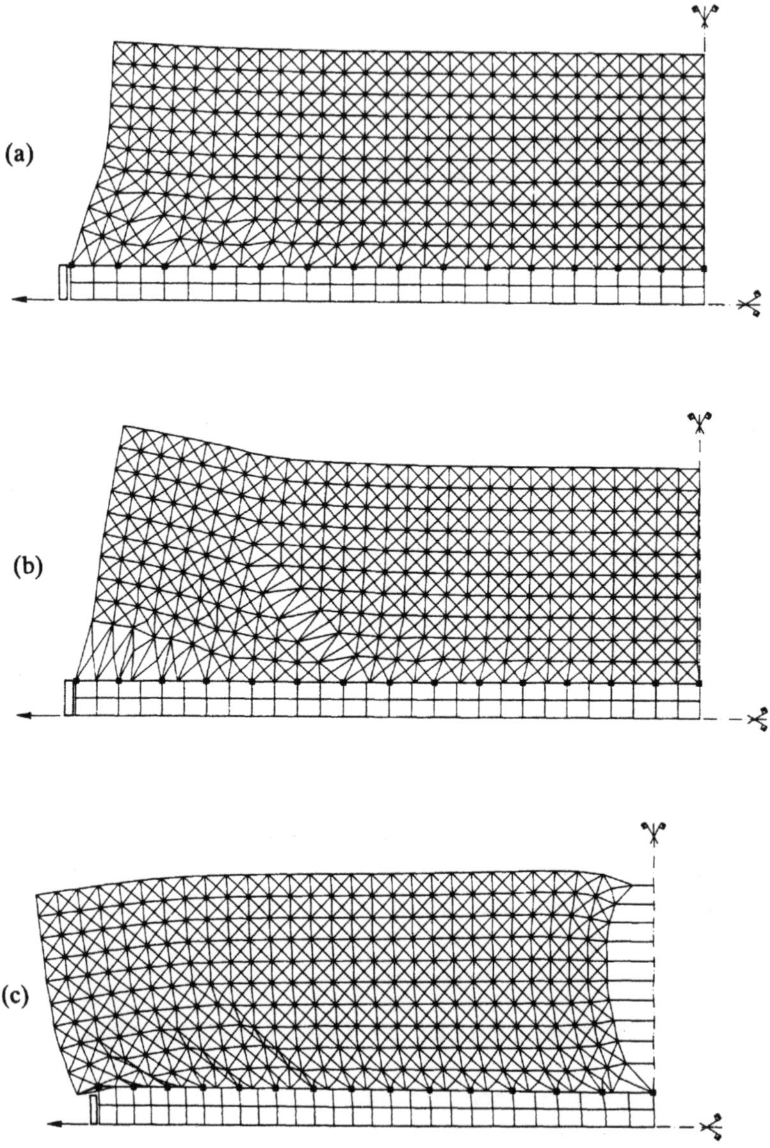

Fig. 2 Incremental deformations of tension-pull specimen: (a) early stage ($F = 32{\cdot}7$ kN); (b) at impending primary crack formation ($F = 51{\cdot}4$ kN); (c) at primary crack formation ($F = 51{\cdot}6$ kN).

(a)

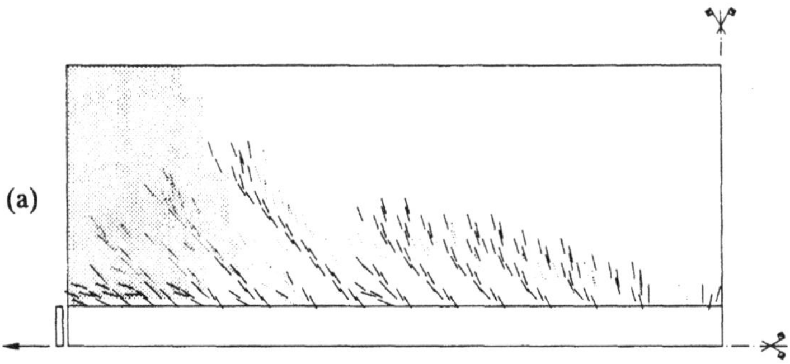

(b)

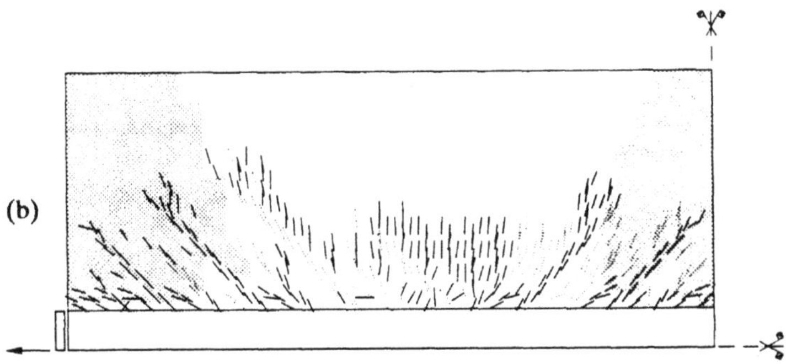

Fig. 3 Crack formation in tension-pull specimen (active cracks drawn, inactive cracks dotted, longitudinal cracks shaded): (a) at impending primary crack formation ($F = 51{\cdot}4$ kN); (b) final stage, beyond primary cracking.

behind the ribs and are cone-shaped. On subsequent loading, the early cracks propagate and additional secondary cracks nucleate further from the end-face. Although each rib of the rebar produces a secondary crack, Fig. 2 reveals that only a limited number of them survives while the others are arrested. This corresponds to experimental bond-crack detection [18]. A further observation from Fig. 2(b) is that the earliest cracks close to the rebar show large openings in a direction perpendicular to the rebar. This suggests radial separation between the steel and the concrete, which confirms experimental findings [17] and conclusions from quasi-linear analyses [19].

4.2 Transverse Primary Cracking

Although secondary cracks rapidly diminish the concrete tensile capacity, the present specimen length renders sufficient tension at

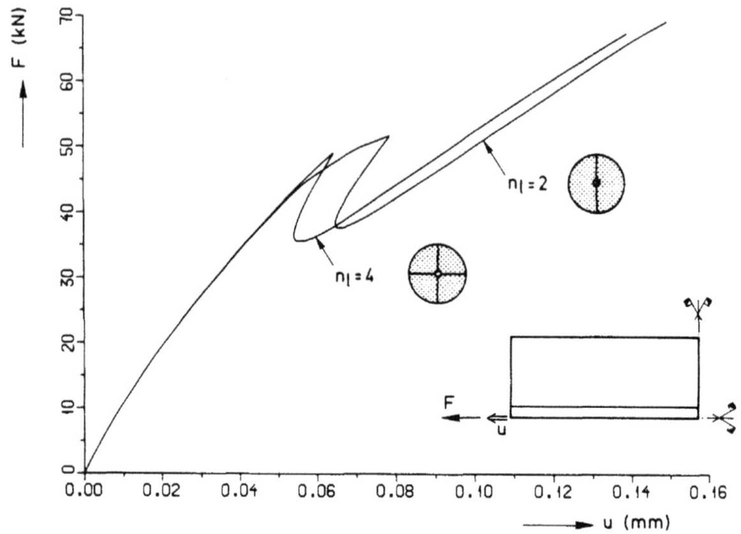

Fig. 4 External load versus end-displacement rebar of tension-pull specimen, for different number n_l of assumed localized longitudinal cracks.

mid-section to trigger the primary crack. Figure 2(c) and the load–elongation curve of Fig. 4 ($n_l = 2$) reveal the catastrophic nature of this crack (snap-back), which is in agreement with previous descriptions that call primary cracks unstable [17]. While traversing the snap-back regime, using indirect displacement control as proposed by de Borst [20], the existing secondary cracks near mid-section showed significant rotation in order to slant towards the free surface provided by the primary crack. When the softening of the primary crack was completed, the load could be incremented again. Secondary cracking and radial separation adjacent to the primary crack proceeded, while the early secondary cracks at the end-face started to re-load in order to finally re-soften.

A major observation is the potential of softening fracture mechanics to accurately capture the stress/strain/crack rotations and redistributions involved in bond-slip cracking. This statement holds especially when the softening models are embedded in a rotating smeared crack concept. The rotating variant of smeared cracking assumes coaxiality between principal stress and principal strain and avoids the overstiff response due to non-zero shear retention factors as adhering to the fixed smeared crack concept (e.g. Willam *et al.* [21], Rots [5]).

Furthermore, the present rotating crack analysis did not noticeably suffer from spurious kinematic modes [22], so that a proper tangent stiffness matrix could be maintained throughout the entire iterative process, giving fast convergence. Preliminary analyses [23] with fixed multi-directional smeared cracks were less successful in this respect, owing to uncontrollable deviations between the leading crack direction and the direction of principal stress, to the shear retention model producing overstiff behavior and to the occurrence of fixed spurious modes.

5. LONGITUDINAL CRACKS AND THE IMPORTANCE OF THEIR BAND WIDTH

Initially, the external force is transferred from the steel into the concrete primarily via axial tensile stresses. On transverse secondary cracking this type of bond action is lost and the transfer of bond forces is subsequently furnished by compressive cones that radiate out from the ribs. The radial components of the compressive cones are balanced by rings of tensile stress, as shown in Fig. 5(a) [7]. When the ring is stressed to rupture a longitudinal crack arises and the balance against the compressive cones is lost, which causes a further breakdown of bond.

Figure 3 presents the extent of longitudinal cracking and Fig. 6 gives a qualitative impression of the associated ring stress with increasing stage of the loading process. Initially, ring stresses are highly concentrated near the end-face, where the steel exits the concrete (Fig. 6(a)). The first longitudinal cracks soon arise and the resulting tangential softening involves significant redistribution as we observe a front of ring stresses that gradually travels away from the end-face (Fig. 6(b)). After primary cracking this phenomenon replicates itself near mid-section, and the final distribution of Fig. 6(c) is stable as further load increments are resisted by the steel rather than by the damaged concrete, the tensile capacity of which is almost exhausted.

Because of axi-symmetry it was impossible to model localization of the longitudinal cracks, which places their band width h in an unclear position. From experimental work [7], [24]–[26], such localization is known to occur, as is exemplified by Figs 5(b) and 5(c) revealing two and respectively four localized longitudinal splitting cracks through the periphery. This information can be transferred into an *a priori*

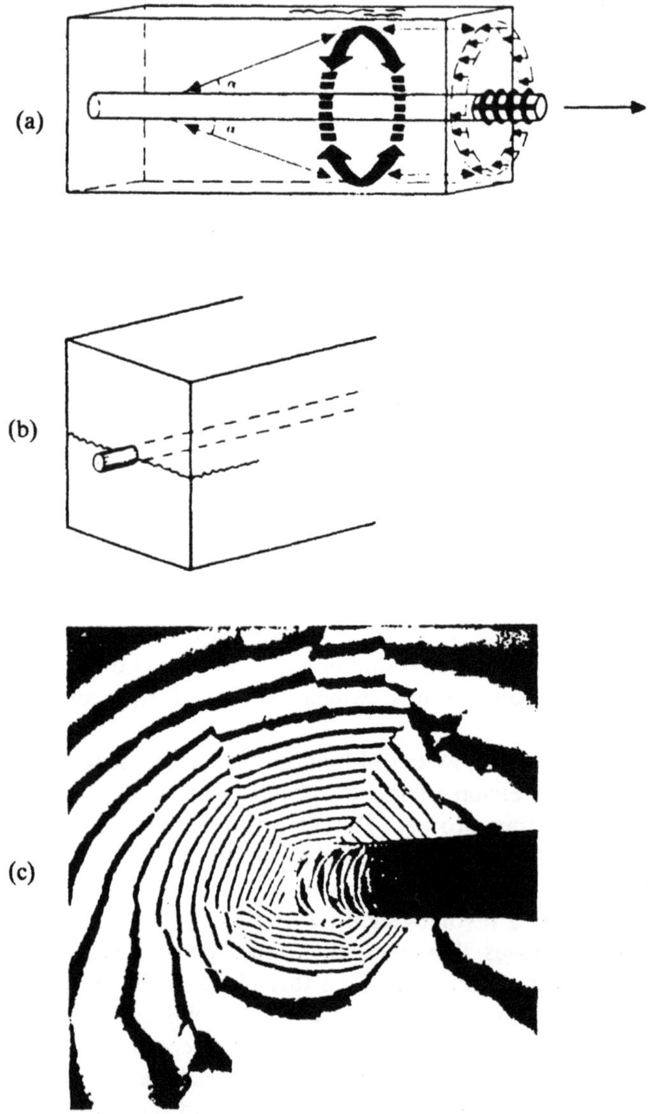

Fig. 5 Bond mechanisms for a deformed rebar, after transverse cracking. (a) Bond action via compressive cones and tensile rings [7]. The tensile ring may give rise to longitudinal cracking (splitting). (b) Bond-splitting with $n_1 = 2$, typical pattern from engineering practice. (c) Bond-splitting with $n_1 = 4$, obtained by holographic interferometry [24].

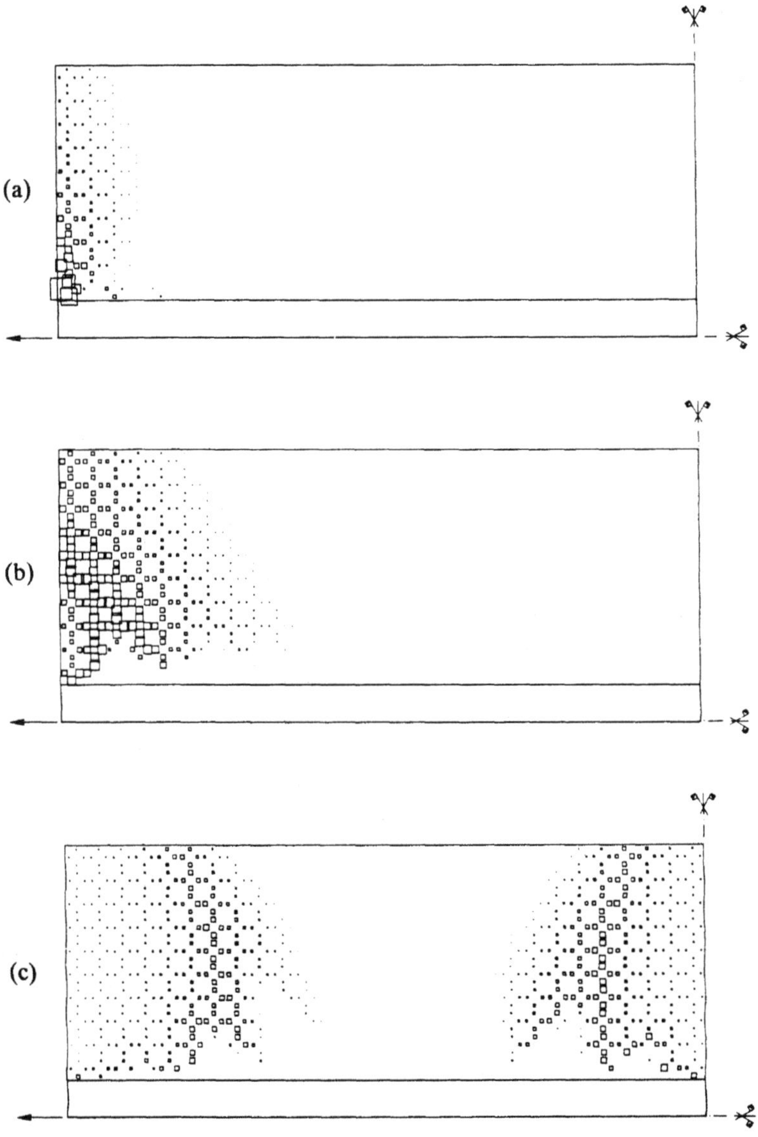

Fig. 6 Redistribution of tensile ring stress in tension-pull specimen (size of square is proportional to the magnitude of the stress): (a) initial stage (linear-elastic); (b) subsequent stage ($F = 37{\cdot}2$ kN); (c) final stage ($F = 68{\cdot}1$ kN), beyond primary cracking.

assumption of the band width according to

$$h = \frac{2\pi R}{n_l} \tag{1}$$

where R is the distance to the axis of rotation and n_l is the imagined number of localized longitudinal cracks. The use of eqn (1) implies n_l times G_f to be consumed in the smeared longitudinal crack.

The above result was obtained for $n_l = 2$ which provides a safe assumption. For the outer integration points with large values of R the band width according to eqn (1) even becomes so large that the local stress–strain relation for the present set of elastic-softening parameters displays a snap-back. To remedy this, the softening model in this study was enhanced with automatic reduction of the strength limit f_{ct} [27]. This steep softening confirms the widespread notion that bond splitting is brittle and dangerous.

An additional analysis was carried out for $n_l = 4$, which doubles tangential toughness. Figure 4 reveals a subtle discrepancy between the load–elongation curves predicted. The previous analysis with $n_l = 2$ not only shows enhanced pre-peak nonlinearity, which is evident from the steeper softening in longitudinal cracks, but it also surprisingly shows the limit load at primary cracking to increase. Obviously, the rapid loss of ring stress transfer for $n_l = 2$ quickly breaks down the bond action via the compressive cones and, hence, it indirectly accounts for a decrease in axial stress transfer to the concrete at mid-section. This postpones primary crack formation and the associated increase of the limit load is explainable from the enhanced contribution of the steel relative to the concrete in resisting the external load. A trial analysis for an even steeper softening ($n_l = 2$ in conjunction with $G_f^I = 37{\cdot}5\ \mathrm{J/m^2}$) underlined this tendency and even prevented the specimen from primary cracking at all, while it snapped on longitudinal cracking. The results reveal the integrated action of axial tensile zones, radial compressive cones and tangential tensile rings.

The bond capacity is dangerously overestimated not only when longitudinal cracks are ignored altogether, as done for example by Ingraffea *et al.* [16], but also when the band width for longitudinal cracks is quantified erroneously, for instance by simply equating it to the band width for transverse cracks (6 mm) which for the outmost sampling points ($R = 69$ mm) would result in a 36-fold (!) overestimation of the tangential softening capacity compared to eqn (1) with

$n_l = 2$. The issue is also relevant with other axi-symmetric problems, like punching shear, or the pull-out of an anchor.

Ideally, one should undertake a fully three-dimensional analysis to remove the uncertainties in the imagined value of n_l. Preliminary results thereof revealed the potential of predicting longitudinal fracture localization [28]. In general, caution must be exercised whenever some kind of symmetry, anti-symmetry or axi-symmetry is introduced in the model. The correct fracture localization may be easily missed and both strength and stiffness may be overestimated. For mirror-symmetry this was exemplified by the direct tensile test [29] and for anti-symmetry an example has been published by de Borst & Rots [22].

6. PREDICTING BOND TRACTION-SLIP CURVES

Let us extract some more quantitative information from the analysis. The bond shear traction has been plotted against the bond-slip in Fig. 7, both for $n_l = 2$ and $n_l = 4$, whereby the curves have been terminated

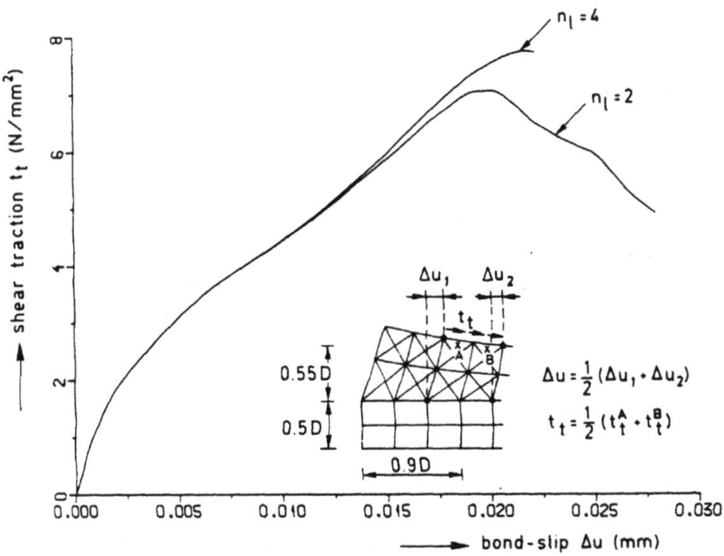

Fig. 7 Local shear traction versus slip in tension-pull specimen for $n_l = 2$ and $n_l = 4$ (distance 18 mm from end-face).

at the onset of primary cracking. The shear traction and the slip have been defined as the average values for a segment in between two ribs. The particular segment underlying Fig. 7 was located at a distance $0{\cdot}9D$ from the end-face, with D denoting the rebar diameter. The traction was determined as the segment-average shear stress of the two integration stations at distance $1{\cdot}05D$ from the axis of rotation, whereafter a surface-correction of $1{\cdot}05D/0{\cdot}5D$ was inserted in order to objectively relate them to the interface surface. The slip was defined as the average of the two node-sets at either side of the segment. These schemes correspond to accepted definitions from experimental research.

The curves show a linear-elastic stage, a stage of decreased stiffness and a softening stage. The stage of decreased stiffness is primarily attributable to transverse secondary cracking and the softening stage to longitudinal cracking. The latter assertion is evident from the fact that $n_l = 2$ yields far more softening than $n_l = 4$. This underlines the importance of the indirect bond action via compressive cones in conjunction with tensile rings. The trilinear curves of Fig. 7 provide a justification for analytical bond-models [30] that were derived on that line of argument. The slip modulus of the second stage of the curves (approximately 300 N/mm^3) falls within experimental scatter [3], [14].

Experimental determinations of traction–slip behavior shows significant scatter [3], [14]. A possible cause is the actual location of the sampling points where traction and slip are recorded. However, there is no consensus on this question since Nilson [13] and Dörr [3] found the curves to depend on the distance from the loaded end-face, while Mirza & Houde [9] report them to be location-independent. To verify this, the curves for the first twelve segments from the end-face have been collected into Fig. 8 for the analysis with $n_l = 2$. Apart from the curve for the first segment, which shows very weak behavior, the curves fall within a relatively narrow band. This suggests that at least the rising portion of the traction–slip curve is location-independent. The only discrepancies relate to the peak traction and to the subsequent softening, which do not show a systematic trend for they depend on local secondary cracking which is nonuniform along the rebar axis.

For cases of bond that are not critical to longitudinal splitting, e.g. flexural bond with sufficient cover, the present results justify application of a unique traction–slip curve. In other cases, the non-local format of traction–slip laws poses a problem as information like

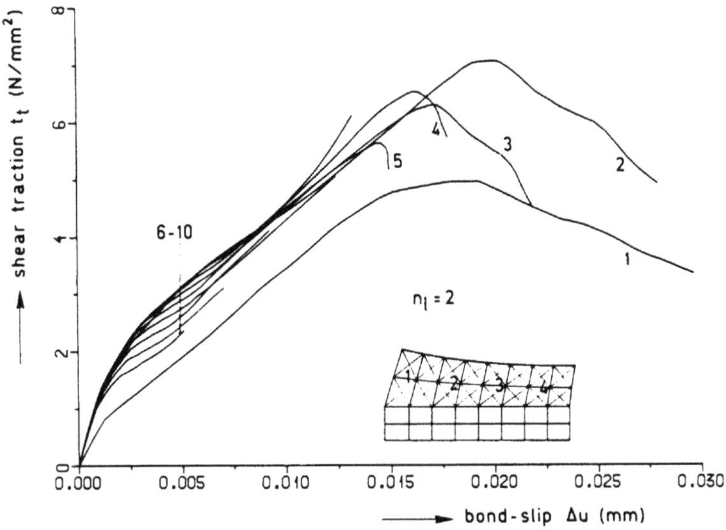

Fig. 8 Local shear traction versus slip in tension-pull specimen, with increasing distance from end-face ($n_1 = 2$).

'distance from the primary crack' is usually not available at local integration point level. For this reason, the search for local parameters like radial pressure [3], [11] should be continued, although the necessity of pursuing high accuracy is doubted since the majority of practical analyses refers to plane-stress configurations where circular bars are replaced by fictitious layers such that identities like radial bond component and lateral pressure are hardly definable.

7. ANCHORAGE STRUCTURE

The second example is a direct application from engineering practice. It concerns the anchorage structure of Fig. 9, which consists of a steel plate embedded in a massive concrete block. The steel plate is pulled out of the concrete by a vertical load *F* which is applied via an anchor bolt. High safety requirements asked for a careful assessment of the failure load, which justified a dedicated nonlinear finite element analysis.

An axi-symmetric schematization with eight-node elements has been adopted. The left boundary of the concrete in Fig. 9 is unconstrained,

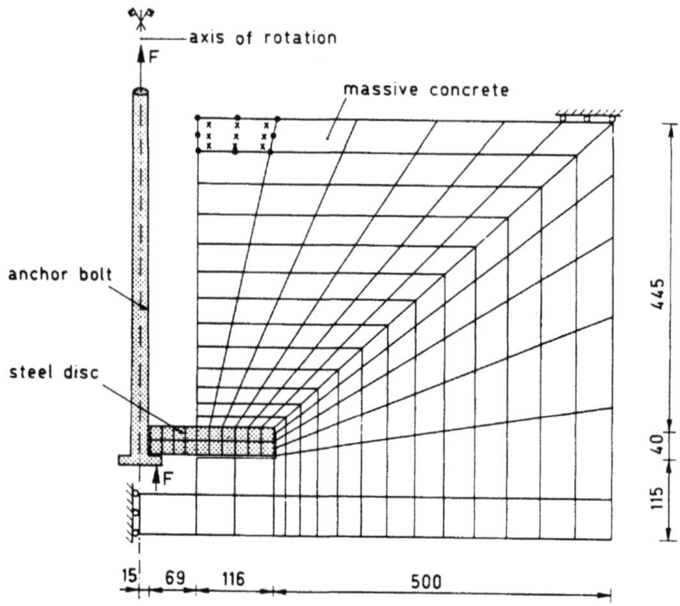

Fig. 9 Axi-symmetric finite element idealization of anchorage structure (dimensions in mm).

which provides a safe assumption. The anchor bolt has been excluded from the analysis, and the load has been directly applied at the bottom of the steel disc. The steel disc has been assumed to be rigidly connected to the concrete only at its top edge, whereas no contact has been assumed along its bottom and side edge. The counterpressure against the pull-out load has been assumed at the upper right corner of the mesh.

The coaxial rotating smeared crack concept has been adopted. The parameters were taken as: $E = 30\,000$ N/mm^2, $\nu = 0{\cdot}2$, $f_{ct} = 2{\cdot}5$ N/mm^2, $G_f = 100$ J/m^2 and a linear softening diagram. The structure included some reinforcement in the direction normal to the plane of Fig. 9. Instead of modeling this reinforcement directly, we preferred to incorporate its effect indirectly, via a somewhat large value $n_l = 10$ for the imagined number of localized longitudinal cracks in eqn (1). This value determines the crack band width and the steepness of the softening for longitudinal cracks. The effect of the parameter will be investigated by an additional analysis for $n_l = 5$. The analysis has been performed under indirect displacement control [20], using the opening displacement over the crack as the control parameter.

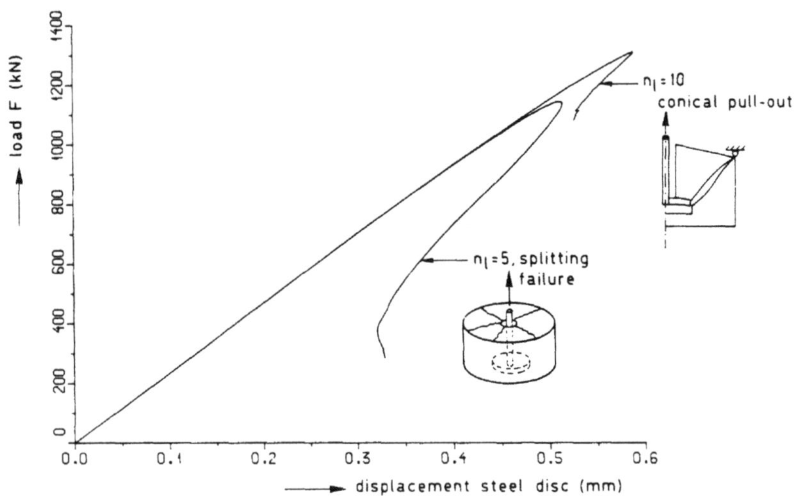

Fig. 10 Load F versus vertical displacement of steel disc of anchorage structure.

Figure 10 presents the response in terms of load versus loading point displacement. Prior to the peak, the nonlinearity in the curve is negligible. The peak load occurs very suddenly, whereafter a rapid drop of the load is observed. This suggests a dangerous, brittle type of fracture. Figures 11 and 12 confirm this suggestion. These figures show the incremental deformation fields and the crack patterns, both at the peak load and at the residual load of Fig. 10. At peak load, the fracture has developed only partially. Beyond peak load it rapidly propagates towards the upper-right corner and shows strong localization in the form of the pull-out of a conical frustum. Figure 12(b) provides an excellent interpretation of the elastic-softening theory. We observe that the steel disc and the concrete that surrounds the fracture surface drastically unload while the fracture propagates, i.e. the available elastic energy stored in the structure at peak load is suddenly released in the softening for creating the conical fracture surface.

The crack patterns of Fig. 11 reveal significant longitudinal cracking. It is recalled that the imagined number of *localized* longitudinal cracks in the underlying 3D situation, which determines the crack band width for the *smeared* longitudinal crack in the axi-symmetric simulation, was assumed to be $n_l = 10$. With this assumption the predicted extent of longitudinal cracking was not critical and on creating the conical

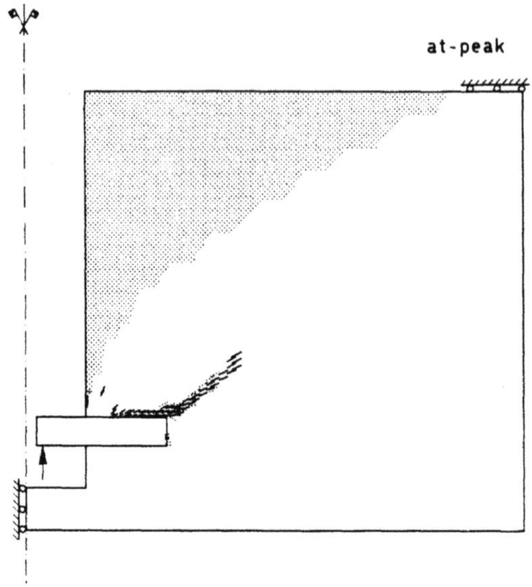

(a)

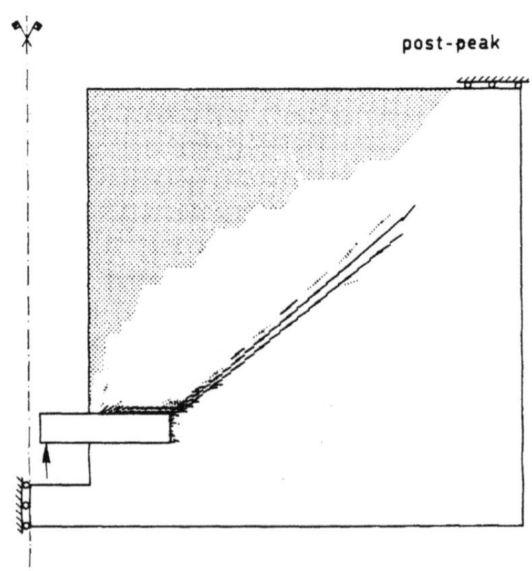

(b)

Fig. 11 Crack patterns at peak load (a) and residual load (b) (longitudinal cracks shaded) for the analysis with $n_l = 10$.

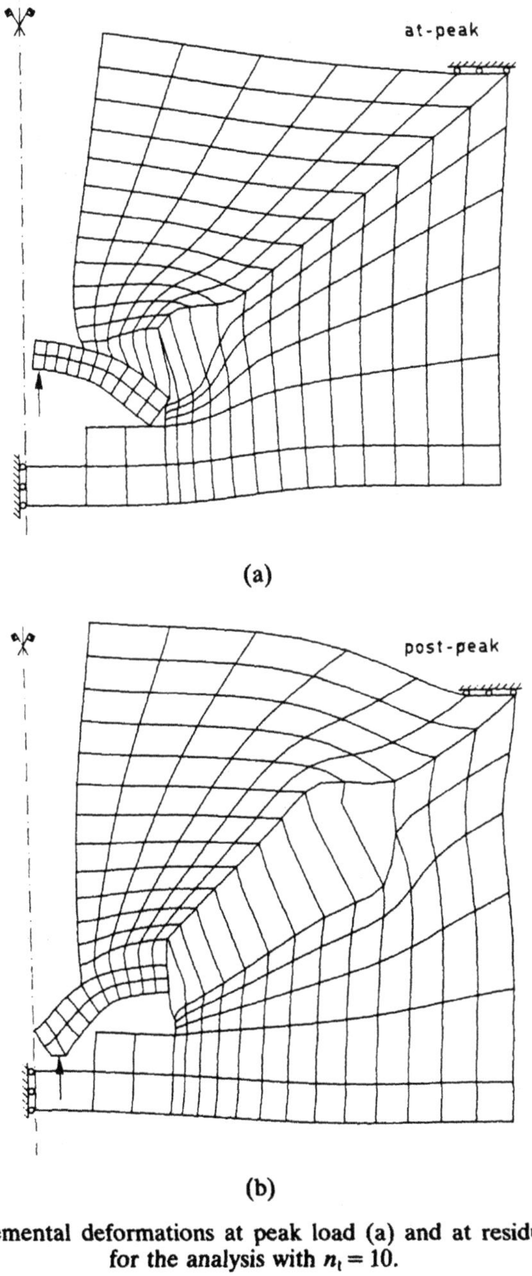

Fig. 12 Incremental deformations at peak load (a) and at residual load (b) for the analysis with $n_t = 10$.

pull-out all existing longitudinal cracks unloaded. An additional analysis was performed for $n_l = 5$, which doubles the crack band width and makes the softening in longitudinal cracking twice as steep. With this analysis the converse occurred as the structure snapped on longitudinal splitting while the conical cracks unloaded. This solution was obtained under standard linearized arc-length control, and is included in Fig. 10. The snap-back is somewhat less sharp as it is for the pull-out type fracture and the limit load has decreased. In analogy to the bond-splitting analysis, further quantitative improvements can be achieved by a fully 3D analysis which eliminates the uncertainties in the imagined number of n_l.

8. CONCLUDING REMARKS

In this chapter bond-slip and anchorage research has been rejuvenated with recent achievements of softening fracture mechanics. It has been demonstrated that computational tools have evolved so far that the final outcome is consistent with the essentials of the elastic-softening theory. The examples correctly reveal that the elastic energy stored in a structure is released for creating the fracture surface.

Accurate simulations of transverse secondary cracks, primary cracks and longitudinal splitting associated with bond-slip around reinforcing bars can be undertaken. This supports the verification of design rules regarding bond shear traction–slip curves. The simulations provide insight into the fundamentals of the bond and give an extension to experimental research. Regarding anchorage research, a direct application from engineering practice has been presented. It demonstrates that softening analyses are able to predict structural behavior associated with dangerous, brittle types of pull-out fracture.

ACKNOWLEDGEMENTS

The computations in this chapter have been performed using the DIANA finite element package of TNO Building and Construction Research (formerly TNO-IBBC). The anchorage study has been originally carried out for Kraftwerk Union AG by TNO-IBBC within a study dedicated to nonlinear analyses of an anchorage for mechanical components. The results are presented by permission of both companies. The bond study was financially supported by the Netherlands Academy of Sciences (KNAW) and the Centre for Civil Engineering Research, Codes and Specifications (CUR).

REFERENCES

[1] Blaauwendraad, J., Supply and demand in computational concrete mechanics—survey of a national attempt, *Proc. 2nd. Int. Conf. Computer Aided Analysis and Design of Concrete Structures*, ed. N. Bicanic *et al.* Pineridge Press, Swansea, 1990.

[2] Schäfer, H., A contribution to the solution of contact problems with the aid of bond elements. *Comp. Meth. Appl. Mech. Engng*, **6** (1975) 335–54.

[3] Dörr, K., Bond behaviour of ribbed reinforcement under transverse pressure. Final Report *IASS symp. Darmstadt*, ed. G. Mehlhorn *et al.* Werner-Verlag, Dusseldorf, 1978, Vol. 1, 13–24.

[4] de Groot, A. K., Kusters, G. M. A. & Monnier, Th., Numerical modelling of bond-slip behaviour. *HERON*, **26**(1B) (1981) 1–90.

[5] Rots, J. G., Computational modeling of concrete fracture. Dissertation, Delft University of Technology, Delft, 1988, 135 pp.

[6] Rehm, G., On the essentials of bond between concrete and reinforcement. *Deutscher Ausschuss fur Stahlbeton*, **138** (1961).

[7] Tepfers, R., Cracking of concrete cover along anchored deformed reinforcing bars. *Mag. of Concr. Res.*, **31**(106) (1979) 3–12.

[8] Bartos, P. (Ed.), *Bond in Concrete.* Applied Science Publishers, London, 1982.

[9] Mirza, S. M. & Houde, J., Study of bond stress–slip relationships in reinforced concrete, *J. Am. Concrete Inst.*, **76** (1979) 19–46.

[10] Gambarova, P. G. & Giuriani, E., Discussion of 'Study of transfer of tensile forces by bond' by D. H. Jiang *et al. J. Am. Concrete Inst.*, **82**(3) (1985) 381–3.

[11] Vos, E., Influence of loading rate and radial pressure on bond in reinforced concrete, Dissertation, Delft Univ. of Techn., 1983.

[12] Dragosavić, M. & Groeneveld, H., Concrete mechanics—local bond, Part I: Physical behaviour and constitutive consequences, Report BI-87-18. Part II: Experimental research, Report BI-87-19. TNO Inst. for Building Mat. and Struct., Delft, 1988.

[13] Nilson, A. H., Internal measurement of bond-slip. *J. Am. Concrete Inst.*, **69** (1972) 439–41.

[14] Lahnert, B. J., Houde, J. & Gerstle, K. H., Direct measurement of slip between steel and concrete. *J. Am. Concrete Inst.*. (1986) 974–82.

[15] Reinhardt, H. W., Cornelissen, H. A. W. & Hordijk, D. A., Tensile tests and failure analysis of concrete. *J. Struct. Engng* (ASCE) **112**(11) (1986) 2462–77.

[16] Ingraffea, A. R., Gerstle, W. H., Gergely, P. & Saouma, V., Fracture mechanics of bond in reinforced concrete. *J. Struct. Engng* (ASCE), **110** (1984) 871–89.
[17] Goto, Y., Cracks formed in concrete around deformed tension bars. *J. Am. Concrete Inst.*, **68**(4) (1971) 244–51.
[18] Jiang, D. H., Shah, S. P. & Andonian, A. T., Study of the transfer of tensile forces by bond, *J. Am. Concrete Inst.*, **81**(3) (1984) 251–9.
[19] Lutz, A. L. & Gergely, P., Mechanics of bond and slip of deformed bars in concrete, *J. Am. Concrete Inst.*, **64**(11) (1967) 711–21.
[20] de Borst, R., Computation of post-bifurcation and post-failure behavior of strain-softening solids. *Computers & Structures*, **25**(2) (1987) 211–24.
[21] Willam, K., Pramono, E. & Sture, S., Fundamental issues of smeared crack models, *Fracture of Concrete and Rock*, ed. S. P. Shah *et al.* Springer-Verlag, New York-Berlin-Heidelberg, 1989.
[22] de Borst, R. & Rots, J. G., Occurrence of spurious mechanisms in computations of strain-softening solids. *Engineering Computations*, **6** (1989) 272–80.
[23] Rots, J. G., Bond-slip simulations using smeared cracks and/or interface elements. Res. Report 85-01, Struct. Mech., Dept. of Civil Engng., Delft Univ. of Techn., 1985.
[24] Beranek, W. J., Experimental techniques for the analysis of deformation. Documentation-Page 119, Inst. TNO for building Mat. and Struct., Delft, 1980.
[25] Losberg, A. & Olsson, P., Bond failure of deformed reinforcing bars based on the longitudinal splitting effect of the bars. *J. Am. Concrete Inst.*, **76**(1) (1979) 5–18.
[26] Schmidt-Thrö, G., Stöckl, S. & Kupfer, H., Influence of the shape of the pull-out specimen and of the arrangement of the slip measurement on the results of pull-out tests. *Deutscher Ausschuss für Stahlbeton*, **378** (1986) 111–69.
[27] Bazant, Z. P. & Cedolin, L., Blunt crack band propagation in finite element analysis. *J. Engng Mech. Div.* (ASCE), **105**(2) (1979) 297–315.
[28] Rots, J. G., Kusters, G. M. A. & Blaauwendraad, J., Significance of crack models for bond-slip studies. *IABSE Reports 54*, Coll. Comp. Mech. of Reinforced Concrete, Delft Univ. Press, 1987 121–9.
[29] Rots, J. G., Hordijk, D. A. & de Borst, R., Numerical simulation of concrete fracture in 'direct' tension. *Proc. Fourth Int. Conf. Numerical Methods in Fracture Mechanics*, ed. A. R. Luxmoore *et al.* Pineridge Press, Swansea, 1987, 457–71.
[30] Dragosavić, M., Bond model for concrete structures. *Computer-Aided Analysis and Design of Concrete Structures 1*, ed. F. Damjanić *et al.* Pineridge Press, Swansea, 1984, 203–14.

Chapter 12

Analysis of Steel–Concrete Bond with Damage Mechanics: Non-linear Behaviour and Size Effect

J. MAZARS, G. PIJAUDIER-CABOT & J. L. CLEMENT
Laboratoire de Mécanique et Technologie, Ecole Normale Supérieure de Cachan, 94235 Cachan Cedex, France

ABSTRACT

The behaviour of reinforced concrete structures is determined by the steel–concrete bond. Numbers of tests have shown that the nonlinear behaviour of the interface is mainly due to a progressive degradation of a concrete layer around the reinforcement. This chapter is a review of recent works done on the application of damage mechanics to this problem, in particular at the LMT where a nonlocal damage model is used to describe this phenomenon. A comparison with pull-out tests shows its capabilities to give information at the global level (nonlinearity and ultimate strength) as well as at the local level (strain evolutions on R bars). On the same kind of tests we show the ability of the model to predict a size effect consistent with experimental observations. The phenomena related to the realization of the anchorage at the bottom of prestressed concrete members are also described. Finally we present the simulation of the behaviour of an anchor bolt (those correspoding to the 1990 RILEM round-robin) in which the sensitivity of the boundary conditions is particularly pointed out.

1. INTRODUCTION

Predicting the response of prestressed and reinforced concrete structures requires three major ingredients: (1) constitutive equations for steel and concrete which fit closely with experimental reality; (2) an objective computational method which allows predictions up to the failure of the structure; and (3) a consistent model for the steel–concrete interface.

The first item has been extensively studied over the past years and many models for concrete are available, e.g. fracturing strain models [1], continuous damage theories [2], or microplane models [3].

Finite element calculations in the presence of strain-softening due to distributed cracking in concrete is still a debated issue. From recent studies [4], it appears that the implementation of localization limiters provides physically realistic results with a finite energy dissipated at failure and objectivity with regard to the analyst's choice of finite element discretization. Results based on nonlocal continuum with local strain formulations agree with several size effect test data [5], [6]. The experimental aspects of the bond strength have been enlightened on numerous occasions [7], [8] but numerical aspects, and in particular the treatment of reinforcing bars in finite element calculations, may appear still unsatisfactory. Reinforcement can be modelled, in a finite element discretization, by unidimensional rods attached to the nodes of the mesh in which 2D or 3D elements represent the concrete. An adequate combination of shape functions provides the continuity of the displacement at the interface between steel and concrete (see Hibbitt *et al.* [9], and references therein). In this approach the steel–concrete interaction is neglected, especially the differential deformations resulting from Poisson's effects. Recently a trend to combine interface elements with regular bidimensional elements representing steel or concrete has emerged [10]–[12]. The nonlinear response of concrete near the steel bar is lumped into a fictitious interface which has special constitutive equations. This method has certainly become a powerful tool for predicting the response of reinforced concrete. In addition, some mechanisms such as friction which depend on the shape of the steel bars (deformed or not), have been included into the behaviour of interface elements. Implementing special bond-slip relations raises the problem of the identification of additional material properties. These characteristics are usually obtained from the analysis of pull-out tests or reinforced concrete tension pull specimens [13], [14]. It is not surprising that the identification is based upon the same lumping method of the damage zone to an interface of zero thickness. In Ref. [13] the thickness of the damaged layer around the reinforcement is equal to the radius of the bar.

Clément justified this technique from homogenization in 1987 (see Ref. [12]). Thus, omitting interface elements and considering that concrete is progressively damaged around the steel bar is strictly equivalent to using interface elements if the bar presents surface deformations, i.e. if the displacements are continuous at the interface. In this chapter, we will restrict the analysis to ribbed bars. In particular, the large relative slip at the interface observed when non-deformed bars are used will be left out of consideration. In this latter case interface elements are needed to represent friction between steel and concrete.

In finite element applications, omitting interface elements should certainly reduce the computational cost. However, it requires the implementation of accurate constitutive equations for concrete and this is sometimes too delicate, although consistent results can be obtained with smeared or discrete crack concepts [10], [15].

A scalar damage model for concrete has been developed over the past years at the Laboratoire de Mécanique et Technologie. It has been applied to the analysis of plain concrete, reinforced concrete and prestressed concrete structures [16]–[19]. Using three-dimensional analysis, Sun [20] proved recently that reinforcing bars could be modelled in a 2D calculation by an equivalent homogenized layer made of steel and concrete in which continuity of the displacements at the interface is assumed. The implementation of strain-softening constitutive equations for concrete raises however a major difficulty in numerical analysis. From calculations on plain and reinforced concrete elements [6], it appears that spurious localization instability and a strong mesh sensitivity renders the use of softening constitutive equations extremely delicate and sometimes physically unrealistic. A nonlocal damage model was developed to circumvent this problem [21]. Its application to reinforced concrete however has never been addressed because analysis must combine local and nonlocal constitutive equations. We will show that this question can be treated in a simple and efficient manner. Our discussion will be mainly focused on the analysis of the pull-out test. In our opinion, this study is a prerequisite before starting on analysis on full structures. Comparisons with experimental data will be presented and the size effect inherent to the fracture aspect of the pull-out failure will be studied from the numerical standpoint.

Finally some additional results are presented. One concerns the anchorage of wire at the bottom of prestressed concrete members for which recent experimental results have been obtained [22] and the other is related to the 1990 RILEM round robin on anchor-bolts [23].

2. CONSTITUTIVE EQUATIONS

The geometry for the pull-out test recommended by ACI or European standards is such that the steel bar embedded in concrete never yields. Therefore, linear elasticity is sufficient to describe the reinforcing bar. In usual computations on reinforced concrete members, elastoplasticity would be required. This refinement, which is rather classical in nonlinear finite element computations, is not of interest in the scope of the present discussion.

Among the possibilities offered by the continuous damage theory [2], we choose the isotropic (scalar) damage model. This choice is rather a simplifying assumption than an accurate description of the behaviour of concrete. Indeed, damage which represents the density of microcracks in the medium develops in specific directions: microcracks propagate perpendicular to the load in tension, and parallel to it in compression. In past studies, it was demonstrated however that the scalar model is sufficient to describe accurately the response of plain concrete notched plates and laboratory-type fracture specimens [6]. The stress–strain relationship is:

$$\sigma_{ij} = (1 - D)C_{ijkl}\varepsilon_{kl} \tag{1}$$

where σ_{ij} and ε_{ij} are the components of the stress and strain tensors respectively ($i, j, k, l \in [1], [3]$), C_{ijkl} are the initial stiffness moduli, and D is the damage variable. The material is initially isotropic with E and ν the Young's modulus and Poisson's ratio respectively. According to this scalar description of damage, concrete is assumed to remain isotropic up to failure. The damage variable D ranges from 0 for the virgin material to 1 at asymptotic failure ($\varepsilon_{ij} \rightarrow \infty$, $\sigma_{ij} \rightarrow 0$). In the presence of strain-softening, spurious localization modes and mesh inobjectivity may occur. To eliminate these unrealistic features, we use the nonlocal damage formulation [21]. The variable that controls the evolution of D is nonlocal. In the model, the positive strains control the growth of damage which is mainly due to microcracks opening in Mode I.

The following norm called 'equivalent tensile strain' is defined [26] as

$$\tilde{\varepsilon} = \sqrt{\sum_{i=1}^{3} (\langle \varepsilon_i \rangle)^2} \tag{2}$$

where $\langle \varepsilon_i \rangle = 0$ if $\varepsilon_i < 0$, and $\langle \varepsilon_i \rangle = \varepsilon_i$ if $\varepsilon_i > 0$; ε_i are the principal strains. $\tilde{\varepsilon}$ represents physically a measure of the amount of tensile strain for any arbitrary state of deformation in concrete. Next, we introduce the nonlocal variable $\bar{\varepsilon}$ which represents the average of $\tilde{\varepsilon}$ over the representative volume surrounding each point $\mathbf{x}$ in the material. $\bar{\varepsilon}$ will be the variable that controls the growth of damage:

$$\bar{\varepsilon}(\mathbf{x}) = \frac{1}{V_r(\mathbf{x})} \int_V \tilde{\varepsilon}(\mathbf{s})\alpha(\mathbf{s}-\mathbf{x})\,\mathrm{d}v \tag{3}$$

where V is the volume of the structure, $V_r(\mathbf{x})$ is the representative volume at point $\mathbf{x}$, and $\alpha(\mathbf{s}-\mathbf{x})$ is a weight function:

$$\alpha(\mathbf{s}-\mathbf{x}) = H(\mathbf{s}-\mathbf{x}) \exp\left[-(4\,|\mathbf{s}-\mathbf{x}|/l_c)^2\right] \tag{4}$$

$$V_r(\mathbf{x}) = \int_V \alpha(\mathbf{s}-\mathbf{x})\,\mathrm{d}v \tag{5}$$

H is the Heaviside function, equal to 0 if $|\mathbf{s}-\mathbf{x}| > l_c/2$, equal to 1 if $|\mathbf{s}-\mathbf{x}| \le l_c/2$.

l_c is the so-called characteristic length of the nonlocal continuum. It is proportional to the smallest size of the damage localization zone. This length was measured experimentally in Ref. [25]; $l_c \approx 3d_a$ in which d_a is the maximum size of the aggregate in concrete.

The evolution of D is specified according to a formalism which is similar to plasticity:

$$F(\bar{\varepsilon}) = \bar{\varepsilon} - K \tag{6}$$

$$\begin{aligned} &\text{If } F(\bar{\varepsilon}) = 0 \text{ and } \dot{F}(\bar{\varepsilon}) = 0 \text{ then } \dot{D} = f(\bar{\varepsilon}) \text{ (loading)} \\ &\text{If } F(\bar{\varepsilon}) < 0 \text{ or } F(\bar{\varepsilon}) = 0 \text{ and } \dot{F}(\bar{\varepsilon}) < 0 \text{ then } \dot{D} = 0 \text{ (unloading)} \end{aligned} \tag{7}$$

Equation (6) defines the damage surface, and $f(\bar{\varepsilon})$ is the evolution law of damage in eqn (7). K is the softening parameter and takes the largest value of $\bar{\varepsilon}$ ever reached at the considered point in the medium. Initially $K = K_0$, in which K_0 is the threshold of damage. K_0 may be regarded as the tensile strain at which damage is initiated. We consider that the threshold K_0 is reached when the stress is maximum in uniaxial tension:

$$K_0 = \frac{f_t}{E} \tag{8}$$

in which f_t is the tensile strength of concrete. From experimental data, the growth of damage is different in tension (microcracks are directly opened by the load) from compression (microcracks open due to Poisson's effects). Thus, damage is decomposed into two parts, D_t for tension and D_c for compression. D is computed from the relation in eqn (9) which combines these two types of damage:

$$D = \alpha_t D_t + \alpha_c D_c \tag{9}$$

The factors α_t and α_c are such that in uniaxial tension $\alpha_t = 1$, $\alpha_c = 0$, and $D = D_t$, while in compression $\alpha_c = 1$, $\alpha_t = 0$, and $D = D_c$; they are functions of $\bar{\varepsilon}$, as we see below. The damage variables D_t and D_c are also functions of $\bar{\varepsilon}$:

$$D_t = 1 - \frac{K_0(1 - A_t)}{\bar{\varepsilon}} - \frac{A_t}{\exp\left[B_t(\bar{\varepsilon} - K_0)\right]} \tag{10a}$$

$$D_c = 1 - \frac{K_0(1 - A_c)}{\bar{\varepsilon}} - \frac{A_c}{\exp\left[B_c(\bar{\varepsilon} - K_0)\right]} \tag{10b}$$

and constants A_c, B_c, A_t, B_t are material parameters. Their range of variation is (see Ref. [26])

$$\left.\begin{array}{l} 1 \le A_c \le 1{\cdot}5 \\ 10^3 \le B_c \le 2 \times 10^3 \\ 0{\cdot}7 \le A_t \le 1{\cdot}2 \\ 10^4 \le B_t \le 5 \times 10^4 \end{array}\right\} \tag{11}$$

The material parameters E, ν, K_0, A_t, A_c, B_t, B_c can be identified from standardized experiments: the response of a standard cylinder (diameter 16 cm, height 32 cm) in compression yields E, ν, A_c, B_c and data from three-point bending beams (40 cm × 40 cm × 10 cm) provide the remaining unknowns. The identification procedure was detailed by Saouridis [6]. Figures 1(a) and (b) show the stress–strain law in tension and compression, respectively. The material parameters are:

$$E = 40\,000\ \text{MPa}, \quad \nu = 0{\cdot}2, \quad K_0 = 1{\cdot}1 \times 10^{-4},$$
$$A_t = 1, \quad B_t = 2 \times 10^4, \quad A_c = 1{\cdot}2, \quad B_c = 1515.$$

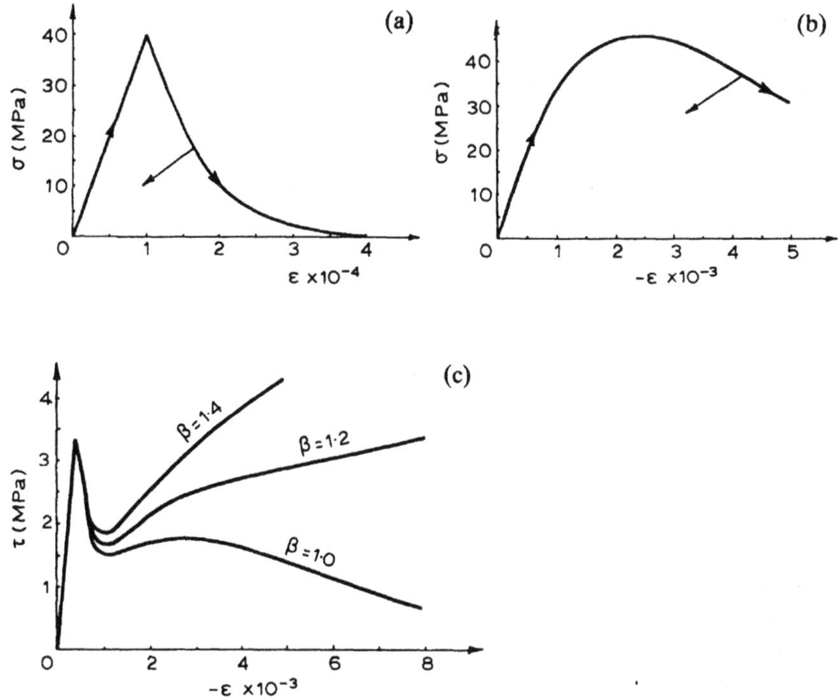

Fig. 1 Responses of the model to (a) tension, (b) compression, and (c) shear.

The factors α_t and α_c are expressed as non-dimensional functions of the principal strains. We introduce the notation:

$$\langle \varepsilon_i \rangle = \varepsilon_{ci} + \varepsilon_{ti} \tag{12}$$

in which ε_{ti} is the positive strain due to positive stresses and ε_{ci} is the positive strain due to negative stresses (Poisson's effect). α_t and α_c are defined as (see Ref. [26]):

$$\alpha_t = \sum_{i=1}^{3} \left(\frac{\varepsilon_{ti}(\varepsilon_{ti} + \varepsilon_{ci})}{\bar{\varepsilon}^2} \right)^{\beta} \qquad \alpha_c = \sum_{i=1}^{3} \left(\frac{\varepsilon_{ci}(\varepsilon_{ti} + \varepsilon_{ci})}{\bar{\varepsilon}^2} \right)^{\beta} \tag{13}$$

If all the stress components have the same sign, we obtain $\alpha_t = 1$ and $\alpha_c = 0$ or $\alpha_c = 1$ and $\alpha_t = 0$. If the stress state is a combination of tension and compression, the value of α_t and α_c depends both on the relative value tension–compression and on the value of β. The first formulation of this model [26] was with $\beta = 1$, but in this case the results obtained for the shear strength were underestimated which is a crucial problem for the steel–concrete bond description. Studies on this problem [24] have shown that $\beta > 1$ was necessary for a good response in the case of shear problems. For similar reasons, the shear retention factor was introduced in smeared crack models [10], [27]. In these constitutive equations the effect of damage is unidirectional, i.e. damage affects the longitudinal stiffness of a specimen loaded in compression or tension. The shear modulus is kept constant in the original smeared crack model and the shear retention factor was introduced to account for the effect of damage on the shear modulus of concrete. It is appropriate to address the question of the identification of β. From the model, experiments on concrete subjected to shear stresses would be required. Such tests are not usual and are very difficult to set up properly. An indirect procedure of identification from pull-out tests may be preferred. In the fourth part of this chapter, we will show that numerical simulations on pull-out tests are extremely sensitive to values of β which can be easily identified then. Figure 1(c) shows the response of the model to pure shear for various values of β. Once the peak stress has been reached, damage starts to grow, especially D_t. Accordingly, the response exhibits first a strain-softening portion, then hardening (regarded as the effect of internal friction and aggregate interlock) may occur.

3. FINITE ELEMENT IMPLEMENTATION: MESH OBJECTIVITY

These constitutive equations have been implemented in the finite element code CESAR developed at LCPC [28]. A secant stiffness matrix algorithm is used to solve the nonlinear equations of equilibrium. This algorithm ensures convergence of the iterative process on points located beyond the maximum load of a structure, on the softening portion of the response. The nonlocal model does not require any special finite element formulation (see Ref. [29]). For each integration point, the other gauss points which belong to its representative volume are identified before the calculation starts. This information which is needed in the calculation of the average effect strain, is stored once and for all in a connectivity table.

In reinforced concrete applications, we need to combine local (steel) and partially nonlocal (concrete) stress–strain relations. In the boundary layer of concrete touching steel, regions of the representative volume of concrete attached to an integration point lie in steel. A similar problem is faced for the boundary layer located near a free surface: regions of the representative volume lie outside the structure. In these cases the averaging of the equivalent strain in eqn (2) must be reconsidered: the shape of the ribs and the material properties of steel should not influence the calculation of $\bar{\varepsilon}$ in the interface region. Therefore the presence of steel can be regarded as a special boundary condition applied to concrete which does not modify the constitutive equations of the material. The computation of $\bar{\varepsilon}$ near the interface is identical to the computation of $\bar{\varepsilon}$ near a free boundary, and regions of the representative volume protruding outside concrete are deleted from the integrals in eqns (3) and (5). In a recent work, Pijaudier-Cabot & Berthaud [30] showed from the study of an interacting crack system that damage should be nonlocal. Justifications for the calculation of the nonlocal damage in the boundary layer near a free surface require the study of the interaction of a system of cracks with a free surface. This work is currently in progress.

Computations on a pull-out test with various meshes were carried out to investigate mesh sensitivity. The specimen is a cylinder of concrete cast in a rigid metal hollow cylinder of internal diameter $\varnothing = 20$ cm and height $h = 15$ cm (Fig. 2). A steel bar of diameter $d_b = 40$ mm is embedded at the centre of the specimen. The pull-out force is applied at one end of the bar and displacements on the circumferential surface of the cylinder of concrete are prevented. The material parameters for concrete are $E = 29\,000$ MPa, $\nu = 0{\cdot}20$, $A_c = 1{\cdot}07$, $B_c = 1512$, $A_t = 1$, $B_t = 2 \times 10^4$, $\beta = 1$ and $l = 60$ mm. For steel, we took $E = 210\,000$ MPa and $\nu = 0{\cdot}3$. The finite element discretizations are two-dimensional axisymmetric. Five meshes made of four noded elements were considered (15, 35, 120, 460, 810 elements, see Fig. 2(b)(c)(d)). Computations with the local ($l_c = 0$) and nonlocal damage model were compared.

We have plotted on Fig. 3 the evolution of the energy dissipated due to damage, W, with the number of finite elements in the mesh. This energy corresponds to the area under the load–displacement curves and it represents the amount of energy consumed when the steel bar is completely pulled out. From this plot, we can see that in local computations W goes to zero when the number of elements increases. It means that bond-breaking occurs without energy consumption, which is physically unrealistic. The nonlocal formulation remedies to

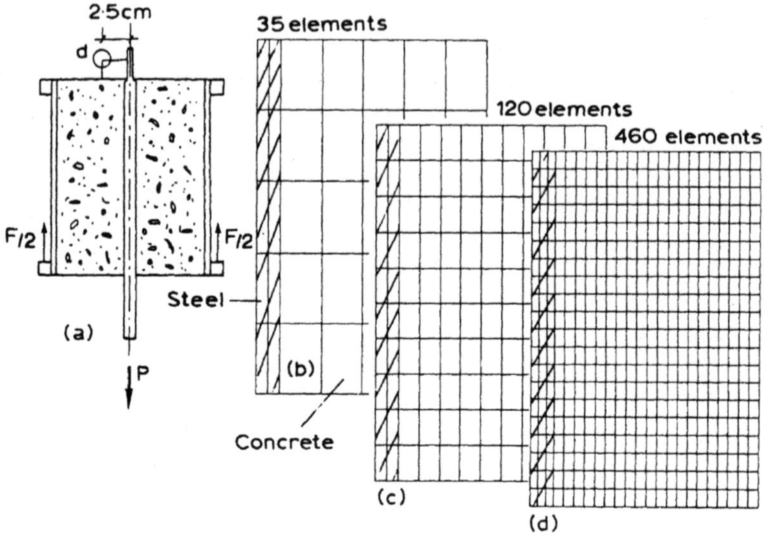

Fig. 2 Pull-out test: (a) specimen geometry; and finite element meshes (b, c, d) for the convergence study.

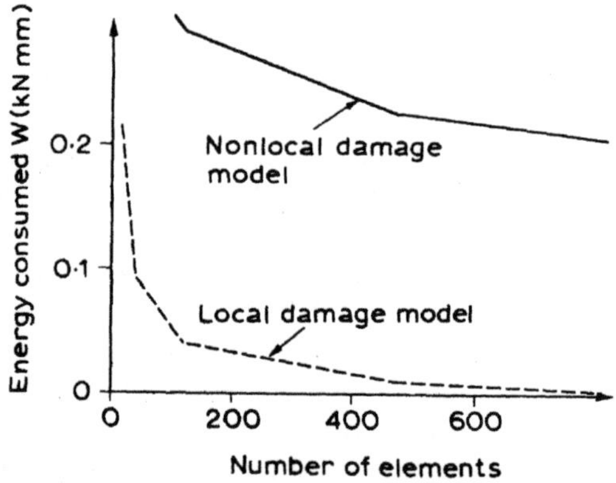

Fig. 3 Convergence of the energy dissipated at failure with mesh refinement—comparison for the local and nonlocal calculations.

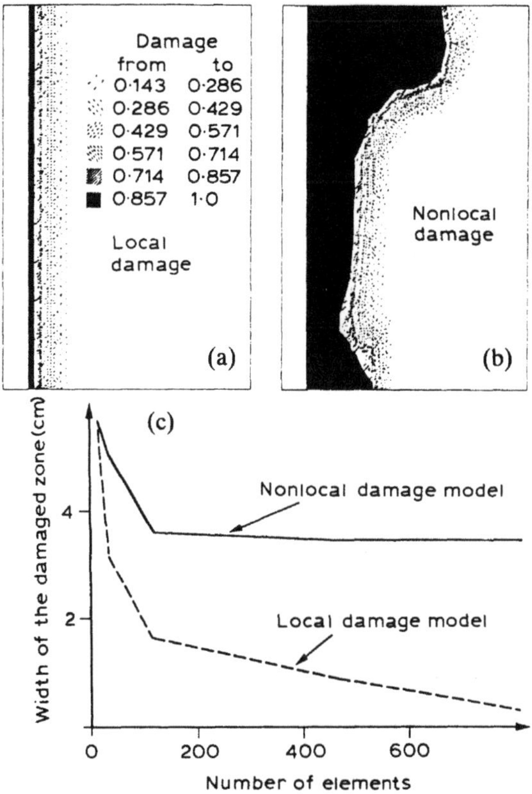

Fig. 4 Damage zone at the peak load (a, b); comparison of the local and the nonlocal calculation for the mesh with 120 elements; (c) width of localization zone at failure versus the number of elements.

this problem and a proper convergence towards a non-zero energy dissipated at failure is observed. Figure 4 shows the local and nonlocal results for the mesh with 120 elements. The dark region represents concrete which has failed ($0{\cdot}85 < D < 1$). In the local calculation, damage remains confined to the row of elements located next to the steel bar. This is not the case for the nonlocal computation. We see that damage has propagated over a large number of elements and the shape of the damage zone is consistent with previous computations and experimental observations [10]. We have reported on Fig. 4(c) the evolution of the thickness of the damage zone at failure with mesh refinement. h is defined as the minimum thickness of the damaged layer around the bar in which $D > 0{\cdot}85$. As expected, h tends to zero when the element size goes to zero in the local analyses, and the width of the band is clearly controlled by the size of the finite elements. h is almost constant in the nonlocal calculation for meshes made with 120 elements and more. A three-dimensional solution to this damage localization problem would be required to establish accurately the relation between the width of the localization zone and the characteristic length of the material.

4. COMPARISON WITH TEST DATA

Adrouche & Lorrain [31] performed a series of pull-out tests on specimens similar to Fig. 2(a). The diameter of the concrete cylinder was reduced to $\varnothing = 85$ mm, with $h = 25$ cm and $d_b = 12$ mm. Tests were conducted at an early age (3 days). Strain gauges were glued on the steel bars at regular intervals (see Fig. 5), and the strain profiles along the bar were recorded at different load levels. From compression and bending tests, the following material properties were measured: $f_c = 16$ MPa (compressive strength), $f_t = 3$ MPa, $E = 20\,909$ MPa. In the present model we assumed that f_t was reached at the onset of damage in tension (for $\tilde{\varepsilon} = K_0$), thus we obtained $K_0 = 1{\cdot}1 \times 10^{-4}$. The parameters A_c, B_c were fitted from the reported data in compression: $A_c = 1{\cdot}07$, $B_c = 2000$, with $\nu = 0{\cdot}2$. The experimental stress–strain curve of concrete in tension was not reported and usual values of A_t and B_t were selected: $A_t = 0{\cdot}8$, $B_t = 2 \times 10^4$. Finally, the maximum aggregate size was $d_a = 3{\cdot}15$ mm which yields $l_c = 9{\cdot}45$ mm.

We used a mesh with 160 degrees of freedom for the axisymmetric computation and fitted the coefficient β so that numerical results would give the same strain at gauge A when the maximum load applied in the experiments is reached. We obtained $\beta = 1{\cdot}1$.

Figure 5 presents the comparison between the calculated and measured strain profiles for the loads $P = 2$, 14, 18, 24 kN. Considering the lack of information provided, especially the response of concrete in tension at early ages, the agreement is good. Figure 5 shows also the progression of damage with increasing loads. In these experiments (see also Ref. [32]) damage was first noticed on the global response at $P = 14$ kN (onset of nonlinear behaviour). Our calculation agrees with this observation. The experiments were not conducted up

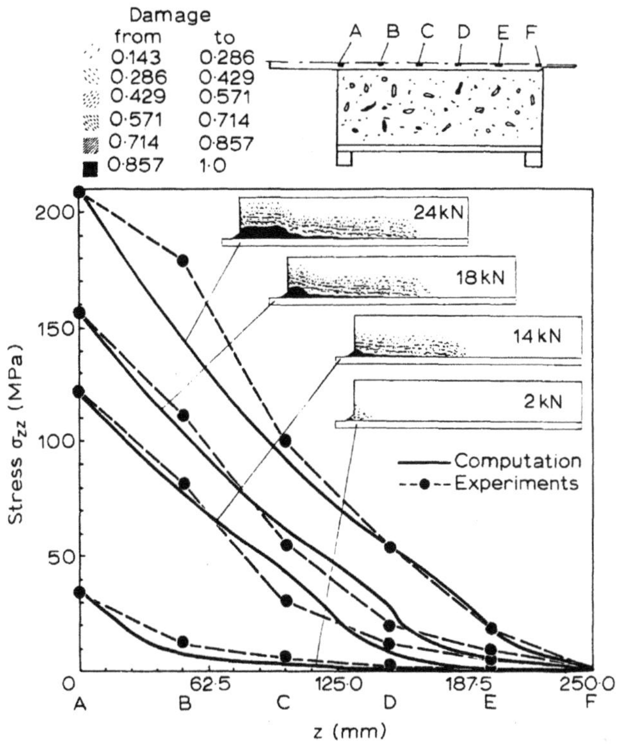

Fig. 5 Comparison of the predicted strain distribution in the reinforcing bar with Adrouche's results [32].

to the maximum load, we calculated it at approximately 28 kN. Our comparision further suggests that it is possible to obtain β with reasonable accuracy from load versus slip (or strain in the steel bar) relations.

5. SIZE EFFECT AND PULL-OUT TESTS

In most building codes, ACI [33], CEB [34], the steel–concrete bond is characterized by a strength criterion. On the other side, various experiments (see Ref. [14]) point out the fracture aspect of the pull-out failure which can be decomposed into two mechanisms:

—The splitting failure is caused by the radial pressure applied by the lugs of the bar to the concrete; this pressure is balanced by circumferential tensile forces causing cracking in concrete.

—The shear failure occurs if the concrete cover is much larger. Concrete is crushed in front of the lugs and cracking occurs in a band (the 'shear band') located near the interface. Debonding does not appear however; the lugs prevent any steel–concrete relative slip. Large slip is the consequence of microcracking in the damage zone around steel. This crack opening is represented by a shear strain (fracturing strain) in the homogenized material equivalent to damaged concrete [35].

Since cracking does exist, a size effect should certainly be expected in both types of failure. Bazant & Sener [36] studied the size effect on the pull-out strength from experiments on geometrically similar specimens. They proposed an approximate prediction formula:

$$\frac{V_u}{C_1} = k_1\left(1 + \lambda_0^{-1}\frac{d_b}{d_a}\right)^{-1/2} \tag{14}$$

where:

$$C_1 = \left(1{\cdot}23 + 3{\cdot}23\frac{c}{d_b} + 53\frac{d_b}{l_d}\right)\sqrt{f_c} \tag{15}$$

V_u is the maximum shear stress, d_b is the diameter of the steel bar, d_a is the maximum aggregate size, c is the minimum cover and k_1, λ_0 are material properties. The constant C_1 in eqn (14) is computed in psi (1 psi = 6895 N/mm^2) and corresponds (from Ref [8]) to empirical formula (15). V_u is computed from test data:

$$V_u = \frac{P_{max}}{S l_d} \tag{16}$$

where S is the nominal surface area of the embedded bar; P_{max} is the maximum pull-out load. Equation (15) may be rewritten in a form suitable for linear regression from which k_1 and λ_0 are identified:

$$\frac{C_1^2}{V_u^2} = \frac{1}{k_1^2} + \frac{1}{\lambda_0 k_1^2}\frac{d_b}{d_a} \tag{17}$$

The specimens tested were concrete cubes of side length 38·1 mm, 76·2 mm, and 152·4 mm, in which a steel bar of diameter 1·6 mm,

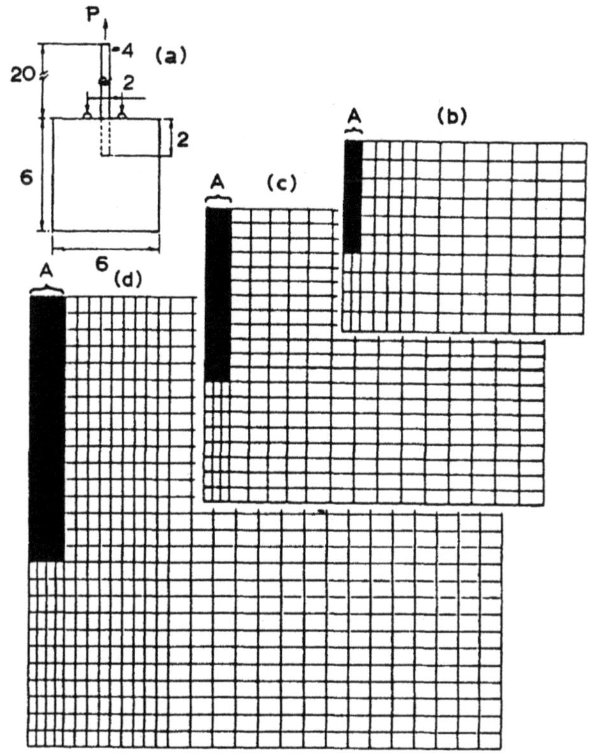

Fig. 6 Bazant & Sener pull-out tests [36]: (a) specimen geometry; (b, c, d) finite element meshes.

3·2 mm, and 6·4 mm, respectively, was embedded. The embedment length l_d was 12·7 mm, 25·4 mm and 50·8 mm respectively. The load was applied on the steel bar, and the cube was held down by a square sleeve (see Fig. 6(a)). The side length of the sleeves respected also the geometrical similarity (12·7 mm, 25·4 mm, 50·8 mm). The compressive strength of concrete reported was $f_c = 45{\cdot}8$ MPa (measured on companion cylinders), and the maximum aggregate size was $d_a = 6{\cdot}4$ mm. The material parameters which were not reported were fitted so that the closest possible prediction of the response of the *medium size specimen alone* was achieved. We obtained: $f_t \approx f_c/10 = 4{\cdot}5$ MPa, $E = 40\,000$ MPa, $\nu = 0{\cdot}2$, $K_0 = 1{\cdot}1 \times 10^{-4}$, $A_t = 1$, $B_t = 2 \times 10^4$, $A_c = 1{\cdot}2$, $B_c = 1515$ and $\beta = 1{\cdot}065$ (with $l_c = 3$, $d_a = 19{\cdot}2$ mm). It must be stressed that this set of material properties was obtained by fitting the model with results on specimens of one size. *Results on the two other sizes are simple predictions.* The same procedure was followed by Saouridis [6] and yielded a size effect on the tensile strength of concrete consistent with experimental data.

For the three specimen sizes, axisymmetric finite element computations were carried out. The concrete block was represented as a cylinder of radius R ($R = 21{\cdot}5$ mm, 43 mm and 83 mm) such that the cross-sections were kept identical to those in the experiments. Variations of R certainly affect the cover of the bar. However, we did not observe substantial differences between computations in which R is equal to half the side of the square and our case, with identical cross sections. The finite element meshes are presented in Fig. 6(b), (c) and (d), the numbers of degree of freedom being 160, 880 and 3200, for the small, medium and large size specimens, respectively. In the calculation, displacements of the steel bar (at points A on Fig. 6) were controlled.

Figure 7 shows the force versus displacement curves for the three sizes and Fig. 8 shows the identification of the size effect relation in eqns (14)–(17) and its comparison with experimental data. We found $k_1 = 4{\cdot}47$ and $\lambda_0 = 0{\cdot}303$. On the usual $\log(V_u/C_1)$ versus $\log(d_b/d_a)$ plot (see Fig. 8(b)), the numerical predictions are consistent with the experiments. Due to the size effect, the maximum shear strength V_u varies from 17 MPa for the smallest specimen to 14 MPa for the largest

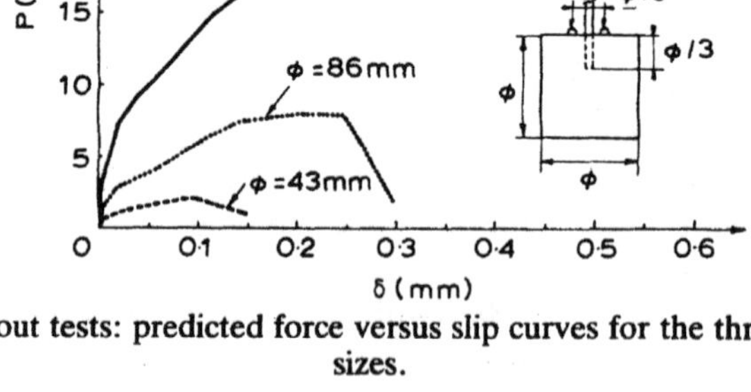

Fig. 7 Pull-out tests: predicted force versus slip curves for the three specimen sizes.

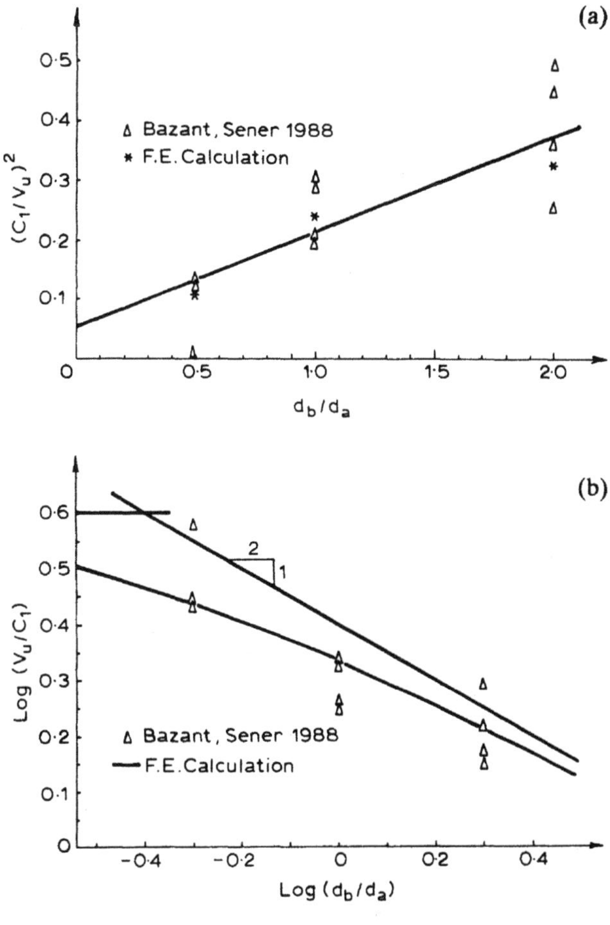

Fig. 8 (a) Identification of the size effect law; (b) size effect in the pull-out test, comparison with experimental data.

specimen. This 20% variation supports the applicability of the size effect formula.

6. ANCHORAGE ZONE FOR PRESTRESSED CONCRETE MEMBERS

This work is part of a French research programme [17] on prestressed concrete members and on the characterization of phenomena linked to the anchorage of wires tightened before moulding the concrete and progressively released after hardening.

Experiments have been done at CERIB [22] on a cylindric concrete specimen (diameter = 15 cm, length = 80 cm, see Fig. 9). The wire (diameter = 7 mm) was placed at the centre of the cylinder and tightened at 4·8 kN. Several measures have been performed:

—The drive in of the wire, at the tip of the member, during its release;
—The evolution of the load in the concrete specimen when prestressing;
—The strain evolution, measured by gauges, along the wire in the anchorage zone.

Calculations have been performed with the same model and computer code as for the pull-out test calculations. An axisymmetric geometry and adapted boundary conditions were used, and material parameters have been determined from previous tests (compression and flexion bending for concrete and traction for steel). The set of parameters used is $E = 29\,000$ MPa, $\nu = 0{\cdot}2$, $K_0 = 1{\cdot}1 \times 10^{-4}$, $A_t = 0{\cdot}5$, $B_t = 9 \times 10^3$, $A_c = 1{\cdot}3$, $B_c = 1400$, $\beta = 1$ and $l_c = 30$ mm (for steel $E_s = 2{\cdot}1 \times 10^5$ MPa, $\nu = 0{\cdot}3$). We present in Fig. 9 two types of result obtained, and for each of them the comparison experiment-calculation. Figure 9(a) shows the evolution of the load as a function of the drive in of the wire during prestressing. Figure 9(b) shows the evolution of the tension of the wire (obtained from strain gauge measurements) at different locations of the anchorage zone, as a function of the load released at the tip of the wire. This last figure is interesting because it relates the effects linked to the progressive damage of concrete along the wire during the anchorage process. This degradation is due to the combination of two effects: the drive in of the wire and its expansion which creates a high 3D stress state on concrete. Outside concrete, at the tip of the wire, the load decreases from 4580 daN to 0 (gauge J13). Inside the specimen, the load decreases more or less quickly and depends on the evolution of the damage zone around the wire:

(1) At 5 cm of the tip (gauge J12) the load seems constant during a short moment and decreases after, as quickly as outside concrete.

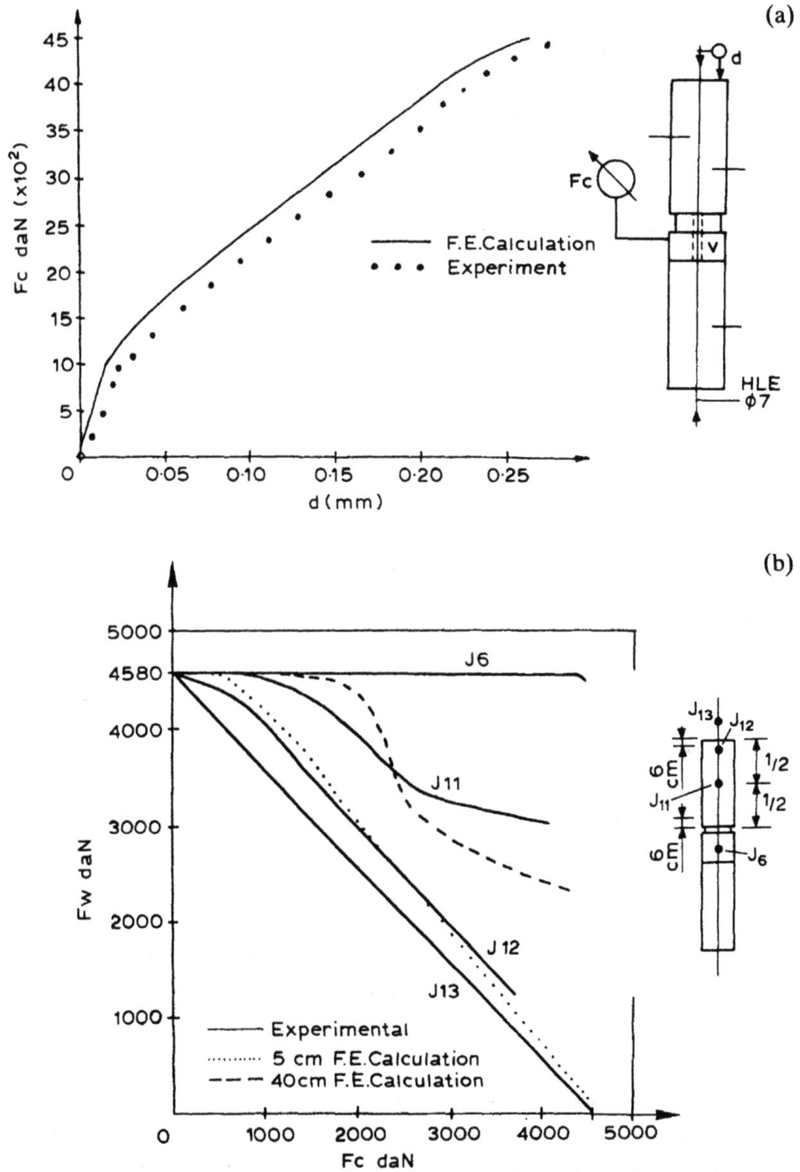

Fig. 9 Prestressed concrete members: (a) evolution of the load inside the specimen with the drive in of the wire; (b) evolution of the load along the wire, obtained from gauge measurements, during the realization of the anchorage.

(2) At 40 cm (gauge J11) the decrease appears later and a residual value exists, even after the complete release of the wire.
(3) In the middle of the specimen (gauge J6) the load is not affected by the release, because this point is outside the anchorage zone.

The evolution given by calculation is of good quality for the global behaviour (Fig. 9(a)). For the local behaviour (Fig. 9(b)), trends are the same and values are not very different. This is important in order to demonstrate the ability of the model to describe well active anchorages, in the case of prestressed concrete, and passive anchorages in the case of the pull-out test. The main difference between the two cases is the state of stress applied to concrete around the wire, which in the first case expands with the release of the tension and then applies a combination of shear and compression, while in the second case the stress around the bar is mainly shear.

7. SIMULATION OF THE BEHAVIOUR OF AN ANCHOR BOLT

Within the RILEM round-robin we have predicted the behaviour of an anchor bolt using the nonlocal damage model [23]. The calculation is two-dimensional, the experimental results being unknown at the time of the calculation. The following presentation is restricted only to the effects of boundary conditions. Figures 10(a) and (b) give the deformed meshes of the problems studied (steel bolt embedded in a concrete specimen) with two boundary conditions, which differ only by the fact that point K at the top of concrete specimen is free in the first case and fixed in the second case.

Figure 10(c) shows the results obtained for the global behaviour in the two cases. When K is fixed, the strength of the anchor bolt is more than twice the strength computed when K is free to move. The case in which K is free to move yields a ductile behaviour while the case in which K is fixed leads to an instability. The analysis of the evolution of the damage zone shows that the plateau in the behaviour of the case in which K is free corresponds to the evolution of two main 'cracking zones' at the same time, and also, at the end, to the damage zone in concrete located under the support. When K is fixed, only one main 'cracking zone' is formed and the instability is due to the sudden crushing of concrete under high compressive stress above the bolt (see Fig. 10(a)).

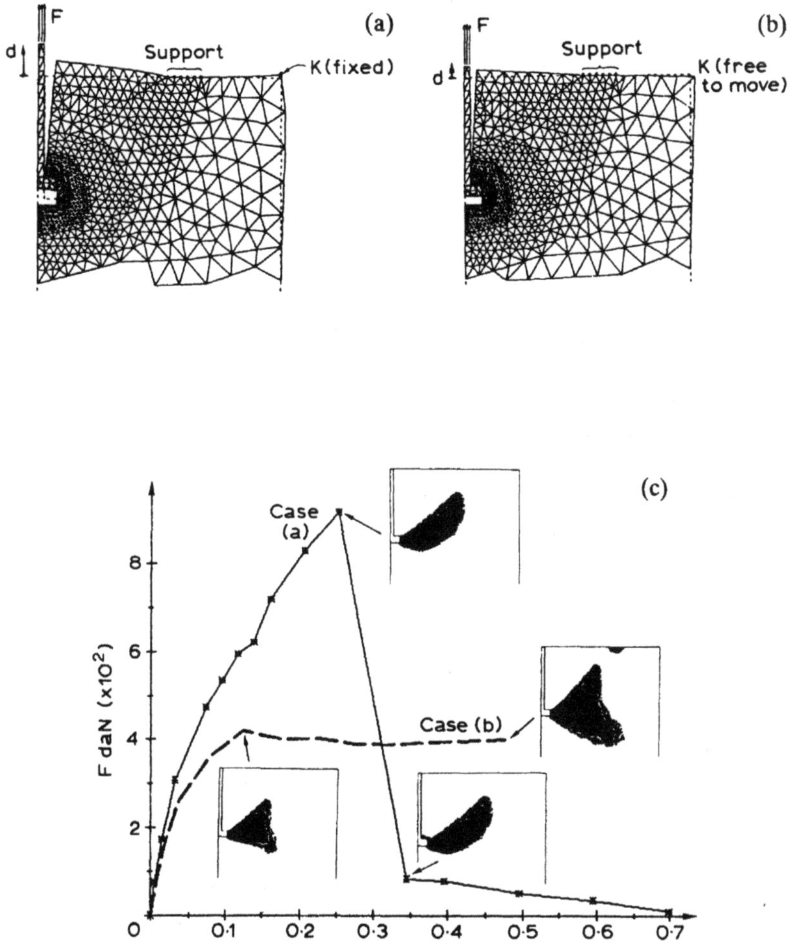

Fig. 10 Simulation of the behaviour of an anchor bolt: (a), (b) deformed mesh for two different boundary conditions; (c) corresponding global behaviours.

These examples show that boundary conditions are very important because they can lead to different failure mechanisms and therefore to different global behaviours. The calculation shows exactly this and, more importantly, it indicates that it is necessary to be extremely careful in the analysis of experimental conditions.

8. CONCLUSIONS

(1) The proposed method of analysis of the response of reinforced and prestressed concrete elements is based on two assumptions: (a) there is no discontinuity of displacements at the steel–concrete interface; (b) the isotropic (scalar) damage model is accurate enough to model the response of concrete under shear stresses developed around bars and wires.

(2) The implementation of the nonlocal damage model in the analysis of reinforced concrete elements must combine local and nonlocal constitutive relations. This is achieved by modifying the computation of the nonlocal damage variable. When the domain of integration intersects the steel bars, the zone lying in steel is chopped off so that the average is computed from the local effective strain in concrete only. Finite element computations on an axisymmetric pull-out test show that the distribution of damage at failure is not sensitive to mesh refinement. The width of the damaged layer of concrete around the bar does not depend on the size of the finite elements and should be controlled by the characteristic length of concrete. Finally, the energy dissipated at failure (i.e. when the bar is completely pulled out), converges toward a non-zero value with mesh refinement. These properties are not observed in calculations when local strain-softening constitutive equations are used and they are essential in the scope of future applications to reinforced concrete beams or frames because computations should not be affected by a spurious damage localization due to the finite element discretization.

(3) The size and shape of the damage zone are consistent with experiments and with previous calculations with smeared or discrete crack models. Comparisons with existing test results show that the present technique allows accurate predictions of the bond strength and of the evolution of damage near the interface. The formulation is also able to predict a size effect consistent with experimental observations.

(4) The description of active anchorages in the case of prestressed members is of good quality even if the stress state around the wire includes a high compressive stress due to its expansion during the realization of the anchorage.

(5) Through the simulation of the behaviour of an anchor bolt, we have shown the sensitivity of the model to different, but close, boundary conditions which can activate different mechanisms of failure.

REFERENCES

[1] Dougill, J. W., On stable progressively fracturing solids. *J. of Applied Math. and Phys.*, (ZAMP), **27** (1979) 423–46.

[2] Mazars, J. & Pijaudier-Cabot, G., Continuum damage theory: application to concrete. *J. Eng. Mechanics* (ASCE), **115**(2) (1989) 345–65.

[3] Bazant, Z. P. & Prat, P. C., Microplane model for brittle plastic material, Parts 1 & 2. *J. Eng. Mechanics* (ASCE), **114**(10) (1988) 1672–1702.

[4] Mazars, J. & Bazant, Z. P. (Eds), Strain localization and size effect due to cracking and damage. *Proc. of the CNRS/NSF Workshop* held in Cachan, Sept. 1988. Elsevier, London, UK, 1989.

[5] Bazant, Z. P. & Lin, F. B., Nonlocal smeared cracking model for concrete fracture. *J. Struct. Eng.* (ASCE), **114**(11) (1988) 2493–2511.

[6] Saouridis, C., Identification et numérisation objective des comportements adoucissants: une approche multiéchelle de l'endommagement du béton. Thèse de Doctorat, Université Paris 6, France, 1988.

[7] Somayaji, S. & Shah, S. P., Bond stress versus slip relationships and cracking response of tension members. *J. of ACI*, **78**(3) (1981) 217–25.

[8] Orangun, C. O., Jirsa, J. O. & Breen, J. E., A reevaluation of test data on development length and splices. *ACI Journal*, **74** (1977) 114–22.

[9] Hibbitt, H. D., Karlsson, B. I. & Sorensen, S. E. *Abaqus*: *Theory manual*, Vers. 4.6, HKS Inc. Publishers, Providence, RhI, p. 4.6, 1987.

[10] Rots, J. G., Computational modeling of concrete failure. PhD thesis, Delft University of Technology, The Netherlands, 1988.

[11] Rots, J. G., Bond slip simulation using smeared cracks and/or interface element. Research Report 85.01, Structural Mechanics, Delft University of Technology, The Netherlands, 1985.

[12] Clément, J.-L., Interface acier-béton et comportement des structures en béton armé—caractérisation—modélisation. Thèse de Doctorat, Université Paris 6, France, 1987.

[13] Dragosavic, M. & Groeneveld, H., Bond model for concrete structures. *Proc. of the Int. Conf. on Computer Aided Analysis and Design of Concrete Structures*, ed. F. Damjanic *et al.* Pineridge Press, Swansea, 1984, pp. 203–14.

[14] Goto, Y., Crack formed in concrete around deformed tensile bars. *J. of ACI* **68**(4) (1971) 244–51.

[15] Ingraffea, A. R. & Saouma, V., Numerical modeling of discrete crack propagation in reinforced and plain concrete. In *Fracture Mechanics of Concrete*: *Structural Application and Numerical Calculation*, ed. G. C. Sih & A. Di Tomaso. M. Nijhoof Pub. Dordrecht, The Netherlands, 1985, pp. 171–225.

[16] Mazars, J., A description of micro and macroscale damage of concrete structures. *Eng. Frac. Mechanics*, **25**(5/6) (1986) 729–37.

[17] Mazars, J., Modélisation de l'endommagement et de la rupture du béton, du béton armé et du béton précontraint. Rapport final du programme de recherche AFREM-MRT-DAEI, Paris, 1988.

[18] Breysse, D., Clément, J. L., Mazars, J. & Saouridis, C., Prévision du comportement à la ruine d'une poutre non classiquement feraillée. *Materials and Structures,* (RILEM), (22) (1989) pp. 420–8.
[19] Clément, J.-L., Mazars, J. & Zaborski, A., A damage model for concrete reinforcement bond in composite concrete structures. *Proc. Euromech 204, Structures and Crack Propagation in Brittle Matrix Composites,* ed. A. M. Brandt. Elsevier Pub, New York, USA, 1985, pp. 443–54.
[20] Sun, Z. F., Une théorie 3D des poutres élastiques hétérogènes et analyse non linéaire des structures en béton armé. Thèse de Doctorat de l'Université Paris 6, France, 1989.
[21] Pijaudier-Cabot, G. & Bazant, Z. P., Nonlocal damage theory. *J. Eng. Mechanics* (ASCE), **113**(10) (1987) 1512–13.
[22] Dardare, J. & Chevalier, T., Essai de caractérisation de l'ancrage en précontrainte par prétension. Report CERIB, Epernon, France, 1987.
[23] Clément, J. L. & Mazars, J., Analysis of anchor-bolts by a non-local damage model. RILEM Round Robin—Internal Report of the Technical Committee 90-FMA, 1990.
[24] Pijaudier-Cabot, G., Mazars, J. & Pulikowski, J., Steel–concrete bond analysis with non-local continuous damage. *J. Struct. Eng.* (ASCE), **117** (1991) 862–82.
[25] Bazant, Z. P. & Pijaudier-Cabot, G., Measurement of characteristic length of nonlocal continuum. *J. Eng. Mechanics* (ASCE), **115**(4) (1989) 755–67.
[26] Mazars, J., Application de la mécanique de l'endommagement au comportement non linéaire et à la rupture du béton de structure. Thèse de Doctorat ès-Sciences, Université Paris 6, France, 1984.
[27] Schnobrich, W. C., The role of finite element analysis of reinforced concrete structures. *Proceedings of the Seminar on Finite Element Analysis of Reinforced Concrete Structures,* ed. C. Meyer, Tokyo, Japan, May ASCE, New York, 1985, pp. 1–24.
[28] *César: A finite element code. User and Theory Manuals,* Vers. 2.0. Laboratoire Central des Ponts et Chaussées, 75732 Paris Cedex 15, France, 1986.
[29] Bazant, Z. P. & Pijaudier-Cabot, G., Nonlocal continuum damage, localization instability and convergence. *J. Applied Mechanics* (ASME), **55** (1988) 287–93.
[30] Pijaudier-Cabot, G. & Berthaud, Y., Effects des interactions dans l'endommagement d'un milieu fragile. Formulation non locale. CRAS, T 310, Série II, 1990, pp. 1577–82.
[31] Adrouche, K. & Lorrain, M., Influence des paramètres constitutifs de l'association acier-béton sur la résistance de l'adhérence aux chargements cycliques lents. *Materials and Structures,* **20**(118) (1987) 315–20.
[32] Adrouche, K., Contribution à l'étude de l'endommagement de la liaison acier-béton sous chargement cyclique de faible fréquence. Thèse de Docteur Ing., Institut National des Sciences Appliquées de Toulouse, France, 1987.
[33] ACI, *Building Code Requirements for Reinforced Concrete.* ACI publication 318–83, American Concrete Institute P.O. Box 19150, Redford Station, Detroit, Michigan 48219, USA, 1983.

[34] CEB, *Code Modèle pour les Structures en Béton*. Bulletin No. 124/125-F, Vol. 2, Comité Euro-International du Béton, 6 rue Lauriston-75116 Paris, 1978.
[35] Ladeveze, P., Sur une théorie de l'endommagement anisotrope. Report No. 34, Laboratoire de Mécanique et Technologie, Cachan, France, 1983.
[36] Bazant, Z. P. & Sener, S., Size effect in pull-out tests. *ACI Materials Journal*, **85** (1988) 347–51.

Chapter 13

Splitting Failure of a Strain-softening Material due to Bond Stresses

HANS W. REINHARDT

Inst. für Werkstoffe im Bauwesen, Universitat Stuttgart, Pfaffenwaldring 4, D-7000 Stuttgart 80, Germany

&

CORNELIS VAN DER VEEN

Delft University of Technology, Division of Mechanics and Structures, Stevinweg 1, NL-2628 CN Delft, Netherlands

ABSTRACT

The bond between reinforcing bars and concrete is an essential requirement of reinforced concrete. The maximum bond stress will vary depending on the thickness of the concrete cover. Analyses with an elastic-brittle and plastic material assumptions lead to a lower and an upper bound, respectively, to experimental data. It is shown that modelling concrete as a strain-softening material yields best agreement between analytical prediction of bond failure due to splitting and test results.

1. INTRODUCTION

The bond between steel and concrete relies upon three mechanisms: adhesion, friction, and mechanical interlock. Adhesion between the steel surface and the concrete matrix plays a part at small bond stresses at first loading, while friction forces develop after the adhesion has failed and relative displacement between steel and concrete occurs. Both mechanisms are important in the case of smooth bars.

The main contribution to bond resistance of deformed bars is due to the mechanical interlock of the ribs of the bars which are embedded in concrete.

If a bar is pulled out of the concrete the steel ribs are supported by concrete. Large compressive stresses are generated at the ribs which spread into the surrounding concrete at a certain angle. These inclined forces cause circumferential forces in the concrete. A splitting crack occurs if the tensile strength of concrete is reached.

Whether this state determines also the maximum bond stress depends upon the stress–strain behaviour of the concrete. An elastic-brittle material will fail whereas an elastic-plastic material and a strain-softening material has still more capacity to sustain an increasing bond stress.

In the following, strain-softening material behaviour will be applied to concrete and the results will be compared with elastic and plastic solutions. The onset of failure is particularly relevant with respect to the required thickness of concrete cover.

2. ANALYTICAL MODEL FOR THE SPLITTING FAILURE MECHANISM

It is generally accepted that bond forces of deformed bars emanate from the steel ribs and radiate out into the surrounding concrete at an

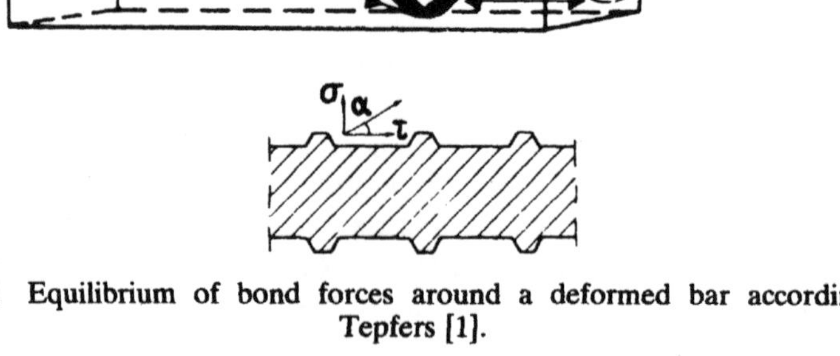

Fig. 1 Equilibrium of bond forces around a deformed bar according to Tepfers [1].

inclination α as shown schematically by Fig. 1. The bond stresses can be split into radial and tangential components. The radial compressive component is balanced by tensile ring stresses in the concrete. Tepfers [1] stated that the radial component of the bond stress, which is proportional to tan α, can be regarded as an hydraulic pressure acting on a thick-walled ring formed by the concrete which surrounds the bar. The wall thickness of the ring is determined by the smallest available dimension which is usually given by the least thickness of the cover, c.

To determine the cracking resistance Tepfers considers three stages: an elastic stage, a partly cracked elastic stage, and a plastic stage. The three stages are illustrated by Fig. 2. Tepfers has derived the maximum bond stress, τ_{br}, as a function of geometry and concrete tensile strength. The appropriate formulas are given as follows:

Elastic stage
$$\tau_{br} = \frac{f_{ct}}{\tan\alpha}\frac{(c + d_s/2)^2 - (d_s/2)^2}{(c + d_s/2)^2 + (d_s/2)^2} \tag{1}$$

Partly cracked elastic stage
$$\tau_{br} = \frac{f_{ct}}{\tan\alpha}\frac{c + d_s/2}{1{\cdot}664 d_s} \tag{2}$$

Plastic stage
$$\tau_{br} = \frac{f_{ct}}{\tan\alpha}\frac{2c}{d_s} \tag{3}$$

The elastic solution is the lower bound, the plastic solution the upper bound, while the partly cracked elastic one lies in between. This means that the elastic solution refers to the onset of cracking but that the loading capacity of the concrete ring is not yet exhausted when cracking starts. The maximum loading capacity is only reached after

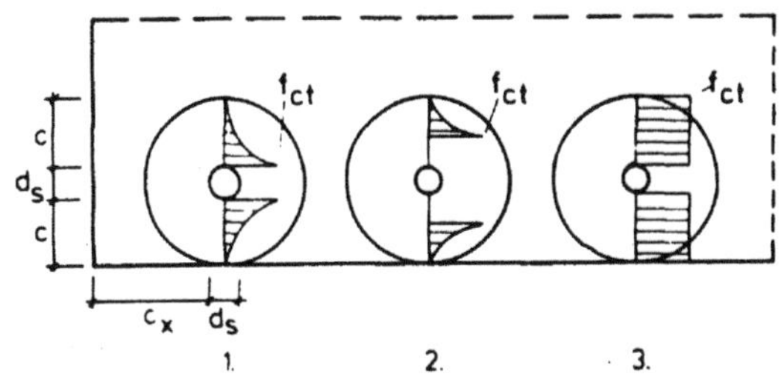

Fig. 2 Distribution of tangential stresses around a deformed reinforcing bar: (1) elastic, (2) partly cracked elastic, (3) plastic. (d_s = steel bar diameter, c = concrete cover, f_{ct} = tensile strength of concrete.)

the cracks penetrate some distance into the concrete. On the other hand, the plastic solution cannot apply to concrete since concrete is certainly not a plastic material.

Generally speaking concrete is neither pure elastic nor is it plastic. It has been shown by several researchers that concrete is a strain-softening material the behaviour of which depends upon the concrete mix, age, curing condition, temperature and moisture content. In the following, concrete will be modelled as a strain-softening material and the relationship between concrete cover and bond strength will be determined.

3. BOND STRENGTH OF A STRAIN-SOFTENING MATERIAL

The idea of a concrete ring which is loaded by hydraulic pressure, as Tepfers [1] has proposed, is also used. It is assumed that the inner part of the concrete ring which touches the steel bar is cracked. Furthermore it is assumed that the cracks transfer some stress according to the softening behaviour of the concrete. Figure 3 shows the geometry of the bar with the surrounding partly cracked concrete ring.

The total resistance of the concrete ring is built up by the outer

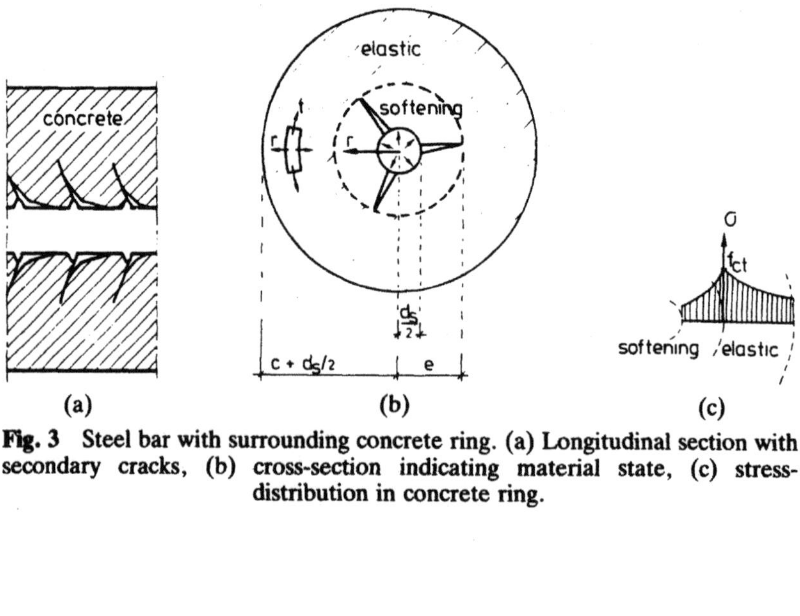

Fig. 3 Steel bar with surrounding concrete ring. (a) Longitudinal section with secondary cracks, (b) cross-section indicating material state, (c) stress-distribution in concrete ring.

non-cracked part and by the softening part. The contribution of the elastic part is given by

$$\tau_{\mathrm{br}} = \frac{f_{\mathrm{ct}} 2e}{\tan \alpha d_{\mathrm{s}}} \frac{(c + d_{\mathrm{s}}/2)^2 - e^2}{(c + d_{\mathrm{s}}/2)^2 + e^2} \tag{4}$$

as derived by Tepfers [1]. The second part is analysed in the following.

To this end, the displacements and stresses have to be considered. At a radial distance r the total tangential elongation δ_{tot} consists of an elastic part and the width of n cracks

$$\delta_{\mathrm{tot}} = 2\pi r \varepsilon_{\mathrm{t}} + nw \tag{5}$$

At a radial distance e, the tensile strength is reached and no cracks exist. There, the total tangential elongation amounts to

$$\delta_{\mathrm{tot}} = 2\pi e \varepsilon_{\mathrm{t}} \approx 2\pi e \frac{f_{\mathrm{ct}}}{E_{\mathrm{c}}} = 2\pi e \varepsilon_{\mathrm{cr}} \tag{6}$$

In this solution Poisson's effect is neglected which makes an error of about 10% in the elastic solution.

The associated radial displacement $u_{\mathrm{r}} = \varepsilon_{\mathrm{cr}} e$ is assumed to be constant for the sake of simplicity. This means that a rigid body translation occurs for the cracked part which is shown in Fig. 4(a). If tensile stresses are transferred by the cracks, Fig. 4(b) applies. In this

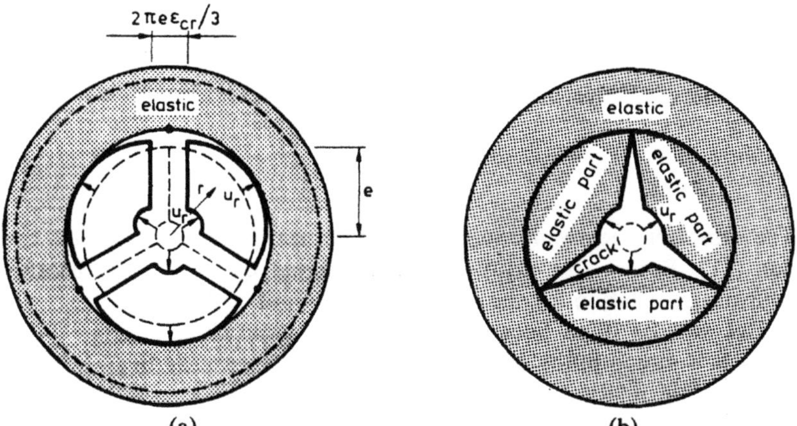

(a) (b)

Fig. 4 Rigid body translation in the radial direction: (a) without tangential stresses in the cracked part; (b) with tensile stresses in the cracked part.

case the tensile stress at $r = e$ reaches just the tensile strength which leads to compatibility between the cracked and uncracked parts of the concrete ring. At a distance r the tangential displacement is given by eqn (5) which is equal to eqn (6) by definition

$$\delta_{tot} = 2\pi r \varepsilon_t + nw - 2\pi e \varepsilon_{cr} \tag{7}$$

In eqn (7) ε_t and w depend on the constitutive relation of the softening material.

Van der Veen [2] has used the power function

$$\frac{\sigma}{f_{ct}} = 1 - \left(\frac{w}{w_c}\right)^k \tag{8}$$

as proposed in Ref. [3] with w = crack opening and w_c = stress free crack opening. Substituting w in eqn (7) into eqn (8) with $\varepsilon_t = \varepsilon_{cr}$ and rearranging leads to the stress in the partly cracked part

$$\sigma_t = f_{ct}\left\{1 - \left[\frac{2\pi\varepsilon_{cr}}{nw_c}(e - r)\right]^k\right\} \tag{9}$$

Note, that it is assumed that $\varepsilon_t = \varepsilon_{cr}$ for $r < e$. Because the stress is calculated from the stress-crack opening curve it follows that only small differences in stress occur. The force developed in the cracked zone is given by the integral from $r = d_s/2$ to $r = e$

$$F = \int_{\tau = d_s/2}^{e} \sigma_t \, dr = f_{ct}(e - d_s/2)\left\{1 - \left[\frac{2\pi\varepsilon_{cr}}{nw_c}(e - d_s/2)\right]^k \frac{1}{k+1}\right\} \tag{10}$$

The bond resistance due to this force is

$$\tau_{br} = \frac{2F}{d_s \tan \alpha} \tag{11}$$

The total bond resistance is given by the superposition of eqn (4) and eqn (11). Before this expression is evaluated quantitatively, another constitutive relation for the softening behaviour will be used.

This will be done because eqn (8) has some shortcomings. First, there is a discontinuity at $w = 0$ between the ascending branch of the stress–strain relation and the softening branch and second, the stress drop at small w is too sharp as has been shown by evaluating test results [4]–[6]. To avoid these deficiencies, another constitutive relation has been proposed [7] which showed best agreement between experimental results and prediction [5]. This relation reads

$$\frac{\sigma}{f_{ct}} = \left[1 + \left(C_1 \frac{w}{w_c}\right)^3\right] \exp\left(-C_2 \frac{w}{w_c}\right) - (1 + C_1^3) \exp(-C_2) \qquad (12)$$

C_1 and C_2 being constants. In analogy to eqn (9), the stresses in the cracked part are given by

$$\sigma_t = f_{ct}\left\{1 + \left[C_1 \frac{2\pi\varepsilon_{cr}}{nw_c}(e - r)\right]^3\right\} \exp\left[-C_2 \frac{2\pi\varepsilon_{cr}}{nw_c}(e - r)\right]$$
$$- \frac{2\pi\varepsilon_{cr}}{nw_c}(e - r)(1 + C_1^3) \exp(-C_2) \qquad (13)$$

The force in the cracked zone follows from the integral similarly, as in eqn (10)

$$F = \int_{\gamma = d_s/2}^{e} \sigma_t \, d\tau = \frac{f_{ct}}{C_2\beta}\{1 - \exp[-C_2\beta(e - d_s/2)]\}$$
$$+ \frac{6C_1^3 f_{ct}}{C_2^4\beta} - \frac{f_{ct}(C_1\beta)^3}{C_2\beta} \exp[-C_2\beta(e - d_s/2)]\Big[(e - d_s/2)^3$$
$$+ \frac{3(e - d_s/2)^2}{C_2\beta} + \frac{6(e - d_s/2)}{(C_2\beta)^2} + \frac{6}{(C_2\beta)^3}\Big]$$
$$- \tfrac{1}{2} f_{ct}\beta(1 + C_1^3) \exp(-C_2)(e - d_s/2)^2 \qquad (14)$$

with the short hand notation $\beta = 2\pi\varepsilon_{cr}/(nw_c)$. Equation (11) holds also for this derivation.

In the following section, the bond resistances for a strain-softening material according to eqns (8) and (12) will be evaluated and will be compared with the elastic, the partly cracked elastic, and the plastic solution, respectively.

4. QUANTITATIVE EVALUATION

The lower bound to eqns (10) and (14) are obtained for one single crack in the concrete ring, i.e. $n = 1$. Therefore, this case will be evaluated. The results will be plotted in a normalized form as τ_{br}/f_{ct}, which shows the direct relationship between shear strength and uniaxial tensile strength (mean value) of concrete. The following quantities will be varied

—tensile strength f_{ct} = 2, 3, 4·5 (MPa)
—fracture energy G_F = 75, 100, 140 (J/m^2)
—bar diameter d_s = 10, 20, 40 (mm)
—Young's modulus E = 15, 30, 45 (GPa)
—concrete cover c = 1–6 times d_s
—angle of inclination α = 45°

From these mechanical quantities, the cracking strain ε_{cr} and the stress-free crack opening w_c follow:

$$\varepsilon_{cr} = f_{ct}/E \qquad \text{and} \qquad w_c = 5{\cdot}14 G_F/f_t \text{ [8]}.$$

The constants are:

$$k = 0{\cdot}248 \text{ [5]}, \qquad C_1 = 3, \qquad C_2 = 6{\cdot}93 \text{ [8]}.$$

Of course, not all combinations are meaningful. The combination $E = 15$ GPa, $f_{ct} = 2$ MPa, $G_F = 75$ J/m^2 may represent a lightweight concrete, the combination 30/3/100 an average normal weight concrete, and 45/140 a high strength concrete. The corresponding values for ε_{cr} are: 133, 100, 100×10^{-6} respectively, while w_c takes the values 193, 171, 160×10^{-6} respectively. Variation in maximum aggregate size would lead to smaller w_c for smaller aggregate and to larger w_c for larger aggregate [5], [9].

The following figures show some results of the calculation in terms of bond stress at onset of failure versus cover to bar diameter ratio, for comparison reasons. Figure 5 contains the predictions for an elastic-brittle material, a plastic material, and a partly cracked elastic material. Comparing predictions with the experimental results [10, 11] it can be seen that the plastic case is an upper bound whereas the elastic case is underestimating all experimental findings. The assumption of a partly cracked ring represents a conservative lower bound. The predictions using a strain-softening material leads to bond stress versus cover to bar diameter relations which agree best with the test results. The power function as applied by van der Veen leads to slightly lower bond stresses than in the present exponential function. This difference can be explained by the shallower softening branch than in the case of the power function. However, since test results show large scatter, the difference between the analytical results is relatively small. Both functions can be used for computing realistic behaviour.

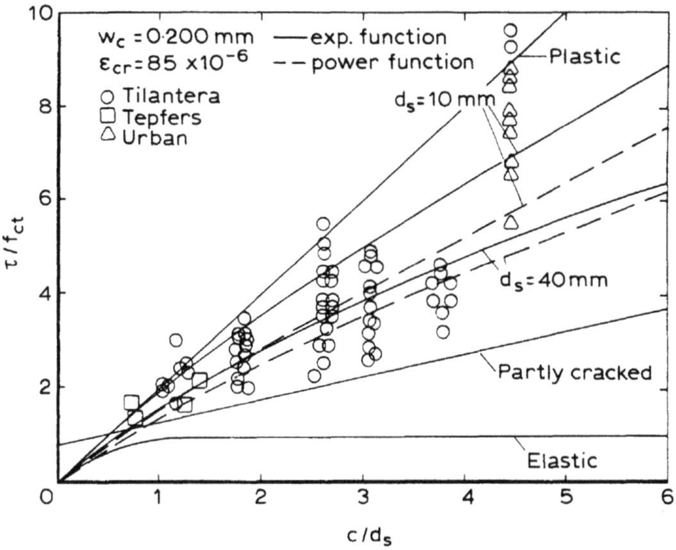

Fig. 5 Bond stresses at onset of failure versus thickness of concrete cover. Experimental and analytical results.

Figure 6 applies to average normal weight concrete with material properties as stated above. Only analytical values are shown. The lines for $d_s = 40$ mm are practically the same. Compared to Fig. 5, all results show lower bond stresses than could be expected since w_c and ε_{cr} are smaller than in Fig. 5.

Lightweight concrete exhibits smaller bond stresses at onset of failure than normal weight concrete, which can be seen from Fig. 7. Although, the differences seem rather small it should be kept in mind that the vertical axis of the figures is normalized with respect to the tensile strength. Since f_{ct} is assumed to be smaller in this case, the bond stress is similarly smaller.

Figure 8 represents high strength concrete. The same as stated for lightweight concrete can be repeated again, however, with the opposite sign. This means that the absolute bond stress is significantly higher than for lightweight and normal weight concrete.

A hypothetical concrete has been evaluated in Fig. 9, i.e. $w_c =$ 0·250 mm and $\varepsilon_{cr} = 75 \times 10^{-6}$. This may represent a low strength concrete with a rather tough crack interlocking mechanism as encountered

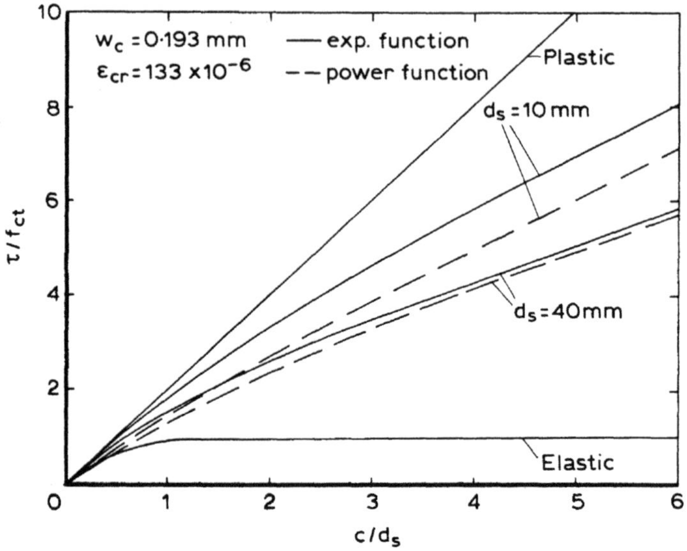

Fig. 6 Bond stress at onset of failure versus thickness of concrete cover. Average normal weight concrete.

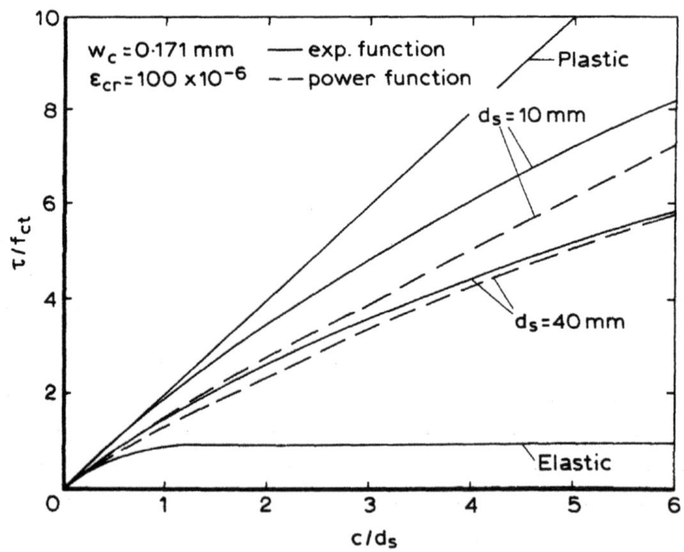

Fig. 7 Bond stress at onset of failure versus thickness of concrete cover. Lightweight concrete.

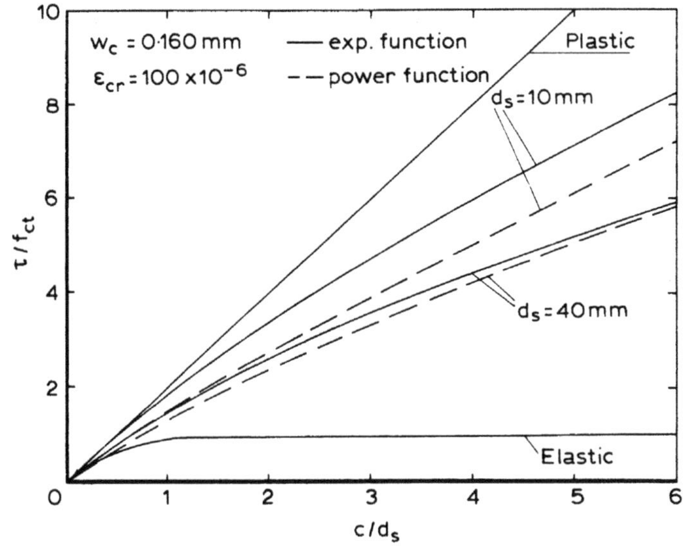

Fig. 8 Bond stress at onset of failure versus thickness of concrete cover. High strength concrete.

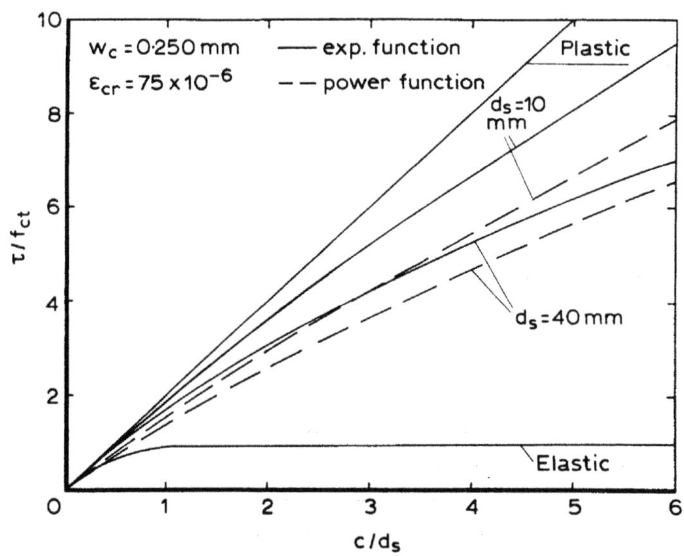

Fig. 9 Bond stress at onset of failure versus thickness of concrete cover. Concrete with large stress-free crack opening w_c.

with large aggregate size. It can be seen that this combination of properties has beneficial consequences for the bond stress at onset of failure. The highest values in normalized form can be found.

5. APPLICATION TO ENGINEERING PRACTICE

The question arises of how the results of the sections above can be used in practical engineering. There are at least two ways. One is to support the code of practice by rational rules and a second is to estimate the influence of various parameters on bonding using known material properties.

All codes of practice contain rules with respect to anchorage of bars, concrete cover, and splice length. These rules were derived from experiments and supported by elastic or plastic analyses. Both theories lead to bounds to real behaviour. Including the softening of concrete has also yielded close agreement. Since the relations could be expressed by analytical formulae it is easy to calculate bond strength by knowing basic material properties without bond testing. If the bond has to be predicted for a new type of concrete this could be done using the basic material properties.

Civilization is progressing and technical development keeps pace with it. This means that new fields of application for reinforced concrete are entered and that unknown parameters appear. For instance high strength concrete and high performance lightweight concrete are being developed, low temperatures are encountered at LNG storage facilities [2] and an even lower temperature range applies to hydrogen storage and transportation facilities. Until now this last field has not been explored as far as concrete structures are concerned.

Another area where experimental results are very scarce is hard impact on concrete structures. High strain rates are the dominating features of such events. The influence of high strain rates on concrete properties has been studied and it has been demonstrated that bonding is influenced by high strain rates in the same way as tensile and compressive strengths are [13]. Thus, knowing the effect on the basic mechanical properties of concrete, bond behaviour can be analytically predicted if the mechanism does not change from low to high strain rate.

6. CONCLUSIONS

The bonding of deformed bars in concrete is a paramount feature and a prerequisite for reinforced concrete. It depends on the geometry of the steel bar, on the mechanical properties of concrete, and on the geometry and loading conditions of the structure. One important parameter is the thickness of the concrete cover since this determines the smallest concrete volume around a steel bar which is available for the transfer of bond stresses.

At a small concrete cover, splitting cracks occur already under small bond stresses. There is a strong relationship between concrete cover, bar diameter, tensile strength of concrete, and bond stress at the onset of failure. It has been shown that these relationships can be predicted realistically if strain-softening of concrete under tensile stresses is taken into account. Two different relations for the softening were used, a power function and an exponential function. Both relations lead to results which agree well with experimental findings. The exponential function is apparently better suited to predict the influence of stress-free crack opening and strain at maximum stress on the bond stress at onset of splitting failure.

The analytical expressions can be used easily to determine bond stress at longitudinal cracking for all types and strength classes of concrete provided that the basic mechanical properties of concrete are known. This procedure enables also the prediction of bonding as influenced by parameters such as temperature and loading rate.

REFERENCES

[1] Tepfers, R., A theory of bond applied to overlapped tensile reinforcement splices for deformed bars. Report 73-2, Chalmers University of Technology, Göteborg 1973, 328 pp.

[2] Van der Veen, C., Cryogenic bond stress-slip relationship. Thesis, Delft University of Technology, Delft 1990, 111 pp.

[3] Reinhardt, H. W., Fracture mechanics of an elastic softening material like concrete. *Heron*, **29**(2) (1984) 42 pp.

[4] Alvaredo, A. M., Hu, X. Z. & Wittmann, F. H., A numerical study of the fracture process zone. In *Fracture of Concrete and Rock, Recent Developments*, ed. S. P. Shah, S. E. Swartz & B. Barr. Elsevier Appl. Sci., London, 1989, pp. 51–60.

[5] Wolinski, S., Hordijk, D. A., Reinhardt, H. W. & Cornelissen, H. A. W., Influence of aggregate size on fracture mechanics parameters of concrete. *Int. J. Cement Composites and Lightweight Concrete,* **9**(2) (1987) 95–103.

[6] Rots, J., Strain-softening analysis of concrete fracture specimens. In *Fracture Toughness and Fracture Energy of Concrete,* ed. F. H. Witt mann. Elsevier, Amsterdam, 1986, pp. 137–48.

[7] Reinhardt, H. W., Cornelissen, H. A. W. & Hordijk, D. A., Tensile tests and failure analysis of concrete. *J. Struct. Div.* (ASCE), **112**(11) (1986) 2462–77.

[8] Hordijk, D. A. & Reinhardt, H. W., Growth of discrete cracks in concrete under fatigue loading. *Toughening Mechanisms in Quasi-Brittle Materials,* ed. S. P. Shah, Kluwer Academic Publ., Dordrecht, 1991, pp. 553–68.

[9] Duda, H., Bruchmechanisches Verhalten von Beton unter monotoner und zyklischer Zugbeanspruchung. Diss. TH Darmstadt, 1990.

[10] Urban, V., Anchorage length and bond stress calculation. Res. Rep. 458, Technical University, Prague, 1980.

[11] Tilanterä, T. & Rechardt, T., Bond of reinforcement in lightweight aggregate concrete. University of Technology, Helsinki, 1977, pp. 1–36.

[12] Tepfers, R. *et al.,* Bond action and bond behaviour of reinforcement, state-of-the-art. CEB Bulletin No. 151, Paris, 1982, 153 pp.

[13] Reinhardt, H. W., Blaauwendraad, J. & Vos, E., Prediction of bond between steel and concrete by numerical analysis. *Materials & Structures,* **17**(100) (1984) 311–20.

Chapter 14

Fracture Mechanics Evaluation of Minimum Reinforcement in Concrete Structures

CRESCENTINO BOSCO & ALBERTO CARPINTERI
Politecnico di Torino, Department of Structural Engineering, 10129 Torino, Italy

ABSTRACT

A refinement of the model proposed in 1981 and 1984 by the second author, is presented in this chapter. Compliance and stress-intensification concepts are used again. However, in this case, the reinforcement reactions are applied directly to the crack surfaces and not as closing forces acting at infinity. Moreover, the congruence condition is locally imposed on the crack opening displacement in correspondence with the reinforcement, and not globally on the cross-section rotation. The theoretical results confirm a transition from ductile to brittle collapse by varying a non-dimensional brittleness number defined in previous contributions. With the present model, yielding or slippage of reinforcement can precede or follow crack propagation in concrete, removing the assumption of steel yielding at incipient fracture. Such a theoretical approach appears to be very useful for estimating the minimum amount of reinforcement for members in flexure, assuming simultaneous concrete cracking and steel yielding (transitional condition). A physically similar flexural behaviour is obtained in cases where the brittleness number is the same. The failure mechanism changes completely when the beam depth is varied, the steel percentage remaining the same. Only when the steel percentage is inversely proportional to the square root of the beam depth, is the mechanical behaviour reproduced. The experimental results show that the transitional value of the brittleness number is tendentially constant for each concrete grade, by varying the beam depth. While the minimum steel percentage provided by Eurocode 2 and ACI are independent of the beam depth, the relationship established by the brittleness number calls for decreasing values with increasing beam depths.

1. INTRODUCTION

The application of Fracture Mechanics methods to reinforced concrete elements and, more generally, to fibre-reinforced materials dates back relatively few years. Nevertheless the importance of the results obtained is increasingly recognized since this approach has provided new ways of understanding and, often, represents the unique possibility of interpreting the real structural behaviour which shows different failure modes in connection with different size scales. Recently it has been proven experimentally that even the minimum percentage of reinforcement that enables the element to prevent brittle failure, depends on the scale [1]. With a classical approach these results should not be found nor be predictable, particularly for reinforced concrete structures.

A simple model to describe the behaviour of a cracked reinforced concrete beam element subjected to bending moment and axial force was proposed by Carpinteri [2]–[4], after early studies applied to cracked columns by Okamura *et al.* [5], [6]. The results, even though the analysis was developed by means of LEFM, revealed some important possibilities for application to static loading as well as to repeated cyclic actions, and for the description of scale effects in reinforced beams. An experimental demonstration of the utility of this approach, and in general, of the utility of the application of Fracture Mechanics concepts to reinforced structures, has recently been provided with regard to the scale dependence of the minimum reinforcement in high strength concrete beams by Bosco *et al.* [7].

It was interesting then to extend the analysis of cracked reinforced elements by comparing different theoretical models and assuming a linear elastic behaviour of concrete and a linear elastic–perfectly plastic behaviour of steel. It was shown how the transition from ductile to brittle failure is not significantly influenced by the assumption of a rotational or a displacement congruence condition of the cracked section, while some other aspects, such as the deformation of the element, seem more affected [8], [9].

In fact, when a rotational congruence condition is imposed, the model does not predict local rotation due to the presence of the cracked section until yielding of reinforcement is reached, whereas, even though the actual phenomenon is in effect very complex due to debonding and slippage of reinforcements, from tests carried out on initially uncracked elements with low reinforcement [7], it generally appears that crack propagation occurs before the yielding limit of reinforcement is reached.

This model does not completely fit the experimental results, although it captures their trends, and some actual structural constitutive relation seems not fully obtainable, such as, for example, the moment–rotation response. The authors discuss these aspects of the problem and present a model where the congruence condition is locally imposed on the crack opening displacement in correspondence with the reinforcement, and not globally on the cross-section rotation. This allows better agreement with the experimental results, without losing any other positive characteristic, already obtained by other similar methods.

The second part of the chapter focuses on the practical applicability of theoretical predictions, given by Fracture Mechanics, to the experimental behaviour of reinforced concrete beams in flexure. In particular the minimum reinforcement requirements to avoid brittle failure are emphasized. An extensive experimental campaign carried out at the Politecnico di Torino is described and a simple relationship to obtain the minimum reinforcement is proposed for standard purposes.

2. THEORETICAL MODEL

2.1 Superimposed Effect of Multiple Loads on the Deformation of a Cracked Beam Element

In order to analyse the behaviour of a structure containing a cracked member it is necessary to know the relation between the load and the deformation of the member. When the cracked member undergoes multiple loads simultaneously and it can be considered to behave elastically, a superimposed effect on the deformation should be considered.

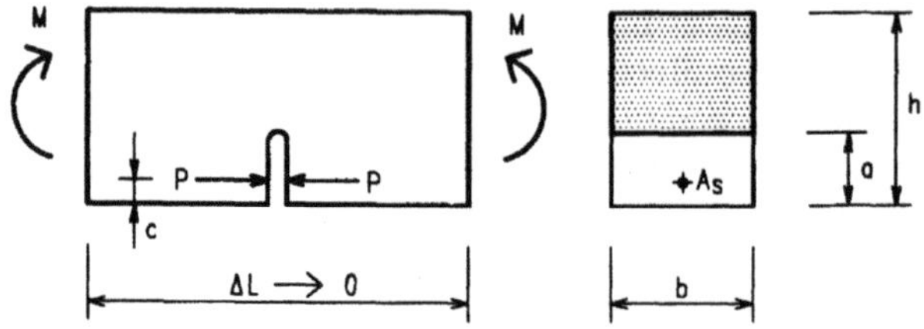

Fig. 1 Cracked element.

Let us consider the cracked member shown in Fig. 1, which undergoes simultaneously the bending moment M and the closing forces P applied on the crack surfaces. We can evaluate the angular deformation $\Delta\varphi_{MP}$ produced by the forces P, together with the crack opening displacement $\Delta\delta_{PP}$, at the point where the forces P are applied, and, at the same time, the crack opening displacement $\Delta\delta_{PM}$ caused by the bending moment M, together with the angular deformation $\Delta\varphi_{MM}$.

By linear superposition it is possible to write

$$\Delta\delta = \Delta\delta_{PM} + \Delta\delta_{PP} = \lambda_{PM}M - \lambda_{PP}P \tag{1a}$$

$$\Delta\varphi = \Delta\varphi_{MM} + \Delta\varphi_{MP} = \lambda_{MM}M - \lambda_{MP}P \tag{1b}$$

where λ_{MM}, λ_{PM}, λ_{MP}, λ_{PP}, are the compliances of the member due to the existence of the crack. The factors λ can be derived from energy methods considering the moment M acting simultaneously with the forces P.

If G and E are, respectively, the strain energy release rate and the Young's modulus of the material (the Poisson ratio ν is considered negligible), it follows that the variation ΔW of the total potential energy is given by

$$\begin{aligned}\Delta W &= \int_0^c G_M b\,dx + \int_c^a G_{(M+P)} b\,dx \\ &= \int_0^c \frac{K_{IM}^2}{E} b\,dx + \int_c^a \frac{(K_{IM}+K_{IP})^2}{E} b\,dx \\ &= \int_0^c \frac{K_{IM}^2}{E} b\,dx + \int_c^a \frac{K_{IM}^2}{E} b\,dx + \int_c^a \frac{K_{IP}^2}{E} b\,dx \\ &\quad + 2\int_c^a \frac{K_{IM}K_{IP}}{E} b\,dx = \int_0^a \frac{K_{IM}^2}{E} b\,dx + \int_c^a \frac{K_{IP}^2}{E} b\,dx \\ &\quad + 2\int_c^a \frac{K_{IM}K_{IP}}{E} b\,dx\end{aligned} \tag{2}$$

where K_{IM} and K_{IP} are the stress-intensity factors due to bending moment M and forces P, respectively.

Using Clapeyron's Theorem, we also have

$$\Delta W = \tfrac{1}{2}M\,\Delta\varphi_{MM} + \tfrac{1}{2}P\,\Delta\delta_{PP} + \tfrac{1}{2}(P\Delta\delta_{PM} + M\Delta\varphi_{MP}) \tag{3}$$

Recalling that Betti's Theorem provides $P\Delta\delta_{PM} = M\Delta\varphi_{MP}$, from eqns (2) and (3) we obtain

$$\tfrac{1}{2}M\Delta\varphi_{MM} = \int_0^a \frac{K_{IM}^2}{E}\,b\,dx \tag{4}$$

$$\tfrac{1}{2}P\Delta\delta_{PP} = \int_c^a \frac{K_{IP}^2}{E}\,b\,dx \tag{5}$$

$$P\Delta\delta_{PM} = M\Delta\varphi_{MP} = 2\int_c^a \frac{K_{IM}K_{IP}}{E}\,b\,dx \tag{6}$$

The stress-intensity factor produced at the crack tip by the moment M, can be expressed as [5], [6]

$$K_{IM} = \frac{M}{h^{3/2}b}\,Y_M(\xi) \tag{7}$$

while the stress-intensity factor produced by the applied forces P acting at the level of reinforcement, i.e. at a distance c from the lower edge of the beam, is equal to [10]

$$K_{IP} = \frac{2P/b}{\sqrt{\pi a}}\,F(c/a,\,\xi) \tag{8}$$

Rearranging the above expression, it is possible to write

$$K_{IP} = \frac{P}{h^{1/2}b}\,Y_P(c/h,\,\xi) \tag{9}$$

where

$$Y_P(c/h,\,\xi) = F(c/a,\,\xi)\frac{2}{\sqrt{\pi\xi}} \tag{10}$$

Function $Y_M(\xi)$ in eqn (7) is given by [5], [6]

$$Y_M(\xi) = 6(1{\cdot}99\xi^{1/2} - 2{\cdot}47\xi^{3/2} + 12{\cdot}97\xi^{5/2} - 23{\cdot}17\xi^{7/2} + 24{\cdot}80\xi^{9/2})$$

for $\xi = a/h \leq 0{\cdot}7$, while, function $F(c/a, \xi)$ in eqn (8) is given by [10]

$$F(c/a, \xi) = \frac{3{\cdot}52(1 - c/a)}{(1-\xi)^{3/2}} - \frac{4{\cdot}35 - 5{\cdot}28c/a}{(1-\xi)^{3/2}} + \left[\frac{1{\cdot}30 - 0{\cdot}30(c/a)^{3/2}}{(1-(c/a)^2)^{-1/2}} + 0{\cdot}83 - 1{\cdot}76c/a\right][1 - (1 - c/a)] \tag{11}$$

for $a/h < 1$, $c/a < 1$.

Substituting eqn (7) into eqn (4) and dividing by M^2, the compliance λ_{MM} (rotation produced by $M = 1$), can be expressed as

$$\lambda_{MM} = \frac{2}{h^2bE}\int_0^{\xi} Y_M^2(\xi)\,d\xi \tag{12}$$

In the same way, from eqns (9) and (5), the compliance λ_{PP} (crack opening displacement produced by $P = 1$) becomes

$$\lambda_{PP} = \frac{2}{bE}\int_{c/h}^{\xi} Y_P^2(c/h, \xi)\,d\xi \tag{13}$$

Eventually, substituting eqn (7) and eqn (9) into eqn (6) and dividing by the product PM, the compliance λ_{PM} (crack opening displacement produced by $M = 1$), or the compliance λ_{MP} (rotation produced by $P = 1$), can be expressed in the form

$$\lambda_{PM} = \lambda_{MP} = \frac{2}{hbE}\int_{c/h}^{\xi} Y_M(\xi)Y_P(c/h, \xi)\,d\xi \tag{14}$$

Function $Y_P(c/h, \xi)$ is plotted in Fig. 2 for $c/h = 0{\cdot}05$, $0{\cdot}10$, $0{\cdot}15$. For each value of the ratio c/h, it is possible to observe that function Y_P tends to infinity for $\xi \to c/h^+$ and $\xi \to 1^-$. As a consequence, the K_{IP} factor presents a minimum for an intermediate value of the crack depth ξ between c/h and 1.

Functions $Y_M(\xi)Y_P(c/h, \xi)$ and $Y_M^2(\xi)$ with the related integrals, are plotted in Figs 3 and 4 respectively. The diagrams in Fig. 4 represent the compliances λ, see eqns (12)–(14).

2.2 Statically Undetermined Reaction of Reinforcement

Let the cracked concrete beam element in Fig. 1 be subjected to the bending moment M, while the reinforcement transmits to the adjacent matrix surfaces an axial force, statically undetermined, equal to

$$P = \sigma_s A_S \tag{15}$$

where A_S is the reinforcement area and σ_s the related stress.

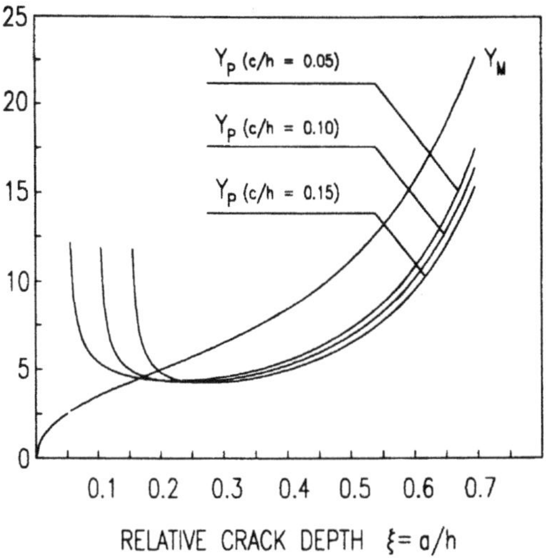

Fig. 2 Shape functions depending on relative crack depth.

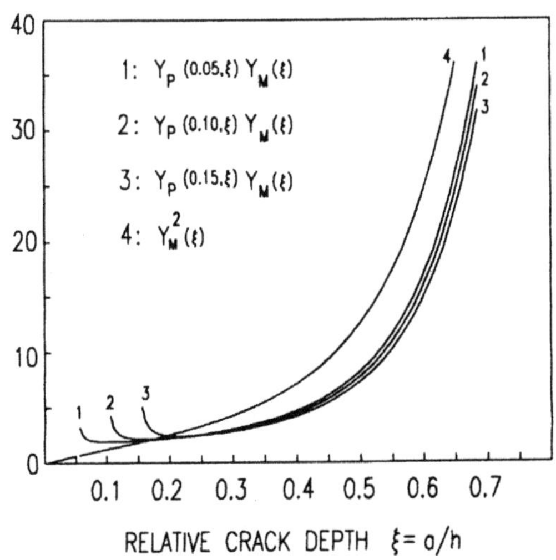

Fig. 3 Products of shape functions.

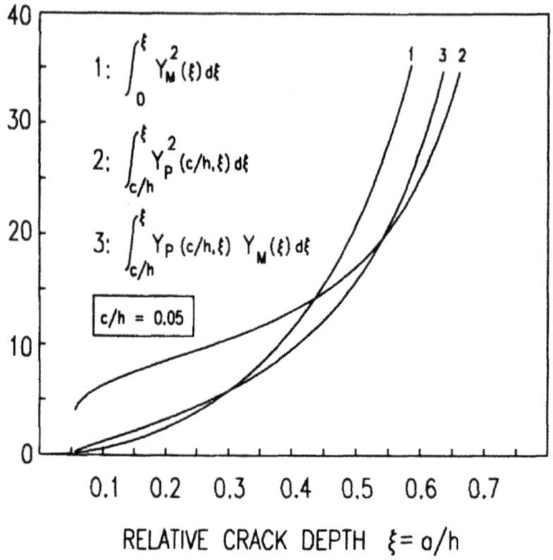

Fig. 4 Compliance functions.

If the displacement discontinuity in the cracked cross-section at the level of reinforcement, is assumed to be zero, up to the moment of yielding or slippage of the reinforcement

$$\Delta\delta = \Delta\delta_{PM} + \Delta\delta_{PP} = \lambda_{PM}M - \lambda_{PP}P = 0 \tag{16}$$

we obtain the displacement congruence condition that allows us to obtain the unknown force P as a function of the applied moment M. In fact, from eqns (12)–(14) and considering eqn (1), it follows that

$$\frac{Ph}{M} = \frac{1}{r''(c/h,\ \xi)}. \tag{17}$$

where

$$r''(c/h,\ \xi) = \frac{\displaystyle\int_{c/h}^{\xi} Y_P^2(c/h,\ \xi)\,d\xi}{\displaystyle\int_{c/h}^{\xi} Y_M(\xi)Y_P(c/h,\ \xi)\,d\xi} = \frac{\lambda_{PP}}{\lambda_{PM}h} \tag{18}$$

Considering a rigid–perfectly plastic behaviour of the reinforcement, the moment of plastic flow or slippage is obtained from eqn (17)

$$M_P = P_P h r''(c/h,\ \xi) \tag{19}$$

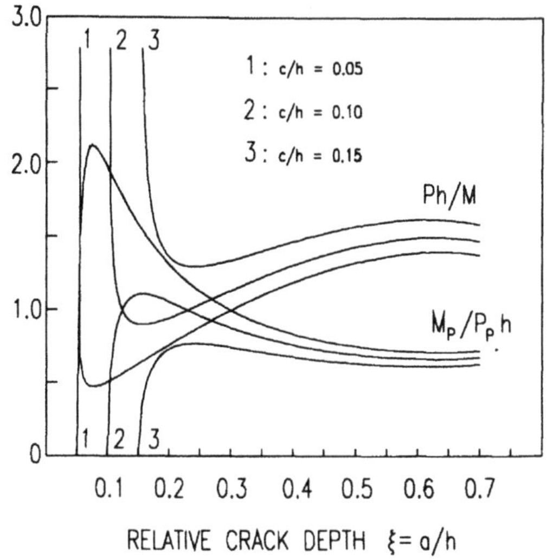

Fig. 5 Statically undetermined reaction of reinforcement versus relative crack depth, varying the position of reinforcement.

where $P_P = f_y A_S$ indicates the yielding (or pulling-out) force, achieved when $\sigma_s = f_y$ (yielding stress of reinforcement).

The statically undetermined reaction of the reinforcement is represented in Fig. 5, against the relative crack depth, for $c/h = 0{\cdot}05$, $0{\cdot}10$, $0{\cdot}15$.

The following remarks can be made:

(a) Function $Ph/M = 1/r''(c/h, \xi)$, i.e. the force transmitted by the reinforcement, is increasing in almost the entire considered range of ξ, i.e. for $c/h \leq \xi \leq 0{\cdot}7$.

(b) For each ratio c/h, function Ph/M always presents an absolute maximum value, over the interval $c/h \leq \xi \leq 0{\cdot}7$. On the other hand, this maximum is very close to the right extreme of the interval so that it is possible to affirm that function Ph/M is increasing monotonically. This means that, by increasing the crack depth (the bending moment M being constant), higher forces P are necessary to satisfy the congruence condition, i.e. to maintain the local crack opening displacement equal to zero.

Functions M_P/P_Ph, according to eqn (19), are also displayed in Fig. 5.

2.3 Stability of the Process of Concrete Fracture and Reinforcement Plastic Flow

The stress intensity factor at the crack tip is (see eqns (7) and (9)):

$$K_I = \frac{M}{h^{3/2}b} Y_M(\xi) - \frac{P}{h^{1/2}b} Y_P(c/h, \xi) \tag{20}$$

for $M < M_P$

$$K_I = \frac{M}{h^{3/2}b} Y_M(\xi) - \frac{P_P}{h^{1/2}b} Y_P(c/h, \xi) \tag{21}$$

for $M > M_P$.

If $M < M_P$, eqn (17) can be put into eqn (20):

$$K_I = \frac{M}{h^{3/2}b} Y_M(\xi) - \frac{1}{h^{1/2}b} Y_P(c/h, \xi) \frac{1}{r''(c/h, \xi)} \frac{M}{h}$$

and, in non-dimensional form

$$\frac{K_I h^{1/2} b}{P_P} = \frac{M}{Ph} \left[Y_M(\xi) - Y_P(c/h, \xi) \frac{1}{r''(c/h, \xi)} \right] \tag{22}$$

In the same way, when $M > M_P$, eqn (21) becomes

$$\frac{K_I h^{1/2} b}{P_P} = Y_M(\xi) \frac{M}{P_P h} - Y_P(c/h, \xi) \tag{23}$$

In order to plot the variations of the stress-intensity factor K_I, against the crack depth, ξ, it is necessary to verify the deformation condition for the reinforcement. This means that we have to verify whether $M/P_P h \geq r''(c/h, \xi)$ or not. In the former case the reinforcement has yielded and we use eqn (23); otherwise eqn (22) must be used as stress σ_s is lower than f_y.

In Figs 6(a) and (b) the stress intensity factor K_I is reported for $c/h = 0{\cdot}05$ and $c/h = 0{\cdot}10$, respectively, against the crack depth ξ and varying the loading parameter $M/P_P h$. First of all, it is necessary to observe that the diagram is divided into two regions, characterized by different conditions of deformation for the reinforcement. The loading conditions and crack depths for which the assumed model predicts non-yielded reinforcement are represented below the separation line.

In Fig. 6(a) all the curves characterized by values $M/P_P h \lesssim 0{\cdot}8$ belong totally to the latter domain, while the curves characterized by

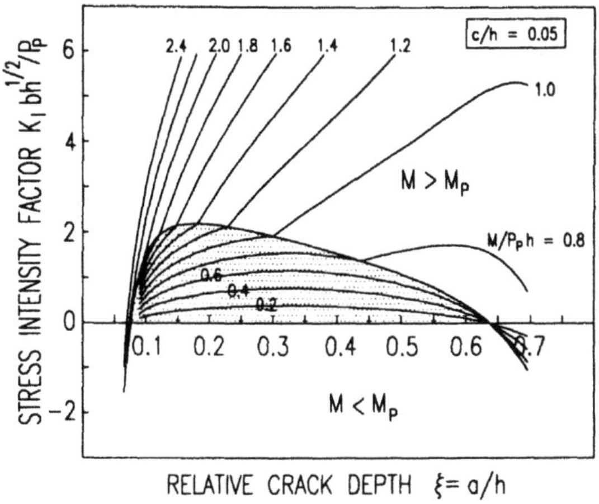

(a)

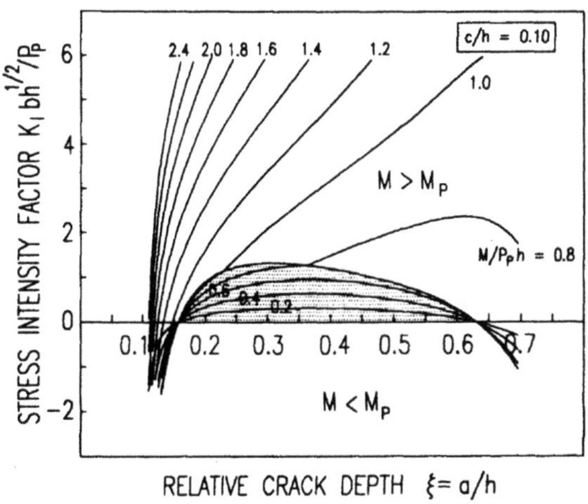

(b)

Fig. 6 Dimensionless stress intensity factor versus relative crack depth ξ, varying the applied bending moment: (a) $c/h = 0{\cdot}05$; (b) $c/h = 0{\cdot}10$.

values $M/P_\mathrm{P}h \gtrsim 0{\cdot}8$ belong only partially. This means that crack propagation is reached with an elastic condition for reinforcement in the entire range of ξ for $M/P_\mathrm{P}h$ ratios lower than 0·8, while the same condition is verified only for limited crack lengths if $M/P_\mathrm{P}h \gtrsim 0{\cdot}8$. In other words, crack propagation is reached before steel yielding either for relatively high content of reinforcement or relatively limited crack depth.

The curves inside the domain where the reinforcement is in the elastic condition show a local or global maximum for $\xi \approx 0{\cdot}35$. The locus of the maxima divides the zone where the cracking process is stable from the other zone where unstable propagation of the crack occurs. Beyond the value $M/P_P h \approx 1$ the curves do not present a maximum and this means that the cracking process is unstable for each crack depth $c/h < \xi < 0{\cdot}7$.

The case $c/h = 0{\cdot}10$ is represented in Fig. 6(b) and the same trends are shown as in Fig. 6(a), whereas the elastic domain shrinks slightly.

2.4 Bending Moment of Concrete Fracture

Assuming that K_I is equal to the matrix fracture toughness K_{IC}, from eqn (20) we have

$$\frac{M_F}{K_{IC}h^{3/2}b} = \frac{1}{Y_M(\xi)} + \frac{P}{K_{IC}h^{1/2}b}\frac{Y_P(c/h,\, \xi)}{Y_M(\xi)} \tag{24}$$

If the force P transmitted by the reinforcement is equal to $P_P = f_y A_S$ or, in other words, if the reinforcement yielding limit has been reached $(M = M_F \geq M_P)$, eqn (24) becomes

$$\frac{M_F}{K_{IC}h^{3/2}b} = \frac{1}{Y_M(\xi)} + N_P\frac{Y_P(c/h,\, \xi)}{Y_M(\xi)} \tag{25}$$

where the *brittleness number*

$$N_P = \frac{f_y h^{1/2}}{K_{IC}}\frac{A_S}{A} \tag{26}$$

is introduced and $A = bh$ is the total cross-sectional area.

In the case $M = M_F < M_P$, i.e. when the reinforcement is in the elastic condition, we can consider the relation

$$\frac{M_F}{K_{IC}h^{3/2}b} = \frac{1}{Y_M(\xi)} + N_P\frac{M_F}{M_P}\frac{Y_P(c/h,\, \xi)}{Y_M(\xi)} \tag{27}$$

since, in that case, it is $\sigma_s/f_y = M_F/M_P$, and therefore $P = \sigma_s A_S = f_y(M_F/M_P)A_S$ in eqn (24).

Equation (27) may be modified by considering eqns (19) and (26):

$$\frac{M_F}{K_{IC}h^{3/2}b} = \frac{1}{Y_M(\xi) - \dfrac{Y_P(c/h,\, \xi)}{r''(c/h,\, \xi)}} \tag{28}$$

Therefore, according to the model when $M_F < M_P$, the moment of crack propagation M_F depends only on the relative crack depth ξ, and is not affected by the brittleness number N_P, i.e. it does not depend on the content of reinforcement but only on its relative position c/h.

The dimensionless fracture moment versus crack depth ξ, is reported in Fig. 7 for $c/h = 0{\cdot}05$ and by varying N_P. The curves $N_P \lesssim 0{\cdot}2$ are descending over the whole range ξ. This means that for low reinforced beams and/or for large cross-sections, the fracture bending moment decreases while the crack extends, i.e. an unstable fracture phenomenon occurs. For higher N_P values, the model predicts a stable fracture process with deep cracks. In particular this occurs for $N_P \gtrsim 0{\cdot}3$.

A more attentive analysis of Fig. 7 can explain the behaviour of cracked sections, varying loading conditions and initial crack length.

(a) Let N_P be equal to $0{\cdot}1$ and the initial relative crack depth $\xi = 0{\cdot}20$. When the applied bending moment M reaches M_F (point A in Fig. 7), the crack propagates and the phenomenon can not be stable anymore, the curve $N_P = 0{\cdot}1$ being always descending. Stable behaviour could only be obtained by reducing the external bending moment. It is worth noting that at point A the reinforcement has already yielded.

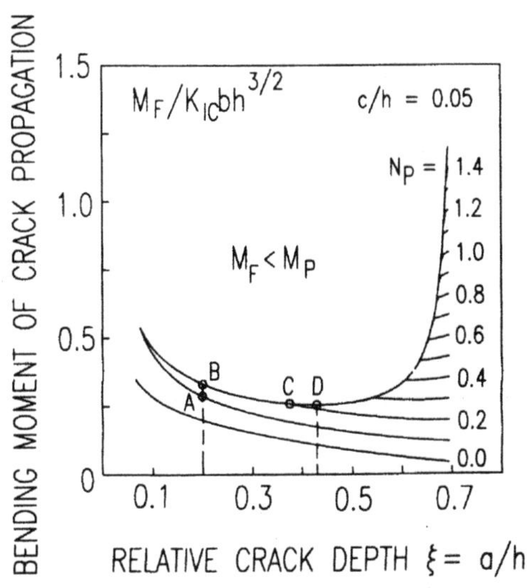

Fig. 7 Dimensionless bending moment of crack propagation versus relative crack depth ξ, varying the brittleness number N_P ($c/h = 0{\cdot}05$).

(b) Now let N_P be equal to 0·2 with the same initial relative crack depth previously assumed. When the bending moment of crack propagation is reached (point B in Fig. 7) the arc B–C of the transitional curve is followed. At point C the reinforcement, previously in elastic condition, yields, but since the curve is still descending, the fracturing process causes instability even in this case and develops up to complete failure of the cross-section.

(c) If N_P is greater than ≈0·25, crack propagation occurs until the transitional curve achieves its minimum, i.e. for $\xi = 0{\cdot}43$ (point D in Fig. 7). Beyond this point it is necessary to increase the bending moment to provoke a crack extension. Then, in these cases, the initial unstable phenomenon becomes stable.

(d) As a final example, let the initial relative crack depth ξ be greater than ≈0·43. In this case for every N_P greater than ≈0·25 the fracturing phenomenon is stable from the beginning. In fact the bending moment of crack propagation is reached in the ascending part of the curves, either in the transitional branch (when the elastic condition for reinforcement prevails) or, for deep cracks, in the plastic branch.

In conclusion, it is shown that the fracture process becomes stable only when N_P is sufficiently high, i.e. when the cross-section of the beam is relatively small and/or the content of reinforcement relatively high, the crack extension being sufficiently deep.

2.5 Moment versus Rotation Response

The local rotation due to the applied loads is given by superimposed effects in the following way

$$\Delta\varphi = \lambda_{MM} M - \lambda_{MP} P \tag{29}$$

where λ_{MM} is obtained from eqn (12) and λ_{MP} from eqn (14). For the relative crack depth ξ, let $\Delta\varphi_F$ be the local rotation due to the presence of a crack when the applied bending moment reaches the value M_F. On the other hand, it is possible to define the local rotation $\Delta\varphi_{F0}$, for the initial relative crack depth ξ_0, at the moment of crack propagation.

For $M = M_F$, eqn (29) can be written as

$$\Delta\varphi_F = \lambda_{MM} M_F - \lambda_{MP} P_P \alpha \tag{30}$$

with $\alpha = M_F/M_P = P/P_P < 1$ if $M_F < M_P$ and $\alpha = 1$ if $M_F \geq M_P$.

Equation (30) can be expressed as a function of the relevant geometrical parameters of the cracked element. To this purpose, assuming $K_I = K_{IC}$ in eqn (20), it is

$$K_{IC} = \frac{M_F}{h^{3/2}b} Y_M(\xi) - \frac{P}{h^{1/2}b} Y_P(c/h, \xi) \qquad (31)$$

and considering that $P = P_P\alpha$, it follows that

$$M_F = \frac{K_{IC}h^{3/2}b}{Y_M(\xi)} + \frac{Y_P(c/h, \xi)}{Y_M(\xi)} P_P h\alpha \qquad (32)$$

Taking into account the expression above, eqn (30) is rewritten as

$$\Delta\varphi_F = \lambda_{MM}\left[\frac{K_{IC}h^{3/2}b}{Y_M(\xi)} + \frac{Y_P(c/h, \xi)}{Y_M(\xi)} P_P h\alpha\right] - \lambda_{MP}P_P\alpha \qquad (33)$$

while, recalling the relationship [8]

$$r'(c/h, \xi) = \frac{\displaystyle\int_{c/h}^{\xi} Y_M(\xi)Y_P(c/h, \xi)\,d\xi}{\displaystyle\int_0^{\xi} Y_M^2(\xi)\,d\xi} = \frac{\lambda_{MP}}{\lambda_{MM}h} \qquad (34)$$

and considering eqn (26), becomes

$$\Delta\varphi_F = \lambda_{MM}\frac{K_{IC}bh^{3/2}}{Y_M(\xi)}\left[1 + [Y_P(c/h, \xi) - r'(c/h, \xi)Y_M(\xi)]N_P\alpha\right] \qquad (35)$$

In eqn (35) we should always consider that $\alpha < 1$ if $M_F < M_P$ or $\alpha = 1$ if $M_F \geq M_P$. We cannot explicitly obtain either M_F or M_P, on the other hand, to obtain the coefficient α, only the knowledge of their ratio is needed. Therefore, from eqns (32) and (19), we have

$$\alpha = \frac{M_F}{M_P} = \frac{K_{IC}bh^{3/2}}{P_P h Y_M(\xi) r''(c/h, \xi)} + \frac{Y_P(c/h, \xi)}{Y_M(\xi) r''(c/h, \xi)} \frac{M_F}{M_P} \qquad (36)$$

Then, considering eqn (26), it follows that

$$\frac{M_F}{M_P}\left[1 - \frac{1}{r''(c/h, \xi)} \frac{Y_P(c/h, \xi)}{Y_M(\xi)}\right] = \frac{1}{N_P r''(c/h, \xi) Y_M(\xi)} \qquad (37)$$

and finally

$$\alpha = \frac{M_F}{M_P} = \frac{1}{[Y_M(\xi)r''(c/h, \xi) - Y_P(c/h, \xi)]N_P} \qquad \text{for } M_F < M_P \qquad (38)$$

while $\alpha = 1$ when $M_F \geq M_P$.

From eqns (35) and (38) it should then be possible to obtain the dimensionless rotation $\Delta\varphi_F / K_{IC}bh^{3/2}$ varying the brittleness number N_P for a given relative crack depth ξ. On the other hand, it is possible to consider the ratio $\Delta\varphi_F/\Delta\varphi_{F0}$ between the values given by eqn (35) for $\xi > \xi_0$ and $\xi = \xi_0$, respectively, and to plot a diagram where $M_F/K_{IC}bh^{3/2}$ given by eqns (25) or (28), is considered as the vertical axis. In this case in fact, if crack propagation is seen as an evolutionary phenomenon, the parameter ξ giving the subsequent positions of the crack tip, the normalized bending moment versus rotation diagram can be obtained until the cracking phenomenon develops up to the complete disconnection of the cross-section.

Therefore, for defined geometry and toughness characteristics, the constitutive behaviour of a cracked section is well described by the brittleness number N_P. This is exactly the same conclusion already found when the congruence condition in terms of rotation is adopted [8]. The difference is that the local rotation is allowed by the present model before reinforcement yielding.

In Figs 8(a)–(e) the moment–rotation diagrams are reported for $N_P = 0$ (no reinforcement), 0·1, 0·26, 0·53 and 0·87, corresponding to the values planned for an experimental campaign on reinforced concrete beams [7]. The same figures also show the ultimate carrying capacity and the asymptotical behaviour (horizontal dotted lines) when the cross-section is completely cracked and the reinforcement has yielded. Since the resultant tensile force is located at $(h - c)$ from the upper edge of the cross-section, the ultimate bending moment becomes $M_u = P_P(h - c)$. Considering that $N_P = P_P h/(K_{IC}bh^{3/2})$, it is possible to obtain $M_u/(K_{IC}bh^{3/2}) = N_P(h - c)/h$, which is the limit value to which the normalized moment–rotation curves tend, for each defined value of N_P. It is worth noting that, for sufficiently high concrete strength and sufficiently low percentage of steel, this type of failure precedes the crushing of the matrix.

Figures 8(a)–(e) also illustrate the failure mechanisms given by the model by varying the brittleness number N_P. The following remarks can be made.

(a) In Figs 8(b) and (c) (where $N_P = 0{\cdot}1$ and $0{\cdot}26$ represent both a very low percentage of reinforcement or a deep cross-section) the contribution of the reinforcement does not involve any appreciable increase in the ultimate load-bearing capacity with regard to the plain cross-section. Then, the phenomenon of crack propagation is markedly unstable. The first cuspidal point represents first cracking, whereas the second represents steel yielding.

(b) In Figs 8(d) and (e) (where $N_P = 0{\cdot}53$ and $0{\cdot}87$, respectively) the benefit of the reinforcement is manifest and the resistant bending moment increases, becoming greater than M_F. The phenomenon

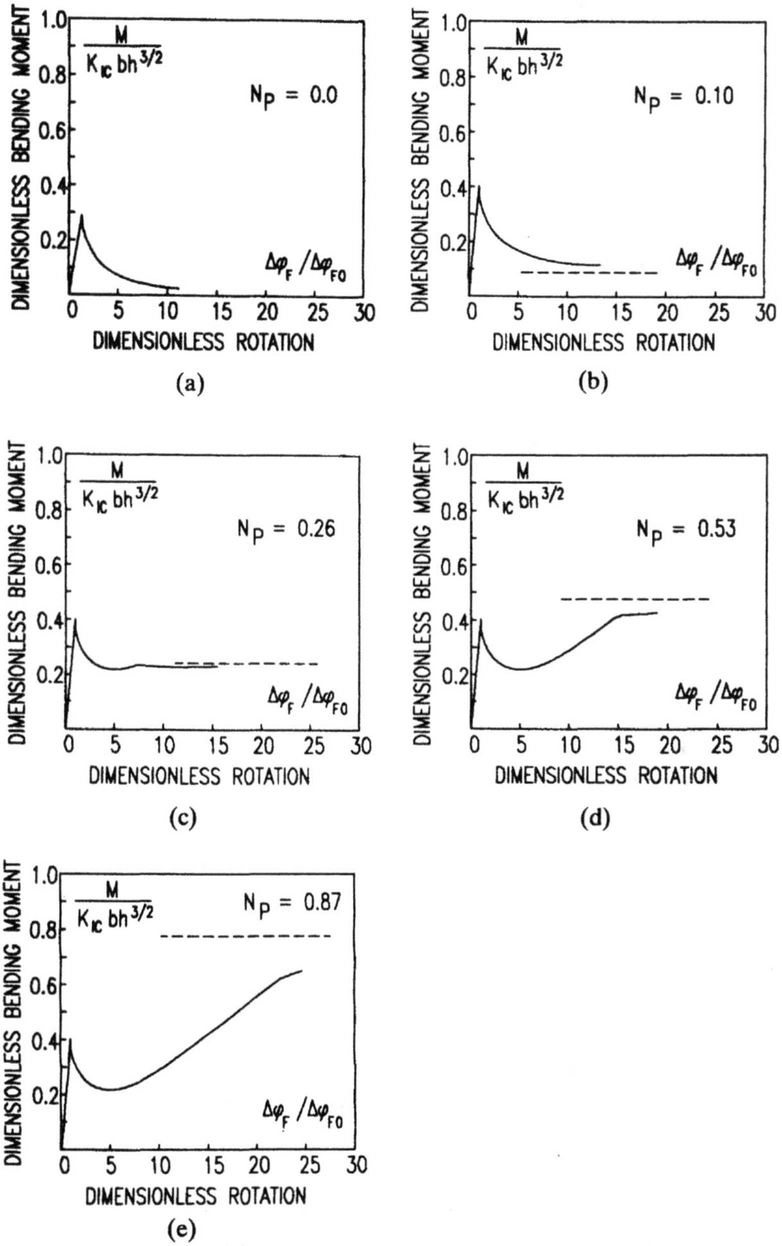

Fig. 8 Dimensionless bending moment versus normalized rotation, varying the brittleness number N_P (analytical response).

changes from unstable to stable and the failure mode does not become brittle if it is strain-controlled. On the other hand, a snap-through instability is predicted, if the process is load-controlled.

2.6 Brittle or Ductile Collapse

For practical purposes it is interesting to determine when M_F (bending moment of crack propagation) is equal to M_P (moment of reinforcement yielding). This condition allows the determination of minimum steel reinforcement [7] which guarantees the stability of the beam behaviour, when the first cracking bending moment is reached.

When $M_F > M_P$, elastic behaviour ($0 < M < M_P$) is followed by linear hardening behaviour ($M_P < M < M_F$). From eqn (29) it is possible to obtain

$$\Delta\varphi = \lambda_{MM}(M - M_P) \qquad \text{for } M > M_P \tag{39}$$

Such linear behaviour stops when crack propagation occurs. At that point, if the fracture phenomenon is unstable, function M versus $\Delta\varphi$ presents a discontinuity and drops to a value obtainable considering the complete ligament disconnection. If the fracture phenomenon is stable, the discontinuity disappears and a continuous hardening response is obtained [2], [3].

The transitional value N_{PC} can be determined from eqn (25) if we impose $M_F = P_P(h - c) \approx P_P h$, i.e. absence of discontinuity in the diagram M versus $\Delta\varphi$.

In nondimensional form

$$\frac{M_F}{K_{IC}h^{3/2}b} = \frac{1}{Y_M(\xi)} + N_P\frac{Y_P(c/h,\,\xi)}{Y_M(\xi)} \approx \frac{P_P h}{K_{IC}h^{3/2}b} \tag{40}$$

and then, recalling eqn (26)

$$N_{PC} = 1/(Y_M(\xi) - Y_P(c/h,\,\xi)) \tag{41}$$

It is then possible, simply on the basis of the cross-sectional geometrical characteristics, to distinguish between the cases of unstable fracture ($N_P < N_{PC}$) from those of stable fracture ($N_P > N_{PC}$).

3. EXPERIMENTAL INVESTIGATION

The experimental tests were carried out on unnotched reinforced concrete beams at the Department of Structural Engineering of the Politecnico di Torino.

TABLE 1 Concrete grades and related compressive strengths, dimensional classes and related beam depths

Concrete grade	*Cylindrical strength f_{cm} (N/mm^2)*	*Dimensional class*	*Beam depth h (mm)*
I	16	A	100
II	30	B	200
III	40	C	400
IV	65	D	800
V	76		

A first series of 30 high strength concrete beams was planned with the main purpose of verifying the existence of size effects in those structural elements. Subsequently, 46 beams of four different concrete grades were planned in order to determine, for each grade, the transitional brittleness number between ductile and brittle failure.

The considered concrete grades and dimensional classes are listed in Table 1.

The steel content depends on the beam size and the selected brittleness number, see eqn (26). The K_{IC} values are determined for each concrete grade on the basis of experimental results, through the values of fracture energy G_F and Young's modulus E

$$K_{IC} = (G_F E)^{1/2} \tag{42}$$

The longitudinal steel reinforcement, arranged at a distance $h/10$ from the beam edge, is constant over the entire length of the beam, so that the effective depth to the total depth ratio is 0·9 for all the beams tested. There is no transverse shear reinforcement and the maximum aggregate diameter is 15 mm.

The three-point bending tests on the initially uncracked beams were realized by a servo-controlled machine. The beams were supported by a cylindrical roller and a spherical connection, respectively, at the two extremities (span = 6 times the depth h). The load was applied in the midspan through an hydraulic actuator and the loading process was controlled by a strain gauge type DD1, placed on the lower beam edge, symmetrically with respect to the force and parallel to the beam axis; its length was equal to the beam depth. The three-point bending scheme and the cross-section of the beam are shown in Fig. 9.

The strain rate was imposed at a constant and very low value, until

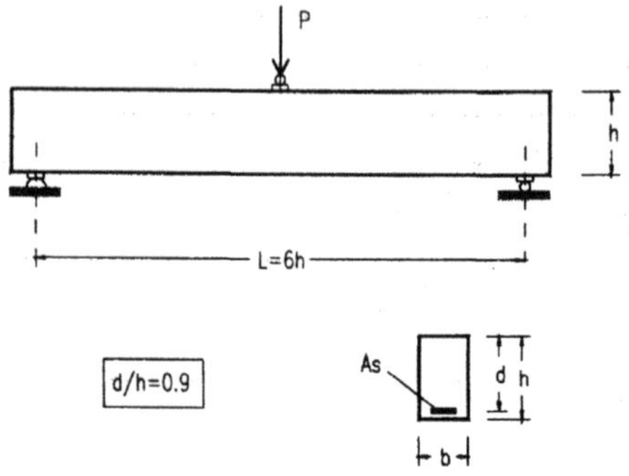

Fig. 9 Schematic of testing apparatus and cross-section of beam specimens.

the crack formation in the middle of the beam was achieved. After cracking the loading process was controlled by the crack mouth opening displacement. Even the compressive and tensile strains on the upper and lower edges of the beam were recorded. The measured central deflection was referred to a bar, connected with the concrete beam at the middle of the depth and in correspondence with the two supports. Deflection and strain gauge deformations were plotted automatically as functions of the applied load.

The transducer at the lower edge of the beam and the deflection reference bar connected to the beam, are shown in Figs 10 and 11 respectively.

The fracture energy G_F was determined on three specimens for each concrete grade according to the method specified in the RILEM draft recommendation 'Determination of the fracture energy of mortar and concrete by means of three-point bend tests on notched beams'. The value of the normal elastic modulus was determined according to the method specified in ISO 6784. Concrete strength was measured on 16 cm × 16 cm × 16 cm cubic specimens for each concrete grade. The cylindrical strength values f_{cm} of concrete are reported in Table 1.

The yield strength of the high bond steel, considered in eqn (26), was experimentally determined for each series of beams. For bars of low diameter (5 mm and 6 mm) where no yielding limit is evident, the stress at 0·2% of permanent deformation was assumed as yield strength.

A general view of the four beam size-scales is given in Fig. 12.

Fig. 10 Strain gauge on the lower beam edge.

3.1 Minimum Reinforcement

As is well known, minimum reinforcement is nowadays an 'open question', and Code provisions are not definitively settled about this parameter. On the other hand, our conclusions deal with statically determined structures, without considering imposed deformation effects (of shrinkage, drying, temperature, etc.) nor regarding maximum crack opening, but only taking into account the avoidance of brittle failure due to directly applied load.

In this case we must satisfy the following equation (Fig. 13):

$$M_Y \geq M_F \tag{43}$$

Fig. 11 Deflection reference bar connected to the concrete beam.

with M_Y being the bending moment of steel yielding and M_F the bending moment of first concrete cracking. For statically undetermined structures this condition is not strictly required, because it is possibly a redistribution of moments upon first cracking.

Typical experimental load–deflection curves, obtained by controlling tensile strain or crack opening displacement on the lower edge of the beam, are represented in Fig. 13, for low reinforced beams. These curves are characterized by a local maximum moment M_F immediately before crack formation, followed by a decrease and then by another peak M_Y at steel yielding. If the bending moment M_F of first cracking is greater than the moment capacity M_Y, a brittle failure occurs under load control. The opposite situation occurs when the content of reinforcement is sufficiently large. The condition in between ($M_F = M_Y$) allows us to determine the minimum amount of reinforcement.

3.2 Fracture Mechanics Role

Linear Elastic Fracture Mechanics theoretically allows the prediction of the transition between brittle and ductile behaviour of reinforced

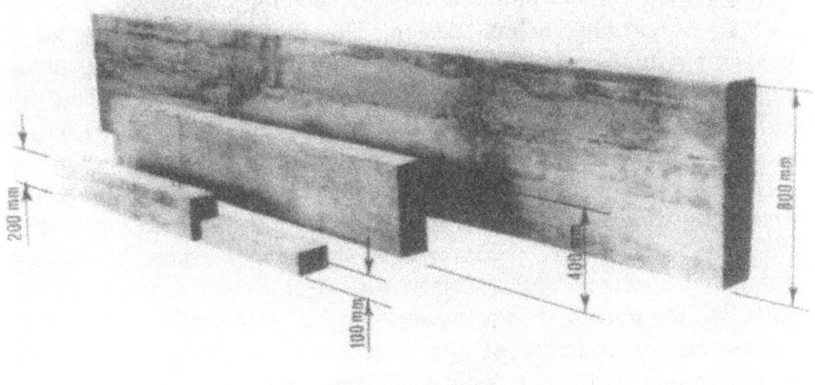

Fig. 12 View of the four beam size-scales.

concrete sections, as described previously. In fact a particular value N_{PC} of the brittleness number, given by eqn (41), divides the two failure modes (brittle and ductile), for determinate concrete grade (G_F and E) and steel characteristics (f_y). The more relevant predictions were obtained for cross-sections with low content of reinforcement or large beam depth, and then those results can be significantly extended to practical applications, as is the case under consideration of minimum reinforcement.

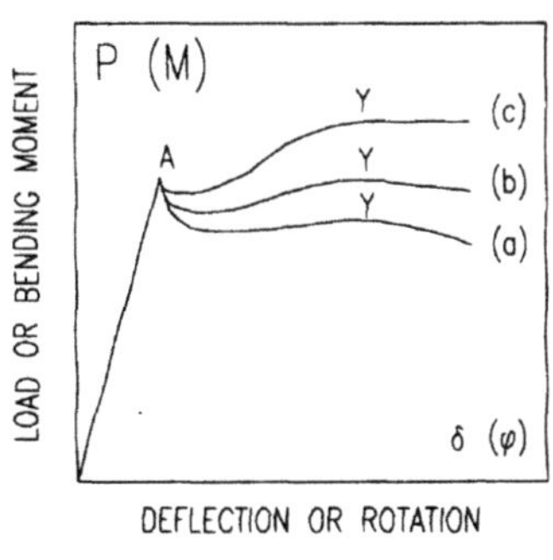

Fig. 13 Load–deflection diagram obtained by controlling the tensile strain on the lower edge of the beam.

3.3 Experimental Results and Discussion

The behaviour of concrete elements with a low steel percentage may experimentally present moment peaks exceeding the theoretical values calculated according to limit design. This difference, formerly called 'hyperstrength', is hard to explain, all the more so as the phenomenon is obviously seen to occur when the steel yield limit is reached and the evaluation of the internal lever arm presents a modest degree of indeterminacy, due to the low steel percentage.

The phenomenon has been justified by referring to a number of aspects which are usually neglected and yet play a significant role in structures with a low reinforcement percentage. Namely, the experimental results are influenced by nonlinear effects: in particular the value M_F is affected by the nonlinearity of the tensile stress–strain relationship of concrete at the onset of first cracking, while the bending moment M_Y is influenced by the stress versus crack opening displacement relationship.

The proposal for the 1990 edition of the CEB Model Code [11] takes into account both these aspects and suggests the schemes shown in Fig. 14, where the crack openings w_c and w_1 are functions of maximum aggregate diameter and concrete grade, respectively. This approach allows not only the provision of a numerical justification of hyperstrength values, but also a description of the intermediate stages of the curves, between the two peaks shown in Fig. 13.

On the other hand, even though the influence of the mentioned nonlinearities exists, the experimental results on the first series of 30 high strength concrete beams revealed that, maintaining constant the scale and varying the steel content, the transition exists and occurs at an order of magnitude as predicted by LEFM.

The load–deflection diagrams for the first series of beams are plotted in Figs 15(a), (b) and (c), for each beam size and by varying the brittleness number from zero to 0·87.

As it is possible to verify, the peak of the first cracking load is decidedly lower than the steel yielding load only in the Cases 3 and 4, i.e. for high brittleness numbers N_P. In Cases 0 and 1, the opposite result is clearly obtained. On the other hand, Case 2 demonstrates a transitional condition between brittle and ductile collapse, the two critical loads being very close. Therefore, the same brittleness transition theoretically predicted in Fig. 8, is also indicated by the experimental diagrams in Fig. 15.

The same experimental results, however, also revealed that the

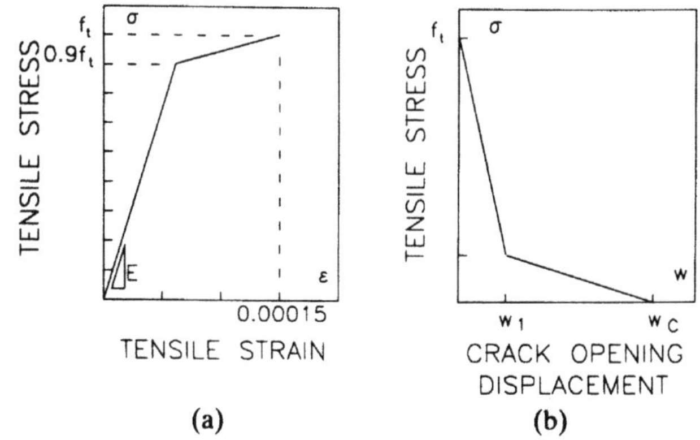

Fig. 14 (a) Stress versus strain diagram in the uncracked zone of concrete. (b) Stress versus crack opening diagram in the fracture zone of concrete.

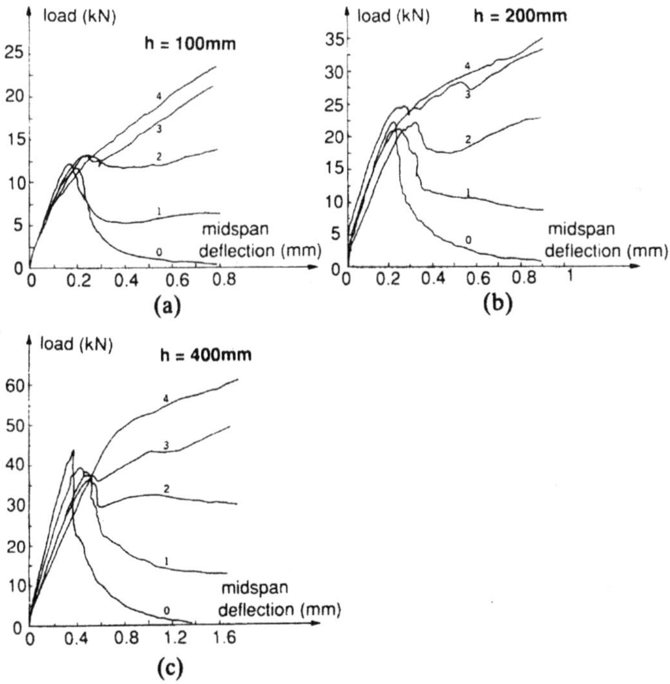

Fig. 15 Load versus deflection diagrams for high strength concrete beams.

transition occurs substantially at the same level, varying the scale. Then it is possible to group the curves in brittleness classes and pay particular attention to that class for which the transition between brittle and ductile failure occurs.

Again, for the first series of beams, the dimensionless bending moment versus rotation diagrams are plotted in Figs 16(a)–(e), for each brittleness class and by varying the beam size. The local rotation is normalized with respect to the value $\Delta\varphi_{F0}$ recorded at the first cracking, and is related to the central beam element of length equal to the beam depth h. The bending moment, on the other hand, is non-dimensionalized with respect to the critical value of stress intensity factor of concrete, K_{IC}, and the beam depth h.

The diagrams in Fig. 16 are significant only for $\Delta\varphi_F/\Delta\varphi_{F0} > 1$, the strain softening and curvature localization occurring only after the first cracking. The dimensionless peak moment does not appear to be the same when the brittleness class is the same and the beam depth is varied. This occurs because reference is made to typical Fracture Mechanics parameters while the cross-section is initially unnotched. On the other hand, the post-peak branches are very close to each other and present the same shape for each selected brittleness class. The size-scale similarity seems then to govern the post-peak behaviour, especially for low brittleness numbers N_P.

The transitional brittleness number N_{PC} for the beams of the first series was found to be approximately equal to 0·26 and then of an order of magnitude as predicted above, see eqn (41) and Fig. 8.

Even the other tests demonstrate that the size-scale effect, theoretically predicted by the model, is experimentally confirmed not only for high strength concrete, but even for concrete strengths normally used in civil structures. In other words, the transitional brittleness number N_{PC} is substantially constant, for each concrete grade, varying the scale between 100 mm and 800 mm.

All the results from the 73 useful tests out of the 76 beams under investigation are reported in Fig. 17. It is possible to observe the monotonic increasing trend of N_{PC} against the compressive strength of concrete.

It is even possible to obtain a simple relationship that, for safety reasons, is assumed to bound the maxima of the distribution (Fig. 17):

$$N_{PC} = 0{\cdot}1 + 0{\cdot}0023 f_{cm} \tag{44}$$

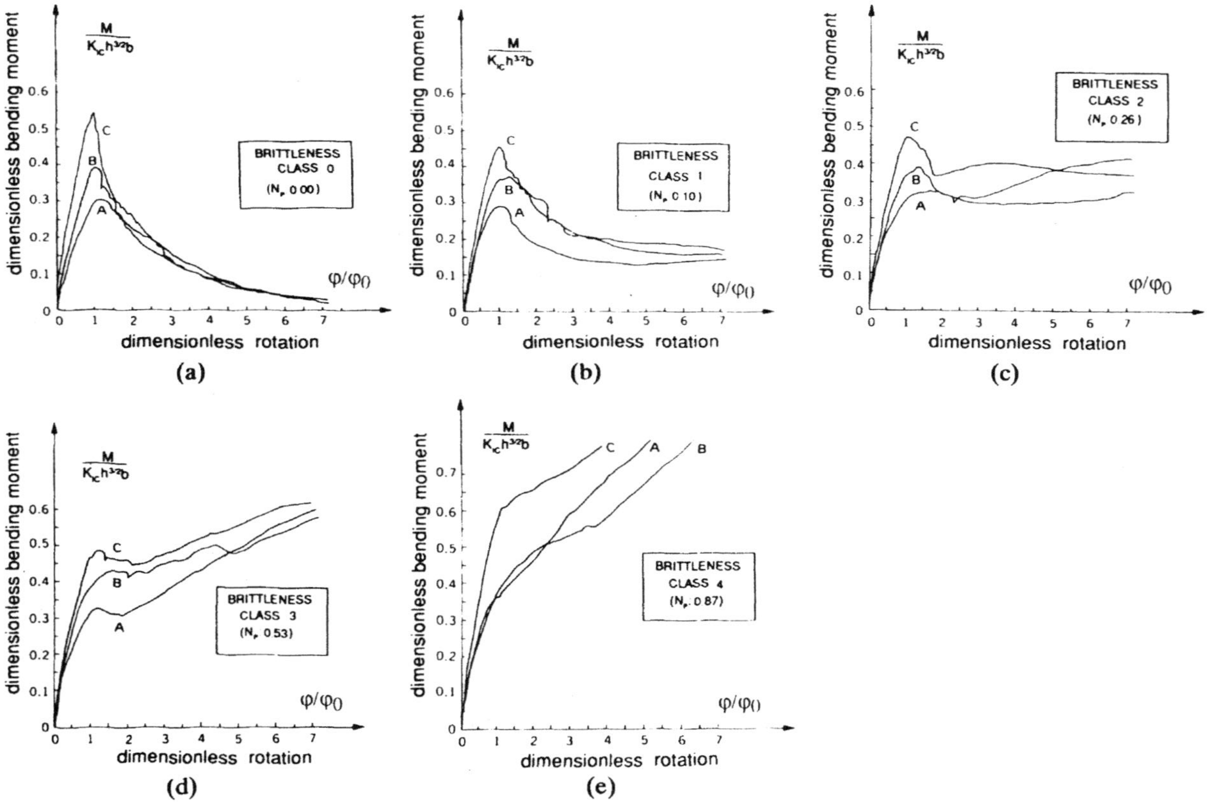

Fig. 16 Dimensionless bending moment versus normalized rotation diagrams for high strength concrete beams.

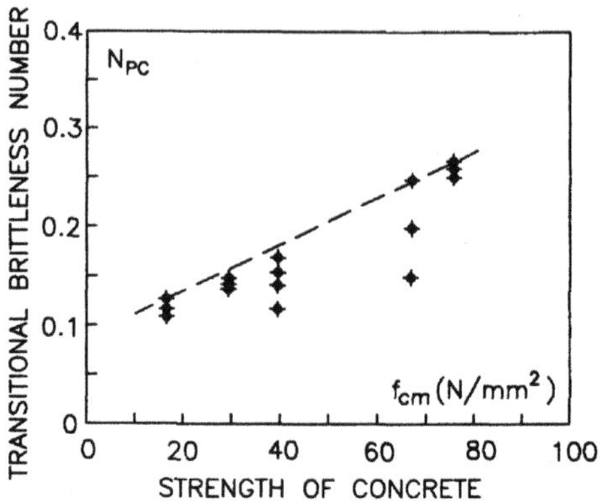

Fig. 17 Transitional brittleness number versus compression strength of concrete.

For a given cylindrical concrete strength f_{cm}, the transitional value of the brittleness number N_{PC} results from the relationship above. Then the minimum reinforcement percentage A_S/A, is obtained from relation (26), versus the beams depth $\boldsymbol{h}$.

The minimum percentage versus the beam depth is reported for four concrete strengths in Fig. 18. In the same figure, the corresponding curves obtained from EC2 [12], are reported, while the ACI 318 Code [13], provides a single percentage of steel.

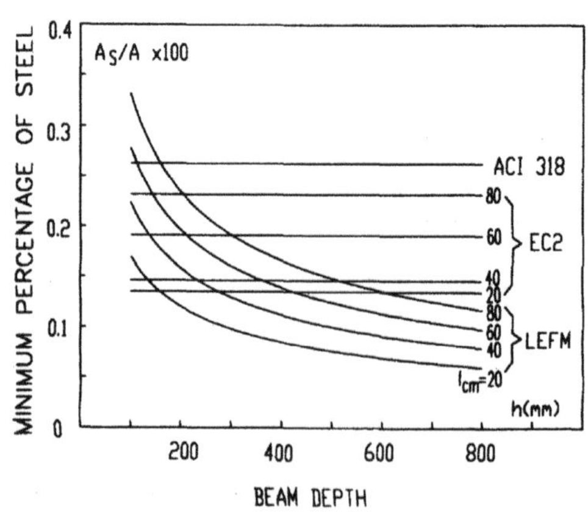

Fig. 18 Minimum steel percentage versus beam depth.

The relationship established with eqn (26) calls for decreasing minimum steel percentage with increasing beam depth, while the values supplied by the Codes are independent of the beam depth. On the other hand, eqn (44) represents the increase of the steel content by increasing the concrete strength, the beam depth being fixed.

In any case, the decrease experimentally found of the minimum steel percentage and represented by relationship (26), has also been verified by the finite element method, assuming nonlinear Fracture Mechanics models [14]. This fact allows us to affirm that the tendency shown by the experimental results obtained so far is sufficiently reliable and can be retained as a sound basis for future code revisions.

4. CONCLUSIONS

(1) The utilization of a theoretical model, where the force transmitted by the reinforcement is directly applied to the crack surfaces, confirms the existence of different behaviours at failure, from ductile to brittle, the latter occurring when the beam depth overtakes certain values and/or the reinforcement content is sufficiently low.

(2) If the crack mouth opening displacement congruence condition is taken into account to determine the force transmitted by the reinforcement, the model gives results in good agreement with the experimental behaviour of low reinforced concrete beams.

(3) The authors discuss the possibility of determining a transitional brittleness number, separating brittle from ductile collapses. Such a theoretical approach appears to be very useful for estimating the critical condition between stable and unstable crack propagation and the minimum reinforcement, using the simple formulae (26) and (44), where not only the traditional mechanical and geometrical parameters appear but even the toughness properties of concrete.

(4) In addition to the well known result that brittleness of plain concrete is a function of size, the experimental results presented above show that, even in the case of reinforced concrete members in flexure, the brittleness increases by increasing the size and/or decreasing the steel content.

(5) For each concrete grade, a physically similar behaviour is revealed in the cases where the brittleness number N_P is the same. Only when the steel percentage is inversely proportional to the square root of the beam depth, is the mechanical behaviour reproduced.

Thus, a criterion to evaluate the minimum percentage of reinforcement can be introduced. Such a percentage, assumed to be provided by the condition of simultaneous first concrete cracking and steel yielding, is found to be scale-dependent, whereas the current standard codes suggest, for direct loading, values independent of the beam depth.

(6) On the basis of the experimental results, the authors suggest the utilization of a simple relationship, theoretically derived, to calculate the minimum percentage of reinforcement for concrete structures, see eqn (44).

ACKNOWLEDGEMENTS

The authors gratefully acknowledge the financial support of the Department for University and Scientific and Technological Research (MURST), and of the National Research Council (CNR).

REFERENCES

[1] Bosco, C., Carpinteri, A. & Debernardi, P. G., Fracture of reinforced concrete: scale effect and snap-back instability. Presented at the International Conference on Fracture and Damage of Concrete and Rock. 4–6 July 1988, Vienna, Austria. *Engineering Fracture Mechanics,* **35**(4–5) (1990) 665–77.

[2] Carpinteri, A., A fracture mechanics model for reinforced concrete collapse. *IABSE Colloquium on Advanced Mechanics of Reinforced Concrete,* Delft University of Technology, Delft, 1981, pp. 17–30.

[3] Carpinteri, A., Stability of fracturing process in RC beams. *Journal of Structural Engineering* (ASCE), **110** (1984) 544–58.

[4] Carpinteri, Al. & Carpinteri, An., Hysteretic behaviour of RC beams. *Journal of Structural Engineering* (ASCE), **110** (1984) 2073–84.

[5] Okamura, H., Watanabe, K. & Takano, T., Applications of the compliance concepts in fracture mechanics. *ASTM STP 536* (1973) 423–38.

[6] Okamura, H., Watanabe, K. & Takano, T., Deformation and strength of cracked member under bending moment and axial force. *Engineering Fracture Mechanics,* **7** (1975) 531–39.

[7] Bosco, C., Carpinteri, A. & Debernardi, P. G., Minimum reinforcement in high-strength concrete. *Journal of Structural Engineering* (ASCE), **116** (1990) 427–37.

[8] Bosco, C. & Carpinteri, A., Fracture behaviour of beam cracked across reinforcement. *Theoretical and Applied Fracture Mechanics* (to appear).

[9] Bosco, C. & Carpinteri, A., Softening and snap-through behavior of reinforced elements. *Journal of Engineering Mechanics* (*ASCE*) (to appear).
[10] Tada, H., Paris, P. & Irwin, G., The Stress Analysis of Cracks Handbook. Del Research Corporation, St. Louis, Missouri, 1963, Part 2, 16–17.
[11] Comité Euro International du Beton (CEB), CEB-FIP Model Code 1990, First Draft, Mars 1990. CEB, CH-1015 Lausanne, Switzerland.
[12] Commission of the European Communities, Industrial Processes—Building and Civil Engineering. Eurocode No. 2. Design of Concrete Structures. Part 1: General Rules and Rules for Buildings. Commission of the European Communities, Brussels, October 1989.
[13] American Concrete Institute, Building Code—Requirements for Reinforced Concrete (ACI 318-89). Detroit, Michigan, 1983.
[14] Hjorteset, K., Minimum reinforcement requirements for concrete flexural members determined using nonlinear fracture mechanics. MSc thesis in Civil Engineering, University of Washington, June 1990.

Chapter 15

Minimum Reinforcement Requirements for Concrete Flexural Members

N. M. HAWKINS* & K. HJORTESET
University of Washington, Seattle, Washington, 98195, USA

ABSTRACT

Minimum reinforcement requirements for concrete flexural members are examined using fictitious crack models (FCM), four different crack closure stress (CCS) distributions, and the results of physical tests on slabs and beams of varying depth. It is concluded that the use of a CSS Straight-Line model with an associated G_F value based on measured material properties is appropriate provided account is taken of shrinkage effects and the way those effects influence cracking. Minimum reinforcement requirements should be based on gross section properties and the condition that the yield moment be greater than the cracking moment. Fracture mechanics principles then demonstrate, as specified in Norwegian Standard NS3473-1989, that minimum ratios depend on member depth, ratios of concrete to steel strength and concrete material properties. However, steel elongation requirements must be added to those Standards if hard drawn wires are used as reinforcement.

Most codes for concrete structures have minimum reinforcement requirements for flexural members [1]–[4]. Those requirements are intended: to prevent excessive crack widths at service loads for aesthetic and durability reasons; to tie the structure together and

* *Present address*: Dept. of Civil Engineering, University of Illinois at Urbana–Champaign, Urbana, Illinois 61801, USA.

ensure that it acts as assumed in the design; and to provide a ductile response and ensure adequate warning of an impending failure at overloads. While such minimum reinforcement requirements are empirical there is little pressure for change since they have been used satisfactorily for many years.

Recent analytical advances make it possible to use fracture mechanics concepts to rationally define the minimum reinforcement requirements necessary to avoid a brittle failure. This chapter defines such requirements through analysis of the results of two laboratory investigations, one concerned with requirements for beams [5] and the other with requirements for slabs reinforced with welded wire fabric [6].

Minimum reinforcement requirements to prevent excessive crack widths depend on bar anchorage characteristics and environmental factors, as well as fracture considerations. An examination of those requirements is not included in the scope of this chapter.

1. INTRODUCTION

Most models codes, with the exception of the Norwegian [3], specify as shown in Fig. 1, minimum reinforcement requirements that are depth independent, although ACI [1] does have different requirements for beams and slabs. By contrast Carpinteri's experiments [5] show

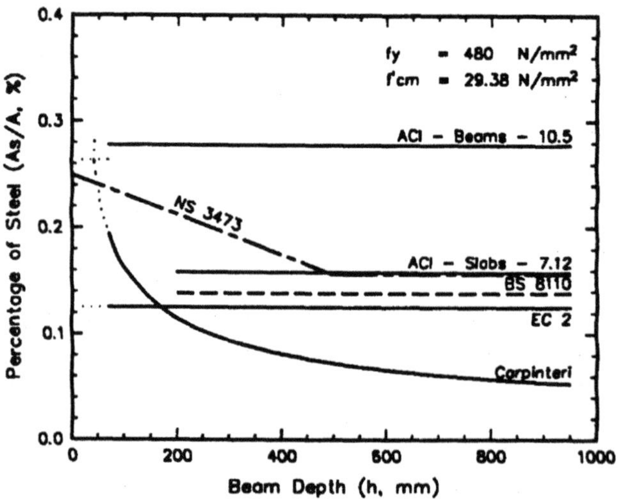

Fig. 1 Minimum reinforcement requirements as a function of beam depth.

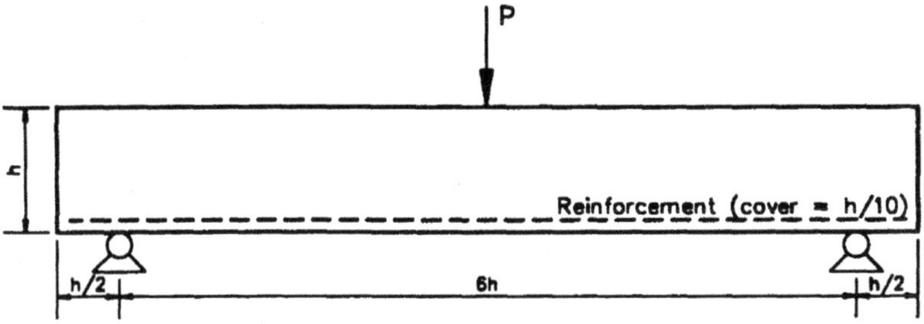

Specimen	Depth, h mm	Beam Span mm	Thickness, b mm
A	100	600	150 or 200
B	200	1,200	150 or 200
C	400	2,400	150
D	800	4,800	200

Fig. 2 Beam geometries.

that minima should be depth dependent while Hu's experiments [6] show they should also be wire size dependent.

Carpinteri conducted three-point bend tests on initially unnotched and uncracked reinforced and plain concrete rectangular beams varying in depth from 100 to 800 mm. His 100-mm depth specimens are studied here and they had the geometries shown in Fig. 2. Those specimens were made with the two different concrete grades and two different steel grades as shown in Table 1. Test results for Carpinteri's

TABLE 1 Material properties for Carpinteri's beams [5]

Concrete grade	2	4
Cube strength f (MPa)	29·38	75·7
Modulus E (MPa)	23 150	34 300
Fracture energy G (N/mm)	0·134	0·090
Tensile strength f (MPa)	2·28	5·30(4·03)
Fic. crack length l (mm)	597	110(190)

	Steel used in Concrete							
	Grade 2				*Grade 4*			
Diameter (mm)	5	6	8	10	4	5	8	10
Tensile strength f (MPa)	673	668	704	562	680	622	675	580
Yield strength f (MPa)	633[a]	489	480	456	637[a]	569[a]	441	456
Elongation %	24	32	33	30	20	22	30	32

[a] Determined as 0·2% offset stress.

TABLE 2 Cracking loads (P_{cr}), steel yield loads (P_y) and ultimate loads (P_u) for beams containing concrete grade 2 [5]

Case	*h* (*mm*)	*b* (*mm*)	*Steel*	A_s/bh (%)	f_y (*MPa*)	P_{cr} (*kN*)	P_y (*kN*)	P_u (*kN*)
1	100	150	1φ5	0·131	633	11·26	11·49	11·49
2	100	150	2φ5	0·261	633	11·10	19·80	21·21
3	100	150	3φ5	0·392	633	11·25	27·40	29·23
1	200	150	1φ6	0·094	489	11·34	10·12	12·26
2	200	150	2φ6	0·188	489	14·05	17·69	23·35
3	200	150	3φ6	0·283	489	12·11	24·91	33·00
1	400	150	1φ8	0·083	480	19·94	20·06	23·65
2	400	150	2φ8	0·167	480	21·73	31·31	37·99
3	400	150	3φ8	0·250	480	18·75	47·50	55·94
1	800	200	1φ10	0·049	456	35·24	27·78	29·10
2	800	200	2φ10	0·098	456	43·37	45·33	44·40
3	800	200	3φ10	0·147	456	45·60	60·60	97·65

specimens studied here in terms of cracking loads, P_{cr}, yield loads, P_y, and ultimate loads, P_u, as reported by Carpinteri, are shown in Tables 2 and 3. Load versus midspan deflection diagrams for typical grade 4 beams are shown in Fig. 3.

Carpinteri [7] has suggested that minimum reinforcement requirements be characterized using a brittleness number, N_P, derived from linear elastic fracture mechanics concepts, LEFM, as

$$A_s/A = N_p K_{IC}/(f_y h^{0.5}) \tag{1}$$

where A_s/A is the steel ratio based on the beam's gross section, K_{IC} is the critical stress intensity factor calculated as $(G_F E_C)^{0.5}$ where G_F is the fracture energy and E_C the modulus determined by standard methods, f_y is the yield strength of the steel and h is the beam's overall depth. Based on the aggregate of his experimental results, and not just those studied here, Carpinteri suggested that N_P be taken as 0·14. For his grade 2 concrete that limitation results in the curve 'Carpinteri' in Fig. 1.

Mirza and MacGregor [8] have pointed out that for cold drawn wires, the drawing process results in the ultimate elongation decreasing as the wire size decreases. Based on theoretical analyses, and an examination of available data, they proposed that for members reinforced with such wires or with fabric made from such wires, minimum reinforcement ratios should be made a function of wire size

TABLE 3 Cracking loads (P_{cr}), steel yield loads (P_y) and ultimate loads (P_u) for beams containing concrete grade 4

Case	*h (mm)*	*b (mm)*	*Steel*	A_s/bh (%)	f_y (MPa)	P_{cr} (kN)	P_y (kN)	P_u (kN)
0	100	150	0	0·000	—	11·38	0·00	0·00
						12·16	0·00	0·00
1	100	150	1ϕ4	0·085	637	11·77	6·43	6·01
						11·77	7·61	5·55
2	100	150	2ϕ5	0·256	569	12·07	14·71	11·77
						12·95	15·70	10·79
3	100	150	2ϕ8	0·653	441	13·73	26·98	22·48
						13·34	28·90	21·64
4	100	150	2ϕ10	1·003	456	15·30	34·14	48·79
						14·52	34·88	46·83
0	200	150	0	0·000	—	23·08	0·00	0·00
						22·03	0·00	0·00
1	200	150	1ϕ5	0·064	569	21·40	10·42	6·10
						17·66	10·18	5·50
2	200	150	3ϕ5	0·190	569	19·50	23·04	17·69
						22·19	23·16	16·59
3	200	150	3ϕ8	0·490	441	21·19	40·79	57·53
						23·54	42·06	55·91
4	200	150	3ϕ10	0·775	456	25·74	64·01	77·31
						27·62	65·89	75·81
0	400	150	0	0·000	—	44·12	0·00	0·00
						36·28	0·00	0·00
1	400	150	2ϕ4	0·043	637	36·67	15·69	8·40
						—	—	—
2	400	150	4ϕ5	0·128	569	39·97	33·08	25·50
						37·49	31·64	23·29
3	400	150	4ϕ8	0·327	441	45·50	55·69	66·54
						40·79	52·16	63·46
4	400	150	4ϕ10	0·517	456	49·03	87·66	99·18
						48·83	81·18	96·12

as follows:

$$\rho_{min} = 0{\cdot}0012/'A_n^{0.5} \tag{2}$$

where ρ_{min} is the steel reinforcement ratio, and A_n is the nominal area of the reinforcing wire.

To explore the appropriateness of proposal [8], Hu [6] conducted elongation tests in air on wires of several different sizes and then on fabric made with those same wires. He measured elongations on a 51-mm gage length with readings for fabric taken both across the weld

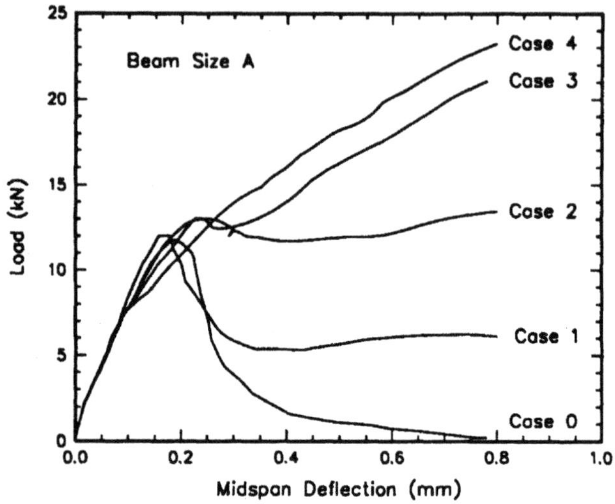

Fig. 3 Load versus deflection diagrams for the beams containing concrete of grade 4.

and between welds. He then conducted four-point bend tests on nine slabs reinforced with those same fabrics and with steel ratios equal to

(A) the lower limit for A_S/A permitted by ACI 318-83 Sec. 7.12.2.1, namely 0·0018 (60 000)/f_y;
(B) the limit for A_s/bd of 200/f_y specified for beams in ACI 318-83 Sec. 10.5.1; and
(C) A_s/bd equal to (0·9)(0·0012)/'A_n which is 90 percent of the permissible limit recommended by Mirza & MacGregor.

The test set-up used by Hu is shown in Fig. 4, and his wire arrangements in Fig. 5. Properties of his test specimens are summarized in Table 4 and his test results in Table 5. In Table 4, h is the slab's overall depth, b is its width, f_{cc} the cylinder strength, f_{sp} the split cylinder strength, d is the effective depth for the slab's reinforcement in its longitudinal direction, ρ_r is A_s/bd, and f_u is the ultimate strength of the wires. The letters and numerals in a specimen's designation, for example SD 7.3A, have the following significance: D indicates a deformed wire; 7.3 is the cross-sectional area of the wire in hundredths of a square inch and A indicates the target ρ value A, B, or C as stated above. In Table 5, M_d is the equivalent deadload moment which includes the weight of the slab and the loading apparatus; $M_{cr,obs}$

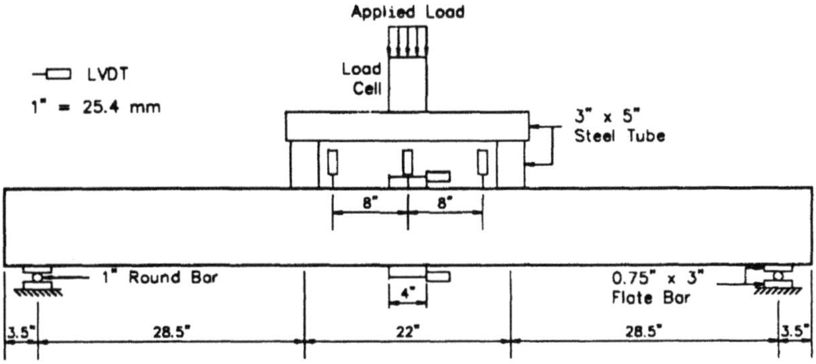

Fig. 4 Test set-up of Hu's slab test (1 in. = 25·4 mm) [10].

is the observed cracking moment and $M_{u,obs}$ is the observed ultimate moment. For all of Hu's specimens ultimate moments were greater than the cracking moments.

2. ANALYSIS OF TEST RESULTS

Analysis of Carpinteri's and Hu's results were conducted using two methods: hand calculations based on methods developed by Hillerborg

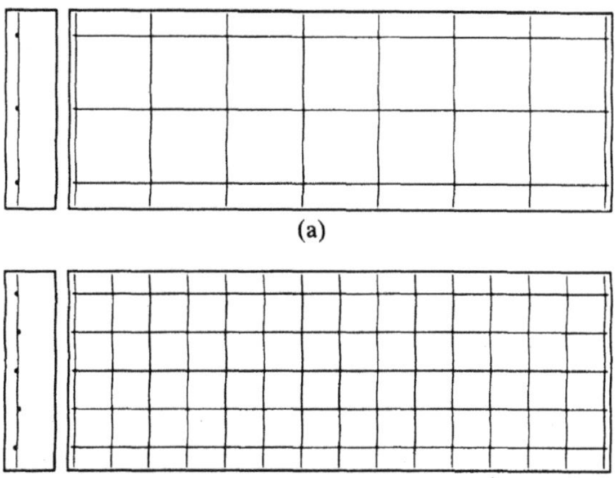

(a)

(b)

Fig. 5 Wire arrangement for Hu's slab specimen [10]. (a) Arrangement; (b) Arrangement.

TABLE 4 Properties of Hu's test specimens

Specimen number	*h (mm)*	*b (mm)*	f_{cc} *(MPa)*	f_{sp} *(MPa)*	*d (mm)*	*Arr type*	ρ_r *(%)*	f_u *(MPa)*
SD7.3A	125	814	51·1	3·9	103·12	1	0·1684	826·85
SD7.3B	144	813	45·4	3·4	120·65	2	0·2401	803·47
SD7.3C	102	813	45·4	3·4	69·60	2	0·4163	819·75
SD9.5A	148	822	35·6	3·3	124·71	1	0·1793	797·54
SD9.5B	158	811	35·6	3·3	131·83	2	0·2865	711·56
SD9.5C	136	814	48·3	3·9	107·70	2	0·3498	776·86
SD11.3A	191	822	43·3	3·2	175·51	1	0·1515	835·40
SD11.3B	153	815	51·1	3·9	132·08	1	0·2032	835·40
SD11.3C	113	814	48·3	3·9	87·88	1	0·3059	835·40

and finite element method (FEM) studies. Shown in Fig. 6, for three characteristic crack closing pressure stress distributions, are variations in the ratio of the flexural strength, f_f, to the concrete tensile strength, f_t, as a function of the beam's overall depth h and the characteristic fictitious crack length l_{ch} equal to $E_c G_F / f_t^2$ [9]. The three stress distributions are those corresponding to a Dugdale model, M, a straight line model, SL, and a Petersson model, C, with all distributions having the same G_F value. The appropriate cracking closing stress, σ, crack width, w, relationship for each distribution is shown alongside the appropriate curve on Fig. 6. Those three curves are used to predict

TABLE 5 Hu's test results

Specimen number	M_d *(k-in.)*	$M_{cr,Obs}$ *(k-in.)*	$M_{cr,Exp} = M_d + M_{cr,Obs}$		$M_{u,Obs}$ *(k-in.)*	$M_{u,Exp} = M_d + M_{u,Obs}$	
			(k-in.)	*(kNm)*		*(k-in.)*	*(kNm)*
SD7.3A	12·74	61·67	74·41	8·41	86·53	99·27	11·22
SD7.3B	14·32	72·97	87·29	9·86	186·77	201·09	22·72
SD7.3C	10·01	34·71	44·72	5·05	104·02	114·03	12·89
SD9.5A	14·61	66·28	80·89	9·14	143·24	157·85	17·84
SD9.5B	15·47	71·20	86·67	9·79	231·20	246·67	27·87
SD9.5C	13·62	65·19	78·81	8·91	217·61	231·23	26·13
SD11.3A	18·21	124·54	142·75	16·13	261·89	280·10	31·65
SD11.3B	15·05	93·61	108·66	12·28	197·51	212·56	24·02
SD11.3C	11·69	48·04	59·73	6·75	127·18	138·87	15·69

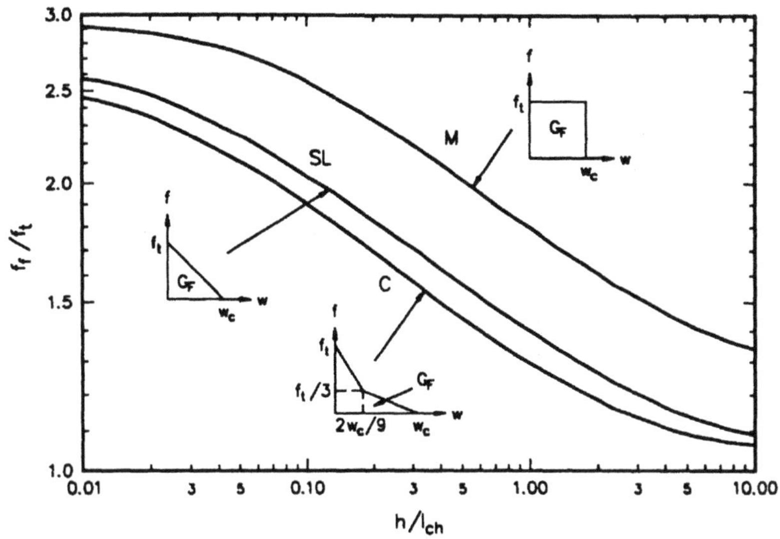

Fig. 6 Theoretical variation in flexural strength f_f for an unnotched beam based on metal model (M), Straight-Line model (SL) and Petersson or Concrete model (C).

variations in cracking loads with beam or slab depth as a check on more detailed FEM calculations.

2.1 FEM Calculations—Unreinforced Beams

Carpinteri tested unreinforced beams of grade 4 concrete only. That material was very brittle with an l_{ch} value of 190 mm. Therefore, the behavior predicted for grade 2 concrete for which the l_{ch} is 597 mm, is also examined. However, only the behavior of the former can be correlated with Carpinteri's data.

The FEM analysis was made using the mesh of Fig. 7. Only one half of the beam was modeled. There were 41 nodes on the centerline with node 1 at the bottom of the beam and node 41, where the 'load' was applied as a constrained displacement, at the top of the beam. The analysis was performed using ANSYS and STIF42 elements. The fracture region was assumed to be confined to the symmetry line and the concrete properties used were those of Table 1. The σ–w relations of Fig. 6 were applied in the FEM analysis as springs on the symmetry line so that for nodes 2 through 40 the force, F_S, crack width, w,

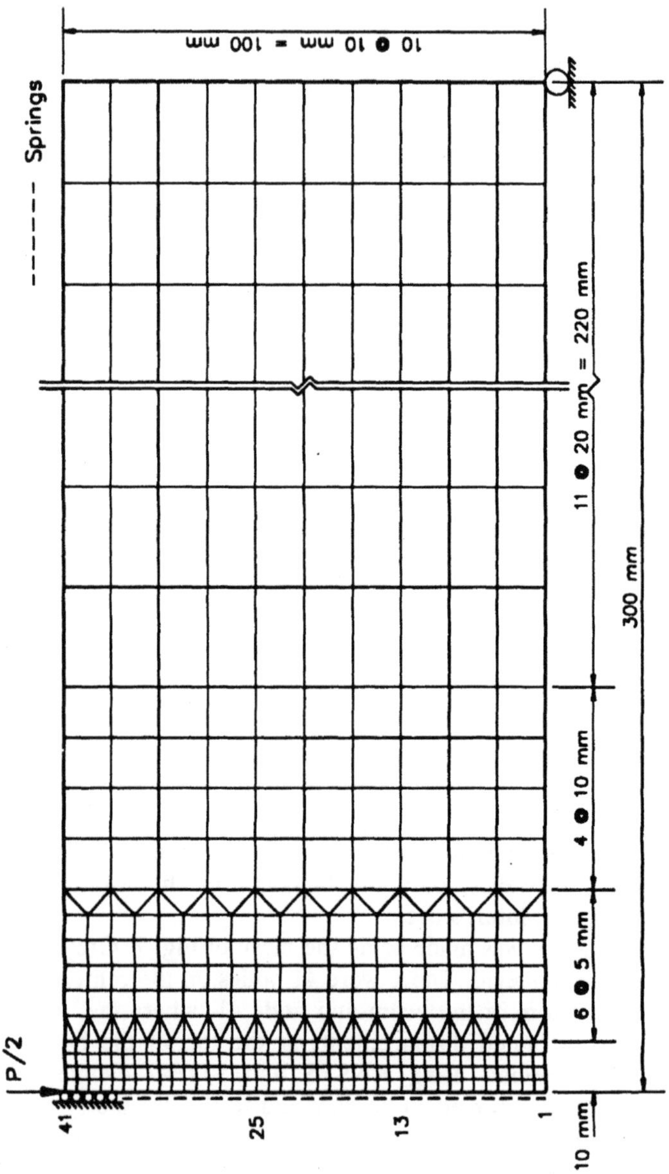

Fig. 7 Finite element mesh used for Carpinteri's beams.

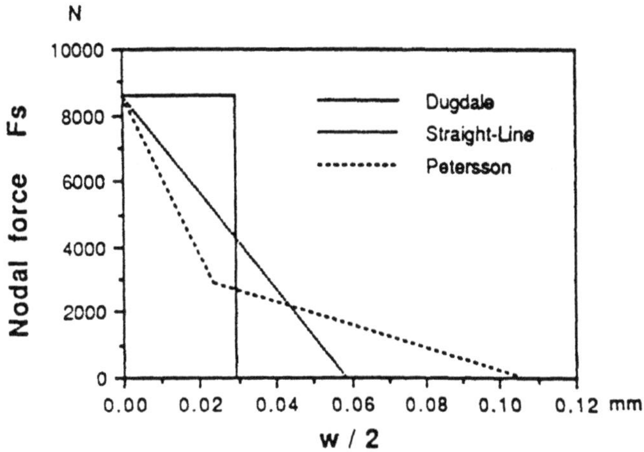

Fig. 8 F_s–w curves for springs applied to the nodes 2, 3, . . . , 40.

relations were as shown in Fig. 8. Because the crack opens on both sides of the symmetry line the width corresponding to a given F_s is half the w value of Fig. 6.

Shown in Fig. 9 are the predicted P–δ curves normalized by dividing P by the predicted elastic cracking load, $P_{cr,E}$, (3·90 kN for the f_t value of 2·28 MPa) and the deflection by $\delta_{cr,E}$ corresponding to that load. The points N1, N4 and N8, etc., in Fig. 9 indicate the

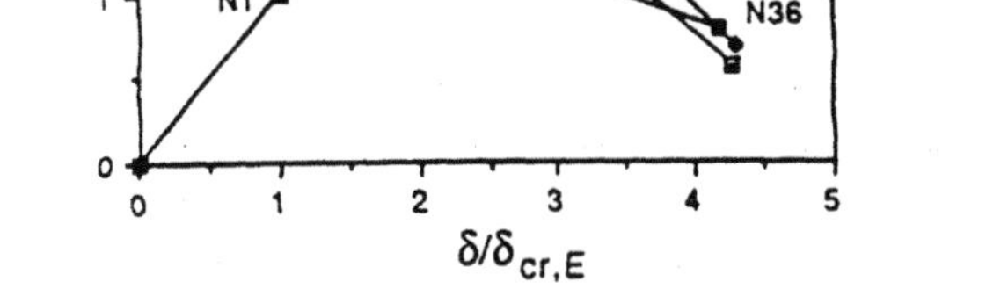

Fig. 9 Normalized P–δ curve for concrete grade 2, $h = 100$ mm. $P_{cr,E} =$ 3·90 kN and $\delta_{cr,E} = 0{\cdot}069\,95$ mm. FEM-calculations using Dugdale, Straight-Line and Petersson model.

conditions at which the stress on nodes 1, 4, 8, etc. first reached the value f_t. Deviations in the responses for the three relationships first became apparent at N16. The maximum $P/P_{cr,E}$ ratios in Fig. 9 are 2·22, 1·79, and 1·65 for the Dugdale, Straight-Line and Petersson models, respectively. The h/l_{ch} value for that beam and concrete grade is 100/597 or 0·17. Those maximum ratios are close to the values of 2·37, 1·86 and 1·73 predicted using Fig. 6 and that h/l_{ch} value. For the Dugdale model snap-back behavior occurs between conditions N28 and N32 because the linear elastic bulk material outside the fracture zone releases more energy than it takes to create a traction free crack surface between nodes 28 and 32. Figure 10 shows the crack width and

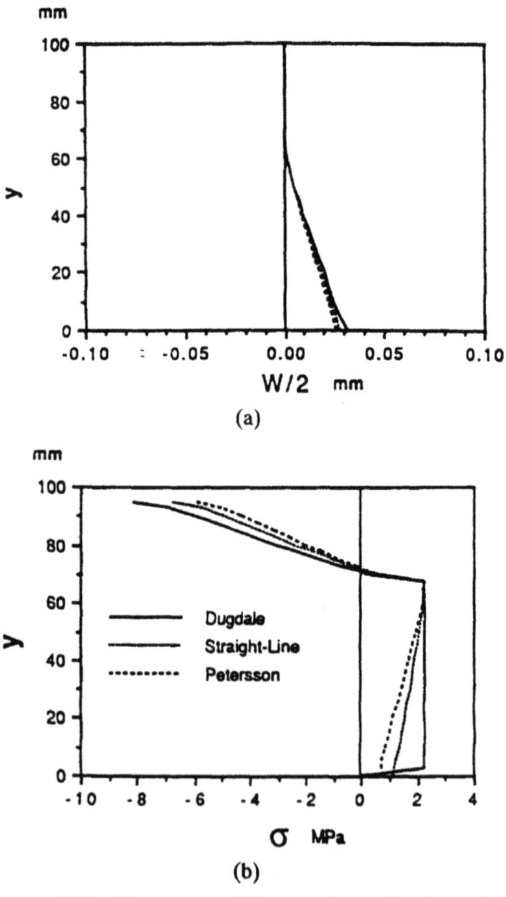

Fig. 10 Load state N28. (a) crack width, w, along the line of symmetry; (b) stresses along the line of symmetry.

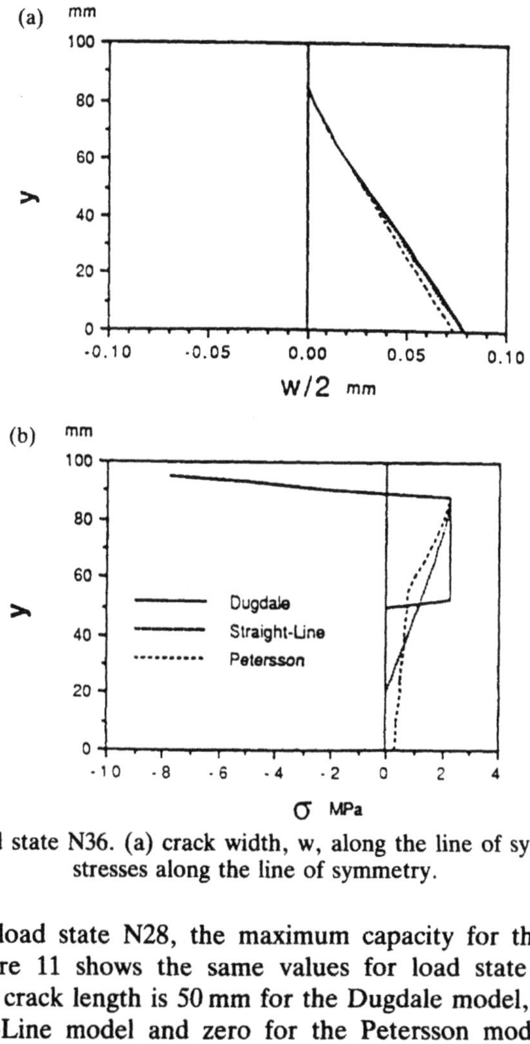

Fig. 11 Load state N36. (a) crack width, w, along the line of symmetry; (b) stresses along the line of symmetry.

stresses for load state N28, the maximum capacity for the Dugdale model. Figure 11 shows the same values for load state N36. The traction free crack length is 50 mm for the Dugdale model, 20 mm for the Straight-Line model and zero for the Petersson model at N36. Clearly, in addition to the maximum capacity, the major difference in the three models lies in the predictions of the immediate post-peak behavior.

Carpinteri has reported P–δ results for unreinforced concrete grade 4 beams. Stress versus crack width relationships of that concrete for the three models examined previously, plus the Kobayashi–Hawkins' model [10], are shown in Fig. 12. For the K–H model, fracture properties were calculated from the concrete compressive strength so

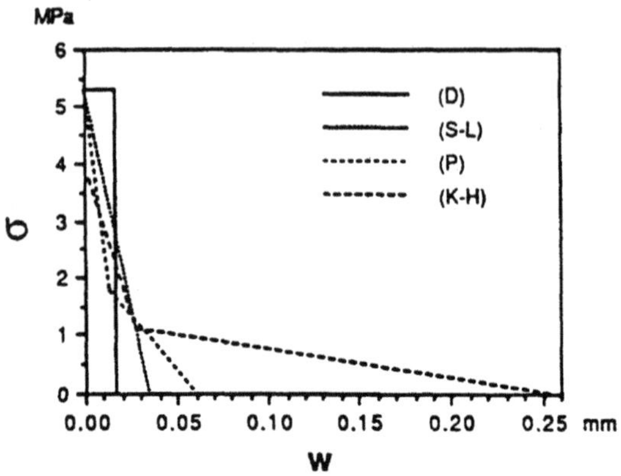

Fig. 12 σ–w curve for concrete grade 4 (Dugdale, Straight-Line, Petersson and Kobayashi–Hawkins model).

that while the tensile strength and fracture energy were 3·72 MPa and 0·196 N/mm for the K–H model they were 5·3 MPa and 0·09 N/mm for the other three models. Measured and computed P–δ curves are shown in Fig. 13. Clearly, while the best overall agreement is with the K–H model there is also a wide discrepancy between the test result and any of the predictions. There are two shortcomings for all four analyses. First, the material away from the centerline was modeled as a linear elastic. Use of a bilinear response with a turning point at $0{\cdot}9f_t$

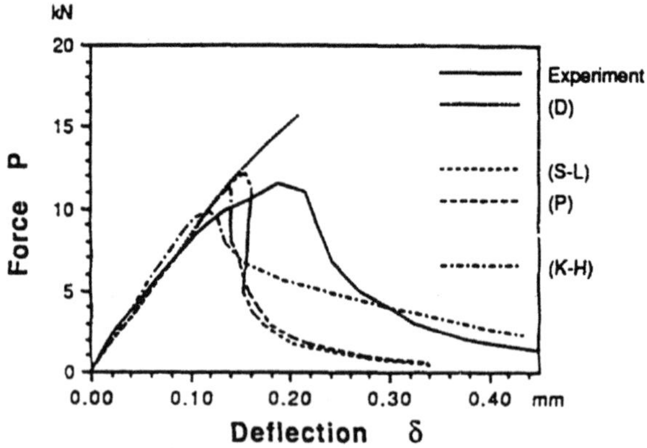

Fig. 13 Measured and calculated P–δ curves for concrete grade 4, $h =$ 100 mm.

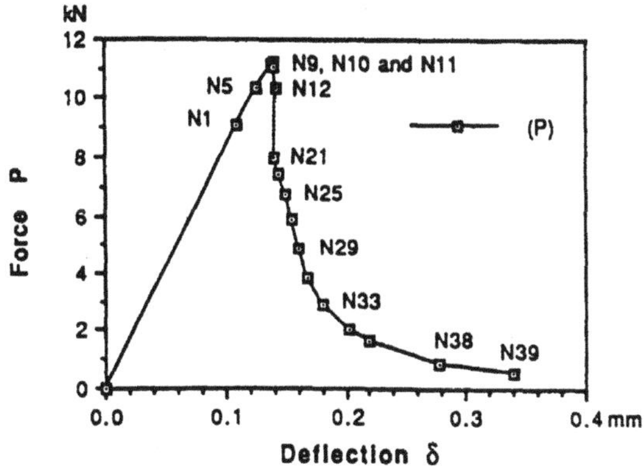

Fig. 14 Calculated P–δ curve, Petersson model, concrete grade 4, $h = 100$ mm.

is more appropriate [11]. Second, constriction of cracking to the centerline only was not consistent with the FEM predictions. Shown in Fig. 14 is the P–δ curve for the Petersson model with conditions N1 through N39 shown for the crack tip at nodes 1 through 39. Shown in Fig. 15 are the corresponding stresses along the bottom of the beam. The curve N1 should, of course, be a straight line, and its unevenness reflects the limits of this FEM analysis. However, the situation for N5 through N10 is of more concern because for those states a peak stress

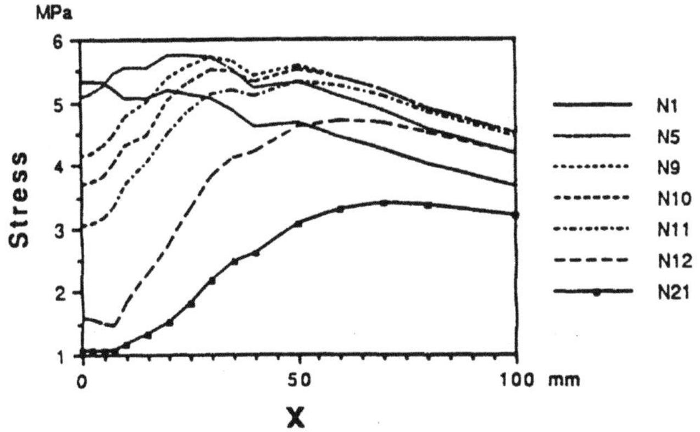

Fig. 15 Stresses along lower edge of beam, Petersson model, concrete grade 4, $h = 100$ mm.

greater than f_t is developed at 30 mm from the centerline. Obviously, a complete analysis would have to consider the possibility of cracking at that location and such behavior would undoubtedly contribute to the actual deflections for states N5 and higher being greater than predicted, as reflected by Fig. 13.

Obviously there are limitations to using discrete crack models such as those of Fig. 12 to predict the complete response of unreinforced beams. However, the maximum to N_1 loads predicted by the FEM calculations for these grade 4 beams are 1·73, 1·34 and 1·23 in the case of the D, S–L and P models respectively. The h/l_{ch} value for that concrete is 100/110 or 0·91 and for that value the corresponding predictions from Fig. 6 are 1·80, 1·41 and 1·29 respectively. Obviously, Hillerborg's f_f/f_t diagram is sufficiently accurate for design purposes for predictions of maximum capacity of unreinforced beams for a wide range of concrete strengths.

2.2 FEM Calculations—Reinforced Beams

FEM analyses were performed on Carpinteri's reinforced specimen, Case 1 Table 3, that had the same dimensions as his unreinforced beams also examined here. The mesh used is shown in Fig. 16. That mesh is the same as the mesh of Fig. 7 adjusted for the presence of the reinforcing bar. The bar was assumed to be elastic–plastic with the plastic stress equal to the yield strength of 637 MPa listed in Table 1 for grade 4 concrete and a 4-mm diameter bar. The bilinear Petersson model was used for concrete fracture properties.

The strain and stress in the reinforcement on the centerline were calculated as shown in Fig. 17, as a function of the load *P*. Loads were divided into seven levels covering the full range of behavior. Two approaches were used to provide bounds on bond effects. The first approach (a) assumed perfect bond, a steel strain equal to the surrounding concrete strain, and was used for load stages. The second approach (b) assumed plane sections remained plane after cracking. Before cracking both approaches are the same.

In Fig. 17 at stage 5, when the crack tip has crossed the steel, the strain in that steel tends to infinity, and therefore the bar's yield stress, when approach a is used. That rigid bond approach gives the maximum load-carrying capacity (*P-MAX*) for the beam (stages A5, 6 and 7 of Fig. 17). Approach A can be expected to predict reinforcement stresses larger than those likely in reality.

Approach b, plane sections remaining plane, is the normal assump-

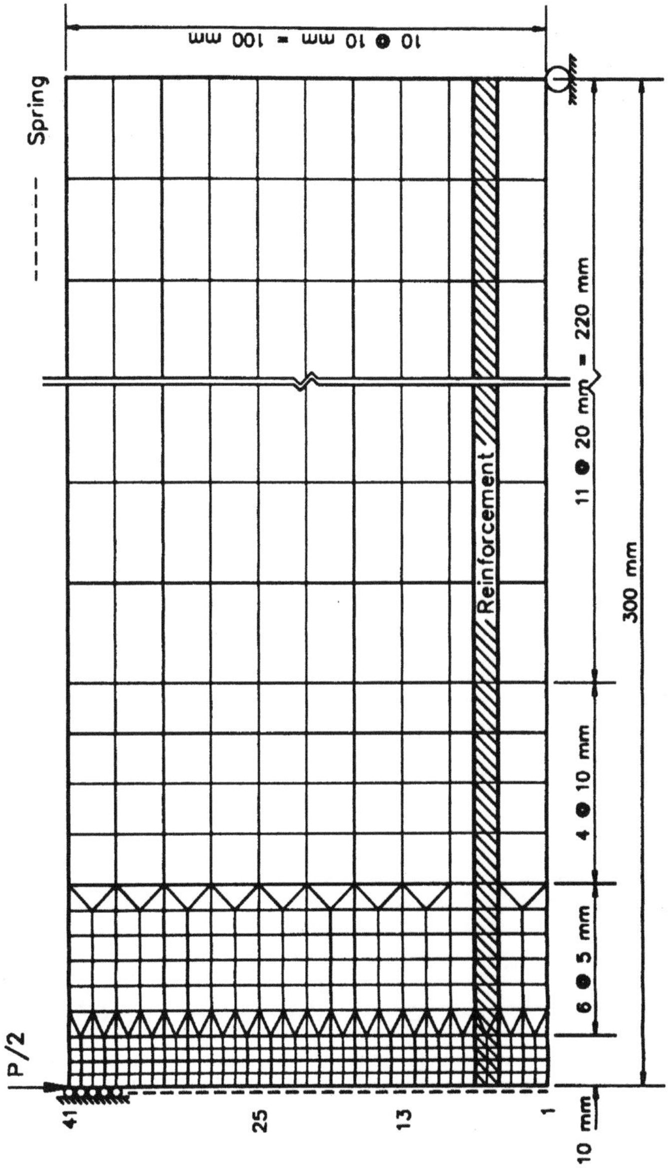

Fig. 16 Finite element mesh used for Carpinteri's reinforced concrete beams.

tion for post-cracking behavior of concrete. This is termed the *P-MIN* approach and the accompanying stages are shown as b5, 6, and 7 in Fig. 17. The predicted capacity for load stages a7 and b7 is the same since the reinforcement is yielding for both concepts.

Shown in Fig. 18 are the responses predicted by the two FEM approaches and in Fig. 19 those responses are compared with the measured response. All three responses are the same through condition N5. After the crack tip crossed the bar the *P-MAX* approach predicted a steel stress of f_y for condition N9 while the *P-MIN* approach yielded values of steel stress of 5·2%, 6·3%, 9·2%, 14·1%, 25·5% and 53·3% of f_y for crack tip at nodes 9, 11, 21, 25 and 29 respectively. Not until the crack tip reached node 36 did the *P-MIN* approach predict steel yielding and the same strength as the *P-MAX* approach.

Shown in Fig. 20 are stresses over the section's depth, with the stresses in the reinforcement distributed to nodes 5, 6, and 7 and 'smeared' through the reinforcement thickness so that node 6 carries 50% of the steel force. Only after the traction free crack has extended a significant distance up the beam, between stages N29 and N36, does the steel stress increase markedly.

The comparison of the experimental and predicted results in Fig. 19 shows, as expected, that the *P-MAX* approach predicts too high a

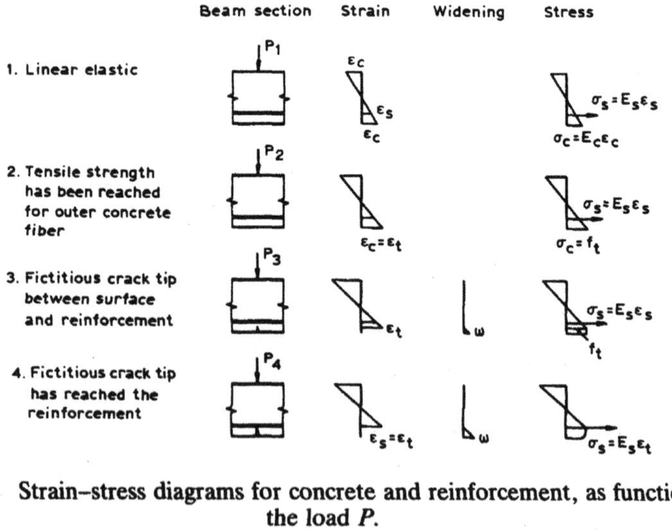

Fig. 17 Strain–stress diagrams for concrete and reinforcement, as function of the load *P*.

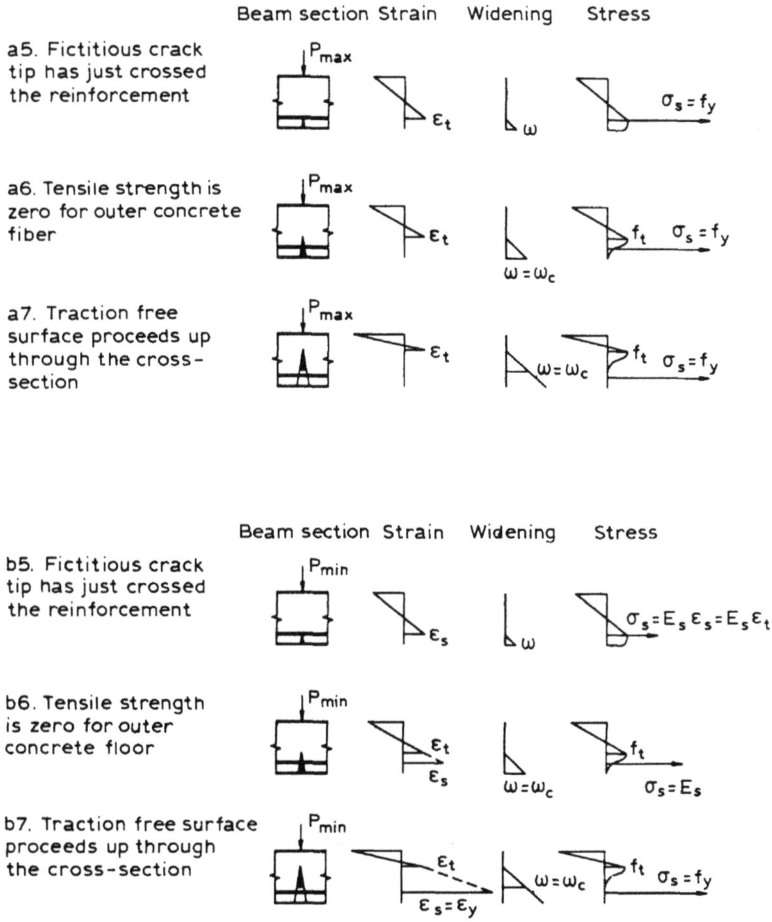

Fig. 17—*contd.* (a) perfect bond is assumed so that reinforcement yields before the crack propagates through the cross section (maximum load-carrying capacity); (b) plane sections are assumed to remain plane for the beam as a whole (minimum load-carrying capacity).

maximum capacity, while the *P-MIN* approach, yields a reasonable prediction of that capacity. However, in the post-maximum range the *P-MAX* approach soon gives a prediction paralleling the experimental result. That may not be unreasonable since perfect bond is to be expected for an unloading situation. However, that result may also be due to the unrealistic assumlption of cracking on the centerline only.

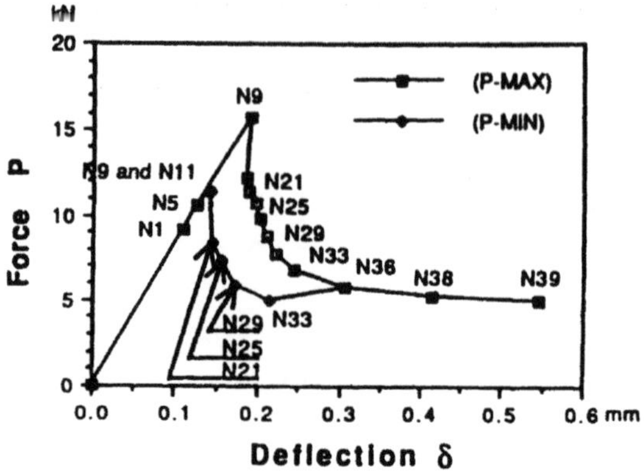

Fig. 18 Calculated P–δ curve using *P-MAX* and *P-MIN* approaches.

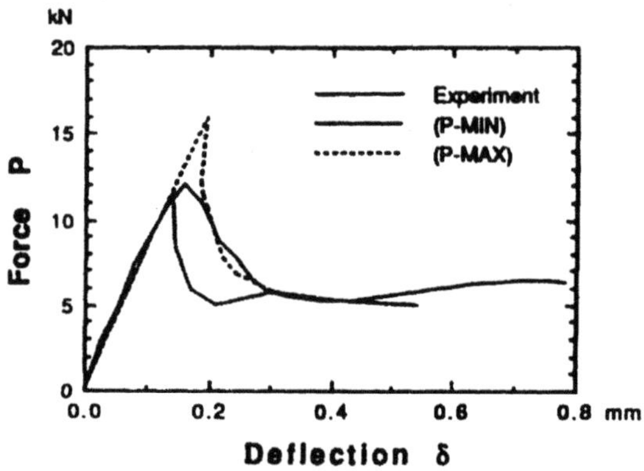

Fig. 19 Calculated P–δ curve. Reinforced concrete beam, 104. Concrete grace 4. $h = 100$ mm.

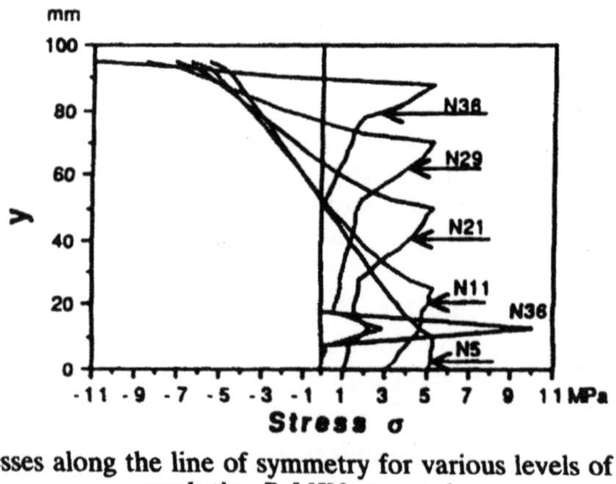

Fig. 20 Stresses along the line of symmetry for various levels of the fictitious crack tip, *P-MIN* approach.

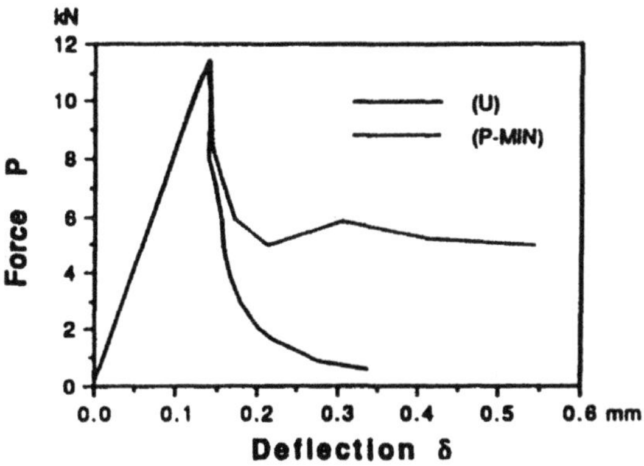

Fig. 21 P–δ curve for unreinforced (U) and reinforced concrete beam. Concrete grade 4. $h = 100$ mm.

At large deflections the predicted and experimental capacities all approach the yield load of 4·54 kN. The tensile contribution of the cracked concrete decreases as the deflection increases. Thus the normal assumption of zero tensile strength for the concrete becomes increasingly realistic with increasing deflections.

In Fig. 21 the P–δ curve calculated using the *P-MIN* approach is compared with that of Fig. 13 for the Petersson model. The cracking loads are the same for the reinforced, (*P-MIN*), and unreinforced, (U) beams because of the very light reinforcement in the former.

3. DISCUSSION

3.1 Carpinteri's Beams

The foregoing FEM calculations demonstrate that for practical purposes minimum reinforcement requirements can be calculated by specifying that:

$$M_y \geq M_{cr} \tag{3}$$

where M_y is the yield moment for the reinforced section and M_{cr} is the cracking moment for the gross section calculated by multiplying the elastic cracking load $M_{cr,E}$, by the ratio f_f/f_t calculated for a given h/l_{ch} value from Fig. 6. Thus, values for M_{cr} and M_y are given by

$$M_{cr} = M_{cr,E}(f_f/f_t) = f_t(bh^2/6)(f_f/f_t) \tag{4}$$

and

$$M_y = A_s f_y d(1 - \alpha/3) \tag{5}$$

where

$$\alpha = (r_1^2 + 2r_1)^{0.5 - r_1} \qquad \text{where } r_1 = (E_s A_s)/(E_c db) \tag{6}$$

In eqn (3) the yield, rather than the ultimate moment, is used as the strength which must exceed the cracking moment. Although the ultimate moment will be greater than the yield moment, in many cases considerable inelastic behavior must occur before it can be achieved.

Tables 6 and 7 compare measured $P_{cr,EXP}$ and computed $P_{cr,FCM}$ capacities for the grade 2 and grade 4 concrete specimens tested by Carpinteri. The measured P_{cr} values are the same as those listed in Tables 2 and 3. For comparisons with predictions, the load, P_d, associated with the beam's dead weight must be added to $P_{cr,EXP}$. $P_{cr,E}$ is the cracking load for linear elastic behavior and the f_t values listed in Table 1, and $P_{cr,FCM}$ is that capacity calculated from eqn (4) for (f_f/f_t) values determined as the straight line value from Fig. 6. The last four columns of Tables 6 and 7 list the ratio q of the calculated capacity to the 'measured' capacity $(P_{cr,EXP} + P_D)$. The results for the 100 mm depth, grade 2 concrete in Table 6 are anomalous. There is no good reason why the cracking capacity for those beams should have been exactly the same as that for the 200 mm deep beams. If those 100 mm results are ignored because their strengths are inconsistent with the trends for deeper beams and other concrete strengths then, for the grade 2 concrete, measured to computed capacities average 1·01 and for the grade 4 concrete they average 1·00.

It is unrealistic to expect measured and computed yield loads to be similar for Carpinteri's beams since the FEM analyses clearly showed that the yield capacity is enhanced by concrete tension stiffening effects. At the same time, however, it is also relatively meaningless to compare computed ultimate capacities calculated using the steel's yield stress with measured ultimate capacities since Carpinteri provided no data on the stress–strain curves for his steel, but reported tensile strengths significantly greater than the yield strengths (Table 1). Shown in Table 8 is a comparison of the calculated and 'measured' yield capacities for those grade 4 concrete beams having 'measured' yield capacities closest to the measured cracking capacities (i.e. beams with a response like that of 2 in Fig. 3). The FEM calculations for the 100-mm deep case showed that the concrete tension effects increased

TABLE 6 Theoretical predictions of the cracking load and comparison with experimental results for concrete grade 2

h (mm)	b (mm)	$P_{cr,Exp}$				P_d (kN)	$P_{cr,E}$ (kN)	h/l_{ch}	f_f/f_t	$P_{cr,FCM}$ (kN)	q			
		(1*) (kN)	(2*) (kN)	(3*) (kN)	(4*) (kN)						(1*)	(2*)	(3*)	(4*)
100	150	11·26	11·10	11·25	11·18	0·10	3·80	0·17	1·86	7·07	0·62	0·63	0·62	0·63
200	150	11·34	14·05	12·11	12·70	0·42	7·60	0·34	1·66	12·62	1·07	0·87	1·01	0·96
400	150	19·94	21·73	18·75	20·84	1·68	15·20	0·67	1·44	21·89	1·01	0·94	1·07	0·97
800	200	35·24	43·37	45·60	39·31	8·97	40·53	1·34	1·30	52·69	1·19	1·01	0·97	1·09

($f_t = 2{\cdot}28$ MPa, $E_c = 23\,150$ MPa, $G_F = 0{\cdot}134$ N/mm, $l_{ch} = E_c G_F / f_t^2 = 597$ mm.)
(1*) = Specimen 1, 1 reinforcement bar.
(2*) = Specimen 2, 2 reinforcement bars.
(3*) = Specimen 3, 3 reinforcement bars.
(4*) = Average of (1*) and (2*).

TABLE 7 Theoretical predictions of the cracking load and comparison with experimental results for concrete grade 4

h (mm)	b (mm)	$P_{cr,Exp}$				P_d (kN)	$P_{cr,E}$ (kN)	h/l_{ch}	f_f/f_t	$P_{cr,FCM}$ (kN)	q			
		(1*) (kN)	(2*) (kN)	(3*) (kN)	(4*) (kN)						(1*)	(2*)	(3*)	(4*)
100	150	11·38	12·16	—	11·77	0·10	8·83	0·91	1·37	12·10	1·05	0·99	—	1·02
200	150	23·08	22·03	—	22·56	0·42	17·67	1·82	1·27	22·44	0·96	1·00	—	0·98
400	150	44·12	36·28	—	40·20	1·68	35·33	3·64	1·17	41·34	0·90	1·09	—	0·99
800	—	—	—	—	—	—	—	—	—	—	—	—	—	—

($f_t = 5{\cdot}30$ MPa, $E_c = 34\,300$ MPa, $G_F = 0{\cdot}090$ N/mm, $l_{ch} = E_c G_F / f_t^2 = 110$ mm.)
(1*) = No reinforcement, test specimen 1.
(2*) = No reinforcement, test specimen 2.
(3*) = Not used.
(4*) = Average of (1*) and (2*).

TABLE 8 Comparison between calculated and experimental yield load, concrete grade 4. $\varepsilon_y = f_y/E_s$

h (*mm*)	*b* (*mm*)	*Steel* *No. of bars*			α	$P_{y,Exp}$ (1*) (*kN*)			P_d (*kN*)	$P_{y,E}$ (*kN*)	$P_{cr,Exp}$ (*kN*)	*q* (1*)		
		No. of bars	*Diam.* (*mm*)	f_y (*MPa*)		(1*) (*kN*)	(2*) (*kN*)	(3*) (*kN*)				(1*)	(2*)	(3*)
100	150	2	5	569	0·17	14·71	15·70	15·21	0·10	12·28	11·77	0·83	0·78	0·80
200	150	3	5	569	0·15	23·04	23·16	23·10	0·41	18·83	22·56	0·80	0·80	0·80
400	150	4	5	569	0·13	33·08	31·64	32·36	1·63	25·52	40·20	0·74	0·77	0·75

(1*) = Test Specimen 1; (2*) = Test Specimen 2; (3*) = Average of (1*) and (2*).

the yield capacity by about 20 percent and therefore the average value for that contribution of 22 percent, as apparent from the ratios in the last three columns of Table 8, is reasonable.

It is realistic to conclude that minimum reinforcement requirements should be calculated from eqn (3) since M_y then represents the capacity once cracks have opened sufficiently to offset concrete tension effects but the steel has not strained sufficiently for the development of strain hardening effects.

3.2 Hu's Slabs

Hu's slabs were of very different proportions to Carpinteri's beams being much wider than they were deep. In Table 9 the cracking moments observed in Hu's tests are compared with those predicted using eqn (4). Column (2) lists the modulus of elasticity, E_c, calculated by the customary formula of ACI Code 318. Column (4) lists the tensile strength f_t, calculations based on prior University of Washington laboratory experience for similar mixes, as (5·5/6·7) f_{sp}. Column (5) lists the theoretical cracking moment, $M_{cr,E}$, for elastic behavior. Column (6) lists the characteristic length, l_{ch} calculated as $(E_c G_F) f_t^2$ where G_F is taken as 0·14 N/mm based on experience from fracture mechanics testing at Washington for specimens made with similar concretes to those used in Hu's tests. Listed in column (8) is the f_f/f_t multiplier calculated from Fig. 6 for a straight line model. All those multipliers are larger than the customary value of about 1·5 assumed for that value in ACI 318, although use of the Petersson model in Fig. 6 rather than the Straight-Line model would reduce values in column (8) by about 10 percent. The resultant $M_{cr,FCM}$ values calculated from eqn (4) are listed in column (9). Listed in the tenth column is the cracking moment measured in the tests, and the ratio of those two cracking moments is listed in column (11).

Obviously use of eqn (4) overestimates the observed cracking moments. It is believed that overestimation is due primarily to neglect of shrinkage effects and the presence of the fabric's transverse wires. After Hu cast his specimens, he covered them with plastic sheets for one week to cure the concrete, and then kept them at room temperature and humidity until the time of test at about three months. Considerable drying shrinkage must have occurred and been partially restrained by the fabric. By contrast Carpinteri stored his beams in a constant temperature/humidity room until test at 20–40 days. For his beams shrinkage effects should have been minimal. Further, even if

TABLE 9 Theoretical predictions of moment for Hu's experimental results

Specimen number (1)	E_c (*MPa*) (2)	f_{sp} (*MPa*) (3)	f_t (*MPa*) (4)	$M_{cr,E}$ (*kNm*) (5)	l_{ch} (*mm*) (6)	h/l_{ch} (7)	f_f/f_t (8)	$M_{cr,FCM}$ (*kNm*) (9)	$M_{cr,Exp}$ (*kNm*) (10)	q (11)	$M_{u,E}$ (*kNm*) (12)	$M_{u,Exp}$ (*kNm*) (13)
SD7.3A	33 831	3·93	3·23	6·86	455	0·28	1·71	11·73	8·41	1·40	11·86	11·22
SD7.3B	31 880	3·45	2·83	7·98	557	0·26	1·72	13·73	9·86	1·39	22·26	22·72
SD7.3C	31 880	3·45	2·83	3·98	557	0·18	1·82	7·24	5·05	1·43	12·84	12·89
SD9.5A	28 231	3·31	2·72	8·14	535	0·28	1·71	13·92	9·14	1·52	17·85	17·84
SD9.5B	28 231	3·31	2·72	9·20	535	0·30	1·69	15·55	9·79	1·59	27·76	27·87
SD9.5C	32 882	3·93	3·23	8·08	442	0·31	1·68	13·57	8·91	1·52	24·80	26·13
SD11.3A	31 145	3·17	2·60	13·06	643	0·30	1·69	22·06	16·13	1·37	31·50	31·65
SD11.3B	33 831	3·93	3·23	10·28	455	0·34	1·63	16·75	12·28	1·36	23·66	24·02
SD11.3C	32 882	3·93	3·23	5·54	442	0·25	1·73	9·58	6·75	1·42	15·56	15·69

Hu's slabs and Carpinteri's beams had been subject to the same curing practice, the cracking moments for Hu's slabs would have been more sensitive to drying shrinkage effects than Carpinteri's beams because of the section's shape. Hu observed that the cracks in his specimens always propagated along the transverse wires, and did not occur between wires. Obviously those wires were crack initiators and were probably in a region where shrinkage had caused significant tension. Those effects demonstrate clearly some practical limitations to any theoretical procedure for calculating cracking moments.

The wires used for the fabric in Hu's specimens had very limited elongations at the maximum load and as observed in Ref. [8], that elongation decreased as the wire size decreased. Further, for a given wire size and multiple samples the range in elongations at maximum load was as much as 100%. In general, but not necessarily for a given wire, the elongation measured on a 2-inch gage length that contained a transverse wire was less than that for a 2-inch gage length between transverse wires. Average stress–strain curves for the latter case for the three different wires sizes used by Hu and the range in elongations at maximum load are shown in Fig. 22. Note that the strain in Fig. 22

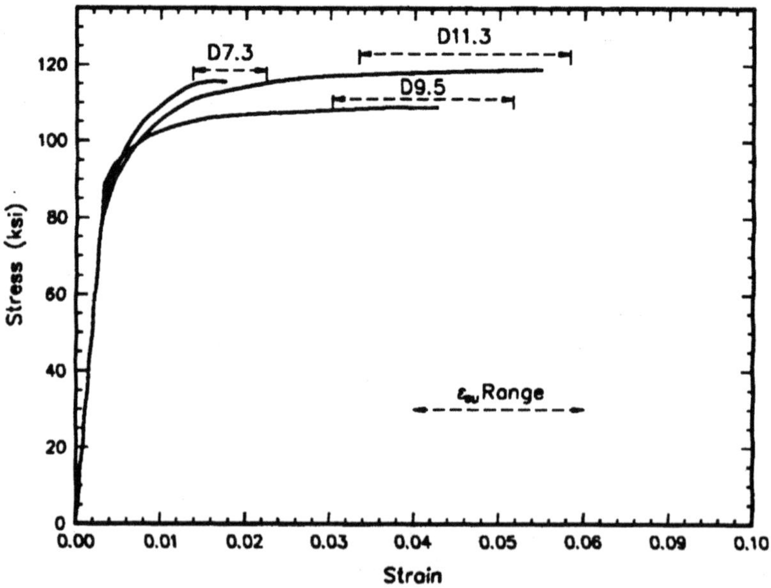

Fig. 22 Stress–strain curves for Hu's wires.

at which the curves terminate is not the ultimate strain. It is the strain at which the instrumentation was removed because the wire had started to neck down. The stress at fracture was about 20% less, and the corresponding strain much larger, than the stress and strain at the wire's tensile strength. Complete details for the stress–strain curves are given in Ref. [6].

From the form of those stress–strain curves it is clear that in any calculation of the yield moment it is more appropriate if the tensile strength of the steel is used for the yield stress, rather than the 0·35% or 0·50% offset stress. Shown in columns (12) and (13) of Table 9 are the 'yield' moment calculated from eqn (5) with f_u of Table 4 used for f_Y, and the measured ultimate moment, respectively. The two values are in close agreement.

For all specimens except SD7.3A the $M_{cr,FCM}$ values are less than the $M_{u,EXP}$ values so that the minimum reinforcement condition of eqn (3) was met. For SD7.3A the $M_{cr,FCM}$ value is only five percent greater than $M_{u,EXP}$. However, that finding does not reveal the complete picture. All the slabs with reinforcing ratios in the A and B category failed when one of the longitudinal wires fractured, and before there was any evidence of concrete crushing in the compression zone. Further, for wire sizes 7.3 and 9.5, and C category reinforcement ratios the crack, at the final failure section, had, for about 50% of its length, penetrated through the full depth of the slab before failure occurred due to wire fracture. Only slab SD11.3C had a normal 'tensile' failure with yielding of the steel followed by crushing of the concrete.

The instrumentation for Hu's slabs, (Fig. 4), allowed calculation of curvatures from either the LVDTs horizontally positioned on the top and bottom surfaces of the slab at midspan, or from the three LVDTs positioned vertically 203 mm (8 inches) apart in the constant moment region. Since the wires in the slabs were at, or close to, their tensile strength at failure of the slabs, the strain in the wires at failure, ε_{su}, could be calculated approximately from the curvatures at failure, ϕ_u, by the formula

$$\varepsilon_{su} = \phi_u(d - c) \tag{7}$$

where c is the neutral axis depth $d\alpha/3$ (eqns (5) and (6)). Values of the wire strains at failure, calculated from eqn (5), together with the ulitimate strains determined from direct tensile tests are plotted versus wire size in Fig. 23. The solid line represents the mean prediction from

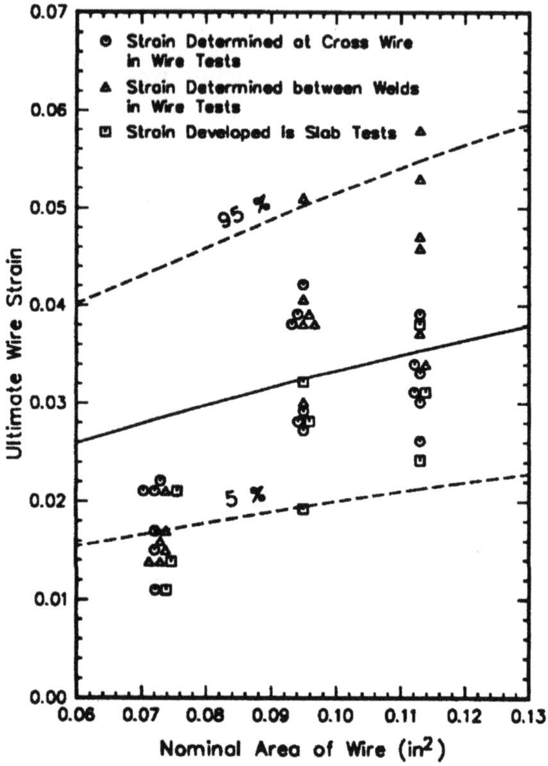

Fig. 23 Ultimate wire strain versus wire size.

Ref. [8] for ultimate wire strains and the two broken lines the predicted 95% and 5% limits, respectively. Effectively, the prediction of Ref. [8] is consistent with Hu's results with ultimate strains decreasing with wire size.

Consistently, wire strains at failure were less for the A slabs than B or C slabs. This result is due to eqn (7) being representative of 'average' conditions in the constant moment region of the slabs and not of conditions in the slabs where the wires fractured. Fractures were always at transverse wires, and since cracks occurred only at transverse wires in lightly reinforced slabs, the spacing between cracks increased as the slab's reinforcement ratio decreased. Thus, for those lightly reinforced slabs, strains calculated from 'average' conditions

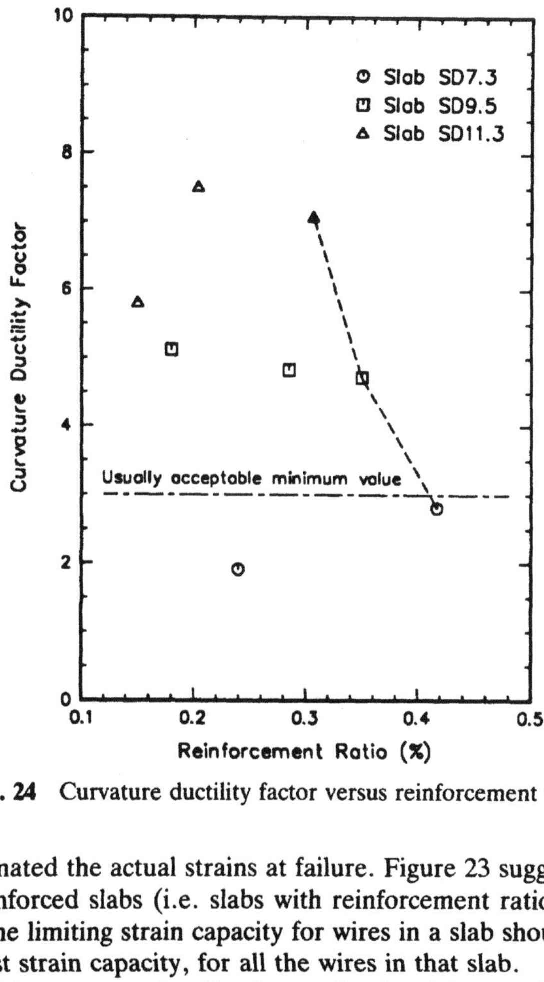

Fig. 24 Curvature ductility factor versus reinforcement ratio.

underestimated the actual strains at failure. Figure 23 suggests that for lightly reinforced slabs (i.e. slabs with reinforcement ratios of type B or less), the limiting strain capacity for wires in a slab should be taken as the least strain capacity, for all the wires in that slab.

In Fig. 24, curvature ductility factors for the slabs are plotted versus reinforcement ratio. Those ductility factors were determined from the measured maximum moment curvature results. The horizontal line in Fig. 24 at a factor of 3·0, indicates the minimum curvature ductility currently acceptable according to ACI 318 principles. That is the result for a grade 60 bar, a tensile reinforcement ratio equal to 0·75 of the balanced ratio, no compression steel, and a concrete strength of 35 MPa. The result for slab SD7.3A is not plotted in Fig. 24 since ductility factors could not be determined from the test results for that slab. Theoretically, the curvature ductility decreases with increasing reinforcement ratio. However, from Fig. 24 it is clear that the ductility also decreases as the wire size used in the section decreases. Obviously, ductility factors for slabs reinforced with small wire sizes may not be able to reach the minimum values commonly acceptable for reinforced concrete construction. In Fig. 24 the results for C slabs are connected with broken lines. That trend suggests that use of the minimum reinforcement ratios recommended by Mirza & MacGregor (eqn (2)) may still not ensure adequate ductility for members reinforced with small wire sizes.

For minimum reinforcement requirements the practical implications that can be drawn from an analysis of Hu's results are twofold. First, for slabs, while shrinkage and reinforcement patterns can significantly reduce the cracking moment, M_{cr}, they have little effect on the yield moment M_y. Therefore, even when those effects are marked, use of eqn (3) is still reasonable. Second, by contrast, however, if a reasonable ductility prior to failure is also needed, then specific requirements, such as eqn (2), must be added to eqn (3) for steels with limited ultimate elongations. In such cases it may also be necessary to recognize strain concentration effects at cracks.

4. CONCLUDING REMARKS

Minimum reinforcement requirements are best expressed in terms of reinforcement ratios based on gross section properties. From substitution in eqn (3) the minimum ratio, $\rho_r = A_s/bh$, can be expressed as

$$\rho_r = 6(f_f/f_t)(f_t/f_y)(h/d)/(1-\alpha/3) \quad (8)$$

where α is given by eqn (6). Both ρ_r and α in eqn (8) are functions of A_s and therefore theoretically eqn (8) must be solved iteratively. However, for most structures with minimum reinforcement satisfying eqn (8), α varies between 0·1 and 0·2 only and that variation has only a 3% effect on ρ_r. Therefore for practical purposes ρ_r can be expressed as

$$\rho_r = 0{\cdot}175(f_f/f_t)(f_t/f_y)(h/d) \quad (9)$$

where f_f/f_t is taken from Fig. 6.

Equation (9) predicts that the minimum reinforcement ratio is dependent both on depth, (h/d), and strength (f_t/f_y), effects as well as

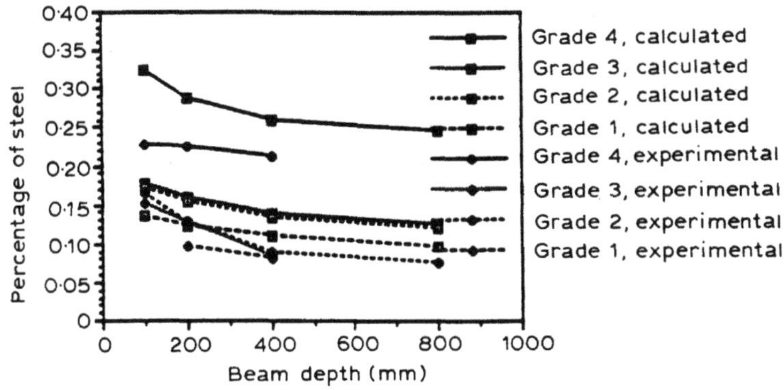

Fig. 25 Comparison of calculated and experimental minimum ratios for Carpinteri's results.

fracture mechanics effects (f_f/f_t). In Fig. 25 minimum reinforcement ratios calculated from eqn (9), using the straight line (f_f/f_t) values from Fig. 6, are compared with ratios established from Carpinteri's experimental results. For the latter case $M_{cr,EXP}$ was approximated as the measured value for beams having one or two reinforcing bars. The value $M_{y,EXP} = M_{cr,EXP}$ was established by linear interpolation from the test results having yield moments immediately above and below the cracking moment. Steel reinforcement ratios were adjusted in proportion to the steel's actual yield strength to the value for f_y of 480 MPa. From Fig. 25 it is apparent that minimum reinforcement ratios calculated using eqn (9) are conservative when compared to test data.

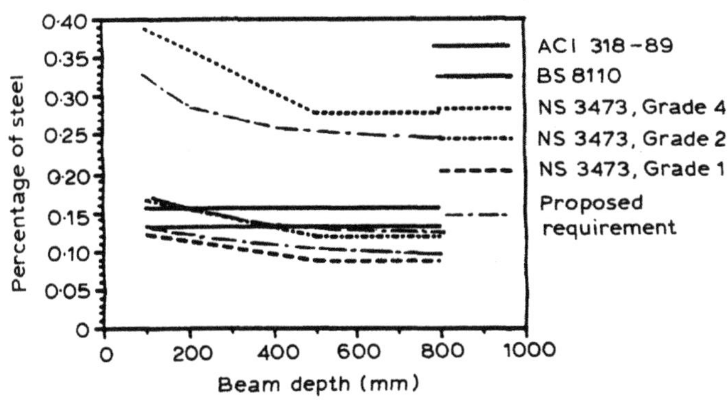

Fig. 26 Minimum reinforcement ratios for yield strength of 480 MPa.

That is understandable because, in eqn (9), in order to allow for ductility demands, the contribution of the tensile capacity of the concrete to the yield moment is ignored. That contribution is, however, included in Carpinteri's results. In Fig. 26 the minimum reinforcement ratio calculated from eqn (9), is plotted versus NS 3473 and other Code requirements for the concrete strengths used by Carpinteri. Clearly, eqn (9) leads to results close to NS 3473 requirements with both requirements being concrete strength dependent. Clearly, from the preceding work, however, requirements such as NS 3473 must be supplemented with steel elongation requirements such as those that result in eqn (2).

REFERENCES

[1] ACI Committee 318. *Building Code Requirements for Reinforced Concrete (ACI 318-89) and Commentary (ACI 318R-89)*. American Concrete Institute, Detroit, MI., USA, 1989, Sec. 7.12, 10.3 and 10.5.

[2] British Standard. *BS8110: Part 1: 1985 Structural Use of Concrete. Part 1. Code of Practice For Design and Construction Standards Association*, London.

[3] Norwegian Standard. *NS3473 E* (English Translation) *Concrete Structures, Design Rules*. Norwegian Council for Building Standardization, Oslo 1989.

[4] Eurocode No. 2. Commission of the European Communities, Paris, 1984.

[5] Carpinteri, A., Minimum reinforcement in reinforced concrete beams. RILEM TC 90-FMA, CODE WORK, Cardiff, 20–22 September 1989, UK.

[6] Hu, C., Ductility of concrete slabs reinforced with welded wire fabric. M.S.C.E. thesis, Department of Civil Engineering, University of Washington, Seattle, WA., USA, 1986.

[7] Carpinteri, A., A fracture mechanics model for reinforced concrete collapse. *IABSE Colloquium on Advanced Mechanics of Reinforced Concrete*, Delft, 1981, pp. 17–30.

[8] Mirza, S. & MacGregor, J. G., Strength and ductility of concrete slabs reinforced with welded wire fabric. *ACI Journal*, **78**(5) (1981) 374–81.

[9] Hillerborg, A., Personal communication with K. Hjorteset. February 1990.

[10] Liaw, B. M., Jeang, F. L., Du, J. J., Hawkins, N. M. & Kobayashi, A. S., Improved nonlinear model for concrete fracture. *J. Eng. Mechanics* (ASCE), **16** (1990) 429–45.

[11] Bazant, Z. P. & Oh, B. H., Microplane model for progressive fracture of concrete and rock. *J. Eng. Mechanics* (ASCE), **111** (1985) 559–82.

[12] Hjorteset, K., Minimum reinforcement requirements for concrete flexural members determined using nonlinear fracture mechanics. MSCE thesis, University of Washington, Seattle, WA., USA, 1990, 118pp.

Chapter 16

Fracture Mechanics Application to Reinforced Concrete Members in Flexure

M. H. BALUCH, A. K. AZAD & W. ASHMAWI

Dept. of Civil Engineering,
King Fahd University of Petroleum & Minerals, KFUPM-Box 1469, Dhahran 31261, Saudi Arabia

ABSTRACT

The work presented in this chapter includes: (i) the introduction of a LEFM based model to predict the residual strength capacity of a RC beam in the presence of an edge crack propagating under flexural loading; (ii) the use of the model in formulating a fracture mechanics based approach to the design of RC members in flexure; and (iii) a description of the minimum reinforcement ratio requirements based on the proposed LEFM model.

1. INTRODUCTION

The well accepted notion that the study of a cracked member or a member with a flaw subject to a stress field should consider the principles of fracture mechanics has given impetus to researchers to apply fracture mechanics to concrete-like materials. However, until now the design and prediction of strength of concrete members, plain or reinforced, have been performed almost exclusively by strength criteria which are based on a stress–strain law and a failure theory or concept.

Although considerable work has been devoted to the application of fracture mechanics to plain concrete, only lately have researchers diverted their attention to reinforced concrete. This was partly due to the fact that reinforced concrete causes considerable difficulty in the formulation of a refined approach due to the heterogeneous nature of the material in which crack growth is controlled by concrete toughness as well as the elasticity of reinforcing steel. The problem becomes compounded by the presence of bond-slip between rebar and concrete. Furthermore, tests of sufficiently reinforced beams [1] have shown existence of stable crack growth, wherein progressive advancement of the crack is always accompanied by load increment, indicating fracture insensitivity. The collapse load of an under-reinforced beam is marginally influenced by the concrete strength, inasmuch as the strength is essentially controlled by the amount of steel and the beam dimensions. The apparent insensitivity of reinforced members to fracture has discouraged the application of the fracture theory to characterize their behavior and failure. Prevailing strength theory seems to have provided satisfactory solutions.

However, it is recognized that for certain situations, a reinforced beam may show fracture sensitivity, in particular those beams which are very lightly reinforced and are made of high strength concrete. Also, the present strength theory would fail to accurately predict the crack growth under a sub-critical load level. Mathematical models adopting fracture criteria are seemingly more accurate and applicable to study crack growth.

In recognition of the fact that the impact of fracture mechanics on concrete can only be felt if reinforced concrete members are considered more rigorously and rationally, researchers have lately diverted their attention to reinforced concrete [2]–[4]. Most of these finite element based analytical works are less appealing to designers, who traditionally prefer to use theoretical approaches.

Carpinteri in a series of papers [5]–[9] has put forward an elegant formulation for predicting the fracture collapse of a reinforced concrete beam, initiated by the presence of an edge crack with its tip subjected to a Mode I stress field. In order to render the problem determinate, this model assumes that the tension reinforcement is in a state of yield at incipient crack propagation [9]. This assumption, although valid for certain combinations of steel area and crack depths, is not generally true, as the early stage of cracking occurs in most cases at steel stress far less than the yield point [1]. This apparent contradiction renders the application of Carpinteri's model inadequate for all stages of crack propagation.†

The primary objective of this chapter is to present a LEFM based model for studying crack propagation in reinforced concrete beams subject to a Mode I stress field. The treatment reconsiders the basic approach of Carpinteri, removing the presumed condition of steel yielding at incipient fracture. The proposed formulation yields the true characteristics of the 'fracture load–crack length' relationship of a beam for the entire spectrum of crack lengths corresponding to the pre-yielding and post-yielding of reinforcement.

In addition to describing the stress intensity factor K_I at the crack tip in terms of the applied bending moment M and the steel force F_s, the formulation makes use of a hypothesized concrete stress distribution based partially on the Euler-Bernoulli assumption of linear strain variation across the depth of the beam. With this assumed stress field, additional equations are made available in satisfying gross equilibrium of the beam. This helps to render the problem determinate in the presence of a steel force $F_S \leq F_p$, where F_p is the force in the reinforcement at yielding.

2. THEORETICAL FORMULATION

2.1 Fracture Moment

In this section, the concept of a fracture moment M_F is presented and it closely parallels the original concept of M_F as developed by Carpinteri [9]. With reference to a reinforced concrete beam element in the presence of a Mode I through-thickness edge crack, the element is subjected to a bending moment M and an eccentric steel force F_s. The beam is assumed to be of rectangular cross-section of size $b \times h$, with cover of the steel being designated by d_s. Crack depth at any instant is symbolized by a (Fig. 1(a)).

Transferring the eccentric steel force F_s to be concentric with the beam axis, the component stress intensity factors K_{IM}, K_{IF} due to net

† *Editor's note*: The new Carpinteri model proposed in Chapter 14 is able to overcome this drawback as crack propagation can be simulated even when steel is in the elastic condition.

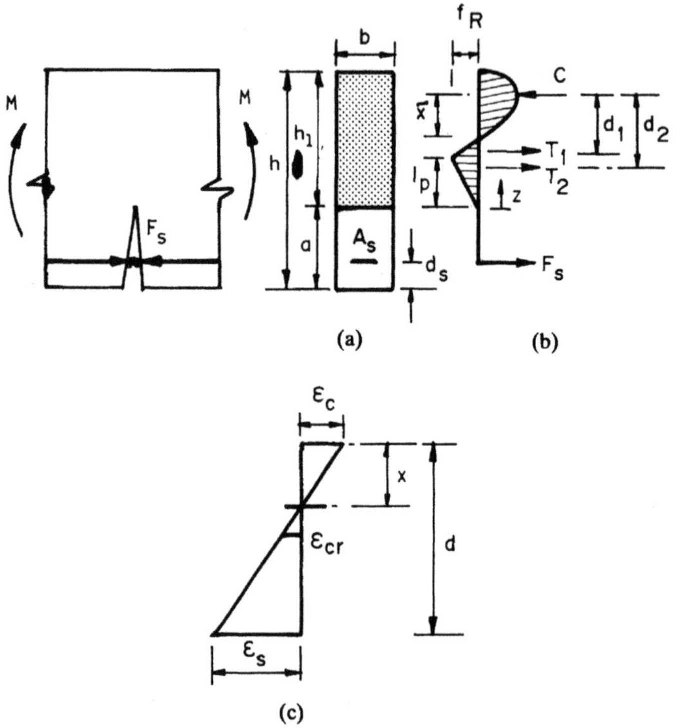

Fig. 1 Beam element and variation of strain and stress across the beam depth.

moment and axial force, respectively, may be written as [10], [11]

$$K_{IM} = \frac{Y_M(\xi)}{h^{3/2}b}\left[M - F_s\left(\frac{h}{2} - d_s\right)\right] \tag{1}$$

$$K_{IF} = -\frac{Y_F(\xi)}{h^{1/2}b}F_s \tag{2}$$

where $\xi = a/h$ is the relative crack depth and Y_M, Y_F are the functions

$$Y_M(\xi) = 6(1{\cdot}99\xi^{1/2} - 2{\cdot}47\xi^{3/2} + 12{\cdot}97\xi^{5/2} - 23{\cdot}17\xi^{7/2} + 24{\cdot}80\xi^{9/2}) \tag{3}$$

$$Y_F(\xi) = 1{\cdot}99\xi^{1/2} - 0{\cdot}41\xi^{3/2} + 18{\cdot}70\xi^{5/2} - 38{\cdot}48\xi^{7/2} + 53{\cdot}85\xi^{9/2} \tag{4}$$

for $\xi \leq 0{\cdot}6$. The contribution of the steel force to K_{IF} and K_{IM} is negative since steel tends to close the crack tip as the loading is applied.

Superposing the component stress intensities K_{IM} and K_{IF}, and setting $K_I = K_{IC}$, the critical stress intensity for concrete, and solving for the moment $M = M_F$ that will cause such a condition, one obtains

$$M_F = \frac{K_{IC} h^{3/2} b}{Y_M(\xi)} + \frac{F_s h}{Y_M(\xi)} \left[Y_F(\xi) + Y_M(\xi)\left(\frac{1}{2} - \frac{d_s}{h}\right)\right] \tag{5}$$

where the moment M_F at incipient fracture is referred to as the fracture moment.

Equation (5) may be recast in the form

$$M_F = \frac{1}{\alpha} K_{IC} + \frac{\beta}{\alpha} F_s \tag{6}$$

where $\alpha = Y_M(\xi)/(h^{3/2} b)$ (7)

$$\beta = [Y_M(\xi)/(h^{3/2} b)]\left(\frac{h}{2} - d_s\right) + Y_F(\xi)/(h^{1/2} b) \tag{8}$$

2.2 Equilibrium Consideration

In this section, the concept of an equilibrium moment is developed, borrowing ideas forming the nucleus of strength theory of reinforced concrete.

With reference to Fig. 1(c), it is assumed that the strain variation across the beam depth will be linear in accordance with the Euler–Bernoulli hypothesis. Shown in Fig. 1(b) is the stress variation across the beam depth consistent with the strain variation. The compressive stress distribution in concrete is assumed in the form given by the 'Madrid Parabola'

$$\sigma = \sigma_0 \left[\frac{2\varepsilon}{\varepsilon_0} - \left(\frac{\varepsilon}{\varepsilon_0}\right)^2 \right] \qquad 0 \le \varepsilon \le 0{\cdot}003 \tag{9}$$

where $\sigma_0 = f'_c$ (uniaxial compressive strength) and $\varepsilon_0 = 2\sigma_0/E_c$, E_c being the modulus of elasticity of concrete.

The tensile stress variation in concrete is assumed to follow the linear elastic law given by $\sigma = E_c \varepsilon$ ($\varepsilon \le \varepsilon_{cr}$) where ε_{cr} is the cracking strain in concrete and computed as $\varepsilon_{cr} = f_R/E_c$, where f_R is the modulus of rupture of concrete. For $\varepsilon > \varepsilon_{cr}$, the concrete stress in the tensile zone is assumed to vary linearly from a maximum of f_R (for

$\varepsilon = \varepsilon_{cr}$) to zero at the crack tip (to allow for micro-cracking) i.e. $\sigma = f_R z/l_p$ where z is measured from the crack tip and l_p is the length of the fracture process zone. The stress in steel is assumed to follow an elasto-plastic constitutive law with $f_s = E_s\varepsilon_s(\varepsilon_s \leq \varepsilon_y)$ and $f_s = f_y(\varepsilon_s > \varepsilon_y)$, where E_s is the modulus of elasticity of steel.

The assumed stress distribution in the vicinity of the crack tip is a quantitative adoption of postulated stress variation as discussed by several researchers [9], [12]. The basic philosophy behind this assumed variation is that as the stress in the vicinity of the crack tip cannot exceed the modulus of rupture f_R, there is a certain zone in the neighborhood of the crack tip where there is redistribution due to softening (fracture process zone). The model simply joins the point of zero stress (crack tip) to the point stressed to f_R by a straight line as an approximation of the actual state of stress in the process zone.

With reference to the stress distribution as illustrated in Fig. 1(b), the following expressions for component tensile stress resultants may be written as

$$T_1 = \tfrac{1}{2}f_R(h_1 - x - l_p)b \tag{10}$$

$$T_2 = \tfrac{1}{2}f_R l_p b \tag{11}$$

Summing moments about the centroid of the concrete compressive stress block yields an expression for the equilibrium moment, M_{eq}, where

$$M_{eq} = T_1 d_1 + T_2 d_2 + F_s d_3 \tag{12}$$

in which d_1, d_2 and d_3 are the moment arms of forces T_1, T_2 and F_s with respect to the centroid of the concrete compressive stress block and are given by

$$d_1 = \tfrac{2}{3}(h_1 - x - l_p) + \bar{x}$$

$$d_2 = (h_1 - x - \tfrac{2}{3}l_p) + \bar{x}$$

$$d_3 = d - x + \bar{x}$$

where $\bar{x}$ is the location of the centroid of the resultant concrete compressive force (Fig. 1(b)).

For a reinforced beam in the presence of a Mode I crack, a unique solution corresponding to incipient crack propagation exists only when the fracture moment M_F from eqn (6) and the equilibrium moment M_{eq} from eqn (12) coalesce to yield an identical value.

3. SOLUTION ALGORITHM

For a given beam, the fracture moment M_F can be determined through a uni-directional search algorithm in which a small value of steel force F_s is assumed initially and then incremented in small steps to check convergence of M_F and M_{eq}. The solution algorithm is subdivided into two parts, corresponding to whether the steel is yielding or not.

3.1 Pre-yield Algorithm ($\varepsilon_s < \varepsilon_y$)

The following steps are prescribed for the determination of M_F.

(i) Fix crack depth a.

(ii) Assume steel force $F_s < F_p$ (where F_p is the steel yield force $= A_s f_y$) and calculate corresponding strain in steel $\varepsilon_s = F_s/(A_s E_s)$.

(iii) Assume depth of the neutral axis x.

(iv) Using $\varepsilon_c = \dfrac{x}{d-x}\varepsilon_s$ and eqn (9), compute using numerical integration the compressive force resultant C and its line of action $\bar{x}$.

(v) Calculate the total tensile force in concrete $T = \frac{1}{2}(h - x - a)f_R$.

(vi) Check force equilibrium condition $C = T + F_s$. If satisfied, go to step (vii). Otherwise, go back to step (iii) and change x until force equilibrium condition is satisfied.

(vii) When equilibrium is attained, find distance l_p at which $\varepsilon = \varepsilon_{cr}$ i.e.

$$l_p = (h_1 - x) - \frac{\varepsilon_{cr}}{\varepsilon_s}(d - x) \qquad (13)$$

(viii) Calculate component tensile forces in concrete T_1 and T_2 from eqns (10) and (11).

(ix) Compute equilibrium moment M_{eq} (eqn (12)) and M_F (eqn (6)).

(x) Check if $M_{eq} = M_F$. If yes, this means that the crack will propagate and the corresponding load P_f can be found from overall equilibrium of the beam. The crack length can now be incremented, and the process repeated to find the accompanying fracture moment and load. If $M_{eq} \neq M_F$, then assume a new steel force F_s and go back to step (ii) and repeat until convergence is attained.

3.2 Post-yield Algorithm ($\varepsilon_s > \varepsilon_y$)

(i) Fix crack depth *a*.

(ii) Steel force is now a known constant $F_s = F_p = A_s f_y$. Assume the steel strain $\varepsilon_s = \varepsilon_p > \varepsilon_y$.

(iii) Follow steps (iii) through (x) of pre-steel yield algorithm, replacing ε_s by ε_p and F_s by F_p in the relevant equations. If $M_{eq} \neq M_F$, change the steel strain $\varepsilon_p > \varepsilon_y$ and start again to seek convergence.

4. EXPERIMENTAL VERIFICATION

A series of three-point bend tests were conducted on concrete beams reinforced with varying amounts of steel to characterize the fracture process [1]. The proposed model was tested against experimental data from some of these beams. Details of two such beams are given in Table 1. The beams were weakly reinforced and tested in three-point bending with a simple span of 1400 mm.

The material properties of concrete were experimentally determined as follows: $f'_c = 23$ N/mm^2 and $K_{IC} = 31$ N/mm$^{3/2}$. The modulus of elasticity of steel was taken as 205 GPa and yield stress as 400 MPa. The modulus of elasticity of concrete, and the modulus of rupture, f_R, were taken in accordance with ACI 318-89 [13] as $E_c = 430\sqrt{f'_c}$ and $f_R = 0{\cdot}62\sqrt{f'_c}$ (f'_c in N/mm^2).

Figures 2 and 3 show the correlation between the experimental P_f–a curves and those predicted by the model for the two beams considered. Also superimposed on the plots is the locus of a load–crack depth relationship assuming yielding of tension steel (Carpinteri model). The present model appears to simulate the actual response of the under-reinforced beams reasonably well, capturing the entire region of the P_f–a curve.

TABLE 1 Beam details

Beam	*Beam width b (mm)*	*Beam depth h (mm)*	*Steel cover d_s (mm)*	*Steel area A_s (mm^2)*	*Initial notch height (mm)*
B1	102	380	25	50·3	38
B2	102	380	10	78·5	38

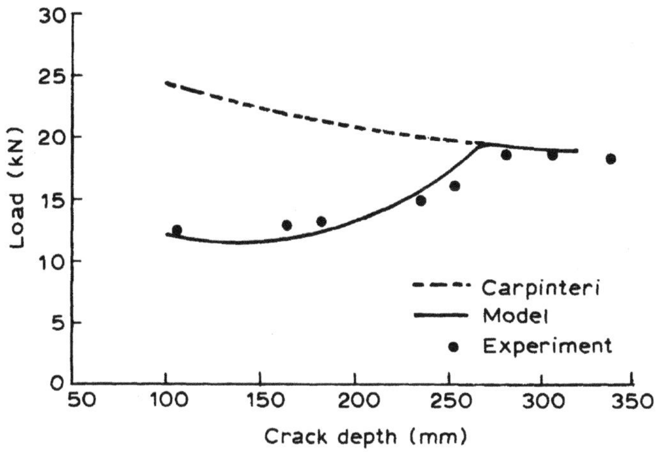

Fig. 2 Comparison of results with test data for Beam B1.

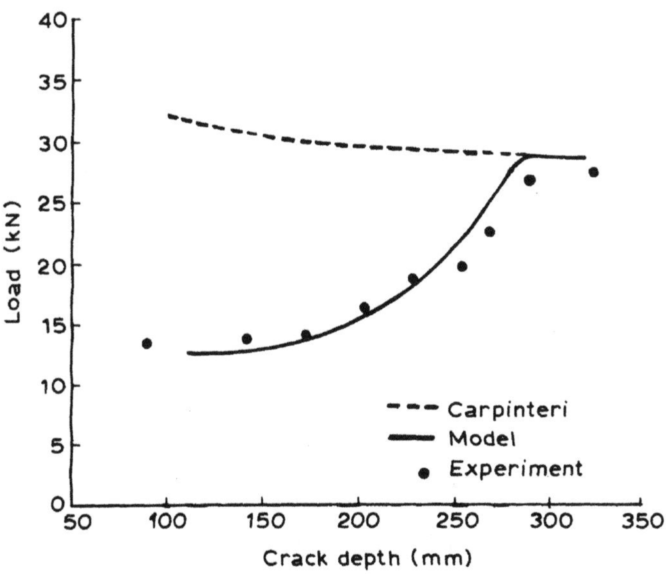

Fig. 3 Comparison of results with test data for Beam B2.

Numerical results for the P_f–a relationship for beams reinforced with a higher amount of steel (sufficiently reinforced) indicate that the crack propagation in Mode I is arrested after its growth to a certain height. This is due to the development of compressive stress in the entire uncracked ligament. For such beams, further load increment would give rise to new cracks at other locations along the beam span under mixed-mode fracture conditions. The arrest of Mode I crack propagation followed by emergence of new diagonal cracks has been observed in bending tests of sufficiently reinforced beams. Both the model and experimental data show clearly that fracture sensitivity can only be witnessed in highly under-reinforced beams. For such beams, experiments show the existence of a strain-softening mode of collapse due to rebar necking following a considerable amount of plastic deformation [1], [14], [15].

5. CRACK-CONTROLLED DESIGN FOR FLEXURE

On many occasions, emphasis on durability, structural integrity and serviceability of a structure dictates a preference for a crack-controlled design in which the crack height is contained within a prescribed limit. By increasing the amount of reinforcement, the crack height can be controlled by virtue of reduced steel stress. Instead of approaching the problem of finding the steel area necessary in a beam by a semi-empirical method based on a strength theory, it seems more rational to propose a fracture based method.

In this section, the proposed LEFM model for computation of fracture moment has been extended to suggest a fracture mechanics based approach to design of a reinforced beam type member for which a maximum permissible crack height (in Mode I stress field) has been prescribed under loading. The model yields the area of reinforcement A_s necessary to satisfy the stipulated loading and crack conditions.

5.1 Algorithm for Solution

In order to obtain the amount of steel reinforcement A_s necessary for a beam subjected to a Mode I crack under a bending moment M_u, an iterative procedure is employed to yield the solution.

The input design parameters include: (a) maximum permissible crack height a; (b) ultimate design moment M_u; (c) concrete fracture and strength characteristics K_{IC}, f'_c and f_R; (d) yield stress of reinforcement f_y; and (e) beam section $b \times h$ (width $\times$ depth).

The procedure for a typical iteration is outlined as follows:

(i) Obtain the steel force F_s by equating the fracture moment M_F (eqn 6) to the design moment M_u

$$F_s = \frac{\alpha}{\beta} M_u - \frac{1}{\beta} K_{IC} \tag{14}$$

(ii) Assume a strain level in steel ε_s.

(iii) Assume the depth of the neutral axis x.

(iv) Compute the compressive force resultant C and its line of action $\bar{x}$ using $\varepsilon_c = \dfrac{x}{d-x}\varepsilon_s$ and the concrete compression constitutive law as given by eqn (9).

(v) Calculate the total tensile force in concrete $T = \frac{1}{2}(h - x - a)bf_R$.

(vi) Check force equilibrium $C = T + F_s$. If yes, then proceed to step (vii). Otherwise, go back to step (iii) and change x until force equilibrium is attained.

(vii) When equilibrium is attained, find distance l_p at which $\varepsilon = \varepsilon_{cr}$ i.e.

$$l_p = (h_1 - x) - \frac{\varepsilon_{cr}}{\varepsilon_s}(d - x) \tag{15}$$

(viii) Calculate M_{eq} using eqns (10)–(12).

(ix) Check if $M_{eq} = M_u$. If yes, then proceed to step (x). Otherwise, go back to step (ii) and increment ε_s until a solution is obtained ($M_{eq} = M_F$).

(x) For a convergent solution, find the stress in reinforcement $f_s = \varepsilon_s E_s$ (if $\varepsilon_s \leq \varepsilon_y$ yield strain) or $f_s = f_y$ (if $\varepsilon_s \geq \varepsilon_y$).

(xi) The area of reinforcement A_s necessary becomes

$$A_s = F_s / f_s$$

5.2 Illustration

Based on CEB Model Code 90 [16], concrete fracture energy G_F can be estimated in N/mm^2 by the empirical equation

$$G_F = \gamma f_c'^{0{\cdot}7} \tag{16}$$

where f_c' is the concrete compressive strength in N/mm^2 and γ is a coefficient dependent on the maximum aggregate size d_{max}. Based on discrete data provided in Ref. 16, the following regressed equation relating γ to d_{max} was obtained

$$\gamma = 2{\cdot}828e^{0{\cdot}0454 d_{max}} \tag{17}$$

where d_{max} is the maximum aggregate size in mm. Also, the critical stress-intensity factor or fracture toughness K_{IC} is given by

$$K_{IC} = \sqrt{G_F E_c} \tag{18}$$

Using eqn (16) in eqn (18), together with the ACI expression for $E_c = 4730\sqrt{f'_c}$ (f'_c in N/mm^2), the following empirical relationship between K_{IC} and f'_c is obtained

$$K_{IC} = 2{\cdot}175\sqrt{\gamma} f_c'^{0{\cdot}6} \tag{19}$$

with K_{IC} in $N/mm^{3/2}$.

As a test example, a simply supported beam of span 1400 mm and rectangular cross-section 102 mm × 380 mm subjected to a concentrated force at mid-span was selected. Compressive strength of concrete was taken as $f'_c = 28{\cdot}27\ N/mm^2$ and fracture toughness K_{IC}

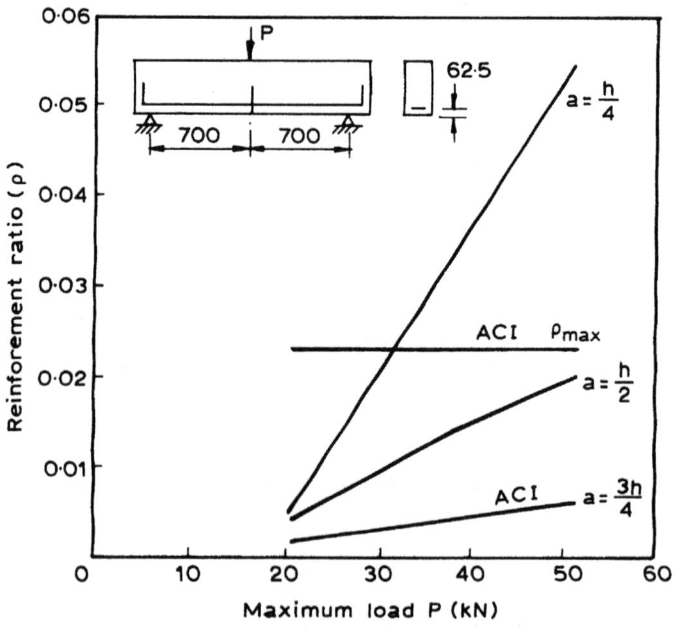

Fig. 4 Load–reinforcement characteristics.

was obtained from eqn (19) as 36·12 $N/mm^{3/2}$. The modulus of rupture f_R was taken as the ACI expression $0{\cdot}62\sqrt{f'_c}$ (f'_c in N/mm^2) = 3·3 N/mm^2. Maximum aggregate size d_{max} used was 12 mm.

Three sets of permissible maximum crack depths were taken, including $a/h = 0{\cdot}25$, 0·50 and 0·75. Shown in Fig. 4 is the variation of steel reinforcement ratio ρ versus ultimate load P_u for the three crack depths considered. Also superposed is the ACI-based reinforcement ratio which remains unaffected by crack size limitations as the reinforcement ratio based on ACI design philosophy corresponds to a crack depth of approximately 0·75h, i.e. if a load corresponding to P_u

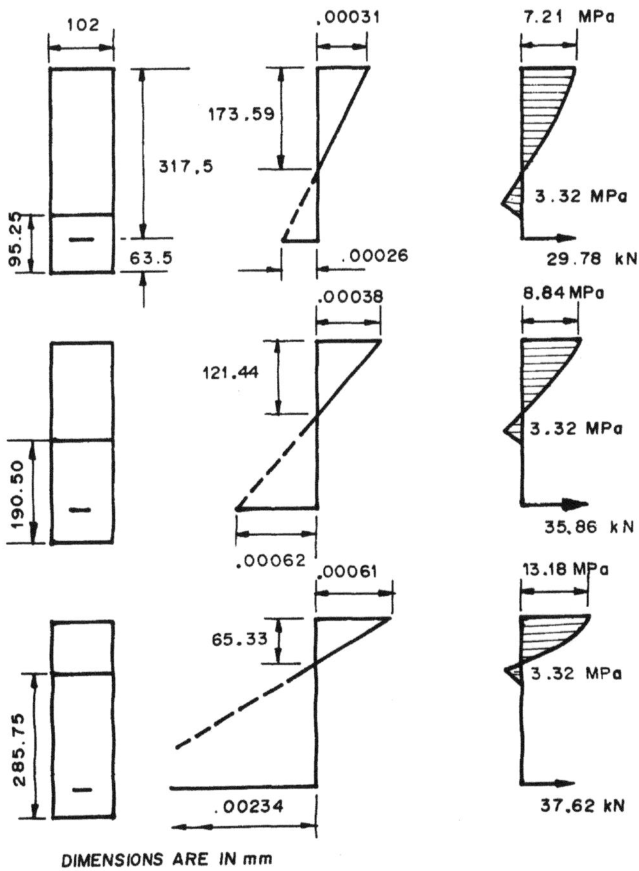

Fig. 5 Stress–strain variation for various crack depths ($P = 30$ kN).

were applied to the member the amount of reinforcement provided by the ACI Code would allow the crack to propagate to an approximate depth in the neighborhood of $0{\cdot}75h$. If the requirement of the structure dictates crack arrests at lower depths ($0{\cdot}5h$ or $0{\cdot}25h$), the rate of reinforcement increase necessary per unit load becomes significantly higher than for the case where the crack is allowed to propagate up to $0{\cdot}75h$.

Shown in Fig. 5 is the stress–strain variation across the depth of the beam for the three crack depths considered and at a maximum load of 30 kN. It is interesting to note that (i) the amount of steel A_s necessary to arrest cracks at lower depths (for the same load) increases, (ii) the steel force F_s increases with increasing crack depth, and (iii) the extension of the microcracked zone decreases with increasing crack depth. Shallow crack depths corresponding to extended net ligaments allow for greater mobilization of tensile forces in concrete, with concomitant low level of steel force. As the specified crack is increased with the same load, the concrete tensile force decreases with steel force tending to increase to the limiting yield value.

6. MINIMUM REINFORCEMENT BASED ON LEFM MODEL

In a lightly reinforced beam, there is a danger that the bending moment causing first cracking of concrete is larger than the moment capacity of the reinforced concrete section, and that the beam would then fail in a brittle fashion (notch sensitive fracture). To safeguard against this sudden failure, a minimum amount of tensile reinforcement has to be used so as to mobilize the moment capacity of the reinforced section and to guarantee that the formation of the first crack will occur at a load level less than that required to yield the reinforcement.

Most building codes provide an empirical prescription of the minimum tension steel to impart ductility and avoid brittle failure. For example, ACI [13] specifies a minimum tension reinforcement ratio, ρ ($\rho = A_s/bd$, d being the effective depth) given by $\rho_{min} = 1{\cdot}4/f_y$, f_y being the yield stress of steel in N/mm^2.

The ACI minimum reinforcement ratio was derived by making assumptions as to the probable tensile cracking strength of the plain concrete section and determining the equivalent amount of reinforcement necessary to balance the tensile force carried by concrete prior to cracking.

6.1 Proposed LEFM Criterion for Minimum Reinforcement

Shown in Fig. 6 is a typical load–crack length diagram for an under-reinforced beam with a fixed value of A_s. It is to be noted that the behavior is a superposition of two continuous P–a diagrams, one representing steel in the elastic state for all values of a and the other for steel being plastic throughout. The intersection of the two continuous curves occurs in a kink, at which crack length the steel yields initially. Ignoring the small load increment that may follow after initial yielding of steel due to redistribution of concrete stresses, it may be hypothesized that the kink or the yield point corresponds to the maximum load (technically pseudo-critical load) that the reinforced beam can sustain. In order to avoid unstable crack propagation, the moment capacity of the beam corresponding to the kink should be greater than the initial cracking moment of the beam behaving essentially as a plain concrete one.

In other words,

$$M_F^{RC} > M_F^{PC} \tag{20}$$

where M_F^{RC} corresponds to the moment at the kink and M_F^{PC} the fracture or cracking moment corresponding to the initial plain concrete behavior of the beam.

The condition

$$M_F^{RC} = M_F^{PC} \tag{21}$$

is the required criterion for definition of the minimum reinforcement ratio ρ_{min} that must be provided to prevent sudden fracture failure.

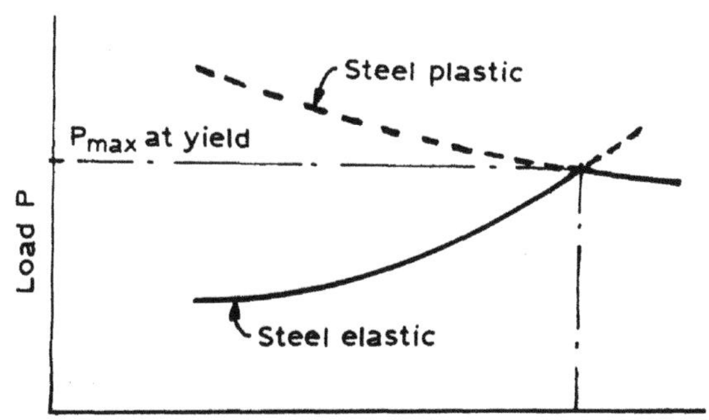

Fig. 6 Typical load versus crack length curve for under-reinforced beam.

The moment corresponding to the kink may be obtained from eqn (6):

$$M_F^{RC} = \left(\frac{1}{\alpha}\right)_{\xi_{max}} K_{IC} + \left(\frac{\beta}{\alpha}\right)_{\xi_{max}} F_p \tag{22}$$

where ξ_{max} = nondimensional crack depth (a/h) for which steel initially yields (at kink).

$(1/\alpha)_{\xi_{max}}$ and $(\beta/\alpha)_{\xi_{max}}$ are functions of ξ_{max} and cross-section dimensions and $F_p = \rho_{min} bdf_y$.

6.1.1. *Fracture Moment Capacity of Plain Concrete Section*

In order to obtain a realistic estimate of M_F^{PC}, use is made of strength theory where the initial cracking moment is controlled by the modulus of rupture f_R. Defining the strength theory cracking moment by M^{PC}, one has

$$M^{PC} = \gamma f_r S \tag{23}$$

when γ = a factor of safety and S = elastic section modulus = $bh^3/6$. Using γ as 2·0

$$M_F^{PC} = f_R \frac{bh^2}{3} \tag{24}$$

The fracture moment of plain section, M_F^{PC}, in the presence of a crack of non-dimensional depth ξ_{min} is given by

$$M_F^{PC} = \left(\frac{1}{\alpha}\right)_{\xi_{min}} K_{IC} \tag{25}$$

where ξ_{min} corresponds to the initial cracks in the reinforced concrete beam prior to loading.

Since the value of M_F^{PC} is dependent on ξ_{min}, which is unknown, for simplicity M_F^{PC} is taken as equal to M^{PC} given by eqn (23).

6.2 Procedure for Computation of ρ_{min}

Since the moment carrying capacity of a reinforced beam in pure flexure reaches a pseudo-critical value at the initial yielding of steel, the corresponding crack is assumed to be of critical height ξ_{max}. For a given set of geometric and material properties, the ξ_{max} of a beam can be calculated for a prescribed value of ρ using the proposed LEFM approach. By varying ρ in steps from a small initial value, M_F^{RC} can be computed from eqn (25) for each assumed value of ρ and then compared with M_F^{PC} for plain concrete. The value of ρ which corresponds to the case when $M_F^{PC} = M_F^{RC}$ is taken as ρ_{min}.

The dependency of ξ_{max} on ρ introduces complexity into the determination of ρ_{min} in a more straightforward way. Furthermore, the ξ_{max}–ρ relationship is valid only for given section and material properties and therefore a unique value of ρ_{min} which is applicable to any flexural member cannot be prescribed by a simple relationship in the same fashion as the ACI stipulation. For the development of a simplified expression, it is essential that an appropriate ξ_{max} be established.

6.3 Upper Limit for ξ_{max}

The complexity present in finding a general form of the ξ_{max}–ρ relationship, valid for a wide range of material and cross-sectional properties, motivated a search for a suitable replacement for the exact ξ_{max}–ρ relationship. This has been achieved by studying the M–a relationship for different values of ρ and the effect of positioning a_{max} or ξ_{max}.

If $\rho < \rho_{min}$, as for a weakly reinforced beam (Fig. 7), the initial crack will propagate instantaneously and the failure is classified as unstable ($M_{ult} < M_{UR}$). By increasing ρ to the value ρ_{min}, it is possible to reach a level where $M_{ult} = M_{UR}$ simulating the critical state. Now, the question is raised whether it is possible for the beam to sustain a moment level equal to M_{UR} at $\bar{a}$ different from a_{max} ($\bar{a} < a_{max}$ or $\bar{a} > a_{max}$)? To answer this, each case is considered independently.

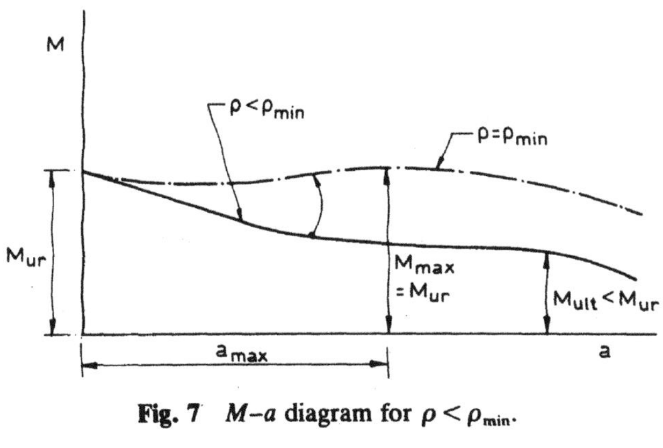

Fig. 7 M–a diagram for $\rho < \rho_{min}$.

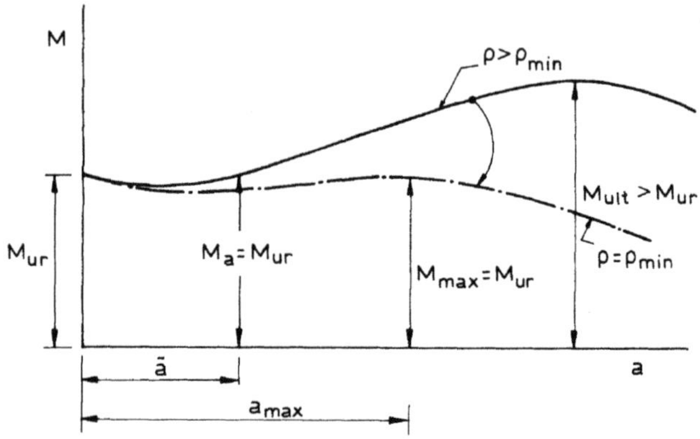

Fig. 8 M–a diagram for $\rho > \rho_{min}$ and $\tilde{a} < a_{max}$.

Referring to Fig. 8, if $\tilde{a}$ is less than a_{max} and M corresponding to $\tilde{a}$ is put equal to M_{UR}, then it can be seen that: (i) the obtained $A_s > \min A_s$; (ii) the steel stress at $\tilde{a}$ will not be yielding; and (iii) $M_{ult} > M_{UR}$. Since steel stress will be different from the yield stress, the case of pursuing $\tilde{a} < a_{max}$ is discarded. If $\tilde{a}$ is chosen so that $\tilde{a} > a_{max}$ (Fig. 8), the following conditions apply: (i) $A_s > \min A_s$, and (ii) $f_s = f_y$.

As a conclusion, it is rational to define an upper limit of ξ_{max} or (a_{max}) as the level at which most or all rebars are likely to yield. The upper limit was found to range from 0·65 to 0·85 and the latter value is used subsequently. Adoption of this value considerably simplifies development of expressions for ρ_{min} using the present formulation, as this eliminates ξ_{max} as an unknown variable.

6.4 Practical Formula for ρ_{min}

Use of eqns (22) and (25) and the condition of $M_F^{PC} = M_F^{RC}$ at ρ_{min} yields

$$\left(\frac{1}{\alpha}\right)_{\xi_{min}} K_{IC} = \left(\frac{1}{\alpha}\right)_{\xi_{max}} K_{IC} + \left(\frac{\beta}{\alpha}\right)_{\xi_{max}} F_p \tag{26}$$

Further, replacing M_F^{PC} by eqn (23)

$$\gamma f_r \frac{bh^2}{6} = \frac{K_{IC}}{(\alpha)_{\xi_{max}}} + \left(\frac{\beta}{\alpha}\right)_{\xi_{max}} F_P. \tag{27}$$

and substituting $F_P = \rho_{min} f_y bd$ into eqn (27) results in

$$\gamma f_r \frac{bh^2}{6} = \frac{K_{IC}}{\alpha_{max}} + \frac{\beta_{max}}{\alpha_{max}} \rho_{min} f_y bd \tag{28}$$

from which

$$\rho_{min} = \tfrac{1}{6}\gamma\left(\frac{\alpha}{\beta}\right)_{\xi_{max}} \frac{h^2}{d}\frac{f_r}{f_y} - \frac{1}{bd}\frac{1}{\beta_{max}}\frac{K_{IC}}{f_y} \tag{29}$$

or

$$\rho_{min} = \frac{1}{f_y}(\kappa_1 f_r - \kappa_2 K_{IC}) \tag{30}$$

where,

$$\kappa_1 = \frac{1}{6}\gamma\left(\frac{\alpha}{\beta}\right)_{\xi_{max}} \frac{h^2}{d} \tag{31a}$$

$$\kappa_2 = \frac{1}{bd}\frac{1}{\beta_{max}} \tag{31b}$$

An alternative form of eqn (30) is now studied in order to highlight the effect of certain salient variables. One first defines a modified steel reinforcement ratio $\bar{\rho}_{min}$ such that

$$\bar{\rho}_{min} = \frac{A_s}{bh}$$

and thus

$$\frac{\bar{\rho}_{min}}{\rho_{min}} = \frac{d}{h}$$

Recasting eqn (29) in terms of $\bar{\rho}_{min}$

$$\bar{\rho}_{min} = \frac{h}{6}\gamma\left(\frac{\alpha}{\beta}\right)_{\xi_{max}} \frac{f_r}{f_y} - \frac{1}{bh}\frac{1}{\beta_{max}}\frac{K_{IC}}{f_y} \tag{32}$$

From the following expressions

$$\alpha_{max} = \frac{Y_M(\xi)_{max}}{h^{3/2}b} \tag{33a}$$

$$\beta_{max} = \alpha_{max}\left(\frac{h}{2} - d_s\right) + \frac{Y_F(\xi_{max})}{h^{1/2}b} \tag{33b}$$

the ratio $\left(\frac{\beta}{\alpha}\right)_{\xi_{max}}$ can be written as

$$\left(\frac{\beta}{\alpha}\right)_{\xi_{max}} = h\left(\frac{1}{2} - \frac{d_s}{h}\right) + h\frac{Y_F(\xi_{max})}{Y_M(\xi_{max})} \tag{34}$$

Substituting for $\left(\frac{\beta}{\alpha}\right)_{\xi_{max}}$ from eqn (34) into eqn (32) yields

$$\bar{\rho}_{min} = \frac{\gamma}{6}\left[\frac{Y_F(\xi_{max})}{Y_M(\xi_{max})} + \frac{1}{2} - \frac{d_s}{h}\right]\frac{f_r}{f_y} - \frac{\gamma}{6}\frac{1}{h^{1/2}Y_M(\xi_{max})\left[\frac{Y_F(\xi_{max})}{Y_M(\xi_{max})} + \frac{1}{2} - \frac{d_s}{h}\right]}\frac{K_{IC}}{f_y} \tag{35}$$

or alternatively

$$\bar{\rho}_{min} = \frac{1}{f_y}(\bar{\kappa}_1 f_r - \bar{\kappa}_2 K_{IC}) \tag{36}$$

where

$$\bar{\kappa}_1 = \frac{\gamma}{6}\frac{1}{\left[\frac{Y_F(\xi_{max})}{Y_M(\xi_{max})} + \frac{1}{2} - \frac{d_s}{h}\right]} \tag{37a}$$

and

$$\bar{\kappa}_2 = \frac{1}{h^{1/2}}\frac{1}{Y_M(\xi_{max})\left[\frac{Y_F(\xi_{max})}{Y_M(\xi_{max})} + \frac{1}{2} - \frac{d_s}{h}\right]} \tag{37b}$$

The ACI-defined ρ_{min} can now be related to $\bar{\rho}_{min}$ as

$$\rho_{min} = \bar{\rho}_{min}\left(\frac{1}{\dfrac{1-d_s}{h}}\right) \tag{38}$$

Equations (36) and (38) can be used to obtain the exact solution for the minimum reinforcement which requires the knowledge of ξ_{max}. This problem is alleviated by using an approximate value of ξ_{max} conservatively taken as 0·85.

A cursory inspection of eqn (36) reveals the dependence of $\bar{\rho}_{min}$ on two parameters, namely a material characteristic parameter depending on a function of the ratio K_{IC}/f_y and a section characteristic parameter as defined by the ratio $\frac{d_s}{h}$.

Extensive computations carried out to identify the functional form

TABLE 2 ρ_{min} for $J_M \sim 0{\cdot}0019$

K_{IC} $(N/mm^{3/2})$	f_y (N/mm^2)	J_M	ρ_{min}
29·94	321	0·001 937	0·004 571
35·59	370	0·001 938	0·004 571
40·69	414	0·001 935	0·004 571
53·94	522	0·001 938	0·004 571

of the material dependency enables us to propose the material characteristic parameter J_M as

$$J_M = \frac{K_{IC}}{f_y^{1{\cdot}21}} \frac{\text{lb/in}^{3/2}}{\text{lb/in}^2} = \frac{14{\cdot}322 K_{IC}}{f_y^{1{\cdot}21}} \left(\frac{\text{N/mm}^{3/2}}{\text{N/mm}^2}\right) \tag{39}$$

and the straightforward relevant sectional characteristic parameter

$$J_S = \frac{d_s}{h} \tag{40}$$

In order to verify the relevance of the adopted indices, fixed values

TABLE 3 ρ_{min} for $J_M \sim 0{\cdot}003$

K_{IC} $(N/mm^{3/2})$	f_y (N/mm^2)	J_M	ρ_{min}
29·94	224	0·002 993	0·006 563
35·58	258	0·002 998	0·006 564
40·69	289	0·003 002	0·006 564
53·94	364	0·002 997	0·006 564

TABLE 4 ρ_{min} for $J_s \sim 0{\cdot}1667$

h (mm)	d_s (mm)	J_s	ρ_{min}
254	42·4	0·1669	0·004 534
508	84·6	0·1665	0·004 594
762	127·0	0·1667	0·004 625

TABLE 5 ρ_{min} for $J_s \sim 0{\cdot}25$

h (mm)	d_s (mm)	J_s	ρ_{min}
254	63·5	0·25	0·005 743
381	95·3	0·25	0·005 781
762	190·5	0·25	0·005 871

of J_M and J_S were chosen by varying their constituent terms and ρ_{min} obtained using eqns (36) and (38), together with $\xi_{max} = 0{\cdot}85$. The results as shown in Tables 2–5 indicate that J_M and J_S may be used as fractural invariants, since ρ_{min} virtually remains unchanged for various combinations of K_{IC} and f_y (such that J_M is constant) and for various values of d_s and h (such that J_s is constant). Thus J_M and J_s can be called *material fractural invariant* and *sectional fractural invariant*, respectively.

Figure 9 illustrates the variation of ρ_{min} with respect to J_M and J_S. On regressing the data of Fig. 9, the following simplified expression

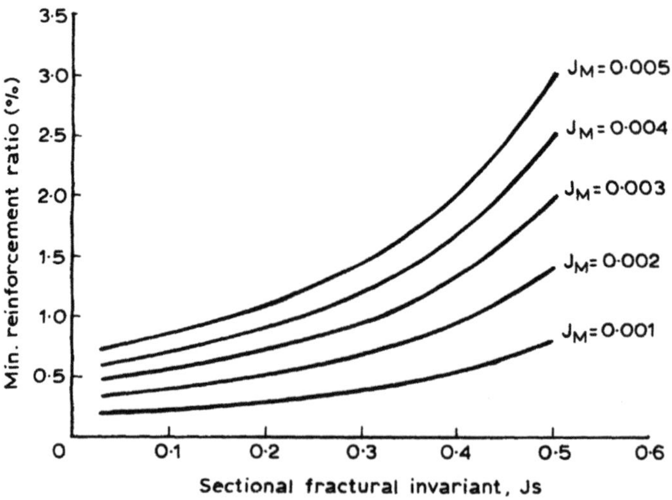

Fig. 9 Minimum reinforcement ratio versus J_S and J_M.

for ρ_{min} is obtained in terms of the two proposed fractural invariants

$$\frac{1}{\rho_{min}} = (1{\cdot}70 - 2{\cdot}6J_S)J_M^{-0{\cdot}82} \tag{41}$$

Equation (41) shows that the minimum reinforcement ratio is sensitive to the member shape (h, d_s) and to the material strength (K_{IC}, f_y) in contrast to the ACI equation ($\rho_{min} = 1{\cdot}4/f_y$) which is independent of the member cross-sectional dimension, steel location and concrete strength characteristics. The recognition of the dependency of ρ_{min} on the characteristic material strengths is important, as there is nowadays a growing trend in the use of high strength concrete ($f'_c > 50$ MPa) for which a simplified form like the ACI one, would be inadequate.

7. CONCLUSIONS

A fracture mechanics based method is proposed to compute fracture moment for a reinforced concrete member subjected to Mode I stress field in the presence of an edge crack. The applicability of the method is not hindered by the condition of steel stress and the crack propagation at both phases of pre-yielding and post-yielding of steel can be predicted with reasonable accuracy.

The proposed LEFM model has the potential of serving as a crack-controlled design in which crack propagation is restricted to a permissible height. The method yields an easy-to-use expression for the minimum reinforcement requirement in flexure (ρ_{min}) in terms of two invariants, which shows dependency on concrete properties (f'_c, f_R and K_{IC}) and beam geometry. This dependency of ρ_{min} on the characteristic materials properties is significant in view of the increased use of high strength concrete.

ACKNOWLEDGEMENT

The support of the King Abdulaziz City for Science and Technology, Riyadh, for this work is acknowledged.

REFERENCES

[1] Azad, A. K. & Baluch, M. H., Fracture characterization of reinforced concrete. Final Report AR7-130, King Abdulaziz City for Science and Technology, Riyadh, Nov. 1990.

[2] Bazant, Z. P. & Cedolin, L., Approximate linear analysis of concrete fracture by *R*-curves. *J. Struct. Eng.*, **110**(6) (1984) 1336–55.

[3] Saouma, V., Finite element analysis of reinforced concrete: a fracture mechanics approach. PhD dissertation, School of Civil and Environmental Eng., Cornell University, USA, 1981.

[4] Ingraffea, A. R. & Saouma, V., Numerical modeling of discrete crack propagation in reinforced and plain concrete. In *Fracture Mechanics of Concrete,* ed. G. C. Sih & A. Di Tommaso. EAFM Series, Martinus Nijhoff Publishers, The Netherlands, 1985, pp. 171–225.

[5] Carpinteri, A., A fracture mechanics model for reinforced concrete collapse. In *Advanced Mechanics of Reinforced Concrete,* IABSE Colloquium, Delft, June 1981, pp. 17–30.

[6] Carpinteri, A., Application of fracture mechanics to concrete structures. *Journal of Structural Division* (ASCE), **108** (1982) 833–48.

[7] Carpinteri, A., Scale effects in fracture of plain and reinforced concrete structures. In *Fracture Mechanics of Concrete.* ed. G. C. Sih & A. Di tommaso. EAFM Series, Martinus Nijhoff Publishers, The Netherlands, 1985, pp. 95–140.

[8] Carpinteri, A., Di Tomasso, A. & Fanelli, M., Influence of material parameters and geometry on cohesive crack propagation. *Fracture Toughness and Fracture Energy of Concrete,* ed. F. H. Wittmann. Elsevier Science Publishers, The Netherlands, 1986, pp. 117–35.

[9] Carpinteri, A., *Mechanical damage and crack growth in concrete.* EAFM Series, Martinus Nijhoff Publishers, The Netherlands, 1986.

[10] Okamura, H., Watanabe, K. & Takano, T., Application of the compliance concept in fracture mechanics. *ASTM STP 536,* American Society for Testing and Materials, 1973, pp. 423–38.

[11] Okamura, H., Watanabe, K. & Takano, T., Deformation and strength of cracked members under bending moment and axial force. *Engineering Fracture Mechanics,* **7** (1975) 531–39.

[12] Petersson, P. E., Crack growth and development of fracture zones in plain concrete and similar materials. PhD thesis, Lund Institute of Technology, Sweden, 1981.

[13] *Building Code Requirements for Reinforced Concrete* (*ACI 318-89*). American Concrete Institute, Detroit, Michigan, 1989.

[14] Azad, A. K., Baluch, M. H. & Ziraba, Y., Fracture characterization of reinforced concrete beams. In *Fracture of Concrete and Rock, SEM-RILEM International Conf.,* eds S. P. Shah & S. E. Swartz, Texas, June 1987, pp. 443–8.

[15] Azad, A. K., Mirza, M. S. & Chan, P., Fracture energy of weakly reinforced beams. *Fatigue and Fracture of Engineering Materials and Structures,* **12**(1) (1989) 9–18.

[16] CEB Model Code, Draft version, 1988.

Chapter 17

Role of Compressive Fracture Energy of Concrete on the Failure Behavior of Reinforced Concrete Beams

KEITETSU ROKUGO & WATARU KOYANAGI
Civil Engineering, Gifu University, Yonagido, Gifu, 501-11, Japan

ABSTRACT

To introduce the description of the relation between compressive fracture energy and the final strain value in tension steel bars, test results of the behavior of reinforcing bars in concrete and the effect of specimen height on the compressive fracture behavior of concrete are presented. The shape of the compressive stress–strain curve after peak load is shown to depend strongly on specimen height. On the other hand, the total compressive fracture energy in specimens of various heights tends to be constant. In the case of deformed bars covered with concrete, the yield plateau does not appear and the strain at rupture decreases. The yield limit reinforcement ratio (similar to the so-called balanced failure reinforcement ratio) and the rupture limit reinforcement ratio are expressed using the compressive fracture energy of concrete. The possibility is pointed out that the rupture limit reinforcement ratio decreases as the effective depth of the RC beam cross-section increases.

1. INTRODUCTION

The fracture behavior of concrete due to cracking can be well described using the parameters defined in Fracture Mechanics of Concrete such as the tension softening diagram and the fracture energy G_F (area under the tension softening diagram). It is accepted in the field of Fracture Mechanics of Concrete that the softening behavior after the peak is size-dependent and causes fracture localization. The energy approach and the softening-localization concept of Fracture Mechanics would be useful tools applicable also to reinforced concrete. Hillerborg [1], [2] has applied the softening–localization concept to the design of reinforced concrete (RC) beams. The authors [3] have reported the application of the fracture energy approach to RC beams.

The properties of RC beams such as strength and ductility are usually evaluated by analysis of the cross-section, assuming suitable stress–strain relationships for concrete and steel under compatibility and equilibrium conditions. The ductility of the RC cross-section is roughly guaranteed in design codes of concrete structures [4], [5] by restricting the tension reinforcement ratio p (or $p-p'$, mentioned later) to be sufficiently lower than the balanced reinforcement ratio p_b.

In this contribution, the upper and lower limits of the tension reinforcement ratio and the ductility of RC beams are expressed using the fracture energy of concrete in compression. To introduce this point, the real failure properties of concrete and steel bars are examined. Test results of the influence of the specimen height on the compressive fracture energy and the compressive stress–strain curve are presented. Test results on the yield strength, the rupture strain and the stress–strain curve of reinforcing steel bars (rebars) covered with concrete are also reported.

2. INFLUENCE OF SPECIMEN SIZE ON COMPRESSIVE FRACTURE OF CONCRETE

The stress–strain curve of concrete in compression is important for the analysis of RC members. However, it is known that the shape of the curve after the peak depends on the size of the specimen or failure region because of softening and localization. The effects of the specimen height on the compressive stress–strain curve and the compressive fracture energy are discussed in this chapter citing data from our report [6]. Compression tests on concrete specimens of different specimen height were performed.

2.1 Outlines of Compression Tests of Concrete

Three kinds of concrete were adopted: mortar ($f_c = 65$ MPa); normal concrete ($f_c = 39$ MPa); and steel fiber reinforced concrete (SFRC,

TABLE 1 Mix proportions of concretes for compression tests

Type of concrete	*W/C (%)*	*s/a (%)*	*Unit content (kg/m³)* W	C	S	G	Fiber [2%]	*Admixture (C × %)*	*Slump (cm)*	*Age at test (days)*
M: mortar	33	—	203	616	1 541	—	—	$3{\cdot}0^a$	—	25 ± 2
P: normal concrete	53	51	190	360	933	925	—		3	18 ± 1
F: SFRC	51	56	215	430	844	839	164	$1{\cdot}0^b$	8	18 ± 1

C: high early strength portland cement; *S*: river sand, F.M. = 2·85; *G*: crushed stone, max. size = 15 mm; fiber: steel cut wire, $\phi 0{\cdot}5 \times 30$ mm; Admixture: [a] superplasticizer, [b] AE water reducing agent.

$f_c = 48$ MPa). The details of the mixes are given in Table 1. The cross-section of the specimens was square (75 mm × 75 mm). Four specimen heights were selected: 75 mm, 150 mm, 300 mm and 450 mm, i.e. the height/width (h/w) ratio equaled 1, 2, 4 and 6, respectively. The compressive deformation was measured with two displacement transducers (stroke: 10 mm) attached to the specimens. The gage length was shorter by 20 mm than the specimen height. The load–deformation curves were recorded on an $X-Y$ recorder. The loading plate of the testing machine was fixed so as not to rotate during the loading test.

2.2 Test Results of Compression Tests and Discussions

Measured load–deformation curves are shown in Fig. 1 with solid lines. The compressive strength, the compressive fracture energy W_q, the applied energy up to the peak load point W_p and the length of the damaged zones are summarized in Table 2. Test results are expressed with the average of more than three specimens for each test condition. The first letter of the series indicates the kind of concrete (M: mortar, P: normal plain concrete, F: SFRC), and the last figure is the h/w ratio. The fracture energy W_q here is defined as the energy dissipated up to the point where the load descends to 1/3 of the peak load, as indicated in Fig. 2. W_q contains a portion of the elastic strain energy. Examples of the length of the damaged zone are sketched in Fig. 3.

For each kind of concrete, the peak loads were almost constant but both the deformations at the peak and the energies up to the peak load W_p were nearly proportional to the specimen height except for the cubic specimens with the h/w ratio equal to 1. The cause of this peculiar result for the cubic specimens would be the confinement due

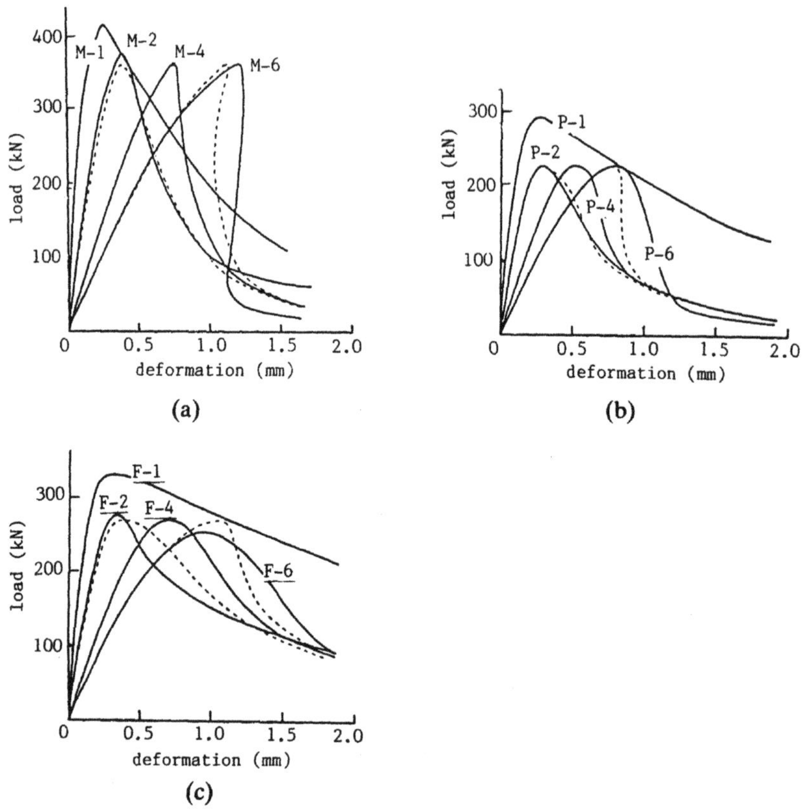

Fig. 1 Measured compressive load–deformation curves. (a) mortar; (b) normal concrete; (c) SFRC.

to friction between the loading plates of the testing machine and the concrete.

Although the compressive fracture energy W_q slightly increased with the specimen height, W_q tended to remain almost independent of the specimen height, except for the cubic specimens. The length of the damaged zone for specimens with a h/w ratio greater than four were shorter than the specimen height and tended to be constant. This means that the fracture does not spread over the full specimen height but concentrates in the case of the taller specimens. It would be possible to make the simple assumption that the compressive fracture energy per unit concrete area (not volume) is constant.

TABLE 2 Results of compression tests

Series	*Maximum load (kN)*	*Compressive strength (MPa)*	*Deformation at peak (mm)*	*Energy at peak W_p (J)*	*Fracture energy W_q (J)*	*Damaged zone (mm)*
M-1	403	71·6	0·26	70	340	75
M-2	364	64·7	0·39	83	174	150
M-4	352	62·5	0·77	168	219	250
M-6	353	62·7	1·22	261	257	306
P-1	279	49·7	0·26	52	437	75
P-2	221	39·2	0·28	39	128	150
P-4	221	39·2	0·54	77	126	235
P-6	219	38·8	0·79	106	160	295
F-1	318	56·4	0·31	71	1 035	75
F-2	269	47·7	0·34	58	287	150
F-4	261	46·4	0·69	112	277	205
F-6	246	43·7	0·93	141	322	227

The dotted lines in Fig. 1 show the curves which are composed of the natural ascending part and the descending part related to the specimens with a h/w ratio of four. In the case of a h/w ratio of six, for example, the ascending curve is 3/2 times that of the deformation before the peak relating to specimens with a h/w ratio of four, and the descending curve is the composition of the ascending curve and the difference between the ascending and descending curve of the specimens with a h/w ratio of four. The composed curves (dotted lines) well fit to the measured curves (solid lines). The ascending part of the load–deformation curve before the peak is dominated by the homogenized behavior over the specimen, whereas the descending part of the load–deformation curve after the peak is governed by the localized phenomena. This is illustrated in Fig. 4. More precisely, the unloading

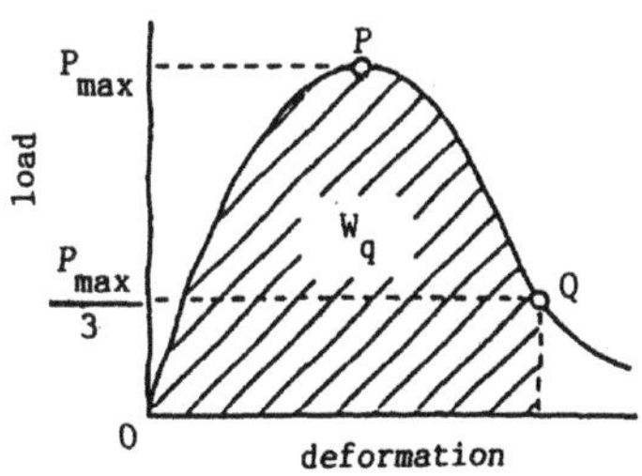

Fig. 2 Compressive fracture energy W_q.

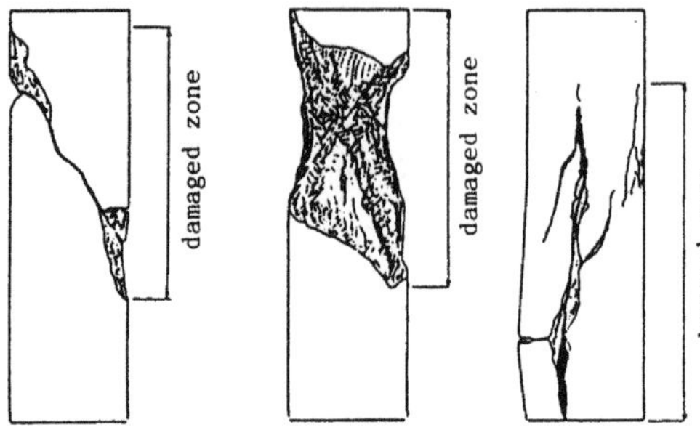

Fig. 3 Length of damaged zone.

path from the peak point outside the damaged zone does not exactly coincide with the loading path.

Figure 5 shows the compressive stress–strain curves which are calculated from the load–deformation curves using the cross-sectional area and the gage length (20 mm shorter than the specimen height). For each kind of concrete, the stress–strain curves before the peak coincide with each other, whereas the curves after the peak are remarkably influenced by the h/w ratio. The post-peak curves become steeper as the specimen length increases. In the case of the mortar specimen with a h/w ratio of six, a snap-back phenomenon was clearly observed.

The conditions in the uniaxial compression tests are different from those in the compression zone of RC beams, where a large strain gradient exists and usually stirrups and compression bars are placed. When applying the stress–strain curve to the analysis of RC beams, the influence of size should be well taken into account. Hillerborg [1],

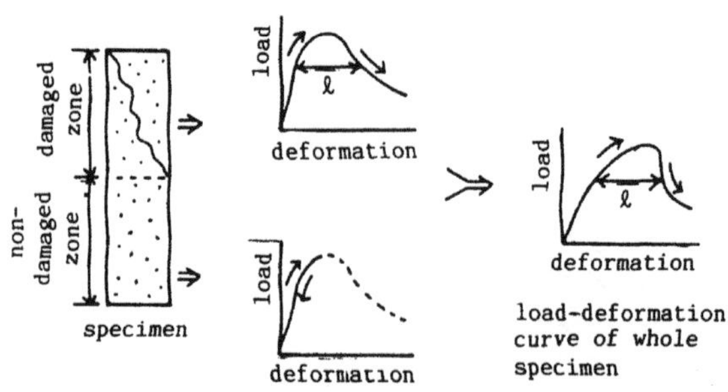

Fig. 4 Composition of load–deformation curve.

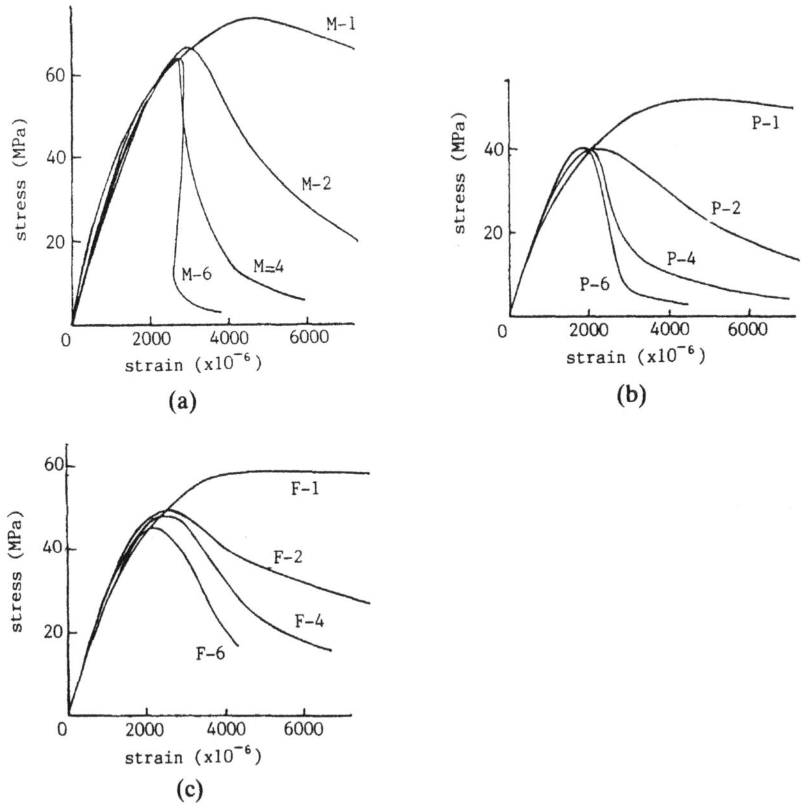

Fig. 5 Compressive stress–strain curves. (a) mortar, (b) normal concrete, (c) SFRC.

[2] proposed that the descending (or horizontal) part of the compressive stress–strain curve should be changed to be dependent on the depth of the compression zone of RC beams. If it can be assumed that the compressive fracture energy to break a unit area of concrete cross-section is constant, similar to the case of the tensile fracture energy G_F, the *difference* between the ascending and descending branches of the stress–strain curve should be changed in inverse proportion to the length of the fractured zone, the length of the hinge zone in the case of RC beams in flexure. According to the assumption above, the compressive fracture energy under the stress–strain curve also changes in inverse proportion to the length of the fracture zone.

The influence of the fracture zone length on the stress–strain curve is discussed in Section 4.

3. BEHAVIOR OF TENSION REINFORCING BARS IN CONCRETE

Since steel is a typical strain-hardening material, the yield zone spreads over the high stress region of a steel bar and then a large amount of energy is consumed. The role of the steel reinforcing bars (rebars) is not only to undertake tension force but also to provide ductility for RC members. However, the mechanical behavior of a deformed rebar covered with concrete is different from the naked one (i.e. without concrete), because the concrete surrounding the bar is a strain-softening material and the bond between the deformed rebar and the concrete is sufficient. The actual stress–strain curves of rebars in concrete should be known for refined structural analysis. The yielding, work-hardening and breaking behavior of rebars in concrete are discussed in this chapter citing data from our report [7]. Tension tests of rebars covered with concrete were carried out to examine the shape of the stress–strain curves and the propagation process of the yield zone.

3.1 Outlines of Tension Tests of Rebars

Figure 6 shows the shape and size of the tensile specimen of a rebar with concrete. A rebar (1000 mm long) was placed at the center of the concrete body (100 mm square section and 600 mm long). Naked bars without concrete were also examined. The conditions of the tests are

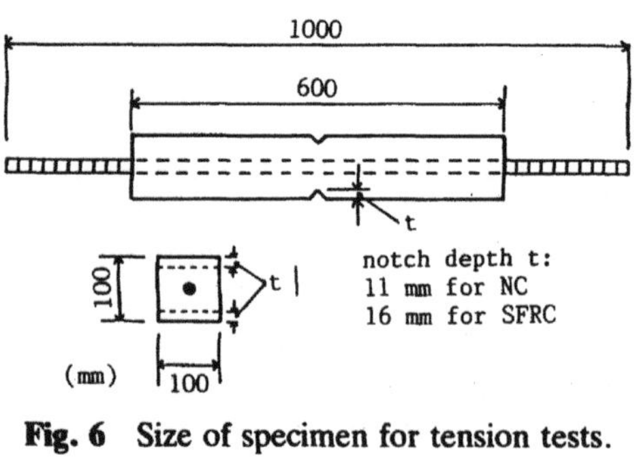

Fig. 6 Size of specimen for tension tests.

TABLE 3 Conditions for tension tests of bars

Specimen		*Concrete type*	*Rebars*			*Gage pitch (mm)*
Name	*Numbers*		*Type*	*Area (mm^2)*	*Diameter (mm)*	
MN16-40	1	NC	Milled	196	15·8	40
MN16-25	1	NC	Milled	196	15·8	25
MF16-40	1	SFRC	Milled	196	15·8	40
MF16-25	1	SFRC	Milled	196	15·8	25
MS16-40	1	—	Milled	196	15·8	40
DS16	3	—	D16	199	15·9	—
DN13	3	NC	D13	127	12·7	—
DF13	3	SFRC	D13	127	12·7	—
DS13	3	—	D13	127	12·7	—
RN13	3	NC	R13	133	13·0	—
RF13	2	SFRC	R13	133	13·0	—
RS13	3	—	R13	133	13·0	—

Yield strength f_{sy}: 372 MPa (D16); 360 MPa (D13); 349 MPa (R13).

reported in Table 3. One kind of round bar (R13), two kinds of deformed bars (D13 and D16) and one kind of milling processed bar (made of D16) were adopted. Normal concrete (NC) and steel fiber reinforced concrete (SFRC) were used. The details of the mixes are listed in Table 4 together with the strengths obtained from standard tests.

As shown in Fig. 7(a), the milled bar was grooved by a channel of 4 mm × 8 mm. First, the channel was grooved into each half section of two bars and then the two halves were glued with epoxy resin into a single bar. The cross-sectional area of the milled bar was 196 mm^2, which is close to the area of D16 bar (199 mm^2). Waterproof type electric resistance wire strain gages (gage length: 2 mm) were attached along the channel at a pitch of 25 mm or 40 mm, as indicated in Fig. 7(b).

For rebars with concrete, the concrete portion was notched on the opposite faces at the middle as shown in Figs 6 and 8, so that the first crack would occur within the gage length (250 mm) of the concrete portion. The notch depth was 11 mm for NC and 16 mm for SFRC, and the reduction ratios of the sectional area were 22% and 32%, respectively. In the case of D13 bars with concrete, both ends of the

TABLE 4 Properties of concretes for tension tests of bars

Kinds of concrete	*Mix proportions*							*Strengths*			
			Unit content, kg/m^3								
	W/C (%)	*s/a (%)*	*W*	*C*	*S*	*G*	*Fiber [2%]*	*Admix-ture (C × %)*	*Compression* f'_c *(MPa)*	*Flexure* f_b *(MPa)*	*Tension* f_t *(MPa)*
NC-1	60	47	183	304	850	954	—	—	40·8	6·34	3·19
NC-2	60	47	183	304	850	954	—	—	44·0	6·24	4·06
SFRC-1	54	69	195	363	1 111	497	157	0·25	46·7	10·47	4·04
SFRC-2	54	69	207	384	1 060	478	157	0·25	53·7	11·87	6·41

NC-1 and SFRC-1 for milled bars, NC-2 and SFRC-2 for the rest. *C*: high early strength portland cement; *S*: river sand, F.M. = 2·70; *G*: crushed stone, max. size = 15 mm; fiber: steel cut wire, ϕ0·5 × 30 mm; Admixture: AE water reducing agent; Testing age: one month.

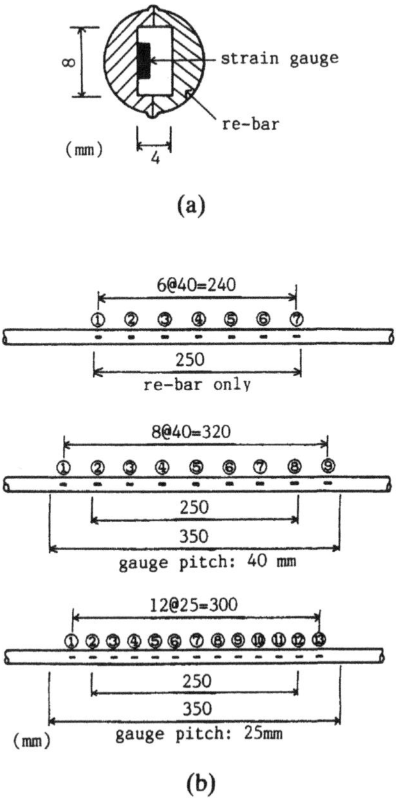

Fig. 7 Milled bar. (a) cross-section, (b) positions of wire strain gages.

D13 bar of 500 mm were welded to D19 bars of 250 mm (total length: 1000 mm) to prevent failure outside the concrete portion. Spiral reinforcement was provided in the end zones of the concrete. For R13 bars, D16 bars and the milled bars with concrete, only spiral reinforcement was provided at the ends of the concrete portion.

For bars with concrete, the deformation of the concrete portion at the gage length of 250 mm was measured with two displacement transducers (stroke: 50 mm) attached to the specimen at symmetrical positions as sketched in Fig. 8. The applied load, the deformation of the concrete portion and the strains in the steel were recorded with a data logger. The displacement of the cross-head of the loading machine was mechanically recorded. For naked bars, the deformation at gage length of 250 mm was measured with attached displacement

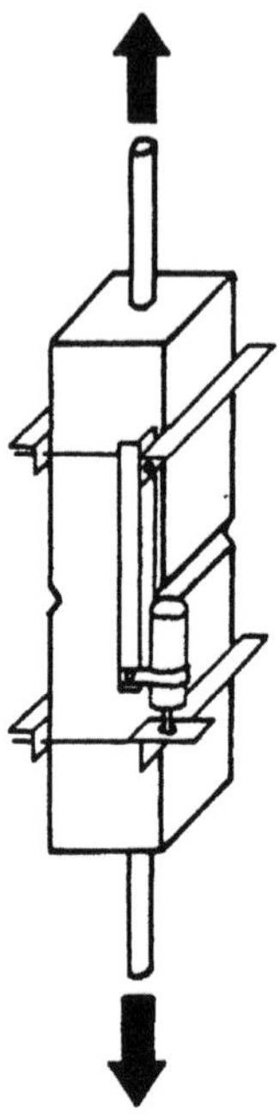

Fig. 8 Equipment for deformation measurement.

transducers on the bar and the relation between the load and the machine displacement was also recorded.

3.2 Test Results of Tension Tests and Discussion

The yield load and the maximum load of each specimen are listed in Table 5. The yield point was defined as the first bend point in a load–deformation curve.

3.2.1 Spreading Process of Yield Zone in Milled Bars

The load–deformation (gage length: 250 mm) curves and the steel strain versus specimen deformation curves of the milled naked bar (gage pitch: 40 mm) are shown in Fig. 9 and those of the milled bars (gage pitch: 40 mm) with concrete are shown in Fig. 10. Figure 11 shows only the load–deformation curves of milled bars with concrete, where the strain in steel was measured at a gage pitch of 25 mm. The circled numbers correspond to the numbers in Fig. 7(b), indicating the positions of strain gages in the bars. As seen from the strain–deformation curves, the strains in steel began to increase

TABLE 5 Results of tension tests of bars

Specimen		*Yield load (kN)*	*Maximum load (kN)*
MN16-40		68·8	103·3
MN16-25		68·1	101·7
MF16-40		88·7	105·0
MF16-25		88·8	102·5
MS16-40		66·6	101·7
DS16	1	73·5	113·0
	2	73·8 [73·9]	112·2 [112·3]
	3	74·4	111·7
DN13	1	47·4	67·6
	2	46·1 [46·7]	67·4 [67·8]
	3	46·6	68·4
DF13	1	60·8	67·6
	2	63·7 [61·8]	67·7 [67·6]
	3	60·8	67·4
DS13	1	45·6	67·6
	2	45·7 [45·6]	67·4 [66·9]
	3	45·5	65·8
RN13	1	46·3	65·7
	2	46·1 [46·3]	65·9 [65·8]
	3	46·4	65·7
RF13	1	46·3	65·7
	2	46·1 [46·2]	65·7 [65·7]
RS13	1	46·3	65·1
	2	46·1 [46·3]	64·7 [65·0]
	3	46·4	65·1

[ ]: average.

suddenly when the strain reached about 0·002 because the yielding initiated. These yield points at each gage position are marked on the load–deformation curves.

In the case of the bar without concrete shown in Fig. 9, immediately after first yielding the yield zone spread to all the gage positions, that is, throughout the whole bar. The work-hardening started and the load increased after the yield spread to the whole bar. The deformation at the initiation of the work hardening was about 4 mm (strain: 0·016).

In the case of the milled bars covered with NC and SFRC, the yield initiated at the notched position (number of gage positions is five in Fig. 10, and six or seven in Fig. 11). The yield zone gradually spread from the first yield zone to both ends. The total deformation when the

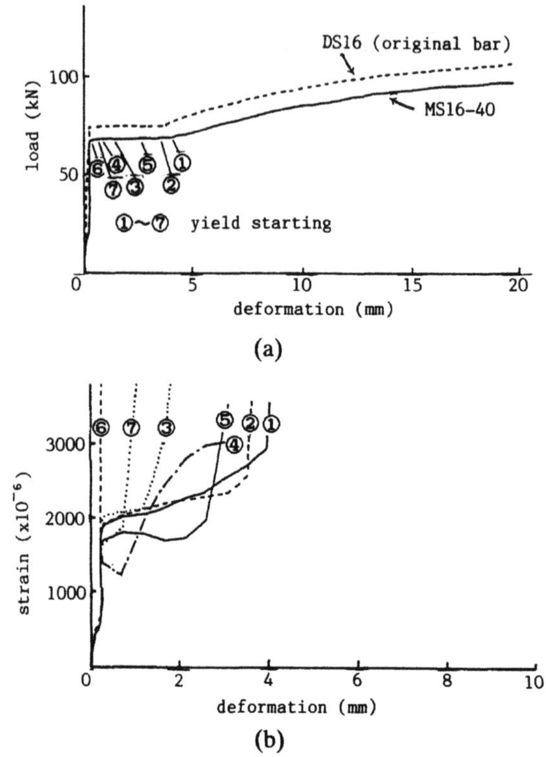

Fig. 9 Yielding behavior of milled deformed bar without concrete (gage pitch: 40 mm) and D16 bar. (a) load-deformation curves (MS16-40, DS16), (b) strain-deformation curves of milled deformed bar without concrete (MS16-40).

yield spread to all gage positions was about 10–20 mm (strain: 0·04–0·08). In the case of the bars with NC, several new cracks formed one after another as the load and deformation increased. When cracks formed, the load dropped slightly and the yield zone propagated.

3.2.2 Yield Behavior of Deformed Rebars

Typical load–deformation curves (gage length: 250 mm) of the deformed bars (D13) with or without concrete are shown in Fig. 12. It is seen from Table 5 and Fig. 12 that the yield load of the deformed bars with NC (specimen: DN13) was 2–4% higher than that of the

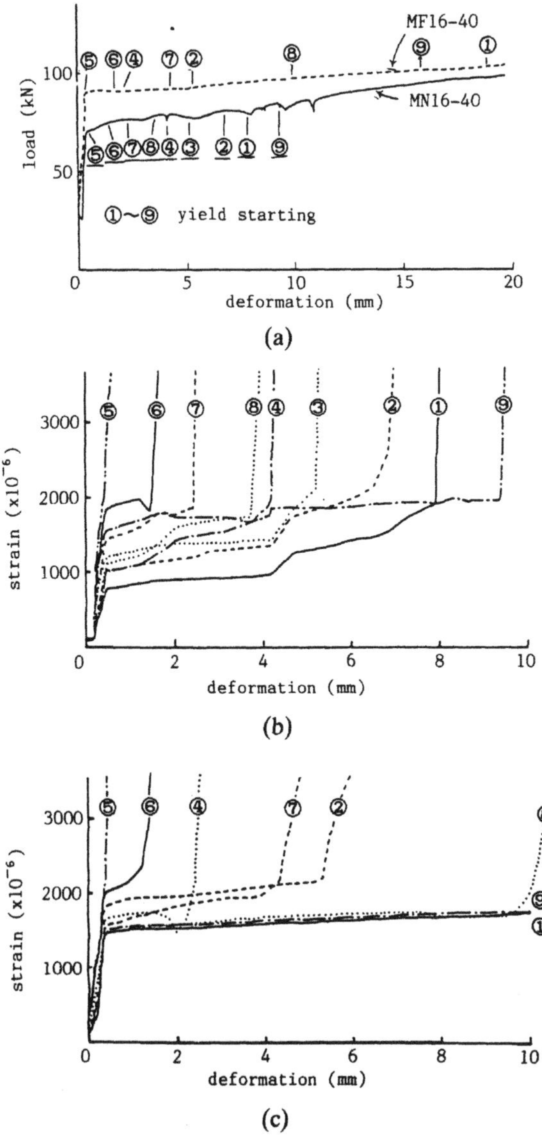

Fig. 10 Yielding behavior of milled deformed bars in normal concrete or SFRC (gage pitch: 40 mm). (a) Load-deformation curves (MN16-40, MF16-40), (b) strain-deformation curves (MN16-40), (c) strain-deformation curves (MF16-40).

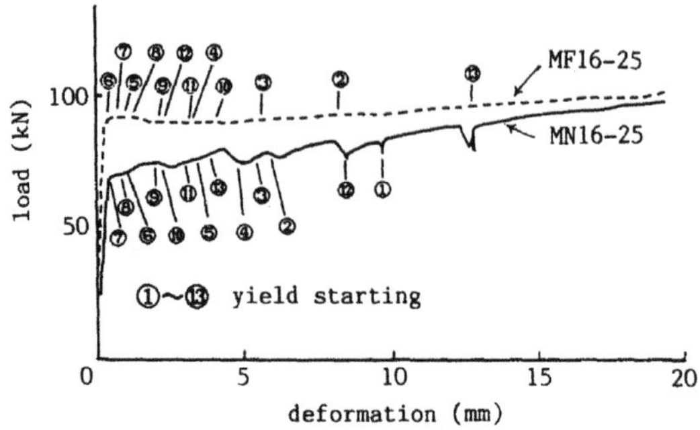

Fig. 11 Load–deformation curves of milled deformed bars in normal concrete or SFRC (gage pitch: 25 mm).

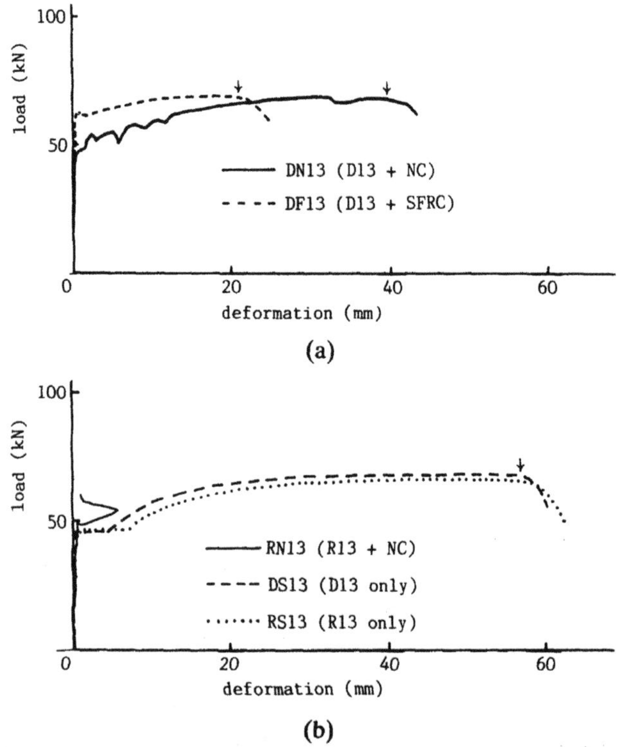

Fig. 12 Load–deformation curves of bars (gage length: 250 mm). (a) D13 bars in normal concrete or SFRC, (b) R13 bar in concrete and naked D13 and R13 bars.

deformed bars (DS13). The ultimate load of DN13 was almost the same as that of DS13.

Though the yield plateau was clearly observed for the naked bars (DS13), the plateau did not appear for the deformed bars with concrete (DN13, DF13) and the load increased gradually with the deformation. This difference can be explained as follows, considering the test results of the milled bars. For the deformed bars with concrete, the plastic deformation in steel due to yield concentrated into the vicinity of the first crack and the work hardening soon initiated in this limited small portion. In the remaining portion, the load applied to the specimen is carried not only by the rebar but also by the surrounding concrete because of the good bonding, and the steel in this portion remains elastic. With the increase in deformation, new cracks are formed one after another, the load drops suddenly and yield occurs at the newly cracked portion. In the newly cracked portions, work hardening starts immediately, and the load level then returns to the level before cracking. As this process continues, the yield region gradually spreads over the whole length. This is the reason why the yield plateau could not be observed and the load increased progressively after yielding in the case of the deformed bars with concrete.

As seen in Table 5, the yield load of the deformed bars with SFRC (DF13) was 30–40% higher than that of the naked ones (DS13). The ultimate loads of both specimens were, however, almost the same. The tensile stress in surrounding SFRC at the initiation of yielding of the specimen was estimated to be about 2 MPa from the difference in the yield loads between DF13 and DS13. In the case of bars with SFRC, a hairline crack was formed along the notched portion prior to the initiation of yield of the specimen. It can be easily supposed that a certain amount of tensile force is transmitted through the cracked section if the steel there has already yielded, because SFRC has large tensile fracture energy G_F in addition to high tensile strength. As a result, the yield strength increased remarkably for the deformed bars with SFRC. As the deformation increased after yielding, no additional crack formed and only the width of the initial crack increased. It is probable that, with the increase of the crack width, the force transmitted through SFRC at the cracked section decreased, while, the force carried by the bar increased due to work hardening. Globally, the load continued to increase.

Since the load of the deformed bars with SFRC at yield was very high, the effect of the surrounding concrete should be taken into account when calculating the yield moment of RC cross-sections. It is an easy way to obtain a higher value for the yield strength of rebars depending on the surrounding concrete, when the tensile resistance of concrete is neglected in calculation. The strain of the deformed bars at rupture is obtained from the curves in Fig. 12. In the calculation of the rupture strain, the descending part of the load–deformation curves after the peak load was omitted because this part is size-dependent and decreases if the bar length increases. Though the rupture strain for the naked bar was 23%, the value for the bar with NC was 16% (about 2/3 of that of the naked bar), and for the bar with SFRC 8·4% (about 1/3).

Figure 13 illustrates the relation between the load and the displacement of the cross head of the testing machine for deformed bars (D13) and for round bars (R13), with or without concrete cover. The displacement before yielding included not only the elongation of the specimens but also some slippage of the gripping portion. In the case of the deformed bars, the yield plateau did not appear, as mentioned above. The total displacement up to the break point for the deformed bars was affected greatly by the concrete cover as compared with the result for the round bars, where there was no effect by concrete cover. For the deformed bars, the effect of SFRC was larger than that of NC.

It can be concluded that the deformation of the deformed rebars is largely influenced by the surrounding concrete cover. The yield strength increases, the yield plateau disappears, and the strain at final rupture decreases. The degree of the restriction depends on many factors, namely, properties of concrete, diameter of bars, area ratio of concrete and steel, surface profiles of bars, and so on.

3.2.3 Yield Behavior of Round Rebars

In the tension tests of the round bars with NC, a crack occurred at the notched concrete section at an early stage of loading. The crack width increased as the load increased. However, no additional cracks occurred. Bonding between the round bar and the surrounding concrete completely disappeared after yielding mainly due to the decrease in the bar diameter (Poisson's effect). In the case of bars with SFRC, no crack formed even at the notched section and the bond between bar and SFRC completely disappeared after yielding.

As shown in Fig. 13(b), the shape of the load versus machine displacement curves was almost the same irrespective of the concrete

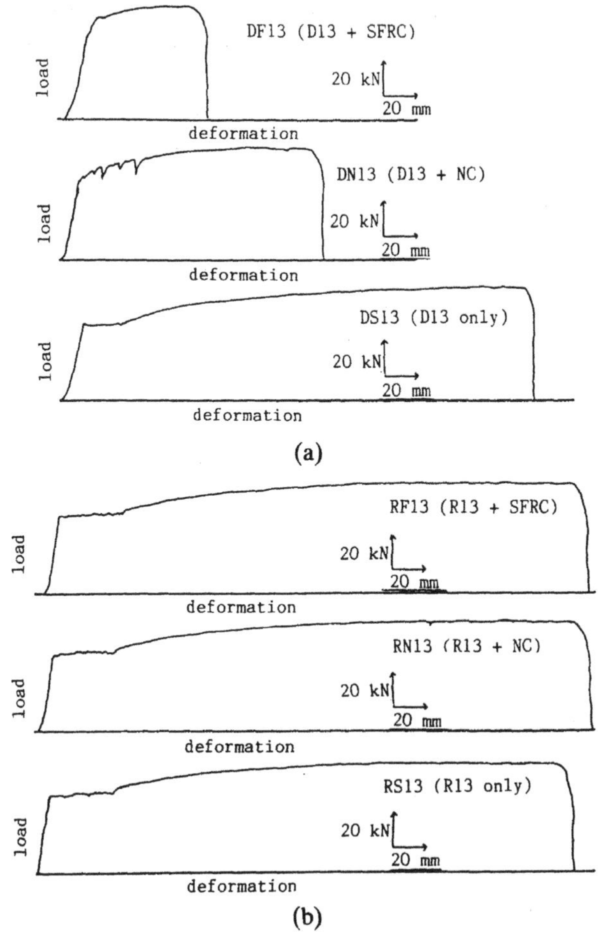

Fig. 13 Load versus machine displacement curves. (a) deformed bars (D13), (b) round bars (R13).

cover. The yield plateau was observed in all cases. The yield loads of the round bars with and without concrete were almost the same as shown in Table 5. In the case of the round rebars, the deformation was not restricted by the concrete cover. It can be concluded that the deformation process of round bars after yielding is not affected by concrete cover.

4. ROLE OF COMPRESSIVE FRACTURE ENERGY OF CONCRETE ON FAILURE OF RC BEAM SECTION

Firstly, our previous research work [3] on the relation between compressive fracture energy and final strain value in tension rebars is reviewed. Then the practical use of this relation and the implications of the real material behavior on the relation are discussed.

4.1 Theoretical Relation between Compressive Fracture Energy and Final Strain in Tension Rebars

The maximum value of strain in tension rebars at the final stage (yield ending point) in the failure process of RC beams, ε_{sr}, can be expressed as follows

$$\varepsilon_{sr} = S(\varepsilon'_{cr})/[pf_{sy} - p'f'_{sy} + N'/(bd)] - \varepsilon'_{cr} \tag{1}$$

where, p and p' are the tension and compression reinforcement ratios, f_{sy} and f'_{sy} are the yield strength of tension and compression rebars, b and d are the width and the effective depth of the RC beam cross-section, and N' is the axial force normal to the cross-section. ε'_{cr} is the strain corresponding to the point where the stress attains the value σ'_{cr} (given by eqn (2)) in the falling branch of the stress–strain curve of concrete as shown in Fig. 14(a). $S(\varepsilon'_{cr})$ is the compressive fracture energy up to the strain ε'_{cr} and is given by the area below the stress–strain curve up to ε'_{cr} as shown in Fig. 14(a). The process to derive eqns (1) and (2) is presented in Section 4.2 together with the explanation about the yield ending point

$$\sigma'_{cr} = pf_{sy} - p'f'_{sy} + N'/bd \tag{2}$$

Equation (1) is convenient for estimating the degree of plastic deformation of tension rebars and for discussing the effects of various

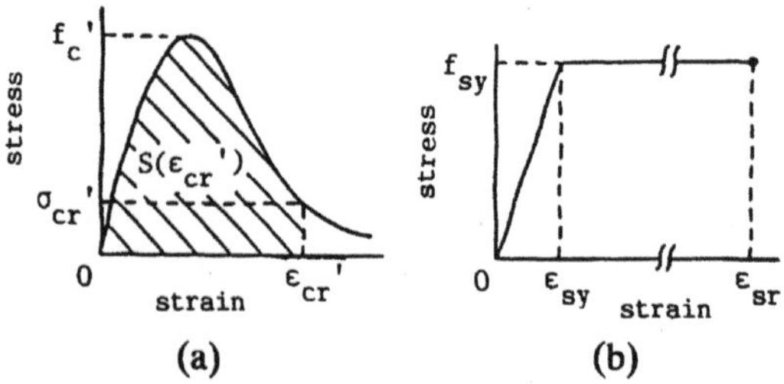

Fig. 14 Stress–strain models for concrete and steel. (a) concrete, (b) steel.

factors on the ductile behavior of RC beams. The yield limit (upper limit) reinforcement ratio and the rupture limit (lower limit) reinforcement ratio can be obtained from eqn (1) as described in Section 4.3. For higher ε_{sr} values, the RC beams show larger ductility and a slightly higher maximum load (load-carrying capacity). This increase in the maximum load is caused mainly by the effect of work hardening in tension rebars at large strain. RC beams with a higher ε_{sr} value are likely to break in the rebars at the final stage.

Equation (1) indicates that the value of ε_{sr} increases in the following cases:

—When the compressive fracture energy up to ε'_{cr}, $S(\varepsilon'_{cr})$, of concrete is large; for example, steel fiber reinforced concrete, confined concrete, high strength concrete.
—When the tension reinforcement ratio p or the yield strength of the tension rebars f_{sy} is small.
—When the compressive reinforcement ratio p' or the yield strength of the compression rebars is large. The effect of the compression rebars is limited if buckling occurs.
—When the axial force N' is small.

4.2 Yield Ending Point and Derivation of Equation (1)

Equation (1) can be derived by introducing the concept of the yield ending point. As shown in Fig. 15, the moment–curvature (M–ϕ) or load–displacement (P–δ) relationship of RC beams in bending can be generally characterized by the cracking point, the yield starting point of tension reinforcing bars, the strength failure (maximum load) point, the spalling initiation point, the yield ending point, the rupture point, and so on. At the yield ending point of tension reinforcing bars, the yielding stops due to crushing in concrete and/or buckling of compression bars and then the stress and strain in the tension reinforcing bars

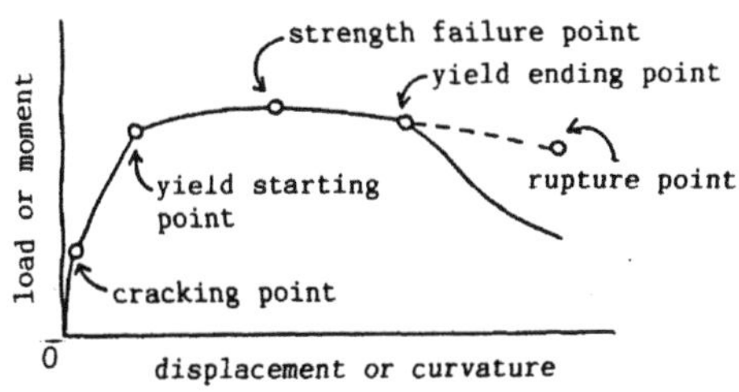

Fig. 15 Failure process of RC beams.

begin to decrease. After the yield ending point, the load of RC beams decreases markedly. Since the greater part of the absorbed energy of RC beams in bending is dissipated in tension reinforcing bars by yielding [3], the yield ending point as well as the yield starting point are considered to be the most suitable points for the evaluation of ductility of RC beams.

Equation (1) has been derived for RC beams with a rectangular cross-section [3]. The stress–strain curves of concrete and steel shown in Fig. 14 are used, where the curve for steel is idealized by two straight lines. When both the tension and compression bars have yielded, the internal tension force T and the compression force C normal to the cross-section are

$$T = pf_{sy}bd \tag{3}$$

$$C = S(\varepsilon_c')bx/\varepsilon_c' + p'f_{sy}'bd \tag{4}$$

where x is the depth of the compression zone, ε_c' is the strain in the extreme fiber of concrete, and $S(\varepsilon_c')$ is the compressive fracture energy, which is expressed as the area below the compressive stress–strain curve of concrete up to ε_c': x is given by

$$x = [\varepsilon_c'/(\varepsilon_s + \varepsilon_c')]d \tag{5}$$

where ε_s is the strain in the tension bars. Equation (4) can be rewritten as

$$C = S(\varepsilon_c')bd/(\varepsilon_s + \varepsilon_c') + p'f_{sy}'bd \tag{6}$$

Axial equilibrium provides

$$N' = C - T \tag{7}$$

and then

$$N' = S(\varepsilon_c')bd/(\varepsilon_s + \varepsilon_c') + p'f_{sy}'bd - pf_{sy}bd \tag{8}$$

Therefore

$$\varepsilon_s = S(\varepsilon_c')/[pf_{sy} - p'f_{sy}' + N'/(bd)] - \varepsilon_c' \tag{9}$$

At the yield ending point, ε_s stops increasing while ε_c' is still increasing

$$\mathrm{d}\varepsilon_s/\mathrm{d}\varepsilon_c' = 0 \tag{10}$$

This is the key point of the derivation process. From eqn (9):

$$[dS(\varepsilon_c')/d\varepsilon_c']/[pf_{sy} - p'f'_{sy} + N'/(bd)] - 1 = 0 \tag{11}$$

where $dS(\varepsilon_c')/d\varepsilon_c'$ is the concrete stress σ_c' corresponding to ε_c' on the descending part of the stress–strain curve. Therefore, the stress in the extreme fiber of concrete at yield ending point σ_{cr}' is given by

$$\sigma_{cr}' = pf_{sy} - p'f'_{sy} + N'/(bd) \tag{2}$$

The concrete strain at the yield ending point ε_{cr}' corresponds to the point where the stress reaches σ_{cr}' in the descending part of the stress–strain curve. Replacing ε_c' by ε_{cr}' in eqn (9), eqn (1) is obtained.

4.3 Upper and Lower Limits of Tension Reinforcement Ratio

There are two characteristic limits for the tension reinforcement ratio. One is the yield limit and the other is the rupture limit. The yield limit (upper limit) reinforcement ratio p_y can be considered as the ratio for which the yield starting point coincides with the yield ending point, and can hence be obtained by substituting the yield strain of steel ε_{sy} for the steel strain at the yield ending point ε_{sr} in eqn (1). Rearrangement of the equation gives

$$p_y = [S(\varepsilon_{cr}')/(\varepsilon_{sy} + \varepsilon_{cr}') + p'f'_{sy} - N'/(bd)]/f_{sy} \tag{12}$$

The value of p_y can be found by iteration using eqns (2) and (12), because ε_{cr}' in eqn (12) is a function of p through eqn (2). When $p' = 0$ and $N' = 0$, it is simplified as

$$p_y = S(\varepsilon_{cr}')/[(\varepsilon_{sy} + \varepsilon_{cr}')f_{sy}] \tag{13}$$

The conventional balanced failure reinforcement ratio p_b is generally defined as the ratio for which the tensile steel reaches the yield strength f_{sy} and the extreme fiber of concrete attains the ultimate strain ε_{cu}' of 0·0035 (or 0·003), simultaneously. Therefore, for $p' = 0$ and $N' = 0$, p_b is

$$p_b = \sigma_c''\varepsilon_{cu}'/[(\varepsilon_{sy} + \varepsilon_{cu}')f_{sy}] \tag{14}$$

where, σ_c'' is the average concrete stress in the compression zone. In the Japanese concrete code [5], σ_c'' is $0{\cdot}68f_{cd}'$ and ε_{cu}' is 0·0035 where f_{cd}' is the design value of f_c'. $\sigma_c''\varepsilon_{cu}'$ is the compressive fracture energy below the stress–strain relation up to ε_{cu}''. Though the fixed value of ε_{cu}' is used in the calculation of p_b, ε_{cr}' for p_y varies with σ_{cr}' $(= pf_{sy} - p'f'_{sy} + N'/(bd))$ and the shape of the stress–strain curve.

The rupture limit (lower limit) reinforcement ratio p_f can be considered as the ratio for which the rupture point coincides with the yield ending point. The following equation for p_f is obtained by substituting the steel rupture strain ε_{sf} for ε_{sr} in eqn (1)

$$p_f = [S(\varepsilon'_{cr})/(\varepsilon_{sf} + \varepsilon'_{cr}) + p'f'_{sy} - N'/(bd)]/f_{sy} \tag{15}$$

When $p' = 0$ and $N' = 0$, it is simplified as

$$p_f = S(\varepsilon'_{cr})/[(\varepsilon_{sf} + \varepsilon'_{cr})f_{sy}] \tag{16}$$

It can be seen from eqns (12) and (15) that p_y and p_f increase with the increase of fracture energy $S(\varepsilon'_{cr})$ and the compression reinforcement p', and decrease with the increase of axial force N'. The energy absorption capacity of a RC beam section and the ductility factor are discussed in our report [3] using eqn (1).

4.4 Simplification of Compressive Fracture Energy

The stress of the extreme fiber in concrete at yield ending point, σ'_{cr}, is smaller than half of the compressive strength f'_c in general cases where the tension reinforcement ratio p is less than 2·5% and f'_c is greater than 20 MPa; for example, in the case of $f_{sy} = 400$ MPa, $p = 2{\cdot}5\%$, $p' = 0$ and $N' = 0$, σ'_{cr} becomes 10 MPa. Generally, the strain in concrete ε'_{cr} is sufficiently smaller than the strain in steel ε_{sr} at yield ending point. Therefore, it is practical to assume both $S(\varepsilon'_{cr})$ and ε'_{cr} to be material constants (S'_c and ε'_u, respectively) for general cases with moderate reinforcement ratios. The compressive fracture energy S'_c should be evaluated from measured compressive stress–strain curves of concrete by appropriate simplification; for example, the area under the curve up to the point $f'_c/3$ on the descending branch.

For design purposes the equivalent rectangular stress block as sketched in Fig. 16(c) has been used instead of the real stress

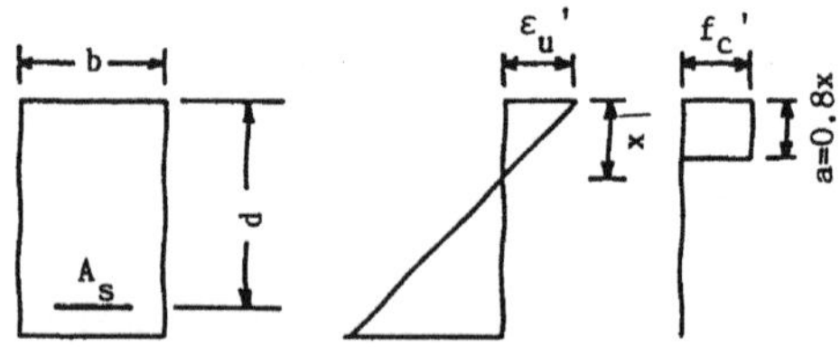

Fig. 16 Cross-section of RC beam and equivalent stress block. (a) cross-section, (b) strain distribution, (c) equivalent stress block.

distribution. In the Japanese code for concrete structures [5], the stress block has a mean stress of $0{\cdot}85f'_c$ and a depth of $0{\cdot}8x$, where x is the neutral axis depth. When the depth of the block is taken to be $0{\cdot}8x$, the ultimate strain ε'_u is given by eqn (17), for example, keeping the area of the block equal to that of the real stress distribution

$$\varepsilon'_u = S'_c/(0{\cdot}8f'_c) \quad (17)$$

The value of ε'_u defined by eqn (17) is suitable, for example, for calculating the ultimate bending strength of RC beams made with high toughness concrete.

4.5 Implications of Real Behavior of Concrete and Rebars Covered by Concrete

The test results reported in Section 3 indicate that the behavior of deformed rebars in concrete is different from that of naked bars as listed below:

—The yield load of specimens of deformed bars with concrete is higher than that of naked bars. This is remarkable for SFRC.
—The rupture strain decreases markedly.
—There is no yield plateau.

In the calculation of yield limit reinforcement ratio p_y as well as the yield moment of the RC cross-section, a higher value of f_{sy} should be taken depending on the combination of concrete and steel and their properties. In the calculation of the rupture limit reinforcement ratio p_f, a smaller value of the rupture strain of steel, ε_{sf}, should be used. For example, in the case of D13 rebars with normal concrete, ε_{sf} was 2/3 of that for naked bars. The yield strength f_{sy} in the calculation of p_f should be replaced by the ultimate strength f_{su}, which is the stress at peak.

According to the results in Section 2, the compressive fracture energy to break a unit area of concrete cross-section seems to be constant. If this is true, the difference in strain between ascending and descending branches of the compressive stress–strain curve should be changed in inverse proportion to the size of the fractured cross-section. Moreover, if the size of the fracture zone of RC beams in the direction of the axis is proportional to the depth of the compression zone x, then the compressive fracture energy under the stress–strain curve $S(\varepsilon'_{cr})$ and the concrete strain ε'_{cr} should be changed in inverse

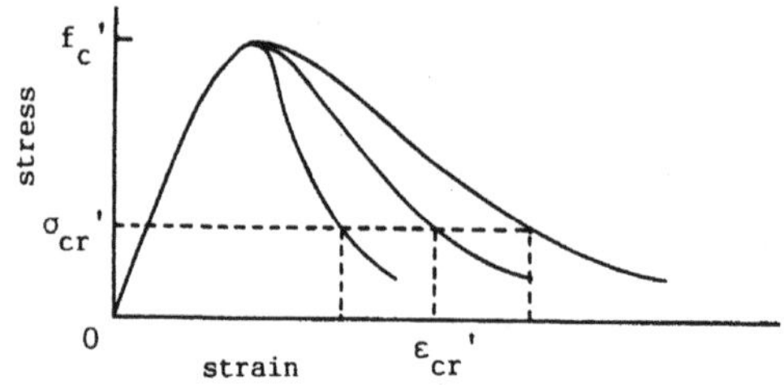

Fig. 17 Compressive stress–strain curve variation with size of fracture zone.

proportion to the depth of the compression zone x.

$$\varepsilon'_{cr} \propto 1/x \tag{18}$$

$$S(\varepsilon'_{cr}) \propto 1/x \tag{19}$$

As shown in Fig. 17, ε'_{cr} is proportional to $S(\varepsilon'_{cr})$. From eqn (5), when ε_{sr} ($= \varepsilon_{sf}$ at rupture point) is sufficiently large compared with ε'_{cr}, x at the yield ending point is given as

$$x = [\varepsilon'_{cr}/(\varepsilon_{sr} + \varepsilon'_{cr})]d \simeq [\varepsilon'_{cr}/\varepsilon_{sr}]d \tag{20}$$

From eqns (18), (19) and (20), the following relations are obtained

$$\varepsilon'_{cr} \propto 1/\sqrt{d} \tag{21}$$

$$S(\varepsilon'_{cr}) \propto 1/\sqrt{d} \tag{22}$$

Since the rupture limit reinforcement ratio p_f is proportional mainly to $S(\varepsilon'_{cr})$, the value of p_f is expected to decrease in inverse proportion to the square root of the effective depth of the RC beam cross-section. Further research work is needed especially to determine the value of S'_c for RC beams and to examine the failure behavior of RC beams with low reinforcement ratios.

5. CONCLUSIONS

It is pointed out that the energy approach and the softening–localization concept of Fracture Mechanics of Concrete are useful for the research and design of RC members.

Compression tests on concrete specimens differing in specimen height ($h/w = 1$–6) were performed. For mortar and SFRC, in addition to normal concrete, the energy up to peak load was nearly proportional to specimen height. However, the compressive fracture energy to break a unit area of the specimen cross-section was almost constant. It is proposed to change the shape of the stress–strain curve and the compressive fracture energy under the curve in inverse proportion to the size of the fracture zone.

The behavior of deformed rebars with concrete cover was different from that of naked bars. The specimens of deformed rebars with concrete cover had a higher yield load and shorter deformation up to rupture compared with the naked bars. The yield plateau was not observed for the deformed bars with concrete cover. These differences were more remarkable for SFRC than for normal concrete. On the other hand, the behavior of round rebars with concrete cover coincided with that of the naked bars. There was no effect of concrete cover for round bars even in the case of SFRC.

The theoretical relation between the compressive fracture energy and the strain in tension rebars is presented. The concept of the *yield ending point* is introduced to obtain this relation. Two characteristic limits for the tension reinforcement ratio are derived, and it is shown that the yield limit (upper limit) reinforcement ratio and the rupture limit (lower limit) reinforcement ratio increase with the increase of fracture energy of concrete and the compression reinforcement ratio, and decrease with the increase of the axial force. A procedure to simplify the treatment of fracture energy is presented for practical purposes. It is emphasized that the real behavior of concrete and rebars in concrete should be used in the calculation. Because of the size-dependent nature of the compressive fracture energy, the rupture reinforcement ratio is expected to decrease as the effective depth of RC beams increases.

REFERENCES

[1] Hillerborg, A., Fracture mechanics concepts applied to moment capacity and rotational capacity of reinforced concrete beams. *Engineering Fracture Mechanics*, **35** (1/2/3) (1990) 233–40.

[2] Hillerborg, A., The compression stress–strain curves for design of reinforced concrete beams. In *Fracture Mechanics: Application to Concrete, ACI SP-118*, ed. V. C. Li & Z. P. Bazant. American Concrete Institute, 1989, pp. 281–94.

[3] Rokugo, K., Iwase, H. & Koyanagi, W., Application of fracture energy of concrete in compression to section design of reinforced concrete members. *Fracture Toughness and Fracture Energy of Concrete,* ed. F. H. Wittmann. Elsevier Science Publishers, Amsterdam, 1986, pp. 587–596.

[4] ACI: *Building code requirements for reinforced concrete.* American Concrete Institute, 1989.

[5] JSCE: *Standard specification for design and construction of concrete structures, Part 1.* Japan Society of Civil Engineers, 1986.

[6] Koyanagi, W., Rokugo, K. & Uchida, U., Compressive toughness of concrete. *Review of the 42nd General Meeting.* Cement Association of Japan, Tokyo, 1983, pp. 268–70, (in Japanese).

[7] Koyanagi, W., Rokugo, K. & Iwase, H., The stress–strain relationship of reinforcing bars in concrete and mechanical behavior of RC beams in flexure. *Proc. JSCE,* **5** (372) (1987) 83–92 (in Japanese), *Concrete Library of JSCE,* (12) (1989) 53–65 (English translation).

Chapter 18

Shear Crack Stability along a Precast Reinforced Concrete Joint

T. P. TASSIOS & E. VINTZĒLEOU
National Technical University of Athens, GR-106 82 Athens, Greece

ABSTRACT

In this chapter, a model previously developed by the authors is used for the prediction of the behaviour of sheared reinforced concrete interfaces. The model consists of adding the contribution of each separate mechanism of load transfer, taking into account their interaction.

The validity of the method is proved by comparing the shear stress, τ, versus shear slip, s, diagrams produced by applying the model with experimental τ–s curves. Subsequently, the model was applied to calculate the total energy produced up to the failure of RC interfaces, as well as the critical slip value corresponding to failure. In the examples reported in this chapter, the influence of several parameters (concrete compressive strength, percentage of reinforcement, embedment length of bars) on the total energy and critical slip was also investigated.

1. INTRODUCTON

The propagation of shear cracks in RC interfaces seems to be less studied than that of cracks due to axial actions. Non-conventional meso-level models of force transfer are needed for this purpose. Their development, calibration and use may be considered as a new trend, since several problems in modern design necessitate a more rational handling of shear cracks. This is, among others, the case for shear transfer along tensional cracks newly developed across a monolithic RC element; since such cracks appear rather early in the loading history of an element, their posterior shear response is of paramount importance. Similarly, the shear displacement along a joint of precast RC elements, needs to be predicted step-by-step by means of rational models other than plasticity models suitable only for ULS considerations.

2. SHORT DESCRIPTION OF THE MODEL OF INTERFACE SHEAR TRANSFER

2.1 Shear transfer mechanisms

As is well known, when a reinforced concrete (precast or precracked) interface is subject to shear, a slip, s, is taking place along the interface, accompanied by lateral dilatancy, Δw, due to the roughness of the interface (see Fig. 1), under zero external normal force. Thus, two mechanisms of load transfer are mobilised, namely:

- dowel action of the reinforcement crossing the sheared interface; and
- friction, generated by the internal compressive stress acting on the interface; such stresses result from the tensile stresses induced in the reinforcing bars when lateral dilatancy is manifested (Fig. 1).

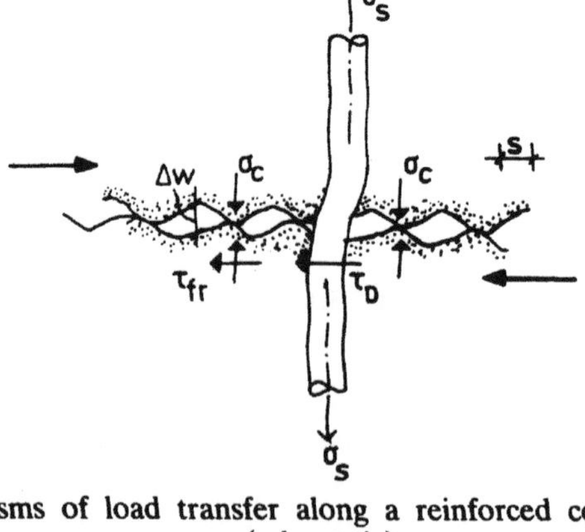

Fig. 1 Mechanisms of load transfer along a reinforced concrete interface (schematic).

Therefore, assuming that analytical models are available for each individual mechanism entering the game, one can predict the shear behaviour of a RC interface as follows (see Fig. 2):

(a) For a given shear slip-value, the lateral dilatancy, Δw, can be read on the s–Δw curve of Fig. 2(a).
(b) Knowing the anchoring characteristics of reinforcing bars, as well as their embedment length into the concrete, one can

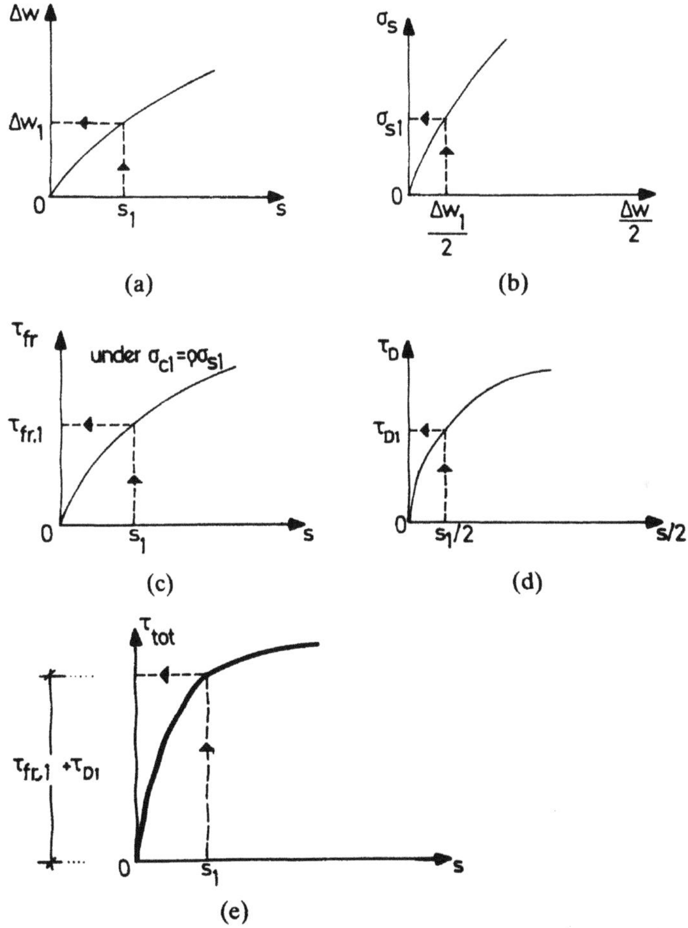

Fig. 2 Generation of the total resistance τ_{tot} versus shear displacement curve (e), on the basis of the models for individual mechanisms (curves (a)–(d)).

determine the tensile stress, σ_s, developed in the reinforcement for a pull-out slip equal to half the lateral dilatancy ($\Delta w/2$).

(c) In the case of reinforcement being practically uniformly distributed across the interface, the tensile stress of the reinforcement can be transformed into compressive stress onto the concrete ($\sigma_c = \rho\sigma_s$, where ρ is the percentage of reinforcement).

Obviously, in the case where there is an initial tensile stress in the reinforcement (due for example to a tension force acting on the interface), this initial tensile stress should not be taken into account when calculating σ_c.

This compressive stress of concrete allows for the development of a frictional resistance, τ_{fr}, to be estimated on the basis of τ_{fr}–s curves of Fig. 2(c).

Where there is a compressive stress acting on the interface due to an external load or to prestressing, this additional compressive stress has to be taken into account when estimating the frictional resistance of the interface.

(d) On the other hand, the shear slip mobilises the dowel resistance of reinforcing bars (Fig. 2(d)). This shear resistance, τ_D, should be added to the frictional one (Fig. 2(e)) taking, however, into account the possible unfavourable interaction of the two mechanisms.

2.2 Friction and dowel action constitutive laws

Some years ago, an extensive experimental investigation of load transfer mechanisms was carried out at the National Technical

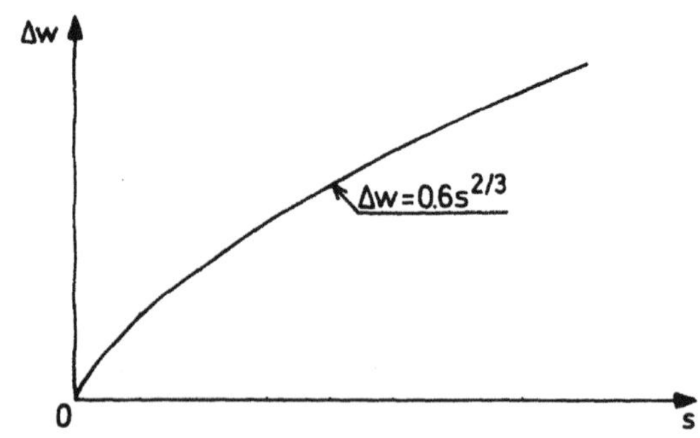

Fig. 3 Lateral dilatancy versus shear displacement curve for 'natural' (rough) cracks.

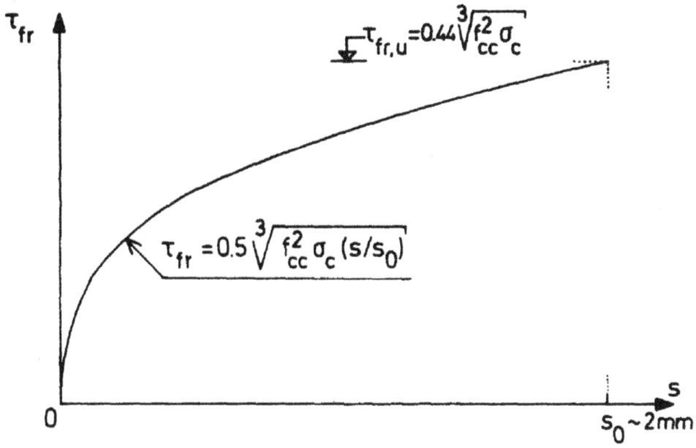

Fig. 4 Frictional resistance of natural cracks.

University of Athens, with the aim of calibrating physical analytical models developed within the project.

The results of this project [1] are presented in Figs 3–5 and they are briefly discussed below.

(a) In the case of 'natural' cracks (rough interface) in concretes with a compressive strength not higher than 56 MPa, tests have shown [2], [3], that the shear slip versus lateral dilatancy relationship is independent of parameters such as maximum aggregate size, normal compressive stress on the interface,

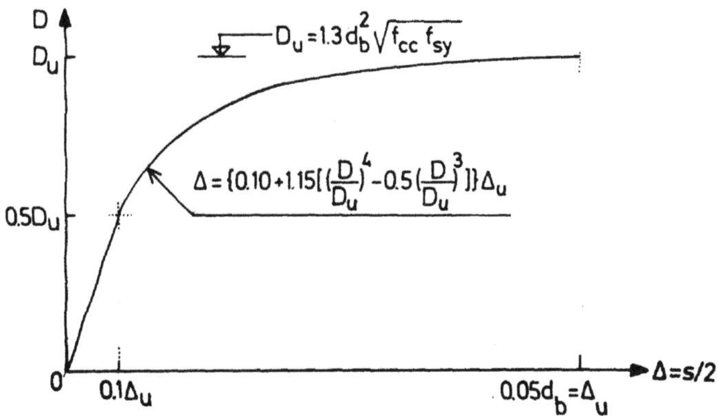

Fig. 5 Dowel force versus shear displacement curve for bars with large concrete cover.

concrete compressive strength, etc. It seems that the following formula can be applied for such rough interfaces (Fig. 3)

$$\Delta w = 0{\cdot}6s^{2/3} \not> 1{\cdot}4\ \text{mm} \quad (\Delta w,\ s \text{ in mm}) \tag{1}$$

(b) The frictional resistance of a rough interface under a constant normal stress, σ_c, can be estimated by means of the following expression

$$\tau_{fr,u} = 0{\cdot}44\sqrt[3]{f_{cc}^2\sigma_c} \tag{2}$$

while the frictional stress versus shear displacement curve is expressed by the following equation (see also Fig. 4):

$$\tau_{fr} = 0{\cdot}50\sqrt[3]{f_{cc}^2\sigma_c\left(\frac{s}{s_0}\right)} \qquad s_0 \cong 2\ \text{mm} \tag{3}$$

(c) The dowel resistance versus shear displacement curve is given in Fig. 5. The physical model of this figure, which was checked and calibrated against all available experimental data, is valid for the case of bars having large concrete cover around them, so that splitting of the concrete is avoided.

(d) It should be noted that for the above case, the presence of axial stresses in the reinforcing bars, simultaneously with shear, leads to a decrease of the dowel resistance, according to the following expression [4]

$$D_u' = 1{\cdot}3d_b^2\sqrt{f_{cc}f_{sy}}(1-\alpha^2) \tag{4}$$

where d_b denotes the bar diameter, f_{cc} and f_{sy} are the concrete compressive strength and the yield stress of steel, respectively, and α $(=\sigma_s/f_{sy})$ denotes the ratio between the axial stress acting on the reinforcing bar and the yield stress of steel.

2.3 Pull-out force-displacement relationship

The mechanism of pull-out was not investigated within the above mentioned programme. Although international literature is rich in experimental results on pull-out, it seems that better physical modelling of the mechanism has still to be developed. However, for the needs of the present work, a rather rough pull-out model was used, as described below.

Let us assume that a reinforcing bar (diameter d_b, yield stress f_{sy}) is embedded in concrete having a tensile strength equal to f_{ct}. The

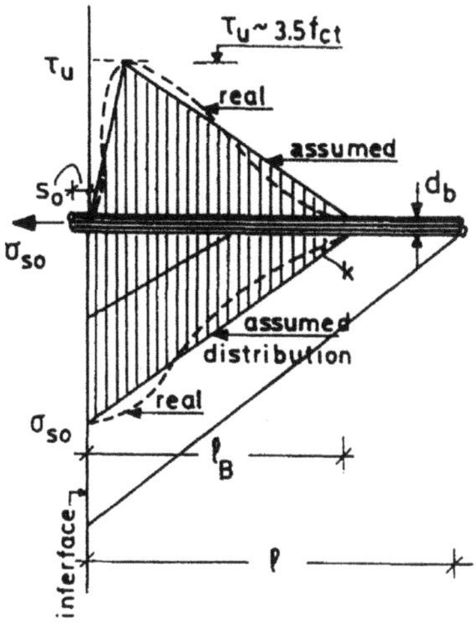

Fig. 6 Steel stresses and bond stresses along a reinforcing bar with insufficient embedment length ($\max\sigma_{s0} < f_{sy}$).

assumed distributions of both tensile stresses and bond stresses along the bar are shown in Fig. 6.

On the basis of the simplification shown in Fig. 6, the σ_{s0}–s_0 curve (s, being the slip between steel and concrete at the interface) can be produced, as follows

$$\sigma_{s0} = 7\frac{l_B}{d_b} f_{ct} \not> f_{sy} \qquad \text{(equilibrium, } l_B \leq l\text{)} \tag{5}$$

$$s_0 = 3{\cdot}5\frac{f_{ct}}{E_s}\frac{l_B}{d_b} = \frac{\Delta w}{2} \qquad \text{(total deformation)} \tag{6}$$

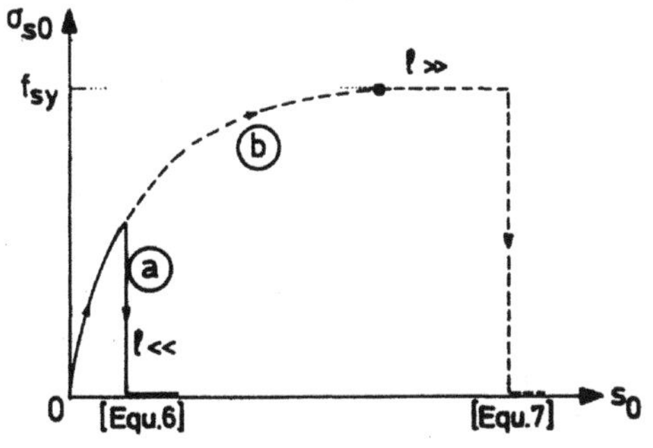

Fig. 7 Axial stress versus pull-out displacement for reinforcing bars with insufficient (a) or sufficient (b) embedment length, respectively.

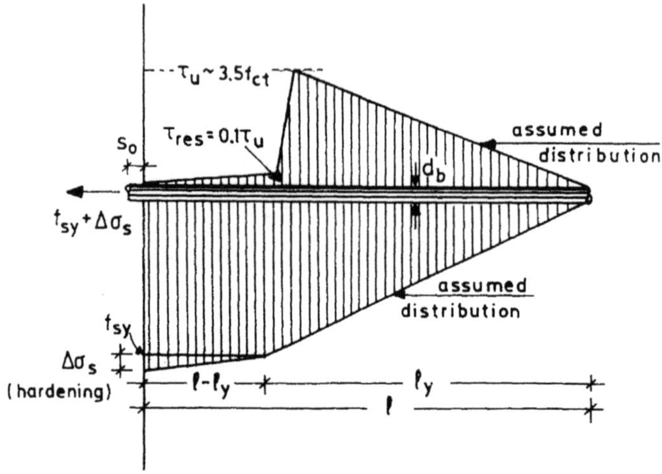

Fig. 8 Steel and bond stresses along a reinforcing bar with embedment length larger than that required for anchoring of the yield force of the bar ($\Delta\sigma_s = 2$ MPa, $\varepsilon_{su} \cong 60‰$). (Thus hardening is practically not taken into account).

For insufficient anchorage lengths, i.e. if l is smaller than required for σ_{s0} to become equal to f_{sy}, the bar will be pulled out for $\sigma_{s0} < f_{sy}$ (see curve 'a' in Fig. 7).

For l-values large enough to allow for $\sigma_{s0} = f_{sy}$, pull-out can still happen for an s_0-value given by the following expression (see also Fig. 8, and curve 'b' in Fig. 7)

$$s_0 = \left(l - \frac{1}{7}\frac{f_{sy}}{f_{ct}} d_b\right)\left[\frac{f_{sy}}{E_s} + 0{\cdot}029\right] + \frac{1}{14}\frac{f_{sy}}{E_s}\frac{f_{sy}}{f_{ct}} d_b \tag{7}$$

provided that the above s_0-value is smaller than that corresponding to the steel strain at failure (ε_{su}).

3. CHECK OF THE MODEL VERSUS EXPERIMENTAL DATA

The procedure described in the previous section, as well as the models for the individual mechanisms, were used to predict the behaviour of sheared reinforced interfaces which were tested by Mattock [5] and Walraven *et al.* [3]. The results of the comparison between experimental and predicted curves are shown in Figs 9 and 10; they seem to be rather satisfactory.

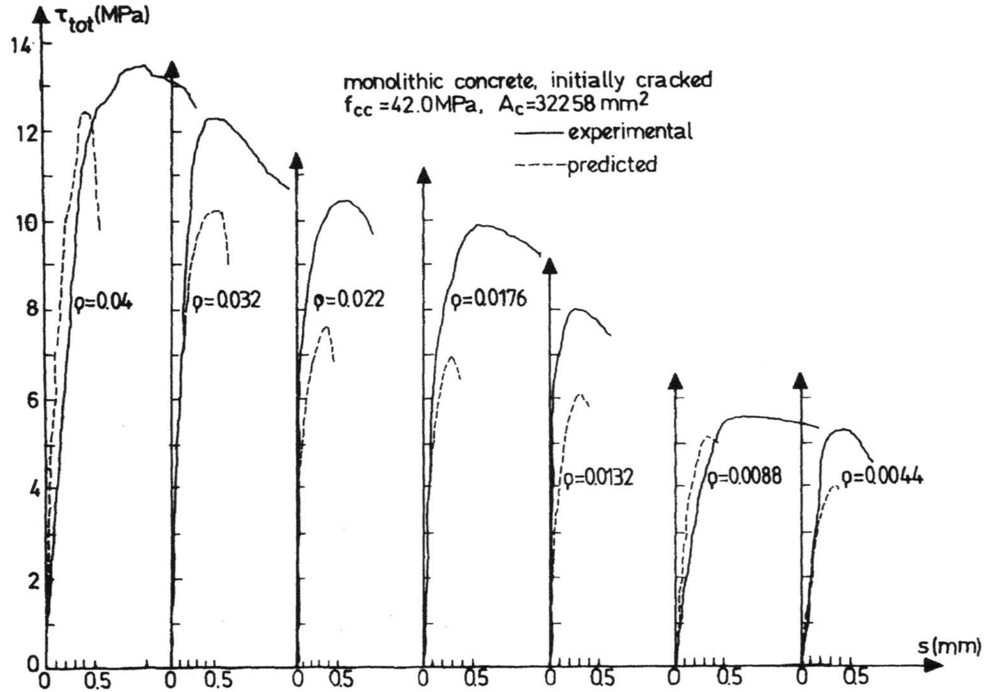

Fig. 9 Comparison between experimental curves 'shear resistance versus shear displacement' (Mattock [5]) and curves predicted by the analytical method described in this chapter.

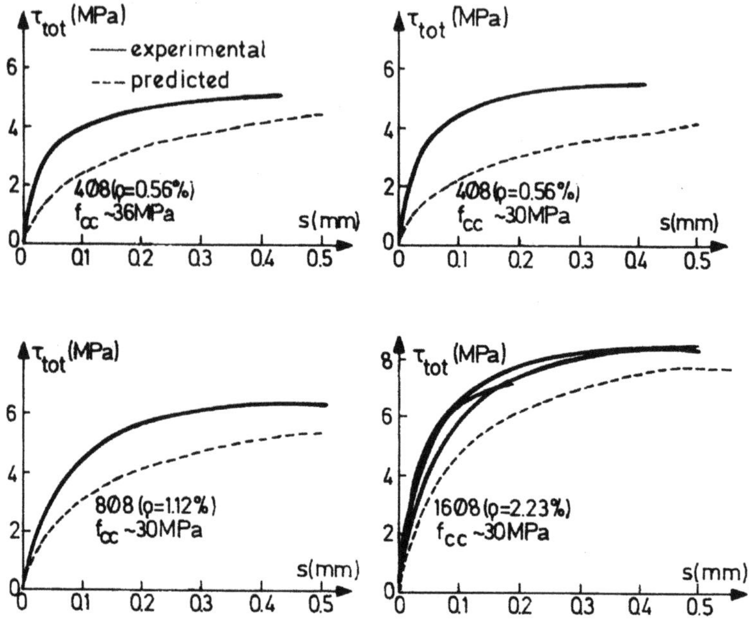

Fig. 10 Comparison between experimental curves 'shear resistance versus shear displacement' (Walraven *et al.* [3]) and curves predicted by the analytical method described in this chapter.

It should be noted that both Mattock and Walraven *et al.*, in their experimental results indicate values of initial crack widths produced during pre-cracking of specimens before any application of shear loads.

The influence even of small w_0-values on the final τ versus s curve is rather important, while the reliability of measuring crack width values as small as 0·01–0·05 mm might be relatively low. This fact may explain a certain deviation of the predicted τ–s curves from the experimental ones.

However, given the scatter inherent in the phenomena related to the behaviour of sheared RC interfaces, it is believed that the procedure described and the models of the related mechanisms proved to be rather satisfactory for the purposes of this chapter. Their somehow more conservative results are compatible with the design-oriented nature of this investigation.

4. 'AXIAL' AND 'SHEAR' ENERGY PRODUCED BY RC INTERFACES

The models described in Section 2, as checked in Section 3, were used to calculate: (i) the energy produced by a sheared RC interface up to its failure; and (ii) the shear slip corresponding to failure (s_{crit}).

4.1 Shear stress-slip relationships for various concrete strengths and rebar diameters

To this purpose, the following calculations were made for the examples of Fig. 11:

(a) The σ_s versus $w/2$ curves were produced (Fig. 12) for four concrete categories.

(b) Using the curves of Fig. 12 and the models of Figs 3–5, the total mobilised shear stress, τ_{tot}, was plotted (Fig. 13) against shear slip for each concrete quality and for four percentages of reinforcement.

(c) For each percentage of steel and for each concrete strength, the total fracture energy of the sheared interface was calculated; both shear displacement and dilatancy energy were taken into account.

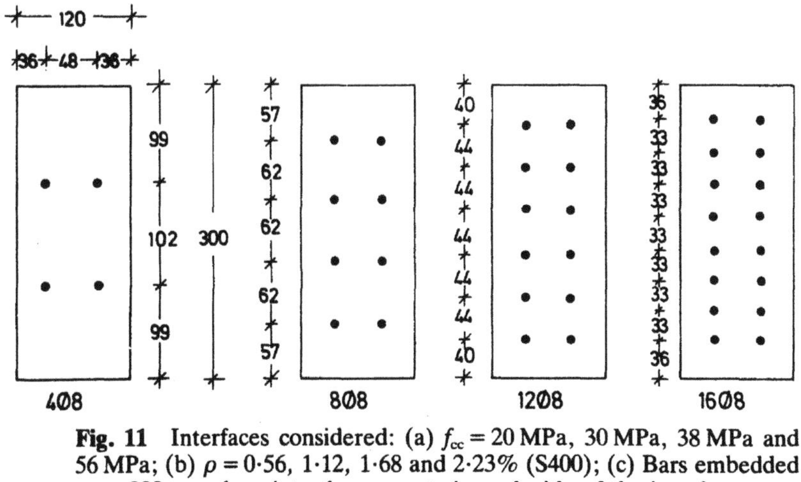

Fig. 11 Interfaces considered: (a) $f_{cc} = 20$ MPa, 30 MPa, 38 MPa and 56 MPa; (b) $\rho = 0{\cdot}56$, 1·12, 1·68 and 2·23% (S400); (c) Bars embedded 200-mm deep into the concrete in each side of the interface.

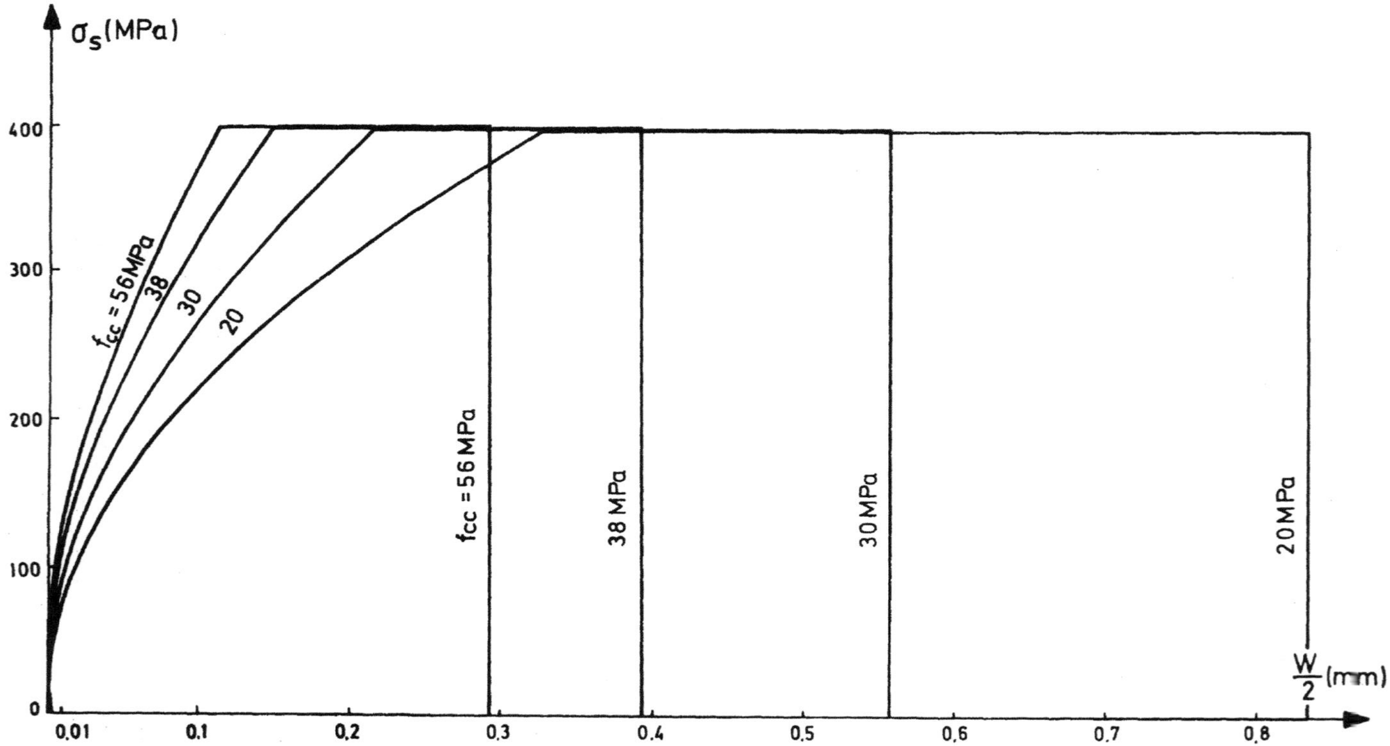

Fig. 12 Steel stress versus lateral dilatancy for an 8-mm diameter steel bar S400, embedded 200-mm deep into concrete with different compressive strengths.

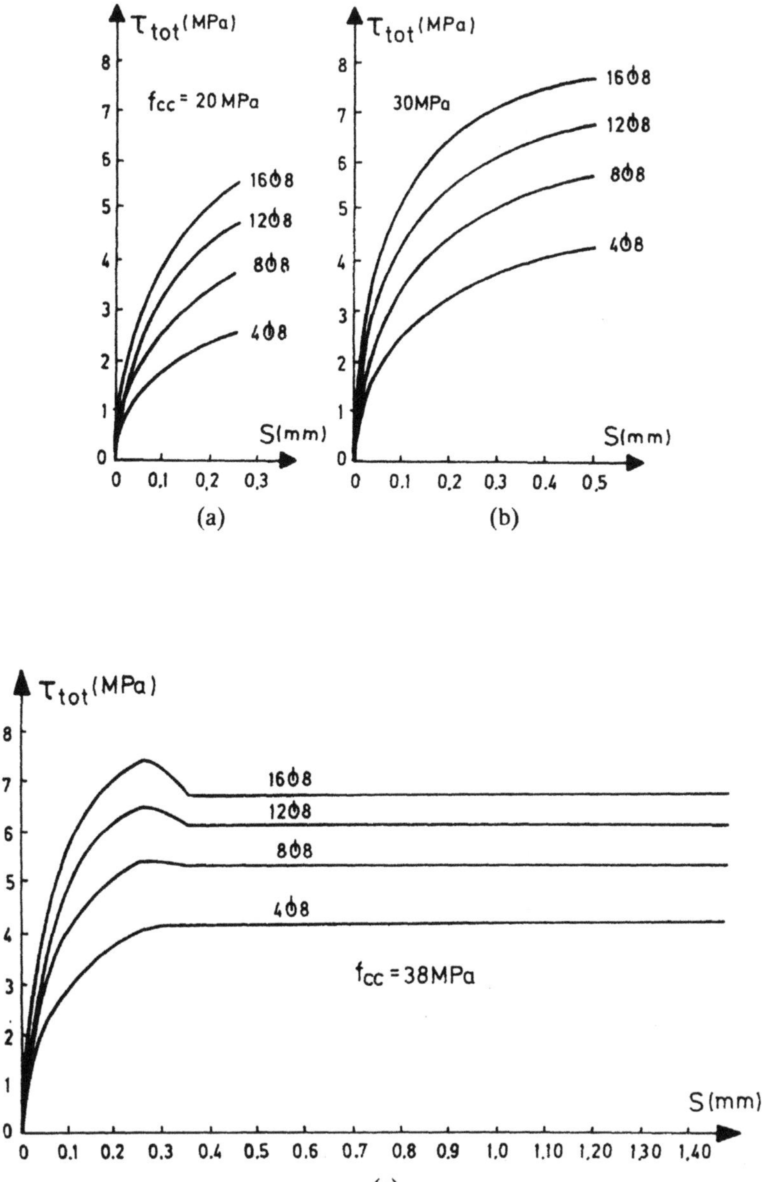

Fig. 13 Shear resistance versus shear displacement curves for the interfaces of Fig. 11.

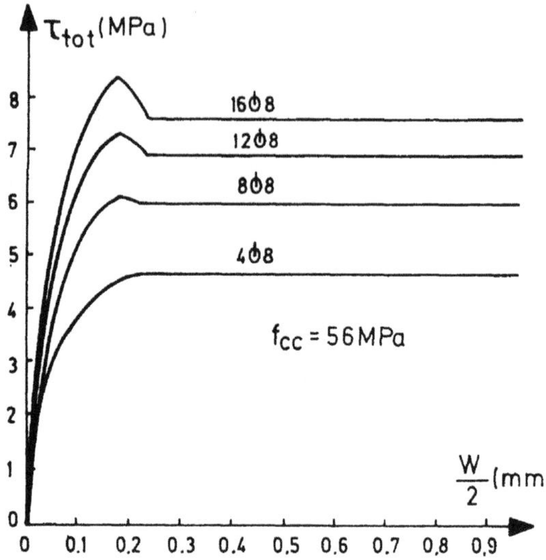

(d)

Fig. 13—(*cont'd*)

In Fig. 14a, the total energy-values are plotted against ρ for the four concrete grades, but only for an embedment length equal to 200 mm. One may observe that for a given embedment length, the fracture energy is linearly increasing with the percentage of steel (the bar diameter was, however, the same in all cases). As shown in Fig. 12, for increasing concrete strength, the crack width (and the shear slip) required for the failure of the pull-out mechanism is decreasing. It is therefore expected that for increasing concrete strength the total fracture energy will be decreasing. However, in this example, for $f_{cc} = 20$ MPa and 30 MPa, the available embedment length of bars is not sufficient for σ_{s0} to become equal to f_{sy}, the interfaces fail at relatively low s-values and the total energy for those two qualities of concrete are lower than for $f_{cc} = 38$ MPa.

In Fig. 14b, one may observe that the shear slip required for bars to be pulled-out (s_{crit}) is decreasing with increasing concrete compressive strength.

4.2 The role of the embedment length

The investigation of the role of the *embedment length* was carried out by means of another example (Fig. 15).

In this case too, τ_{tot} versus s curves were produced using the

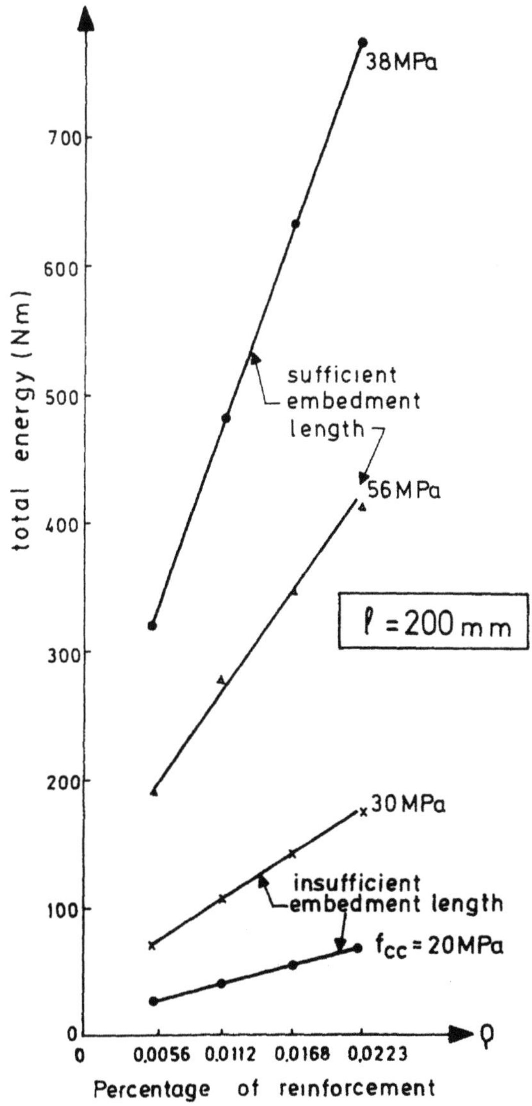

Fig. 14a Total fracture energy as a function of the percentage of reinforcement for various concrete qualities (for constant embedment length 200 mm).

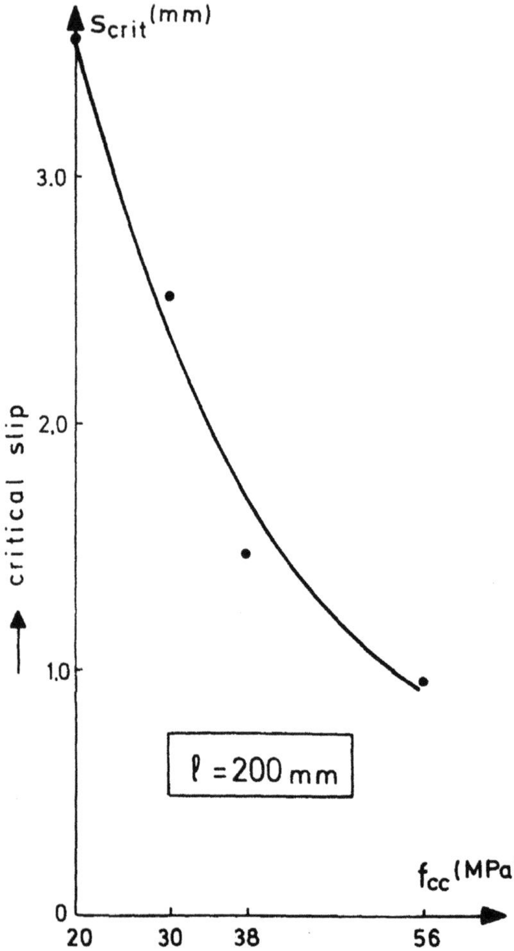

Fig. 14b Critical slip values as a function of concrete compressive strength (for constant embedment length 200 mm).

σ_s–$\Delta w/2$ curves of Fig. 12 for $f_{cc} = 20$ MPa and 30 MPa. The models of Figs 3–5 were also used here. The output of the application of the procedure described in Section 2 is shown in Fig. 17.

As in the case of the previous example, the required energy-values are plotted against the embedment length for two concrete strengths (Fig. 18). In the same figure, the critical shear slip values (shear slip

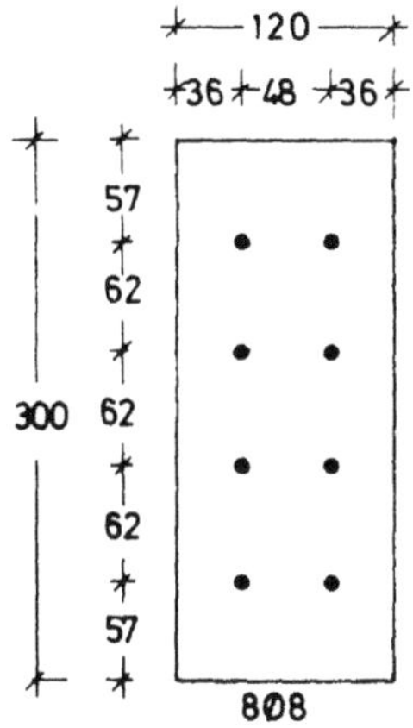

Fig. 15 Interfaces considered: (a) $f_{cc} = 20$ MPa, 30 MPa; (b) $\rho = 1{\cdot}12\%$ (S400); (c) embedment lengths of bars: $6d_b$–$40d_b$, approximately.

required to cause complete pull-out of the reinforcing bars and, thus, failure of the interface) are plotted against embedment length.

One may observe that the energy required for the failure of the interface increases very rapidly with increasing embedment length. This is especially true for embedment lengths larger than those

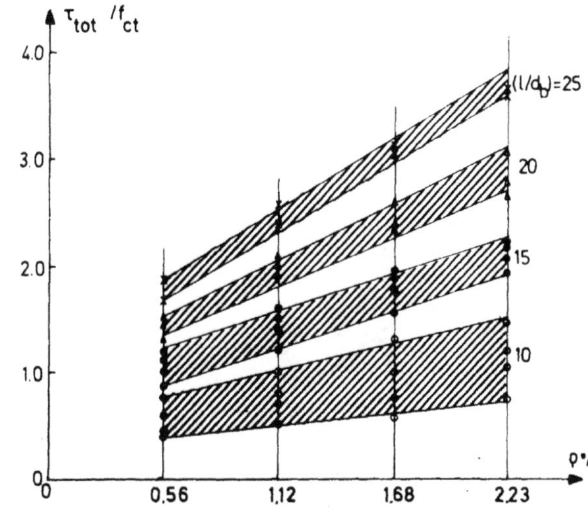

Fig. 16 Shear resistance of interfaces normalised to the tensile strength of concrete, as a function of the percentage of steel crossing the interface, for various embedment lengths of bars (S400).

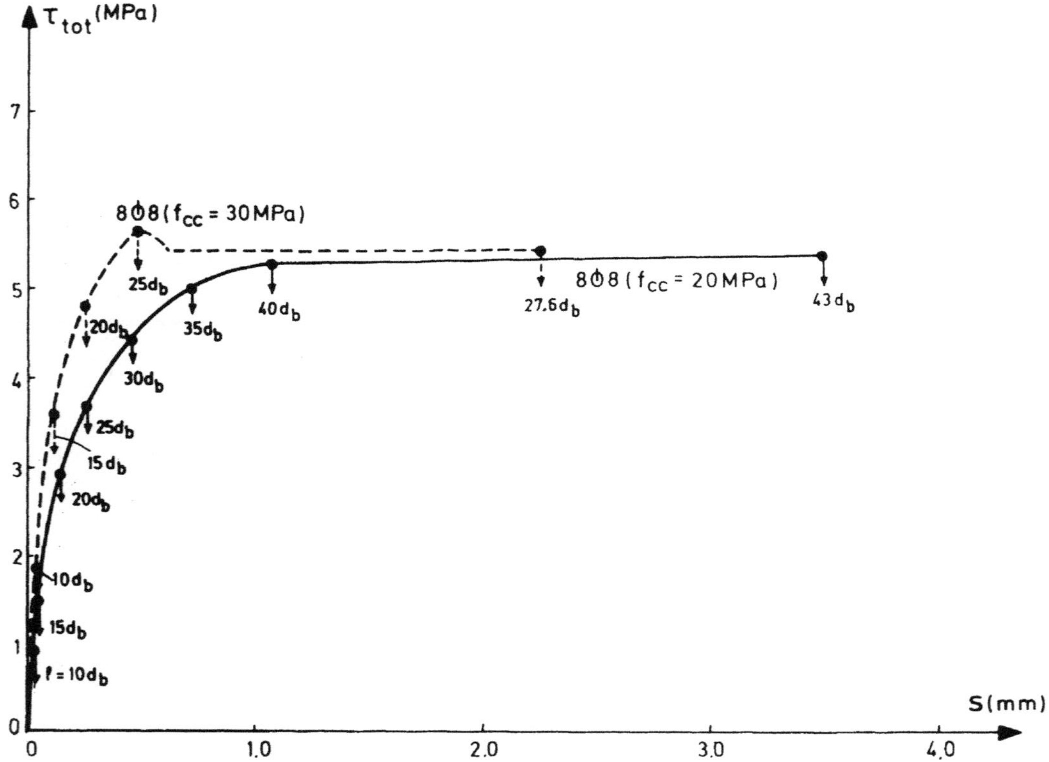

Fig. 17 Shear resistance of interfaces reinforced by means of eight bars each 8 mm in diameter, for two concrete qualities and several embedment lengths.

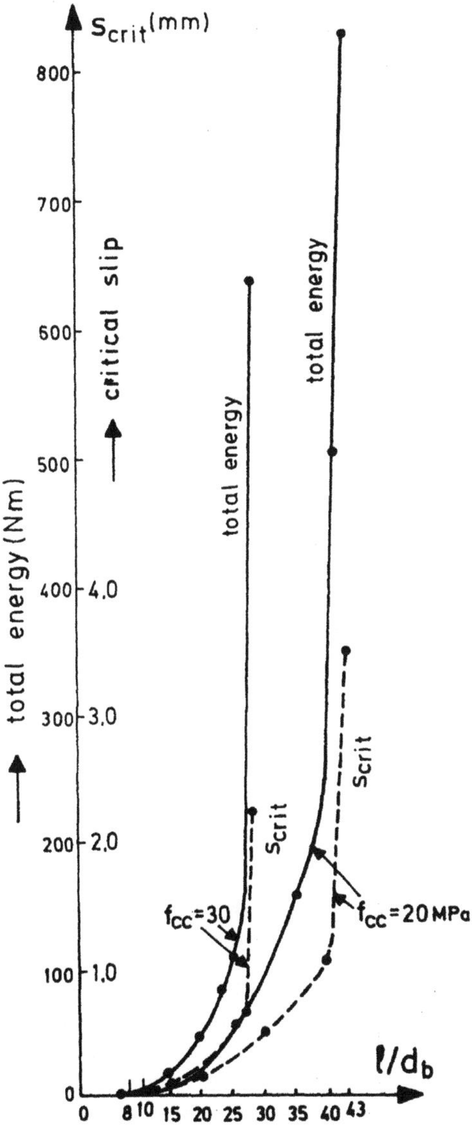

Fig. 18 Total fracture energy and critical slip of interfaces reinforced by means of bars 8 mm in diameter (8Φ8) for different embedment lengths of bars and two concrete qualities.

required for σ_s to become equal to f_{sy}. In fact, for sufficiently large l-values, steel stresses σ_s (Fig. 12) remain practically equal to f_{sy} for increasing shear slip values. Similarly, (Fig. 17), for sufficiently long l-values, total shear stresses τ_{tot} remain practically constant near their maximum value as shear slips increase. This stability of maximum σ_s and τ_{tot} values for increasing slip values explain the considerable increase of total energy under the above mentioned conditions.

The same holds true where the critical slip at failure (s_{crit}) is concerned (Fig. 18).

5. CONCLUSIONS

(a) The proposed analytical model, accounting for all shear transfer mechanisms (such as dowel forces and frictional resistance mobilised by the clamping action of transverse reinforcement) seems to be a versatile tool for the prediction of the stability of Mode II cracks in RC precast joints.

(b) It is quantitatively proven that the critical shear displacement (s_{crit}) and the total fracture energy of such joints is a complex function of structural parameters such as:

—concrete strength
—steel percentage, and
—embedment length of steel

(c) More specifically, it is shown that for all concrete categories, the energy of fracture is a linear function of steel ratio (ρ) and an almost hyperbolic function of the embedment length of transversal reinforcement.

(d) The computational procedure proposed can also be profitably used in studying some practical *design problems*, such as:

—Rational assessment of the minimum steel ratio across the joint. A conventional requirement may be formulated for a shear strength (τ_{tot}) reinstated after cracking, as a percentage of the concrete tensile strength (f_{ct}): subsequently, for given anchorage lengths (l) of transverse bars, a minimum steel ratio (ρ_{min}) may be assessed via analyses such as that shown in Fig. 16.

—An alternative ULS requirement may be expressed in terms of shear–displacement values being lower than a value thought to correspond to a Beyond Limit of Repair (BLR) situation.

—A monitoring-set may be installed, checking s-values under service loads, to be compared with expected values analytically predicted.

REFERENCES

[1] Vintzēleou, E., Load transfer mechanisms along reinforced concrete interfaces under monotonic and cyclic actions. PhD thesis, National Technical University of Athens, 1984, 549 pp. (in Greek).

[2] Tassios, T. P. & Vintzēleou, E., Concrete-to-concrete friction. *Journal Structural Division* (ASCE), **113** (1987) 832–49.

[3] Walraven, J. C., Vos, E. & Reinhardt, H. W., Experiments on shear transfer in cracks in concrete, Part I: Description of results. Report 5-79-3, Delft University of Technology, 89 pp.

[4] Vintzēleou, E. & Tassios, T. P., Behaviour of dowels under cyclic deformations. *ACI Structural Journal,* **84** (1987) 18–30.

[5] Mattock, A. H., Shear transfer under monotonic loading, across an interface between concretes cast at different times. Report SM76-3, University of Washington, Sept. 1976, 62 pp.

Chapter 19

Shear Strength of Reinforced Concrete Beams

ARNE HILLERBORG

Division of Building Materials, Lund Institute of Technology, Box 118, S-221 00 Lund, Sweden

ABSTRACT

The development of cracks leading to shear failure is discussed. Crack development and the failure load can in principle be analysed by means of non-linear fracture mechanics, e.g. the fictitious crack model. Due to numerical difficulties and a limited knowledge of material properties, only rather approximate analyses have so far been performed. The results of these analyses show a good general agreement with test results regarding the relative influence of different factors. The analyses may support some changes in design formulas. With further development of numerical methods and with increased knowledge of material properties the analytical studies may prove very useful as a basis for design rules.

1. GENERAL PRINCIPLES

1.1 The Fictitious Crack Model, FCM

The general principles of the Fictitious Crack Model (FCM) are rather well-known, and they will only be summarized.

In pure tension the material properties are described by means of two relations, Fig. 1, *one stress–strain relation*, valid for all the material, and *one stress–displacement relation*, valid for a fracture zone (a 'fictitious crack'). A fracture zone starts forming when the ultimate strain ε_u is reached. The strain outside the fracture zone will never surpass ε_u, but an increase in total deformation will depend on

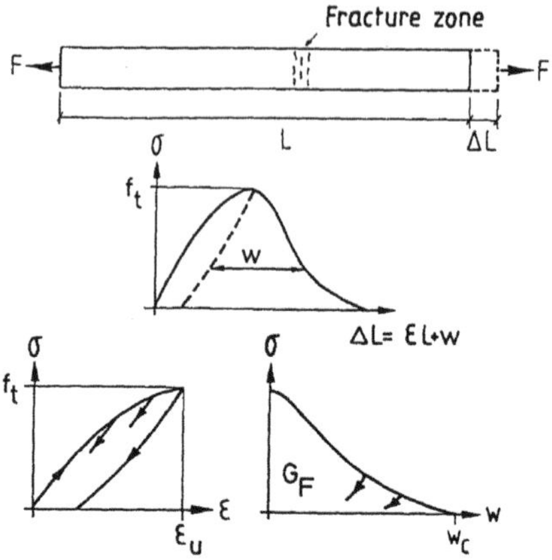

Fig. 1 Illustration of the 'fictitious crack model' for tensile fracture. A stress–strain curve is valid for all material where the strain is smaller than ε_u. The additional deformation within the fracture zone is given by a stress–displacement relation.

the displacement within the fracture zone, whereas the material outside this zone unloads and contracts.

The area below the stress–displacement curve represents the energy absorbed when a unit area of a real crack is formed. This energy is called *fracture energy* and is denoted by G_F.

The fracture zone is formally represented as a crack with the ability to transfer stresses. Such a stress-transferring crack is called a *fictitious crack*, as distinguished from a real crack, which cannot transfer tensile stresses perpendicular to itself.

The term 'fictitious crack' will generally not be used below, but the term 'crack' will be used to cover fictitious as well as real cracks.

The model described in Fig. 1 for the case of pure tension is assumed to be generally applicable to all situations where cracks are formed due to tensile stresses in a structure.

Although the crack starts with a direction which is perpendicular to the main tensile stress direction when it first forms, further deformation of a structure may be such that *sliding displacements* and *shear stresses* also occur in the crack. The displacements perpendicular to the crack are called *opening displacements*.

In the general case of both sliding and opening displacements the transferred normal and shear stresses both depend on the two displacements, and also on the previous displacement history. The relations are thus path dependent. Tests to determine how the shear and normal stresses depend on the sliding and opening displacements and on the displacement path have been reported by Hassanzadeh [1]. At a late displacement stage the behaviour can be described as 'aggregate interlock'.

When shear as well as normal stresses are acting in the crack, the principal tensile stress direction forms an angle with the normal to the crack. This tensile stress may reach a value which causes a new crack to form at an angle to the first crack.

1.2 Formation and Growth of Shear Cracks

The first cracks in a reinforced concrete beam with ordinary dimensions form as bending cracks in the region of maximum moment, where the shear forces are small. As the load increases, further cracks are formed within the region where the shear forces are larger, and where a shear failure may occur. These cracks as a rule also start as bending cracks, approximately normal to the axis of the beam, but later they bend and form a skew angle to the axis, Fig. 2. These cracks are called shear cracks.

Although all parts of a crack start as opening cracks, perpendicular to the main principal tensile stress, sliding displacements may also later occur in shear cracks, as they grow. Thus shear as well as normal stresses occur. The opening displacements are counteracted by the reinforcement crossing the crack (bending as well as shear reinforcement). As a consequence the shear stresses which are transferred in the crack are increased by the reinforcement, and the shear strength of the beam is increased.

The reinforcement also affects the crack growth. The more reinforcement, the shorter the cracks, and the higher the shear force transferred through the uncracked compression zone.

All kinds of reinforcement crossing the crack may be expected to increase the shear strength, also, for example, reinforcement parallel to the beam axis at the centre of the beam section, even though ordinary shear reinforcement is more efficient.

The combination of normal and shear stresses in the crack may lead to a stress state causing a secondary crack to form, which crosses the first crack, Fig. 2(a). The secondary crack often leads to an unstable situation and a sudden collapse of the beam. This is a type of *diagonal tension failure.*

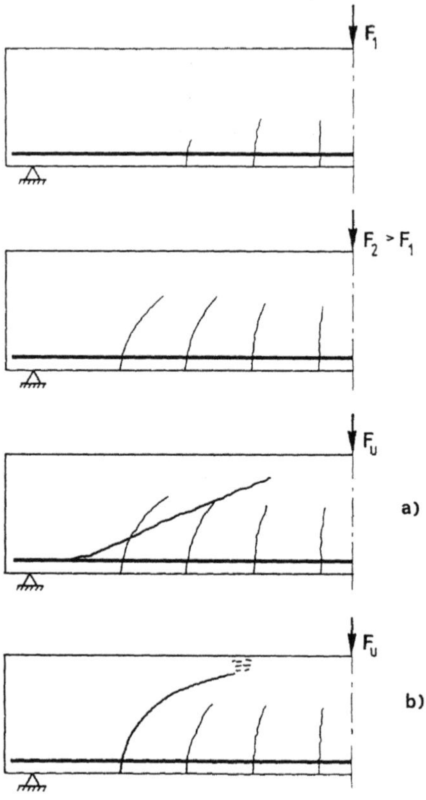

Fig. 2 The successive development of cracks, leading to different types of shear failure.

If no secondary crack is formed, the growth of a shear crack leads to a reduced concrete compression zone, in which shear stresses are also acting. The final failure depends in this case on the failure of the compression zone due to combined compression and shear stresses, Fig. 2(b). This type of failure may be called *shear–compression failure.*

For short beams, beams with an I-shape and prestressed beams, a crack may start as a skew crack in the web due to high shear stresses, Fig. 3. This may lead to a diagonal tension type of failure.

1.3 Dowel Action and Bond Stresses

The sliding displacement in a crack at the level of the tensile reinforcement causes shear forces in the reinforcement. The reinforce-

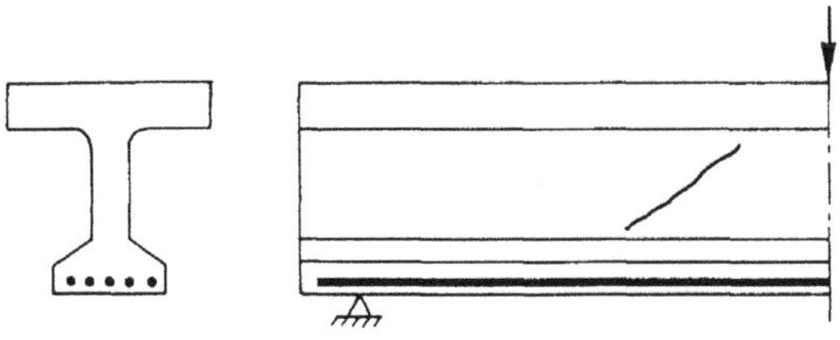

Fig. 3 A shear crack starting in the web of a beam with a strong flange or with prestressed reinforcement.

ment will act as a dowel, carrying part of the total shear force in the beam.

The dowel force has to be carried over to the concrete on both sides of the crack. On the side of the crack where the force acts in the direction towards the concrete surface it will give rise to tensile stresses in the concrete, which may cause spalling of the concrete cover, Fig. 4.

In addition to normal stresses, bond stresses also act between reinforcement and concrete, particularly close to a crack. Besides the bond stress caused by the variation in bending moment, an additional bond stress close to a crack is caused by the bond slip due to the

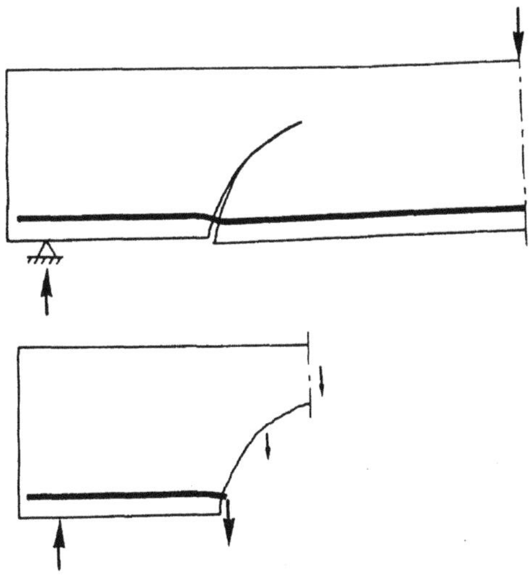

Fig. 4 Sliding deformation in a crack at the level of reinforcement, giving a dowel action.

opening displacement of the crack. High bond stresses can thus be expected close to a crack, at the same position where dowel action may cause spalling. These bond stresses also cause spalling tensile stresses in the concrete, which are added to the stresses from the dowel force.

The stress situation which may cause spalling of the concrete cover is thus rather complex. In principle it may be theoretically analysed if all necessary stress–strain and stress–displacement relations are known, but so far this has not been possible in practice.

Where dowel action has been taken into account in theoretical analyses, the model has been based on some simple assumptions regarding stress distribution between steel and concrete and a determination of some coefficients from tests [2].

2. NUMERICAL APPLICATION

2.1 Assumptions and Approximations

A complete application of the principles outlined above is very complex. So far it has only been possible to perform numerical analyses of shear fracture based on these ideas if rather crude approximations and simplifying assumptions are applied. The only systematic analysis of this type seems to be Gustafsson's [3]. He analysed the shear fracture of a beam without shear reinforcement. His analysis will be summarized below.

In a real beam there are many cracks before failure, bending cracks as well as shear cracks. As a simplification it was assumed that only one crack existed, the shear crack leading to shear failure.

The development of cracks in a beam is to a certain extent of a stochastic nature. There will also be a certain distance between the cracks, because there must be a certain bond length to develop the concrete stresses causing a new crack. For these reasons the shear crack leading to failure may appear at slightly different positions in different beams. For real beams this leads to a scatter in test results, as different crack positions are more or less dangerous for shear failure.

In a numerical analysis it is also necessary to take into account that the critical crack may form in any position. Therefore different crack positions have to be analysed. In the analysis the critical crack is taken as the crack which gives the lowest failure load.

In order to facilitate the analysis, the crack paths considered were

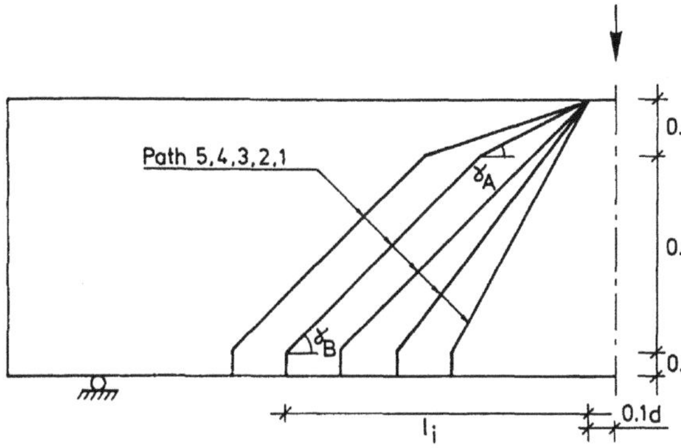

Fig. 5 Examples of crack paths used by Gustafsson [3].

predetermined, with a bilinear or trilinear shape, Fig. 5. The paths were chosen so that the principal stress direction at crack growth would be approximately perpendicular to the crack path.

The details of the assumptions are reported in full by Gustafsson [3] and they are briefly described by Gustafsson & Hillerborg [4]. They may be briefly summarized as follows:

—The stress–strain curves for concrete and steel are linear.

—The descending stress–displacement curve for the stress transfer through the crack is linear.

—The shear stress transfer through the crack (aggregate interlock) has not been taken into account.

—The bond–slip relation is elastic–plastic.

—Dowel action is not taken into account.

—The failure criterion in the compression zone is based on a modified Coulomb–Mohr criterion.

—The shape of the beam is as shown in Fig. 6.

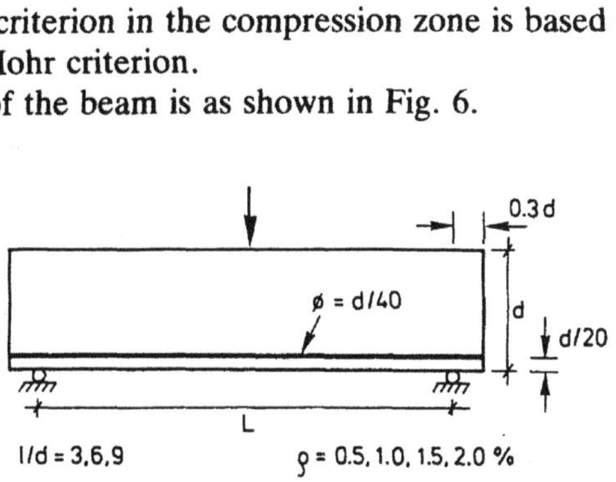

Fig. 6 Geometrical shape of beam analysed by Gustafsson [3].

Normally the crack starts from the bottom of the beam and grows upwards. At first the crack development is stable, i.e. the load increases as the crack grows. At a certain crack length an instability may occur, which means that the load formally decreases as the crack grows. At further crack growth the load may increase again in a stable way, until failure occurs in the compression zone. The relation between crack length and load may thus be of the shape indicated in Fig. 7, with two maxima. The highest of these two maxima is the failure load.

For each beam a number of different crack positions are analysed, and the shear failure load is taken as the smallest of the failure loads for these different crack positions.

2.2 Numerical Results

The numerical results of the analysis performed by Gustafsson are summarized in Fig. 8 [5], which shows how the formal shear strength v_u depends on the following factors:

—concrete tensile strength f_t
—beam depth d
—the 'characteristic length' $l_{ch} = EG_F/f_t^2$, where E is the modulus of elasticity of concrete. For ordinary concrete l_{ch} often has a value around 0·4 m
—reinforcement ratio
—span/depth ratio L/d

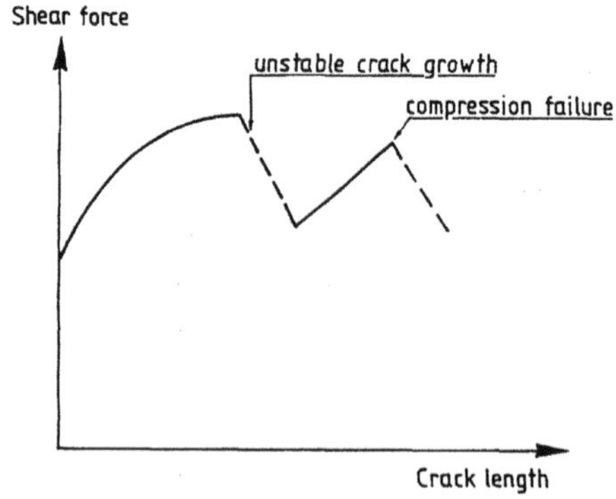

Fig. 7 General relation between shear force and length of shear crack.

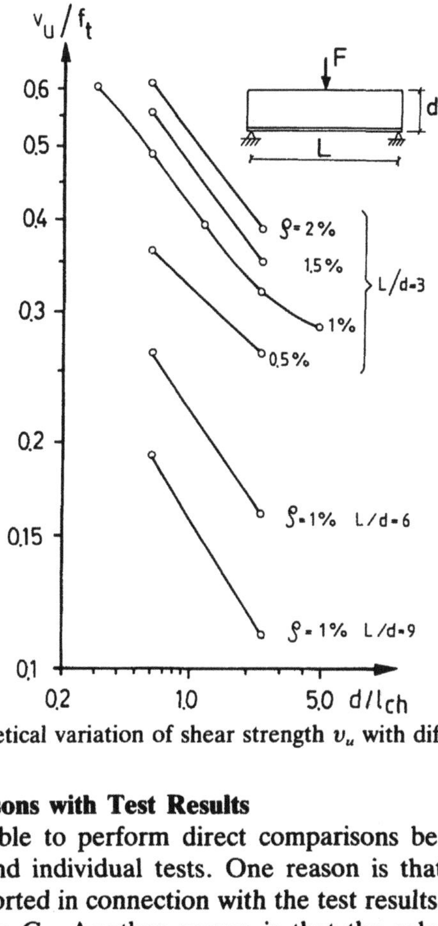

Fig. 8 Theoretical variation of shear strength v_u with different factors [5].

2.3 Comparisons with Test Results

It is not possible to perform direct comparisons between the values from Fig. 8 and individual tests. One reason is that not all relevant values are reported in connection with the test results, in particular the fracture energy G_F. Another reason is that the calculated values are based on too approximate assumptions, where among other things the shear stresses (aggregate interlock) in the crack and also the dowel action have been disregarded. If these factors had been taken into account the formal shear strength values would have been higher.

The only comparison which can be made between the calculated values and test results concerns the relative influence of different factors. Figure 9 shows such comparisons, where the variation of the shear strength with different factors according to Fig. 8 has been compared to empirical formulas based on a large number of test results [6]–[8]. It will seem that there is a reasonable general

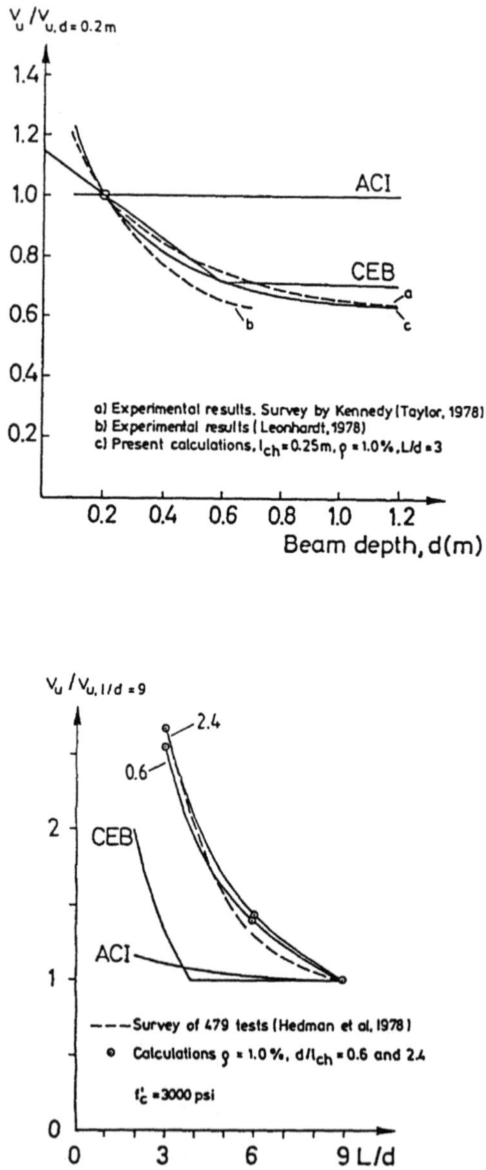

Fig. 9 Relative influence of different factors on shear strength [5]. Comparisons between theoretical results, test results [6]–[8] and code design formulas [9], [10].

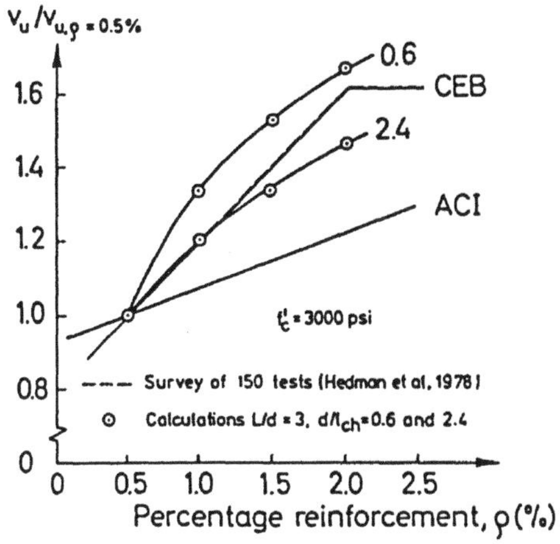

Fig. 9—(*cont'd*)

agreement. Figure 9 also shows a comparison with code formulas according to CEB [9] and ACI [10].

Of particular interest is the influence of the beam depth. The curves in Fig. 8 all have a slope of approximately 1:1 (45°). With the chosen logarithmic scales, where the vertical scale is extended four times in comparison with the horizontal scale, this leads to the following approximate relation:

$$v_u/f_t \sim (d/l_{ch})^{-1/4} \tag{1}$$

This is in full agreement with a comprehensive summary of test results [7], which resulted in the conclusion that the shear strength is inversely proportional to $d^{1/4}$. It may therefore be assumed that the theoretical relation (1) is realistic.

The influence of the beam depth on the shear strength cannot be explained with conventional theories, e.g. the truss analogy or the theory of plasticity. One theoretical explanation of this size dependency can, however, be given by the 'tooth model' [2], Fig. 10, where the aggregate interlock in the cracks plays a dominant role. This model is only valid for beams with a higher span/depth ratio than the beams studied by Gustafsson. It leads to a size dependency of the same order as given above.

3. DESIGN CONSIDERATIONS

As the theoretical values are in reasonable agreement with test results with regard to the relative influence of different parameters, these values may be used as an extended basis for design formulas, which have so far mainly been based on test results. Thus the theoretical analysis may give rise to a change in some design formulas, leading to enhanced safety. One such example will be given below.

The relation (1) may be reformulated in the following way

$$\frac{v_u}{f_t} \sim \left(\frac{l_{ch}}{d}\right)^{1/4} = \left(\frac{EG_F}{df_t^2}\right)^{1/4} \tag{2}$$

$$v_u \sim (EG_F f_t^2/d)^{1/4} \tag{3}$$

From this relation it follows that the shear strength is approximately proportional to the square root of the tensile strength (provided that E and G_F are independent of the tensile strength), and not directly proportional to the tensile strength, as is often assumed. The usual

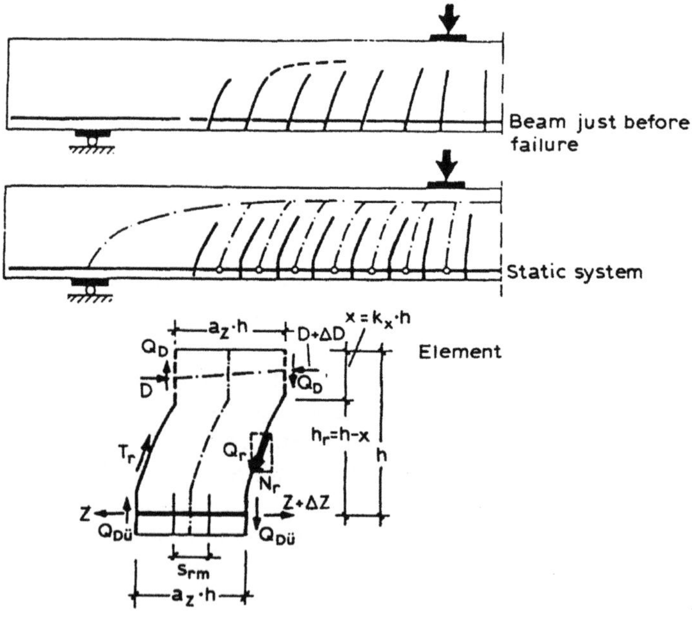

Fig. 10 The tooth model [2].

design formulas can be expected to overestimate the influence of the tensile strength (and indirectly the compressive strength) on the shear strength of beams. There might be a reason to consider a corresponding change in the design formulas. This may be particularly important where high strength concrete is used.

Another conclusion from the above relation is that the fracture energy G_F is an important property for the shear strength of beams. So far this property has not been determined in connection with shear tests. In future shear tests the value of G_F ought to be determined, as it may be a property of about the same significance as the compressive strength.

In the analysis using the 'tooth model' [2], an important parameter is the roughness of the crack surface, which gives rise to the aggregate interlock. This roughness depends on the maximum aggregate size, which in its turn also has an influence on the fracture energy. The roughness of the fracture surface and the fracture energy both tell something about the toughness of the structure. Both theories show that toughness is an important property, which should be taken into account.

4. FURTHER DEVELOPMENT

As the numerical methods develop, and the data about material properties increase, more realistic numerical analyses may be performed, where dowel action and shear stresses in the crack are taken into account, as well as several cracks and crossing cracks.

Furthermore, cases other than the simple beam with a point load and without shear reinforcement ought to be theoretically analysed. Examples of such cases are beams with shear reinforcement, beams with uniform loads, continuous beams, and punching failure of slabs with point loads or point supports.

It would be interesting to perform tests with beams where the fracture energy of concrete is varied within wide limits, while other properties are kept as constant as possible, in order to check if the theoretical analysis gives a realistic estimation of the importance of the fracture energy. The fracture energy may be varied by varying the type of aggregate and the maximum aggregate size.

5. CONCLUSIONS

1. Shear failure can in principle be realistically analysed by means of fracture mechanics, e.g. the fictitious crack model.
2. More detailed knowledge is needed regarding the stress–displacement relations where sliding as well as opening displacements occur in a crack.
3. Numerical difficulties limit the possibilities of performing realistic analyses. As the numerical methods develop these difficulties will decrease.
4. The limited approximate analysis performed by Gustafsson [3] shows that the theoretical values agree well with test results regarding the relative influence of different factors. In particular it gives a realistic prediction of the size dependency, which cannot be explained by conventional theories.
5. The analysis shows that the toughness of concrete, expressed as fracture energy G_F, is an important material property, which should be determined in connection with shear tests on beams and slabs.
6. The theoretical analysis may form an important complementary basis for design formulas, and may give rise to some changes in these formulas in order to achieve improved safety of concrete structures.

REFERENCES

[1] Hassanzadeh, M., Determination of fracture zone properties in mixed Mode I and II. In *Proceedings of 1989 SEM Spring Conference on Experimental Mechanics*. Society for Experimental Mechanics, pp. 521–7.

[2] Reineck, K.-H., Ein mechanisches Modell für den Querkraftbereich von Stahlbetonbauteilen. Dr.-Ing. thesis, Institut für Tragwerksentwurf und -konstruktion der Universität Stuttgart, 1990.

[3] Gustafsson, P. J., Fracture mechanics studies of non-yielding materials like concrete. Report TVBM-1007, Div. of Building Materials, Lund Institute of Technology, Sweden, 1985.

[4] Gustafsson, P. J. & Hillerborg, A., Sensitivity in shear strength of longitudinally reinforced concrete beams to fracture energy of concrete. *ACI Structural Journal*, **85** (1988) 286–94.

[5] Hillerborg, A., Fracture mechanics and the concrete code. In *Fracture Mechanics: Application to Concrete*, ACI SP-118. 1989, pp. 157–69.

[6] Taylor, H. P. J., Basic behaviour in shear and the Model Code. In *Comité Euro-International du Béton Bulletin d'information No 126*, 1978, 125–40.

[7] Leonhardt, F., Shear in concrete structures. In *Comité Euro-International du Béton Bulletin d'information No 126*, 1978, 66–124.

[8] Hedman, O. & Losberg, A., Design of concrete structures with regard to shear forces. In *Comité Euro-International du Béton Bulletin d'information No 126*, 1978, 184–209.

[9] CEB-FIP Model Code for Concrete Structures. In *Comité Euro-International du Béton Bulletin 124/125-E*, 1978.

[10] *ACI 318-89, Building Code Requirements for Reinforced Concrete*, American Concrete Institute, 1989.

Chapter 20

Effect of Fiber Modified Fracture Properties on Shear Resistance of Reinforced Mortar and Concrete Beams

VICTOR C. LI,[a] ROBERT WARD[b] & ALI M. HAMZA[a]

[a] *Advanced Civil Engineering Material Research Laboratory, Department of Civil Engineering, University of Michigan, Michigan 48109-2125, USA*

[b] *Cygna Consulting Engineers, Boston, Massachusetts 02210, USA*

ABSTRACT

The ultimate shear strength of longitudinally reinforced fiber mortar and concrete beams, without shear stirrups, is examined by testing beams under center-point bending. Increases in ultimate shear strength of up to 183% were recorded due to random reinforcement with volume fractions of up to 2% of short steel and polymeric fibers. These experimental results confirm the dependence of structural properties on the fracture property of the material which has a wide range of material characteristic length. It is found that the shear strength can be simply related to the flexural and tensile strength, reinforcement ratio, span-to-depth ratio, and beam depth. These semi-empirical results may have practical implications. They provide an avenue for the use of fracture concepts in structural design without direct measurement of fracture properties. This is plausible because information on the fracture properties is already incorporated in the flexural and tensile properties.

1. INTRODUCTION

Development of fracture mechanics of concrete structures in the last decade has progressed to the point that its applications in reinforced concrete structures and its implications on design codes are now being seriously considered. The introduction of fracture mechanics brings out a fracture parameter as a new independent material property [1], [2] in addition to, and distinct from, the concrete compressive strength f_c' commonly used in design codes and in structural behavior predictions. Perhaps one of the strongest driving forces for the need of a fracture parameter is the result of a series of experimental studies (e.g. Refs [3], [4]) clearly showing the size effect of concrete structures, i.e. the failure strength of geometrically similar concrete structures decreases with the structural size. This relationship is generally not predicted by classical strength theories which form the basis of most conventional concrete structure design codes. Although the Weibull concept of flaw size statistical distribution has been invoked, a consistent mechanical-statistical theory which can explain this size-dependent failure behavior is not available (e.g. Ref. [5]). The size effect may of course be a result of the combined influence by fracture characteristics and flaw size statistics [6], [7].

The material characteristic length L_{ch} originally introduced by Hillerborg *et al.* [8] is generally well accepted as providing a good representation of the quasi-brittleness of concrete. Among various concrete structures which fail in a fracture dominated mode (torsional structures, thin walled pipes, embedded anchor bolts, etc.), two of the most studied are the unreinforced concrete beam which fails in flexure due to tensile stresses in the lower concrete fibers and the longitudinally reinforced concrete beam which fails in shear due to crack propagation caused by diagonal tension stresses in the concrete. Using the fictitious crack model in a finite element analysis, Hillerborg and co-workers predicted that the flexural strength f_f and shear strength f_v decrease with the ratio d/L_{ch}, where d is the effective beam depth. In Fig. 1, the flexural strength for four-point loaded beams is predicted to approach the tensile strength for an ideally brittle material ($L_{ch} \to 0$) and to approach three times the tensile strength for an ideally plastic material ($L_{ch} \to \infty$). In between these limits, the structure may be considered to be quasi-brittle. Similarly, for longitudinally reinforced concrete beams under center point shear loads (Fig. 2), the shear strengths are predicted to decrease (approximately to the power $-1/4$) with the ratio d/L_{ch}, for various span-to-depth a/d and steel reinforcement ratios ρ. These results have been based on numerical analyses assuming the development of a single crack in the beam. The general size effect described above has been amply confirmed by a number of

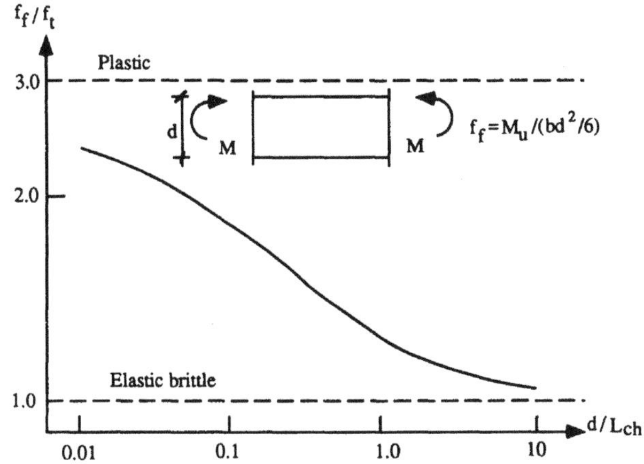

Fig. 1 Numerically predicted size effect on flexural strength based on tension-softening fracture concept [3].

experimental studies usually involving a series of tests on geometrically similar beams with various sizes, but of the same material. In particular, size effect on reinforced shear beams has been confirmed for very large beams up to 3 m in height [9]. These theoretical and experimental studies provide very strong evidence of the validity of the fracture mechanics concept and lay the foundation for a systematic

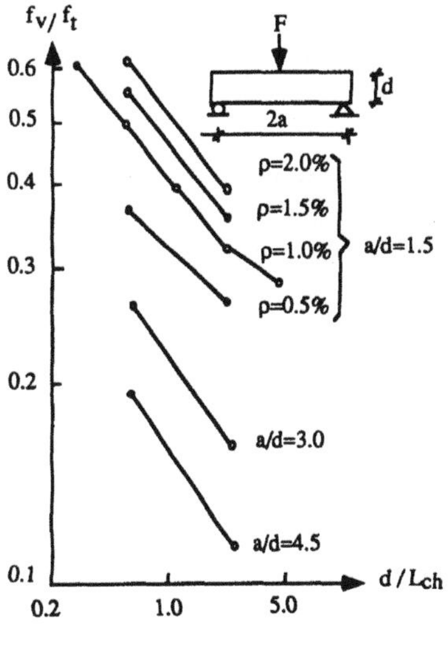

Fig. 2 Numerically predicted size effect on shear strength of axially reinforced beams based on tension-softening fracture concept [18].

upgrading of the concrete structure design codes [1]. Direct experimental confirmation of the size effect based on varying the fracture property L_{ch} of the material (at fixed d), however, has not received as much attention.

The present study has two complementary goals. The dependence of structural properties (flexural and shear strength) on the fracture property of the material is shown experimentally, using materials with a wide range of expected L_{ch}. This is achieved by varying the fiber type, length and volume fraction in both mortar and concrete. The second goal is to directly relate the shear strength of reinforced concrete beams to relatively well known and easily measurable material and geometric parameters. This is achieved by taking advantage of the known fracture study results mentioned above. It is found that the flexural and shear strength of mortar and concrete beams do increase with fiber reinforcement, although a direct correlation with L_{ch} was not attempted partly due to lack of data on the L_{ch} value for the composites used in the present study and partly due to the questionable correctness of using L_{ch} to characterize the fracture property of fiber composites [10]. While providing a qualitative confirmation of the size effect through fracture property change, the present limited findings lie short of the first goal. For the second goal, it is found that the shear strength can be shown to be simply related to the flexural and tensile strength, reinforcement ratio, span-to-depth ratio, and beam depth, for a wide range of materials. These semi-empirical results may have practical implications. In addition to confirming previous findings that fibers may be effective as shear reinforcement in place of stirrups, the simple relationships provide an avenue for the use of fracture concepts in structural design, the importance of which has been already pointed out, without direct measurement of fracture properties such as the fracture energy needed for the calculation of L_{ch}. This is possible because information on the fracture properties is already incorporated in the flexural and tensile properties, as suggested in Fig. 1. For both goals, it is clear that further studies are needed, especially when practical implementations are contemplated.

2. EXPERIMENTAL STUDY OF FRC BEAMS

A series of four-point bend tests and center point shear tests were carried out on mortar and concrete beams reinforced with steel,

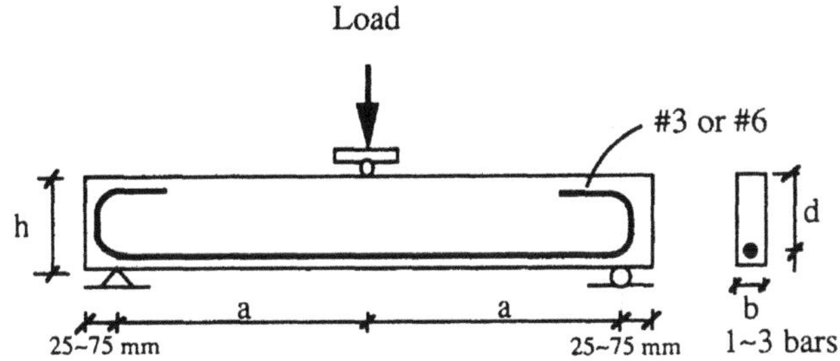

Fig. 3 Loading configuration and specimen details of shear beams. Small section: h = 127 mm, d = 102 mm, b = 63·5 mm, steel #3 bars; large section: h = 204 mm, d = 228 mm, b = 127 mm, steel #6 bars; mortar beams: a/d = 1·0 ~ 4·25 & Concrete beams: a/d = 1·0 ~ 3·0.

acrylic, Spectra 900 (a high modulus polyethylene) and Kevlar 149 (an aramid) fibers with nominal volume fractions V_f of 1% and 2%. The shear beam dimensions and loading configurations are indicated in Fig. 3. The shear beams are also reinforced with #3 and #6 longitudinal steel deformed rebars placed at 102 mm and 204 mm effective depth for small and large beams, respectively, but without shear reinforcements (stirrups). Steel reinforcement ratios vary between 1·1% to 3·3%. Shear span-to-depth ratio varies between 1 and 3 for the concrete beams, and between 1 and 4·25 for the mortar beams.

Two different mortar mixes were used. Mix A had a cement : sand : water ratio of 1 : 1 : 0·5 and Mix B had a ratio of 1 : 1 : 0·4. For the concrete mix, the cement : sand : stone : water ratio was 1 : 1·5 : 2·5 : 0·45. Limestone aggregate and river sand were used for the concrete mix. For mortar, sand passed through a #8 sieve was used. Type III rapid hardening cement was used for both mortar and concrete mix. The properties of the various fibers used are listed in Table 1. Details of the experimental set up can be found in Refs [7] and [11]. Results in first crack and ultimate shear strength are presented in Tables 2a and 2b, for mortar and concrete beams, respectively. Results of flexural strength, split tensile strength, as well as cylinder compressive strength are summarized in Tables 3a and 3b, for mortar and concrete specimens, respectively. Details of the flexural, split tension and compression tests for the mortar specimens have been reported in Ref. [12].

An important observation from the shear test is that the span-to-depth ratio has a significant effect on the failure mode. This is summarized below.

TABLE 1 Properties of fibers

Fiber type	*Fiber length (mm)*	*Aspect ratio (l/d)*	*Tensile strength (MPa)*	*Elastic modulus (GPa)*	*Specific density (g/cc)*	*Surface type*
Steel 25	25	28·5	1000	200	7·9	Crimped
Steel 50	50	57	1000	200	7·9	Crimped
Kevlar 149	6·4	530	2800	130	1·45	Straight
Acrylic	6·4	470	400	6	1·15	Crimped
Spectra 900	12·7	334	2000	100	0·97	Straight
Steel 30[a]	30	60	1172	200	7·85	Hooked
Steel 50[a]	50	100	1172	200	7·85	Hooked

[a] Used only for concrete specimens.

2.1 Beam-action ($a/d \geq 2{\cdot}5$)

In plain mortar and concrete beams with $a/d \geq 2{\cdot}5$, failure occurred suddenly when the first diagonal shear crack appeared. In each case the diagonal crack propagated along the compressive stress path towards the load point and also along the reinforcement towards the support as illustrated in Fig. 4(a). Some flexural cracks formed in the beams before failure with more cracking being observed for higher a/d values and lower reinforcement ratios.

Fiber-reinforced beams with $a/d \geq 2{\cdot}5$ usually exhibited flexure–shear cracking with diagonal shear cracks forming as an extension of a flexure crack. In many cases a number of shear cracks formed along the beam span before ultimate load. As shear cracks propagated and bent over to follow the compressive stress trajectory some cracks began to propagate along the reinforcement as illustrated in Fig. 4(b). Ultimate failure occurred by a breakdown of dowel action due to excessive cracking along the rebars or by failure in the compression zone under combined shear and compressive stresses. It is well recognized that fiber reinforcement greatly improves the resistance to crack propagation and also gives much greater tensile stress capacity across an existing crack. Thus for beams with $a/d \geq 2{\cdot}5$ the major contribution of fibers to increased shear capacity comes through increased tensile capacity across the shear crack and the development of larger dowel forces in the reinforcement.

2.2 Arch-action ($a/d \leq 2{\cdot}5$)

In these beams cracking usually initiated between the load point and the support just below the mid-depth of the beam. The crack then

TABLE 2a First shear crack and ultimate strengths for mortar beams

Fiber properties			Beam properties			First shear crack strength	Ultimate shear strength
Type	*l* (*mm*)	V_f (%)	*d* (*mm*)	*a/d*	*ρ* (%)	f_{vc} (*MPa*)	f_v (*MPa*)
			Mix type A				
		0	102	3·00	2·2	1·20	1·20
		0	204	3·00	2·2	1·02	1·02
Kevlar	6·4	2	102	3·00	2·2	2·48	3·40
		2	204	3·00	2·2	2·22	2·68
Steel	25	1	102	3·00	2·2	1·96	2·55
		1	204	3·00	2·2	1·46	1·95
		2	102	3·00	2·2	2·41	3·21
		2	204	3·00	2·2	1·92	2·56
			Mix type B				
		0	102	3·00	2·2	1·61	1·61
		0	204	3·00	2·2	1·33	1·33
Acrylic	6·4	1	102	3·00	2·2	2·07	2·07
		1	204	3·00	2·2	1·80	1·80
		2	102	3·00	2·2	1·95	2·17
		2	204	3·00	2·2	2·06	2·15
Steel	25	1	102	3·00	2·2	2·41	2·74
		1	204	3·00	2·2	2·11	2·37
		2	102	3·00	2·2	3·11	3·79[a]
Steel	50	1	102	3·00	2·2	2·95	3·55[a]
		1	204	3·00	2·2	2·83	3·45[a]
		2	204	3·00	2·2	2·90	3·63[a]
		0	102	1·00	2·2	3·88	5·17
		0	102	1·50	2·2	2·26	3·37
		0	102	2·00	2·2	1·92	2·31
		0	102	2·25	2·2	1·58	2·04
		0	102	2·50	2·2	1·52	1·52
		0	102	2·75	2·2	1·71	1·71
		0	102	3·75	2·2	1·69	1·69
		0	102	4·25	2·2	1·56	1·56
Kevlar	6·4	1	102	1·00	2·2	6·75	7·76
		1	102	1·50	2·2	4·94	5·25
		1	102	1·75	2·2	3·55	3·99
		1	102	2·00	2·2	3·20	3·68

(continued)

TABLE 2a—*contd.*

Fiber properties			*Beam properties*			*First shear crack strength*	*Ultimate shear strength*
Type	l (*mm*)	V_f (%)	d (*mm*)	a/d	ρ (%)	f_{vc} (*MPa*)	f_v (*MPa*)
		1	102	2·25	2·2	3·06	3·15
		1	102	2·50	2·2	2·92	3·48
		1	102	2·75	2·2	3·03	3·19
		1	102	3·00	2·2	2·94	3·03
		1	102	3·75	2·2	2·99	2·99
		1	102	4·25	2·2	—	2·75[a]
Steel	25	1	102	1·00	2·2	5·24	7·82
		1	102	1·50	2·2	4·02	5·15
		1	102	1·75	2·2	2·92	4·64
		1	102	2·00	2·2	2·59	3·99
Steel	25	1	102	2·25	2·2	2·58	3·62
		1	102	2·50	2·2	2·50	3·17
		1	102	2·75	2·2	2·48	2·75
Spectra	12·7	1	102	1·00	2·2	7·45	7·78
		1	102	2·00	2·2	3·56	5·31
		1	102	3·00	2·2	3·48	3·71[a]
		1	102	3·75	2·2	3·02	2·97[a]
		1	102	4·25	2·2	2·45	2·61[a]
		0	102	1·00	1.1	3·46	4·75
		0	102	2·00	1.1	1·84	2·10
		0	102	2·50	1.1	1·35	1·35
		0	102	3·00	1.1	1·32	1·32
		0	102	3·75	1.1	1·31	1·31
		0	102	4·25	1.1	1·19	1·19
Kevlar	6·4	1	102	1·00	1.1	5·91	5·91
		1	102	2·00	1.1	3·02	3·21
		1	102	2·50	1.1	2·50	2·50
		1	102	3·00	1.1	—	2·48[a]
		1	102	3·75	1.1	—	1·91[a]
		1	102	4·25	1.1	—	1·51[a]
Steel	25	1	102	3·00	1·1	1·98	1·98
Spectra	12·7	1	102	1·00	1.1	5·50	5·70
		1	102	2·00	1.1	3·15	3·87
		1	102	3·00	1.1	2·33	2·48[a]
		1	102	3·75	1.1	2·03	2·03[a]
		1	102	4·25	1.1	—	1·64[a]
		0	102	1·00	3·3	4·10	6·15
		0	102	2·00	3·3	2·04	2·57

TABLE 2a—*contd.*

Fiber properties			Beam properties			First shear crack strength	Ultimate shear strength
Type	l (mm)	V_f (%)	d (mm)	a/d	ρ (%)	f_{vc} (MPa)	f_v (MPa)
		0	102	3·00	3·3	1·61	1·61
		0	102	3·75	3·3	1·63	1·63
		0	102	4·25	3·3	1·54	1·54
Kevlar	6·4	1	102	1·00	3·3	7·40	8·40
		1	102	2·00	3·3	3·42	3·82
		1	102	3·00	3·3	3·05	3·14
		1	102	3·75	3·3	3·13	3·13
		1	102	4·25	3·3	2·74	2·74
Steel	25	1	102	3·00	3·3	2·50	2·75
Spectra	12·7	1	102	2·00	3·3	3·45	5·65
		1	102	3·00	3·3	3·73	3·77
		1	102	3·75	3·3	2·86	3·72
		1	102	4·25	3·3	2·95	3·26[a]
Steel	50	1	102	3·00	3·3	3·24	3·90

[a] Flexural failure.

propagated towards the load point and the support in a manner similar to that observed in a splitting tension test. Fenwick & Paulay [13] postulated that appreciable arch-action develops when the diagonal crack extends to the support, thereby separating the tension and compression zones of the shear span and allowing the relatively large translational displacement associated with arch-action to occur. The various failure modes of the tied-arch mechanism are illustrated in Fig. 5. The plain mortar and concrete beams tended to fail by splitting along the line of the arch compressive force. Many of the fiber beams, with much higher splitting strengths, failed either by crushing of the concrete at the support or by sudden ejection of the upper part of the shear span due to failure of the compression zone under combined shear and compression together with sliding along the diagonal crack faces. Some beams with larger a/d values failed due to the propagation of a flexural tension crack from the top of the beam down to meet the diagonal crack. This was caused by an eccentric line of thrust between the load point and the support. The Spectra fiber beams failed by gradual shearing of the diagonal crack faces leading to

TABLE 2b First shear crack and ultimate strengths for concrete beams

Fiber properties			*Beam properties*			*First shear crack strength*	*Ultimate shear strength*
Type	l *(mm)*	V_f *(%)*	d *(mm)*	a/d	ρ *(%)*	f_{vc} *(MPa)*	f_v *(MPa)*
		0	204	3	2·2	1·56	1·63
		0	102	1	1·1	3·99	4·79[b]
		0	102	2	1·1	2·31	3·14
		0	102	3	1·1	1·32	1·55
		0	102	1	2·2	4·33	4·54[b]
		0	102	2	2·2	2·19	2·64
		0	102	3	2·2	1·59	1·65
		0	102	1·5	1·1	3·92	4·46
Spectra	12·7	1	102	1	1·1	5·07	5·74[b]
	12·7	1	102	2	1·1	3·60	4·17
	12·7	1	102	3	1·1	2·08	2·48
	12·7	1	102	2	2·2	3·79	4·37
	12·7	1	102	3	2·2	2·36	2·82
	12·7	1	102	1·5	1·1	3·93	4·74
Steel	30	1	204	3	2·2	2·27	3·05
	30	1	102	3	2·2	2·78	3·16
	30	1	102	3	1·1	2·14	2·43
	30	1	102	1·5	1·1	4·72	5·64
Steel	50	1	204	3	2·2	2·11	3·05
	50	1	102	3	2·2	2·99	3·55

[b] Bearing failure.

gradual reduction in load capacity after the ultimate. This is due to the ability of these fibers to transfer relatively high tensile stresses even at very large crack openings [14].

For both plain and fiber reinforced mortar and concrete beams, the major distinction between $a/d \geq 2{\cdot}5$ and $a/d < 2{\cdot}5$ as indicated in the discussion above, is that the beams with larger a/d tend to develop delamination along the rebar at increasing applied load as the diagonal crack continues to open up in the web. This suggests that for beams with larger a/d, load transfer along the rebar has occurred and that the rebar has been utilized more efficiently. Delamination is likely to be a subsequent effect of small radial cracks emanating from the deformed rebar surface and which cause the loss of bond between the rebar and the mortar or concrete, as has been observed in rebar pull-out tests by

TABLE 3a Flexural, splitting tensile and compressive strengths for mortar beams

Mix type	Material			Flexural strength f_f^a (MPa)	Splitting strength f_t^b (MPa)	Compressive strength f_c^c (MPa)
	Fiber type	l (mm)	V_f (%)			
A			0	2·2	2·2	51·1
A	Kevlar	6·4	2	6·7	4·0	37·7
A	Steel	25·0	1	4·8	3·4	53·0
A	Steel	25·0	2	6·1	4·1	50·2
B			0	2·6	2·9	57·0
B	Acrylic	6·4	1	4·0	4·3	45·3
B	Acrylic	6·4	2	3·7	3·6	33·0
B	Kevlar	6·4	1	5·1	4·4	50·3
B	Spectra	12·7	1	8·9	3·2	45·7
B	Steel	25·0	1	5·4	3·9	62·6
B	Steel	25·0	2	7·0	4·8	57·0
B	Steel	50·0	1	7·4	4·3	54·1
B	Steel	50·0	2	8·9	5·6	—

[a] Flexural test (Beam 114 mm × 114 mm × 342 mm).
[b] Splitting test (Cylinder 77 mm × 154 mm).
[c] Compressive test (Cylinder 77 mm × 154 mm).

TABLE 3b Flexural, splitting tensile and compressive strengths for concrete beams

Material			Flexural strength f_f^a (MPa)	Splitting strength f_t^b (MPa)	Compressive strength f_c^c (MPa)
Fiber type	l (mm)	V_f (%)			
		0	4·8	3·0	17·8
Spectra	6·4	1	6·1	3·3	19·1
Steel	30	1	10·2	5·2	22·7
Steel	50	1	12·1	5·3	26·0

[a] Flexural test (Beam 100 mm × 100 mm × 300 mm).
[b] Splitting test (Cylinder 100 mm × 200 mm).
[c] Compressive test (Cylinder 100 mm × 200 mm).

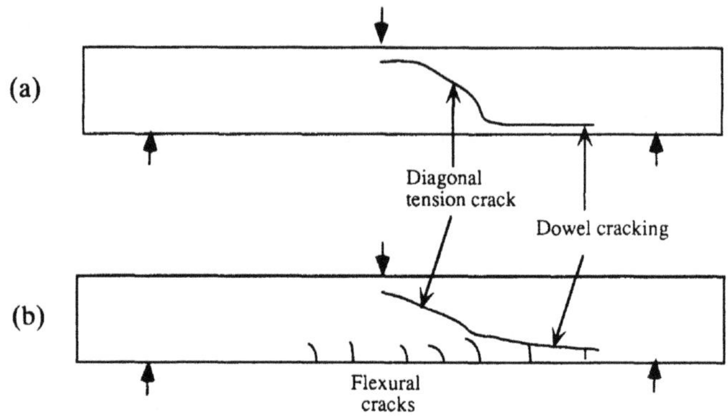

Fig. 4 (a) Typical crack shape in plain mortar and concrete beams with $a/d > 2{\cdot}5$. (b) Typical critical crack shape in fiber reinforced beams with $a/d > 2{\cdot}5$.

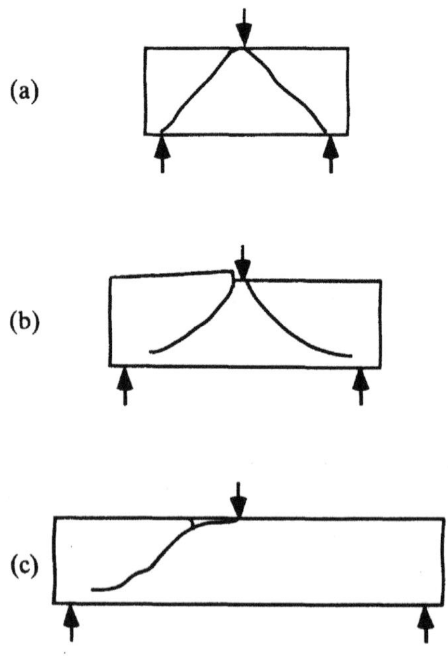

Fig. 5 Typical shear failure patterns in reinforced beams with $a/d < 2{\cdot}5$: (a) splitting failure; (b) shear compression failure; (c) flexural tension failure under eccentric compression force.

Goto [15]. The presence of fibers likely resists the radial cracks and the delamination, with subsequent increase in dowel action and shear strength of the beam.

These observations and postulations are reinforced by Fig. 6 which plots the measured shear strength against a/d, for various fiber reinforced mortars and concrete. For all the steel reinforcement ratios used, the data (clear for mortar but not so clear for concrete, due to lack of concrete data points for $a/d > 2{\cdot}5$) indicate a bend around $a/d \sim 2{\cdot}5$. Fiber reinforcement generally results in larger percentage increases in shear strength for beams with larger a/d, with the magnitude of increase dependent on the fiber type and other factors.

3. PARAMETRIC DEPENDENCE OF SHEAR STRENGTH: $f_v = \text{fcn}\ (f_f;\ f_t;\ \rho;\ a/d;\ d)$

The experimental observations described above suggest that the parametric dependence of shear strength is different for beams with $a/d < 2{\cdot}5$ and for beams with $a/d > 2{\cdot}5$, due to differences in the failure modes. This conclusion was also reached by Zsutty [16] who studied the parametric dependence of shear strength using only plain concrete (with no fibers) beams. Based on empirical fits to experimental data, he concluded that for $a/d > 2{\cdot}5$

$$f_v = 60(f_c' \rho a/d)^{1/3} \tag{1}$$

where f_c' is the concrete compressive strength. Since fiber reinforcement does not significantly increase the compressive strength (10% increase for 1% of 25-mm steel fibers) and in many cases can cause a strength reduction, especially for some synthetic fibers (as much as 20% reduction for 1% of 12·7-mm Spectra fibers in fresh mix with poor workability) it is not possible to use the compressive strength as a material parameter which reflects the strength in shear and flexure. Intuitively it seems that the splitting tensile or the flexural strength should be much better indicators of the improved performance. The flexural strength of unreinforced fiber concrete beams, to some extent, reflects the post-cracking performance of the material. A high stress capacity across a cracked plane allows load increases even after a crack begins to propagate from the bottom of the beam in a flexural test. Thus it may be expected that a material which gives high flexural strengths would also give high stresses across diagonal shear cracks.

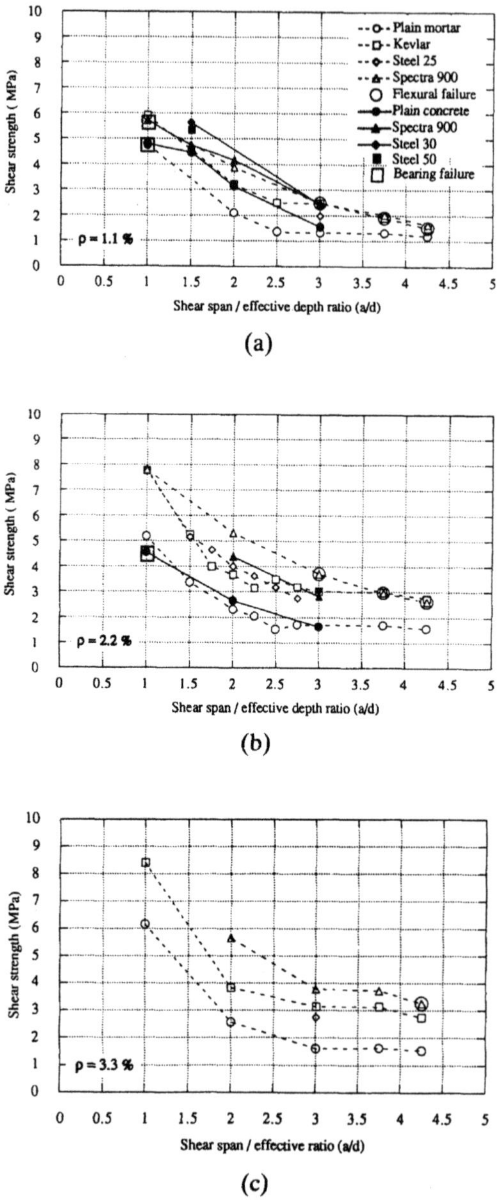

Fig. 6 Influence of shear span/effective depth ratio on shear strength of fiber reinforced mortar and concrete beams for $\rho = 1{\cdot}1 \sim 3{\cdot}3\%$.

After diagonal cracks form in the beam some shear force is transferred to dowel action in the rebars and to direct shear in the compression zone. The ability of the rebars to quickly react to this extra load depends on the stiffness of the load-deformation curve for dowel action. It may be expected that a material with a high cracking tensile strength can effectively resist dowel cracking and thus has a stiff curve. This allows development of relatively high dowel forces before the shear forces across the diagonal crack drop significantly, and thus the overall shear capacity of the beam can continue to increase. From this qualitative discussion, it appears that both the flexural and splitting tensile strengths are important in predicting the ultimate shear strength. For simplicity it was decided to attempt to relate the shear strength to some function of the product $(f_f f_t)$. This product is meant to replace the shear strength dependence on f_c' as in eqn (1).

The dependence on ρ and on a/d is expected not to be different between plain concrete (or mortar) and fiber reinforced concrete (or mortar), and the same functional dependence of $(\rho d/a)^{1/3}$ as in eqn (1) is used for the data of the present study. Finally, the depth (d) size effect so well documented by fracture mechanics analyses and with experimental data, must be included. Shear strength dependence of $d^{-1/2}$ and $d^{-1/4}$ have both been reported in the literature. Our data seem to dictate a dependence of $d^{-1/3}$, together with $(f_f f_t)^{3/4}$ dependence, as shown in Fig. 7. The plot suggests a linear relationship between f_v and $(f_f f_t)^{3/4}(\rho d/a)^{1/3}d^{-1/3}$ both for the plain and fiber reinforced concrete and mortar beams. The best fit lines

$$f_v = A + B[(f_f f_t)^{3/4}(\rho d/a)^{1/3}(d)^{-1/3}] \qquad \text{for } (a/d \geq 2{\cdot}5) \qquad (2)$$

where f_v, f_f, and f_t are in MPa and d is in mm, are also shown in Fig. 7. For the mortar data, $A = 0{\cdot}53$ and $B = 5{\cdot}47$. For the concrete data, $A = 1{\cdot}25$ and $B = 4{\cdot}68$. Equation (2) quantifies how increased fracture tensile properties lead to improved shear strength. The vertical shift in the lines in Fig. 7 (i.e. the difference in the A-value) between the mortar and concrete data probably accounts for the aggregate interlock action associated with shear sliding on the curved diagonal cracks in the concrete specimens. For concrete mixes within practical range, contribution to the shear strength due to the aggregate interlock mechanism is probably about constant. That is, the value of A in eqn (2) may be expected to remain approximately 1·5 for all practical concrete mix types with various fibers. It can be seen that the plain concrete and plain mortar data groups lie to the lower left hand

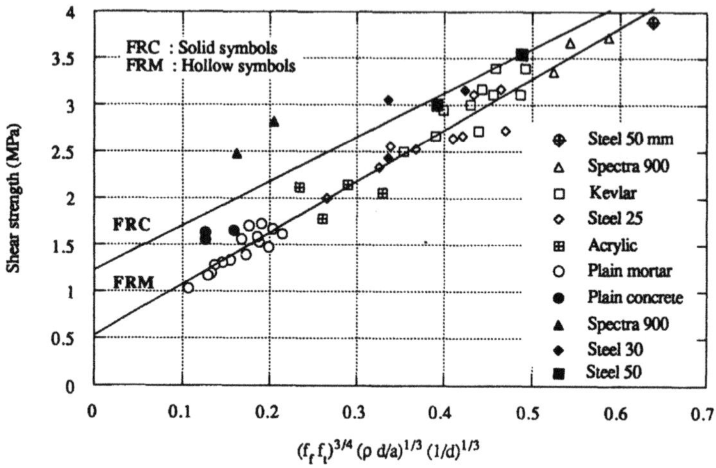

Fig. 7 Semi-empirical relationship between ultimate shear strength and the material and geometrical properties for $a/d \geq 2{\cdot}5$. (FRC = Fiber Reinforced Concrete; FRM = Fiber Reinforced Mortar) (Unit for horizontal axis in $N^{3/2}\ mm^{10/3}$)

corner, while the steel and Spectra fibers data groups generally lie to the upper right hand corner, suggesting the effectiveness of using fibers as shear reinforcement. It is interesting to note that beam depth can be cancelled out of eqn (2) giving dependence of the shear strength on just the material properties, the amount of reinforcement and the shear span.

For $a/d < 2{\cdot}5$, based on empirical fit of concrete data, Zsutty [17] proposed the relation

$$f_v = 150(f_c'\rho)^{1/3}(d/a)^{4/3} \qquad \text{for } (a/d < 2{\cdot}5) \qquad (3)$$

Again, the f_v dependence on f_c' must be modified to reflect the dependence on the tensile capacity of the fiber reinforced materials. The differences between first crack and ultimate shear strength are much greater for beams with $a/d < 2{\cdot}5$ than for beams with $a/d > 2{\cdot}5$. In many cases a crack extends almost all the way from the support to the load point before ultimate load is reached. Thus it was considered reasonable to expect that the cracking tensile strength has little influence on ultimate load and that post-cracking strength is the important material parameter. For this reason it was decided to try to relate the ultimate shear strength to just the flexural strength, as a replacement for f_c' in eqn (3). The reinforcement ratio dependence is judged to be similar for plain and fiber reinforced mortar and concrete, so that the $\rho^{1/3}$ dependence is retained. As to the a/d dependence, the particular increase in shear strength due to arch-action is probably dependent on the width of the loading platen, with greater strength for a wider platen due to distribution of the compressive forces over a greater area. The present data indicate a dependence of (a/d) to the power (-1). A single straight line correlation between f_v and $(f_f)^{2/3}(\rho)^{1/3}(d/a)$ is found for both the plain and fiber reinforced concrete and mortar data (Fig. 8). Data points for beams which suffer from bearing failures at the support points are not included in Fig. 8. Comparison with Fig. 7 suggests that the aggregate interlock mechanism may not be as significant in direct tension transfer as in shear transfer which occurs mainly in the shear sliding of the diagonal cracks in beams with $a/d > 2{\cdot}5$. The best fit line shown in Fig. 8 is given by

$$f_v = 9{\cdot}16[(f_f)^{2/3}(\rho)^{1/3}(d/a)] \qquad \text{for } (a/d \leq 2{\cdot}5) \tag{4}$$

with f_v and f_f being expressed in MPa. For the time being, it is not possible to include the effect of beam size in this expression because all

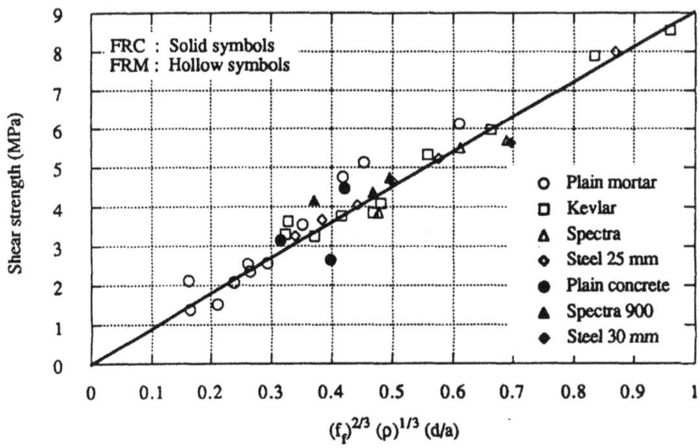

Fig. 8 Semi-empirical relationship between ultimate shear strength and the material and geometrical properties for $a/d \leq 2{\cdot}5$. (FRC = Fiber Reinforced Concrete; FRM = Fiber Reinforced Mortar) (Unit for horizontal axis in MPa$^{2/3}$)

tests for $a/d \leq 2{\cdot}5$ were on a single beam size. However, computational fracture mechanics studies by Gustafsson and Hillerborg, [18] (Fig. 2) seem to indicate similar size effect for all a/d values. If so, an additional factor indicating dependence on d must be included in eqn (4).

4. CONCLUSIONS AND FURTHER DISCUSSIONS

Equations (2) and (4) provide a simple way of quantifying the influence of fracture tensile properties on the shear strength of longitudinally reinforced beams made from various cementitious materials. However, there is usually some discontinuity in the strength versus span-to-depth ratio curves predicted by eqns (2) and (4) at $a/d = 2{\cdot}5$. Problems with this can be avoided if it is realized that eqn (2) is conservative for $a/d < 2{\cdot}5$ and can be used instead of eqn (4) for a/d values just less than 2·5, if it happens to give a higher strength value.

The only property test required in eqns (2) and (4) are standard flexural and split tension tests. As discussed in the Introduction of this chapter, the fracture property L_{ch} is expected to play a strong role in the shear strength f_v. Equations (2) and (4) have indirectly incorporated this information through f_f and f_t. There is admittedly significant amount of empiricism in arriving at eqns (2) and (4). (Part of the empiricism is necessitated by the fact that the relation between f_v and L_{ch} from fracture mechanics analytical studies shown in Fig. 2 is based on a significantly idealized fracture model usually with a single crack and with no delamination or dowel actions.) Furthermore, it would be more satisfying if these equations are cast in terms of nondimensional groups. Nonetheless, it is quite remarkable that the shear strength can be so simply related to easily measurable parameters, and for such a wide range of materials with L_{ch} varying over two orders of magnitude. Certainly guidance from modern fracture mechanics of quasi-brittle material has provided the important hints and fundamental links.

Although it is tempting to use eqns (2) and (4) in structural design, especially for fiber reinforced concrete or mortar beams, it is important to emphasize the need for additional data with increasing range of the parameters studied, especially in relation to concrete quality, reinforcement ratio, span-to-depth ratio and effective beam depth.

With this in mind, we note that the present study does confirm the usefulness of fibers, both steel and synthetic, in potentially replacing stirrups as shear reinforcement in numerous applications. Most importantly, the results may be regarded as a confirmation of the validity of the fracture mechanics concept, albeit indirectly.

ACKNOWLEDGEMENTS

The authors would like to thank the National Science Foundation (Program Manager: Dr K. Chong) and the Air Force Office of Sponsored Research (Program Manager: Dr S. Wu) for their support of this research through grants to the University of Michigan, Ann Arbor.

REFERENCES

[1] Hillerborg, A., Fracture mechanics and the concrete code. In *Fracture Mechanics: Application to Concrete,* ed. V. C. Li & Z. P. Bazant. ACI SP-118, ACI, Detroit, 1989, pp. 157–69.

[2] Li, V. C., Technological implications of concrete fracture research—an overview of tensile failure in cementitious materials and structures. In *Fracture Mechanics: Application to Concrete,* ed. V. C. Li & Z. P. Bazant. ACI SP-118, ACI, Detroit, 1989, pp. 1–16.

[3] Petersson, P. E., Crack growth and development of fracture zones in plain concrete and similar materials. Report TVBM-1006, Div. of Building Materials, Lund Institute of Technology, Sweden, 1981.

[4] Bazant, Z., Fracture mechanics size effect and ultimate load of beams under torsion. In *Fracture Mechanics: Application to Concrete,* ed. V. C. Li & Z. P. Bazant. ACI SP-118, 1989, pp. 171–8.

[5] Bazant, Z., Fracture energy of heterogeneous material and similitude. In *Proceedings SEM/RILEM International Conference on Fracture of Concrete and Rock,* Houston, June 1987 ed. S. P. Shah & S. Swartz. Society for Experimental Mechanics, SEM, Connecticut, USA, 1989 pp. 390–402.

[6] Gustafsson, P. J., Fracture mechanics studies of non-yielding materials like concrete. Report TVBM-1007, Div. of Building Materials, Lund Institute of Technology, Sweden, 1985.

[7] Ward, R., Structural behavior of fiber reinforced mortar related to material fracture resistance. MSc. thesis, M.I.T., 1989.

[8] Hillerborg, A., Modeer, M. & Petersson, P-E., Analysis of crack formation and crack growth in concrete by means of fracture mechanics and finite elements. *Cement and Concrete Res.*, **6** (1976) 773–82.

[9] Shioya, T., Inguro, M., Nojiri, Y., Akiyama, H. & Okada, T., Shear strength of large reinforced concrete beams. In *Fracture Mechanics: Application to Concrete,* ed. V. C. Li & Z. P. Bazant. ACI SP-118, ACI, Detroit, 1989, pp. 259–79.

[10] Hillerborg, A., Determination and significance of the fracture toughness of steel fiber concrete. *Steel Fiber Concrete, U.S. Sweden Joint Seminar,* ed. S. P. Shah & A. Skarendahl, 1985. Elsevier Applied Science Publishers, London, 1986, pp. 257–71.

[11] Li, V. C., Ward, R. & Hamza, A. M., Steel and synthetic fibers as shear reinforcement for mortar and concrete beams. Submitted to ACI American Concrete Institute.

[12] Ward, R. & Li, V. C., Dependence of flexural behavior of fiber reinforced mortar on material fracture resistance and beam size *ACI Journal,* **87** (1990), 627–37.

[13] Fenwick, R. C. & Paulay, T., Mechanisms of shear resistance of concrete beams. *Journal of the Structural Division* (ASCE), **94** (ST10) (1968) 2325–50.

[14] Wang, Y., Li, V. C. & Backer, S., Tensile properties of synthetic fiber reinforced mortar. *J. Cement and Concrete Composites,* **12**(1) (1990) 29–40.

[15] Goto, Y., Cracks formed in concrete around deformed tension bars. *ACI Journal,* **68**(4) (1971) 244–51.

[16] Zsutty, T., Beam shear strength prediction by analysis of existing data. *ACI Journal,* **65**(11) (1968) 943–51.

[17] Zsutty, T., Shear strength prediction for separate categories of simple beam tests. *ACI Journal,* **68**(2) (1971) 138–43.

[18] Gustafsson, P. J. & Hillerborg, A., Sensitivity in shear strength of longitudinally reinforced concrete beams to fracture energy of concrete. *ACI Structural Journal* **85**(3) (1988) 286–94.

Chapter 21

Failure Modes of Longitudinally Reinforced Beams

B. L. KARIHALOO
School of Civil and Mining Engineering,
The University of Sydney, NSW 2006, Australia

ABSTRACT

The role played by the amount of steel in the failure mode transition of longitudinally reinforced beams is studied experimentally. The span, cross-sectional dimensions and the mix properties of the test beams remained unaltered. It is found that the beam with a low amount of reinforcement (just one unanchored bar of diameter 12 mm) failed in the flexural mode, whereas by doubling the amount of reinforcement the failure mode changed to one of diagonal shear. The transition in the failure modes is explained from the viewpoint of fracture mechanics using several existing models. The merits and shortcomings of each fracture mechanics model are pointed out and suggestions put forward for their improvement in the light of the experimental evidence.

1. INTRODUCTION

Reinforced concrete flexural members are often provided with reinforcing steel not only in the longitudinal direction but also in the transverse direction in order to increase their shear capacity. The ultimate failure mode of such members is governed by the amount of steel; a low steel content invariably ensures a tensile mode of failure, whereas a high steel content leads to diagonal shear failure. The latter mode of failure is generally sudden and, if possible, is to be avoided. It is also the least well understood of the failure modes because it cannot be fully explained by conventional failure theories which are based on strength criteria. Moreover, the transition from one mode of failure to another depends not only on the steel content but also on the size of the member itself. Again, the strength based failure theories prove inadequate in explaining this size effect. Therefore, the current design practice is to ignore the size effect and to provide a minimum longitudinal reinforcement content, together with some shear reinforcement in the form of vertical stirrups to prevent the occurrence of sudden shear failure.

Lately though several attempts have been made to gain a better understanding of the role of steel ratio and member size in the failure mode transition using the concepts of fracture mechanics. However, in order to avoid the complications introduced by the shear reinforcement and any anchorages used to secure the longitudinal steel, the fracture mechanics approach has been applied in the first instance to flexural members reinforced only with unanchored longitudinal bars.

This chapter will first present the results of an experimental study of just two longitudinally reinforced beams which were identical in all but one respect, namely the amount of steel. Tests in three-point bending showed that the beam with the lowest steel content (just one unanchored bar of diameter 12 mm) failed in the classical tensile mode, i.e. by the formation and propagation of flexural cracks in concrete, followed by slippage and yielding of steel and eventually the compression crushing of concrete near the applied concentrated load. By doubling the amount of longitudinal steel the failure mode changed to one of diagonal shear.

The chapter will next attempt to explain both modes of failure from the point of view of fracture mechanics using several existing fracture models. The most complete description of the various phases in the tensile mode of failure is provided by the model proposed by Carpinteri [1] (see also Refs [2] and [3]), although some features of this failure mode are also revealed by the model used by Jenq & Shah [4]. The diagonal shear failure mode can be partly explained by the models proposed by Jenq & Shah [4], Bažant & Kim [5], and Gustafsson & Hillerborg [6] (see also Hillerborg [7]). The models for explaining both types of failure also reveal the reasons for the observed size effect. The chapter concludes with a discussion of the merits and shortcomings of each of these models and with suggestions for their improvement in the light of the experimental evidence.

2. EXPERIMENTAL OBSERVATIONS

Two beams of loading span 1·6 m (overall length 1·8 m) were cast with one or two ribbed steel bars (diameter 12 mm) as tensile reinforcement. The beams were 100-mm wide (B) and 150-mm high (W), giving a steel ratio ($\rho = A_s/(BW)$) of 0·75% and 1·5%, respectively. The cover from the centre of the steel bar to the tensile face of the beam was $c = 25$ mm. The concrete mix with a maximum aggregate size $g = 20$ mm had the following mechanical properties: cylinder crushing strength $f'_c = 38$ MPa, $E = 30$ GPa, $f_t = 3{\cdot}4$ MPa. The fracture properties of the mix were not measured, but were estimated from known results for similar mixes [8]. The effective fracture toughness K^e_{Ic} of the mix was estimated to be 1·27 MPa$\sqrt{m}$ giving an effective toughness $G^e_{Ic} = (K^e_{Ic})^2/E$ of 50 J/m^2. The yield strength f_y of the reinforcing steel was 463 MPa.

The beams were tested in three-point bending in a displacement controlled machine. Continuous plots of load–central deflection were obtained. Moreover, the deflections at five points along the span of the beams were measured. Figure 1 shows the actual load–central deflection traces of the beams as plotted on an X–Y recorder, while

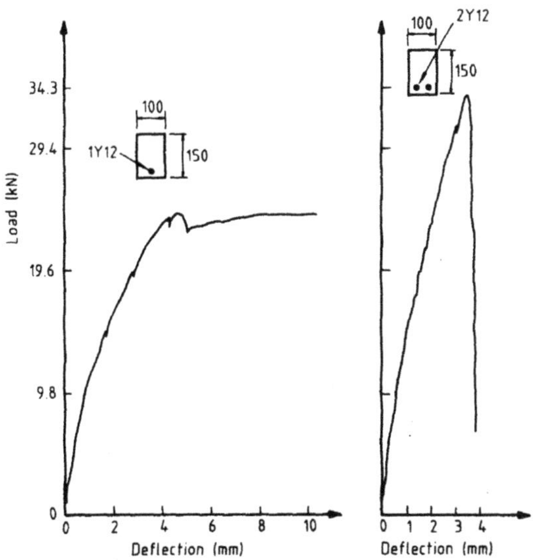

Fig. 1 Load–central deflection traces of the two test beams.

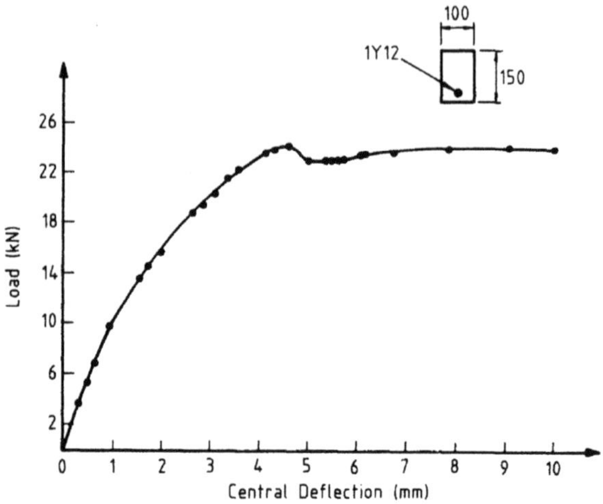

Fig. 2 Load–central deflection plot of the test beam with one reinforcing bar constructed from the load and central deflection readings (filled circles).

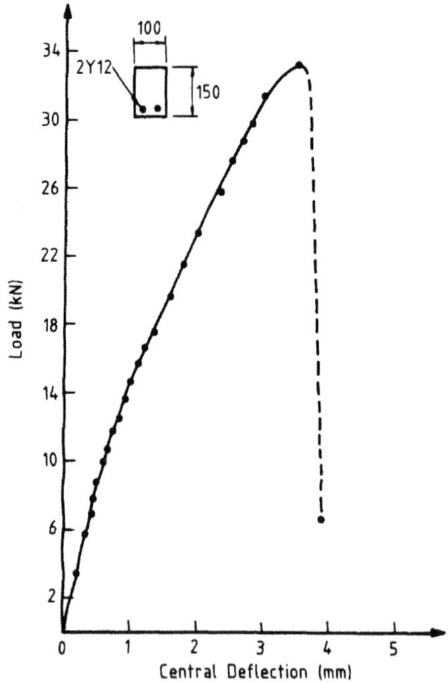

Fig. 3 Load–central deflection plot of the test beam with two reinforcing bars constructed from the load and central deflection readings (filled circles).

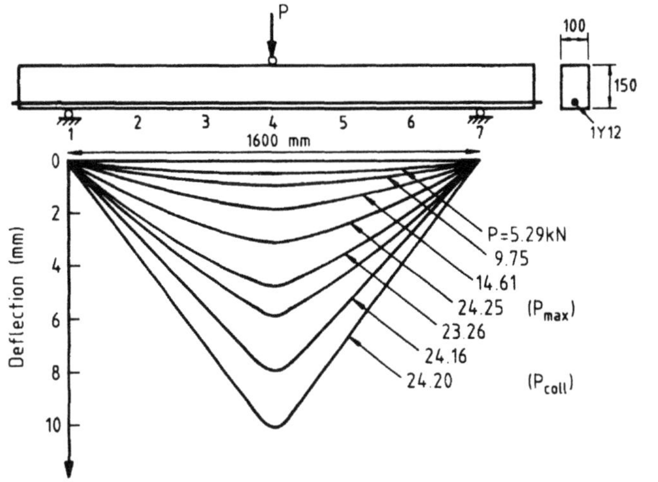

Fig. 4 Deflected shapes of the test beam with one reinforcing bar at various load levels constructed from the deflection readings at the seven indicated locations. The deflection scale is grossly exaggerated for clarity of presentation, but even at this magnification the minor deflections at supports are not visible.

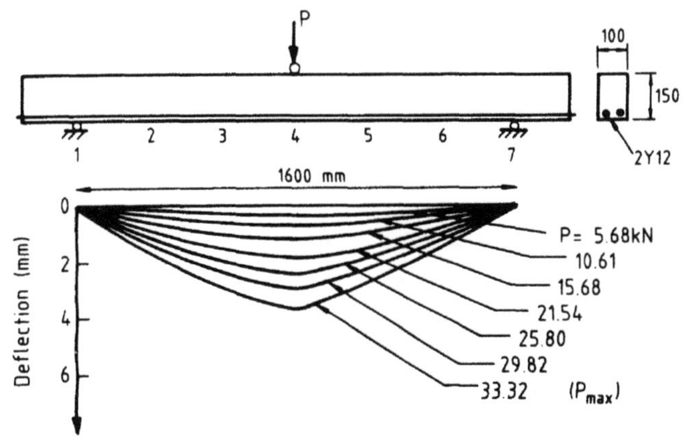

Fig. 5 Deflected shapes of the test beam with two reinforcing bars at various load levels constructed from the deflection readings at the seven indicated locations. The deflection scale is grossly exaggerated for clarity of presentation, but even at this magnification the minor deflections at supports are not visible.

Figs 2 and 3 show the corresponding plots drawn from the load and central deflection readings. Figures 4 and 5 show the deflection patterns of the two beams. These were plotted from the deflections measured at seven locations along the span, including the two support points. The deflection scale has been grossly exaggerated for clarity of presentation.

The beam with a single reinforcing bar reached its maximum load ($P_{max} = 24{\cdot}25$ kN) when the central deflection was around 4·6 mm (Fig. 4). This was followed by a drop in the load to about 23 kN (a drop of approximately 5%) before the load gradually increased to its ultimate value ($P_u = 24{\cdot}2$ kN) at which instant the central deflection had increased to around 10 mm. By this stage several flexural cracks had developed. Moreover, due to excessive flexural cracking a shear crack connecting a flexural crack near the support to the load point formed. This shear crack widened and caused the eventual collapse of the beam simultaneously with the crushing of concrete under the load.

The beam with two reinforcing bars reached the maximum load ($P_{max} = 33{\cdot}32$ kN) when the central deflection was around 3·5 mm (Fig. 5). By this stage several flexural cracks had formed. One of these cracks joined with a debonding crack along the reinforcement and curved towards the load point leading to sudden failure by the opening of the diagonal crack. A trace of the development of diagonal shear failure is reproduced on a reduced scale in Fig. 6, indicating the load

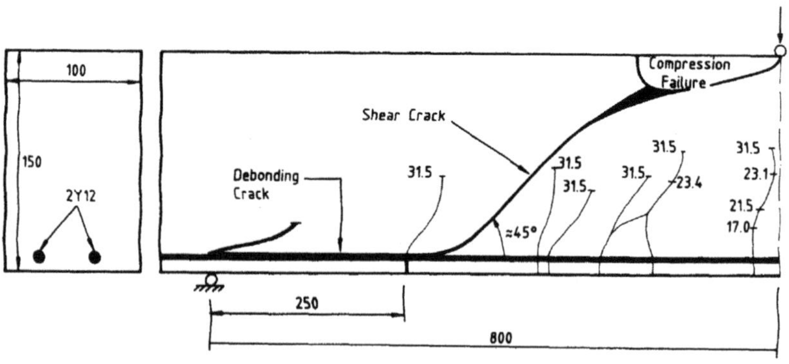

Fig. 6 A scaled-down trace of the actual crack development pattern in the test beam with two reinforcing bars showing load levels in kN. Note that the tensile crack which forms away from the support grows along the reinforcement first as a debonding crack before developing into a diagonal tensile crack (shear crack).

levels (in kN), the position of the flexural crack from which the diagonal crack emanated and its progression towards the load point. The figure shows only that half of the beam in which the diagonal shear crack formed.

3. FLEXURAL FAILURE OF LONGITUDINALLY REINFORCED BEAMS

As mentioned in the Introduction, the most complete description of the various stages in the development of the flexural failure mode is provided in Ref. [1], although several features of this failure mode are also revealed by the model proposed in Ref. [4].

Carpinteri [1] considers a longitudinally reinforced beam section with a through-thickness edge crack which is assumed to have formed in the tensile face of the beam and propagated to the level of the reinforcing steel. This crack will form when the flexural strength of the concrete mix has been reached, and will be practically uninfluenced by the amount of steel. Thus for the three-point beams reported above, the tension crack is likely to have formed at a load level of approximately $P = f_t BW^2/(1{\cdot}5S) = 3{\cdot}2$ kN (S = loaded span = 1·6 m). Once this tension crack reaches the reinforcement, the latter exerts an eccentric axial force on the concrete and resists much of the external bending moment. If it is assumed that the relative rotation of the beam sections straddling the crack is negligible until the steel yields, then it may be shown that

$$\frac{FW}{M} = \left[\left(\frac{1}{2} - \frac{c}{W}\right) + r(\beta)\right]^{-1} \tag{1}$$

where F = force in steel, M = applied bending moment ($= PS/4$), c = cover over steel and

$$r(\beta) = \frac{\displaystyle\int_0^\beta Y_M(x)\,Y_F(x)\,dx}{\displaystyle\int_0^\beta Y_M^2(x)\,dx} \tag{2}$$

Here, $\beta = a_0/W$, and $Y_M(\beta)$ and $Y_F(\beta)$ are the geometrical functions from the linear elastic fracture mechanics relating stress intensity factors to the applied bending moment and axial tensile force, respectively

$$Y_M(\beta) = 6(1{\cdot}99\beta^{1/2} - 2{\cdot}47\beta^{3/2} + 12{\cdot}97\beta^{5/2} - 23{\cdot}17\beta^{7/2} + 24{\cdot}8\beta^{9/2}) \quad (3)$$

$$Y_F(\beta) = 1{\cdot}99\beta^{1/2} - 0{\cdot}41\beta^{3/2} + 18{\cdot}70\beta^{5/2} - 38{\cdot}48\beta^{7/2} + 53{\cdot}85\beta^{9/2} \quad (4)$$

Figure 7 shows the variation in the steel force F with crack growth. F increases linearly with increasing external bending moment M until the yield limit $F_p = f_y A_s$ is reached, where A_s = area of steel and f_y = yield strength. Before the steel yields, cracks grow in concrete only by slip between steel and concrete. Although Carpinteri does not consider slip explicitly, he suggests that it may be simulated by assuming a fictitious yield stress in steel that is less than f_y.

It is assumed that the steel has a rigid plastic behaviour, so that the transmitted force remains equal to F_p after yielding. From eqn (1) it

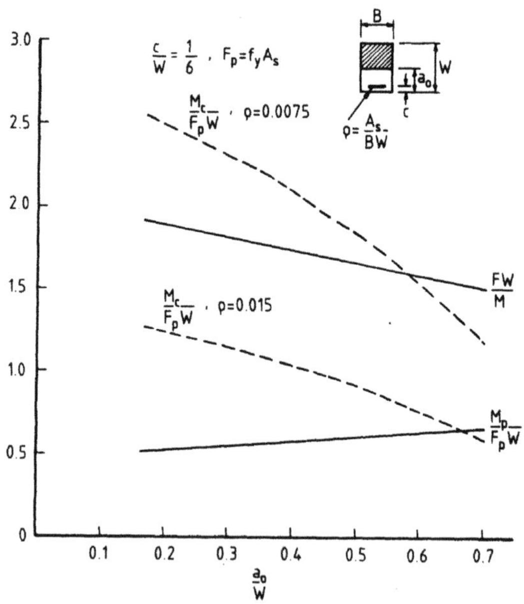

Fig. 7 Variation of steel force (F) with applied bending moment (M) under a growing tensile crack (depth = a_0). Also shown is the variation of the bending moment (M_p) at yielding of steel (F_p) under a growing tensile crack, as well as the moment corresponding to the compression failure of concrete (M_c) for the two steel ratios (broken lines). Note that the M_c curve for $\rho = 0{\cdot}015$ drops below the M_p line at $a_0/W \approx 0{\cdot}65$, whereas that for $\rho = 0{\cdot}0075$ would drop below it well after $a_0/W > 0{\cdot}7$.

follows that the bending moment at the yielding of steel is

$$\frac{M_{\mathrm{P}}}{F_{\mathrm{p}}W}=\left(\frac{1}{2}-\frac{c}{W}\right)+r(\beta) \tag{5}$$

M_{p} is also shown on Fig. 7. It is noted that the increase in plastic moment M_{p} at constant F_{p} is a result of increasing crack depth β. M_{p} cannot however increase indefinitely, because it is quite likely that the compression collapse of concrete will intervene. Assuming a linear stress variation in the compression ligament, it may be shown that the bending moment M_{c} required for the compression collapse of concrete is

$$\frac{M_{\mathrm{c}}}{F_{\mathrm{p}}W}=\frac{f'_{\mathrm{c}}}{f_{\mathrm{y}}}\frac{BW}{A_{\mathrm{s}}}\frac{(1-\beta)(2+\beta-3c/W)}{6} \tag{6}$$

For the two beams tested, the variation of $M_{\mathrm{c}}/(F_{\mathrm{p}}W)$ with β is shown by broken lines on Fig. 7. It is evident that failure by the crushing of the concrete can be expected in the beam with two reinforcing bars ($\rho=A_{\mathrm{s}}/(BW)=0{\cdot}015$) when the fracture has traversed up to about 65% of the depth of the beam, whereas in the beam with only one reinforcing bar ($\rho=0{\cdot}0075$) it may not be expected until the fracture has traversed more than 75% of the beam depth. These results are borne out by experimental observations.

After the onset of plastic flow in steel but before the collapse of concrete by crushing, the crack growth regime in concrete can only be studied by considering the resistance of concrete to crack growth, i.e. its fracture toughness. However, we depart here from the fracture toughness K_{Ic} used by Carpinteri [1] because it is known to be highly dependent on the beam depth W, and so cannot be regarded as a material constant [8]. Instead, we use the effective fracture toughness $K^{\mathrm{e}}_{\mathrm{Ic}}$ which has been shown to be reasonably independent of the depth of the beam in the range $100 \leq W \leq 400$ mm [9]. We could of course also have used the fracture toughness as determined by the two-parameter model ($K^{\mathrm{s}}_{\mathrm{Ic}}$) [10] or the size-effect law ($K^{\mathrm{b}}_{\mathrm{Ic}}$) [11], but these have been shown [8] to be very close to $K^{\mathrm{e}}_{\mathrm{Ic}}$. The bending moment at the tensile fracture of concrete M_{F} ($\geq M_{\mathrm{p}}$) can be calculated by equating the stress intensity factor at the notch tip to the fracture toughness $K^{\mathrm{e}}_{\mathrm{Ic}}$, and so giving

$$\frac{M_{\mathrm{F}}}{K^{\mathrm{e}}_{\mathrm{Ic}}BW^{3/2}}=\frac{1}{Y_{\mathrm{M}}(\beta)}+N_{\mathrm{P}}\left[\frac{Y_{\mathrm{F}}(\beta)}{Y_{\mathrm{M}}(\beta)}+\frac{1}{2}-\frac{c}{W}\right] \tag{7}$$

where the *brittleness number* $N_p=f_{\mathrm{y}}\rho W^{1/2}/K^{\mathrm{e}}_{\mathrm{Ic}}$.

The variation of M_F with β is shown in Fig. 8 for the two beams tested corresponding to $N_p = 1{\cdot}059$ and $2{\cdot}118$, respectively. It will be noted that the curves have a rather weak minimum suggesting that the fracturing process is more or less stable, once the crack has traversed about 35% of the beam depth for the beam with just one reinforcing bar. For the crack growth between $0{\cdot}16 \leq a_0/W \leq 0{\cdot}35$, one can expect instability resulting in a reduced M_F and therefore a drop in the external load. This is borne out by the experiment, as seen from Figs 2 and 3 for the beam with $\rho = 0{\cdot}0075$. Regarding the beam with $\rho = 0{\cdot}015$, Carpinteri's analysis [1] predicts a stable growth of tensile cracks beyond $a_0/W \approx 0{\cdot}2$. In fact, this is quite consistent with the experiment (Fig. 6), but of course in this case another mode of failure eventually takes over, namely the diagonal shear mode that cannot be explained by the above analysis. Before describing the shear mode of failure, it is worth pointing out that the brittleness number N_p introduced above reveals the importance of the size and steel ratio in

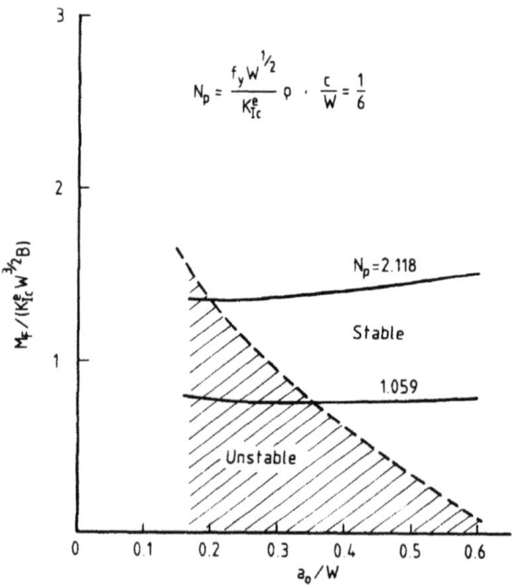

Fig. 8 Bending moment at tensile fracture of concrete against relative crack depth for the two test beams with brittleness number $N_p = 1{\cdot}059$ and $2{\cdot}118$, respectively. The broken line separating the stable crack growth regime from the unstable one is the locus minimorum. These minima are hardly distinguishable on the scale of the figure.

the failure mode transition. The higher the brittleness number the more likely the beam is to fail in an abrupt (brittle) manner due to shear. N_p increases with both the size of the beam (W) and the steel ratio (ρ). However, whilst the brittleness number gives an indication of the role of W and ρ, it alone cannot quantify the failure mode transition. It can however be used, for instance in establishing the minimum flexural reinforcement requirement by demanding that the bending moment at first cracking of concrete coincide with that at, say, the yielding of steel. This condition seems to be especially useful for high strength concrete [3].

4. SHEAR FAILURE OF LONGITUDINALLY REINFORCED BEAMS

As mentioned in the Introduction, several models based on fracture mechanics principles have been proposed for explaining the influence of the beam size and steel ratio upon the shear mode failure of longitudinally reinforced beams. The model described in Ref. [5] is in essence a simple modification of the traditional strength approach to shear failure. The modification is accomplished by invoking the size effect law proposed in Ref. [12]. The second model described in Ref. [6] (see also Hillerborg [7]) is based on the concept of the fictitious crack model [13] and an idealized bond stress against bond-slip relationship. This model can only be implemented numerically in a finite element program. The third model [4] is, in principle, similar to that of Ref. [6], but it permits an analytical estimation of the shear capacity, besides revealing several features of the failure mode transition. Moreover, it uses the fracture toughness value that is reasonably independent of the beam size rather than the fracture energy of the fictitious crack model that is known to vary with the beam size. We shall briefly describe each of these models and apply them to the two test beams.

4.1 The Jenq–Shah model

This model takes a strict fracture mechanics approach and recognizes that a longitudinally reinforced beam can attain the ultimate load either as a result of complete yielding of steel at midspan (tensile failure) or as a result of crack growth in the shear span (diagonal shear failure) (Fig. 9). It transpires that the transition in failure modes is

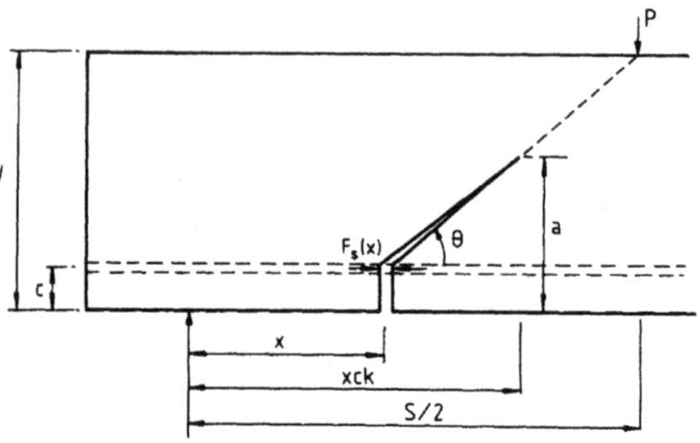

Fig. 9 An idealized diagonal tensile crack that has initiated from a tensile crack at position x and is growing at an angle $\theta(\approx 45°)$ towards the load point. a denotes the location of shear crack at collapse [4].

intimately connected with the distribution of steel force along the reinforcing bars and the depth of the beam. The calculation of the distribution of steel force along the beam is made difficult by the absence of precise information on longitudinal cracking, internal cracking and dowel action. For this reason, it is assumed that it can be represented by a power law

$$F_s(x) = F_{s,max}\left(\frac{2x}{S}\right)^N \leq f_y A_s \tag{8}$$

where $F_{s,max}$ occurs at $x = S/2$. The value of N is adopted after numerical experimentation based on a bilinear approximation to the bond-slip and frictional pull-out behaviour of the reinforcing bars.

Moreover, for simplicity it is assumed that any crack growth in the concrete takes place when the stress intensity factor at the crack tip attains the tensile mode fracture toughness of the material as given by the two-parameter model (K_{Ic}^s) [10]. In other words, any shear and in-plane stresses in the path of a growing crack are ignored. It should be noted that although these stresses force the crack to deviate out of its plane and may even retard its growth it will always grow in a direction in which the local stress field is predominantly one of pure tension (this condition of crack growth is the so-called criterion of local symmetry).

Based on the pull-out behaviour of a reinforcing bar which suggests that the average ultimate bond stress, τ_{av} is inversely proportional to the embedment length, Jenq & Shah argue that for a constant span-to-depth ratio, τ_{av} must be inversely proportional to the beam depth, W. As a result of extensive numerical experimentation, they suggest the following equation for the maximum force in the reinforcing bar at the midspan

$$F_{s,max}\left(x=\frac{S}{2}\right)=19{\cdot}6f_t'\left(\frac{\rho}{W}\right)^{1/2}S \tag{9}$$

where f_t' = direct tensile strength of concrete (in psi). Equation (9) is in Imperial units and is for beams 10 in (≈254 mm) wide having just one reinforcing bar. Careful conversion of this equation into SI units with allowance for a number of reinforcing bars gives

$$F_{s,max}\left(x=\frac{S}{2}\right)=(n\pi B)^{1/2}S\sqrt{\rho W}\tau_{av}$$

$$=205{\cdot}85\sqrt{n\rho}f_t' \tag{10}$$

for the present test beams with $B = 100$ mm, $W = 150$ mm and $S = 600$ mm. Here, n = number of reinforcing bars, ρ = steel ratio and $f_t' \approx 0{\cdot}75f_t = 2{\cdot}55$ MPa for the mix used in the test beams, and $F_{s,max}$ is in kN. For the two test beams with $\rho = 0{\cdot}0075$ and $0{\cdot}015$, respectively, eqn (10) suggests that the maximum force in the reinforcing bars at midspan is approximately 45 kN and 90 kN, respectively. This means that the reinforcing steel is still below the yield limit in both beams (f_yA_s = 52 kN and 104 kN, respectively).

Having estimated approximately the $F_{s,max}$, eqn (8) gives the distribution of steel force along the span of the beam. We are now in a position to estimate the contributions of concrete and steel to the resistance P of the beam for each crack length.

Referring to Fig. 9, the contribution of concrete matrix to the load-carrying capacity is estimated on the assumption that the crack has grown to the current length a_0 at section x such that at its tip the stress intensity factor equals the fracture toughness of the mix K_{Ic}^s. As previously noted, K_{Ic}^s is very nearly the same as K_{Ic}^e for the mix whence

$$K_{Ic}^e=\sigma_N\sqrt{a_0}F_1(\beta) \tag{11}$$

where a_0 is the current length of the crack, $\beta = a_0/W$, σ_N = nominal flexural stress at section $x = (3P_cx)/(BW^2)$, and the geometrical function $F_1(\beta)$ is

$$F_1(\beta)=\frac{1{\cdot}99-\beta(1-\beta)(2{\cdot}15-3{\cdot}93\beta+2{\cdot}7\beta^2)}{(1+2\beta)(1-\beta)^{3/2}} \tag{12}$$

From eqn (11) we obtain the contribution of the concrete matrix to the resistance at the current crack length at section x to be

$$P_c=(K_{Ic}^{e}BW^2)/[3x\sqrt{a_0}F_1(\beta)] \tag{13}$$

(The corresponding equation in Ref. [4] has typographical errors in it.) In both test beams, collapse occurred when the crack had traversed about 125 mm up the beam depth ($a_0=125$ mm at collapse). However, in the beam with just one reinforcing bar the critical tensile crack developed close to the midspan $x\approx 800$ mm, whereas in the beam with two reinforcing bars the critical tensile crack that eventually gave rise to the diagonal shear crack, developed closer to the support at $x=250$ mm (Fig. 6). With these values of a_0 and x and noting that for the mix $K_{Ic}^{e}=1{\cdot}27\ \mathrm{MPa}\sqrt{\mathrm{m}}$, eqn (13) predicts the following contributions of the concrete matrix to the ultimate load carrying capacity of the two beams

$$(P_c)_u=0{\cdot}32\ \mathrm{kN} \qquad (\rho=0{\cdot}0075) \tag{14}$$

$$(P_c)_u=1{\cdot}03\ \mathrm{kN} \qquad (\rho=0{\cdot}015) \tag{15}$$

Equation (14) would seem to suggest that the concrete matrix makes but little contribution to the ultimate load carrying capacity of the beam that fails in the tensile mode. The bulk of the load is carried by the reinforcement and it can be calculated by the conventional strength theory, assuming that the neutral axis is located at the centroid of the uncracked ligament, and so giving ($a_0\approx 125$ mm, $F_{s,max}=f_yA_s=52$ kN)

$$(P_s)_u=\frac{4F_{s,max}}{S}\left(\frac{2W}{3}+\frac{a_0}{3}-c\right)=15{\cdot}2\ \mathrm{kN} \tag{16}$$

Therefore, for the beam with just one bar, $P_u=(P_c)_u+(P_s)_u$ would seem to be only around 15·5 kN, whereas the test showed that the beam collapsed at 24·2 kN. This large discrepancy results from the neglect of strain hardening in steel and of the multiple tensile cracking in concrete which is due to the bond-slip transfer from the midspan towards the supports, as well as that of the dowel action of longitudinal steel.

For the beam with two reinforcing bars the contribution of the concrete matrix to the ultimate load-carrying capacity as predicted by eqn (15) would also be underestimated for a further two reasons in addition to the reasons noted above. First, eqn (13) assumes that the original tensile crack grows in its own plane. Second, any in-plane and shear stresses which would retard the growth of an out-of-plane crack have been neglected. Despite this obvious error in the estimation of the concrete matrix contribution, one may add it to the contribution from steel in order to estimate the ultimate capacity of the test beam with two reinforcing bars. For this beam

$$(P_s)_u = \frac{2F_s(x)}{x_{ck}}\left(\frac{2W}{3} + \frac{a_0}{3} - c\right) \tag{17}$$

where (Fig. 9) $a_0 = 125$ mm,

$$x_{ck} = x + (a_0 - c)\cot 45° = 350\ \text{mm} \tag{18}$$

and so giving

$$(P_s)_u = 60\left(\frac{5}{16}\right)^N \text{kN} \tag{19}$$

(The equation corresponding to eqn (17) in Ref. [4] has typographical errors in it.) From eqns (15) and (19) we get $P_u = (P_c)_u + (P_s)_u$, which may be equated to the experimental value of 33·32 kN in order to estimate the exponent N.

This calculation gives $N = 0{\cdot}53$. Figure 10 shows the distribution of force along the steel bars for several values of N. It is clear from the figure that for $N \approx 0{\cdot}5$ one would expect the ends of the bars to be anchored. In the present test beam the bars were not anchored at the ends, so a value of $N > 1$ would seem to be more appropriate. That the above calculations predicted a value $N < 1$ again points to the fact that the present model underestimates the contribution of the concrete matrix to the ultimate carrying capacity of the beam. It is worth pointing out that in the present model a tensile mode of failure is predicted for $N = 0$ (*cf.* eqns (16), (17)) independently of the size of the beam.

4.2 The Gustafsson–Hillerborg Model

In principle, this model is identical to the Jenq–Shah model described above. In practice though it can only be implemented in a finite

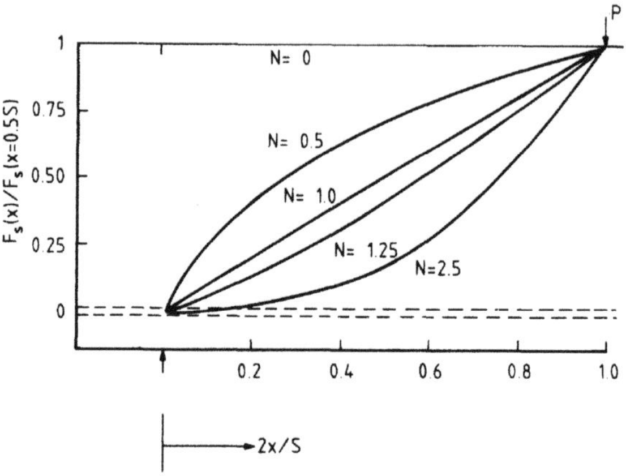

Fig. 10 Power-law representation of the distribution of the steel force along the shear span [4].

element program. The basic assumptions are likewise the same, namely that in the diagonal tensile crack development the aggregate interlock between the fracture surfaces and the shear force carried by the dowel action of reinforcing steel is ignored. However, instead of the fracture toughness K_{Ic}^{s} used in the Jenq–Shah model, the present model describes the concrete fracture resistance through the fracture energy G_F that is known to vary with the beam size [14]. Moreover, a characteristic length parameter $l_{ch} = EG_F/(f_t')^2$ is introduced to characterize the fictitious crack model in an otherwise linear elastic concrete.

The assumptions made in the numerical implementation of the model are summarized in Fig. 11. The reinforcing steel and unfractured concrete are assumed to behave in a linear elastic manner. The interaction between the concrete and steel is modelled by the simple bond-slip relation as in the Jenq–Shah model. Several potential crack paths for the development of a diagonal tensile crack are chosen (Fig. 12) with a view to locating the critical location of the eventual fracture path. The zone ahead of the traction-free crack is modelled by a fictitious crack capable of transferring stresses as described by the stress-crack opening displacement diagram (σ–w) of the concrete.

It is shown that the shear strength of longitudinally reinforced beams is approximately proportional to $f_t'(W/l_{ch})^{-1/4}$. With the above definition of l_{ch} and noting that f_t' is approximately proportional to

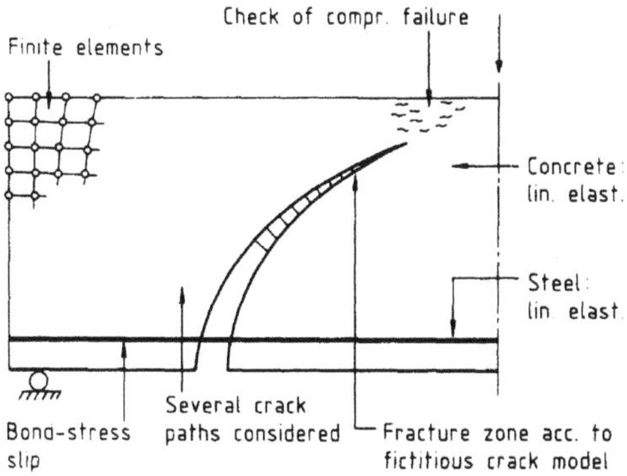

Fig. 11 An overview of the various stages in the shear strength analysis according to Gustafsson & Hillerborg [6].

$(f'_c)^{1/2}$, it follows that the ultimate shear strength is

$$f_s = A(EG_F f'_c / W)^{1/4} \tag{20}$$

where A = constant of proportionality. The influence of beam depth upon the ultimate shear strength as predicted qualitatively by eqn (20) is claimed to be consistent with the results of tests reported by many investigations [15]. It is worth noting that the Jenq–Shah model predicts that f_s scales not as $W^{-1/4}$ but as $W^{-1/2}$ (eqn (9)). It seems likely that this discrepancy is a result of the well-known dependence of G_F itself on W, whereas in arriving at eqn (20), the present model assumes that it is a material constant independent of W.

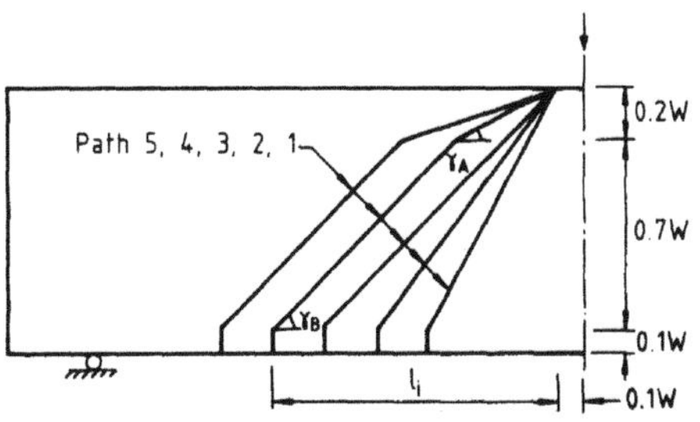

Fig. 12 The various fracture propagation paths tried in the shear strength analysis [6] with γ_A varying between 18° and 61°, γ_B between 45° and 61°, and l_i/W between 1·3 and 0·5.

4.3 The Bažant–Kim Model

This model essentially combines the existing code formulae for diagonal shear failure with the size effect law [12] and establishes the several empirical constants statistically so as to achieve the best fit with the large body of available test data. However, it recognizes the fact that the enormous scatter in the test data is not only due to the size effect, but also a result of how the steel ratio ρ and the shear span ($a = M/V$ where V = shear force at the section on which a bending moment M is acting) are taken into account in these formulae. Therefore the influence of these two factors is reconsidered, together with the size effect.

The mathematical development of the model begins with the conventional notion that the total shear resistance V ($=P/2$ in three-point bending) can be decomposed into two components—a component V_1 resulting from the composite action of concrete and steel through the mechanism of bond-slip and frictional pull-out and a component V_2 due to the so-called arch action (Fig. 13). It is noted that the rate of change of force in steel (T) may be related to the bond stress and the reinforcement ratio, such that

$$V_1 = k_1 \rho^{(1/2-m)} (f'_c)^q BW \tag{21}$$

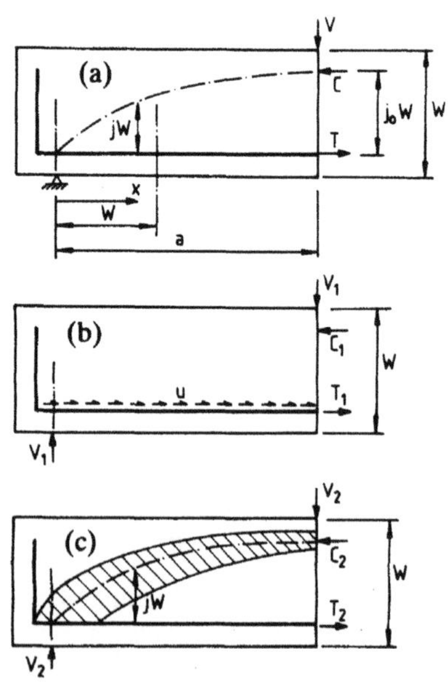

Fig. 13 The two components of the shear capacity (a) result from the composite action between the steel and concrete (b) and from the so-called arch action (c), showing the notation used in the text [5].

where k_1, m and q are empirical constants, with $q \approx 0{\cdot}5$. The arch-action contribution to shear is shown to be (Fig. 13)

$$V_2 = j_0\left(\frac{r}{a}\right)^{r-1} \sigma_s \rho B W^2 \tag{22}$$

where the steel force $T = \sigma_s \rho BW$, and j_0 is a constant defining the location of the compression resultant C at the end of the shear span, $x = a$. Moreover, it is assumed that σ_s is a constant (corresponds to $N = 0$ in the Jenq–Shah model) and that the critical section for arch-action shear is approximately at $x = W$, so that eqn (22) reduces to

$$V_2 = c_2 \frac{\rho^{1-m}}{(a/W)^r} BW \tag{23}$$

Combining eqn (21) with eqn (23) and scaling the result by the size effect law appropriate for the brittle failure condition due to the blunt cracking of concrete [12] gives

$$f_s = k_1 \rho^p \left((f_c')^q + k_2 \frac{\sqrt{\rho}}{(a/W)^r}\right)\left(1 + \frac{W}{\lambda_0 g}\right)^{-1/2} \tag{24}$$

where $k_2 = c_2/k_1$, λ_0 is an empirical constant, and g = maximum aggregate size in the mix. The above approach to the inclusion of size effect implies that both V_1 and V_2 scale according to the same law. Whilst this may be acceptable for the arch-action contribution, it is highly doubtful for the composite action contribution.

From a statistical analysis of all existing test data, the following formula for the mean ultimate nominal shear strength resulted from eqn (24)

$$(f_s)_u = \frac{10\rho^{1/3}}{\sqrt{1 + W/(25g)}} [\sqrt{f_c'} + 3000\sqrt{\rho/(a/W)^5}] \tag{25}$$

where f_c' is in psi (1 psi = 6·895 kPa).

It is further noted that the current codes of practice are based on the philosophy of providing a safety margin not against the ultimate load in diagonal shear failure but against the load at the initiation of such failure. However, since the test data show that the size effect is almost undetectable at diagonal crack initiation whilst it is very pronounced at the ultimate stage, such a philosophy is highly questionable. In fact, a uniform safety margin cannot be assured if the design is based on the diagonal shear initiation concept, because the safety margin would decrease with increasing W/g. In the light of this observation, Bažant & Kim have scaled eqn (25) in such a manner as to ensure a uniform safety margin over a broad range of sizes. This statistical trial and error scaling procedure resulted in the following design formula encompassing all existing test data

$$(f_s)_u = \frac{8\rho^{1/3}}{\sqrt{1 + W/(25g)}}[\sqrt{f'_c} + 3000\sqrt{\rho/\alpha^5}] \tag{26}$$

where $\alpha = a/W$ for the case of a concentrated load. For the test beam with two reinforcing bars that failed in the diagonal shear mode, the parameters appearing in eqn (26) are: $\rho = 0{\cdot}015$, $W = 150$ mm, $g = 20$ mm, $f'_c = (38\,000/6{\cdot}895)$ psi, $a \approx 575$ mm $(=(\frac{S}{2} - x)$ of the Jenq–Shah model). With these geometrical and material parameters (assuming $\alpha \approx 4$), eqn (26) gives the ultimate load at failure $P_u = 2(f_s)_u BW = 7115$ lbs $= 31{\cdot}69$ kN which is very close to the measured ultimate load of 33·32 kN.

The fact that this model predicts a slightly conservative value for P_u confirms the observation made in ref. [5], namely that hardly any data point lies below the shear strength predicted by eqn (26) and that almost all data points lie just slightly above it. However, this confirmation may be just fortuitous, because of the ambiguity inherent in the proper choice of the shear span a. In the derivation of eqn (26) it is measured from the support, whereas in applying eqn (26) to the test beam we measured it from the actual point of initiation of the diagonal crack to its final position at collapse (i.e. the midspan). Had we chosen the shear span between the support and the load as required by the model, the ultimate load at collapse would have been only $P_u = 29{\cdot}56$ kN. It would therefore seem that the accuracy of the ultimate shear capacity predicted by the Bažant–Kim model depends on how best one is able to anticipate the location at which the eventual diagonal crack will initiate.

There is yet another reason to suspect the fortuitous nature of the prediction for the test beam. It will be recalled that in the Bažant–Kim model the steel stress σ_s is assumed to be constant over the shear span. In the notation of the Jenq–Shah model this corresponds to $N = 0$. We have previously shown that N ought to exceed unity for unanchored reinforcement in order to approximate correctly the bond-slip and pull-out behaviour of the reinforcing bar.

Finally, it is interesting to point out that the Bažant–Kim model, just like the Jenq–Shah model, predicts that $(f_s)_u$ scales as $W^{-1/2}$.

5. DISCUSSION AND CONCLUSIONS

The two test beams which differed from each other only by the amount of longitudinal reinforcing steel clearly demonstrated the important role of the steel ratio ρ in the failure mode transition. The beam with a low ρ failed in the tensile mode, whereas by doubling ρ the beam failed in the diagonal shear mode. The various stages in the development of the tensile collapse are satisfactorily explained by the analysis proposed by Carpinteri, including the unstable crack propagation after the attainment of P_{max}. It also explains the observed test result that this instability is only temporary and that stability is restored once the tensile crack has traversed about 35% of the beam depth, whereafter further crack growth occurs stably until the ultimate load P_u is reached. The notion of brittleness number N_p introduced by Carpinteri is helpful in highlighting the role of the size W and the steel ratio ρ in the failure mode transition. But it alone cannot quantify the actual transition levels between flexural and shear failure. Moreover, we proposed that the fracture toughness K_{Ic} (derived from the fracture energy G_F; $K_{Ic} = \sqrt{EG_F}$) used in the definition of $N_p = (f_y p W^{1/2})/K_{Ic}$ be replaced by the effective fracture toughness K_{Ic}^e (or K_{Ic}^s or K_{Ic}^b) that is known to be reasonably independent of W. As it is currently defined, N_p is both explicitly and implicitly dependent on W in view of the known variation of G_F with the beam size W.

Some of the features of the tensile collapse are also revealed by the Jenq–Shah model. However, this model significantly underestimates the ultimate load P_u in this failure mode. This may be because the model neglects multiple tensile cracking in concrete as a result of inadequate representation of the steel pull-out behaviour (i.e. N ought to be positive even in this mode) and the dowel action of longitudinal reinforcement. This model is far more successful in predicting the collapse load in diagonal shear mode of failure. However, even in this mode the contribution of the concrete matrix would seem to be underestimated because the model ignores the frictional interlock between the diagonal tensile crack faces that could significantly retard the crack growth in concrete. It should be noted though that even under mixed-mode conditions the crack growth always takes place along a direction in which the local stress field is predominantly one of pure tension (Mode I). The authors of this model are cognisant of its limitations, and will no doubt overcome these in its future development. In any future revision of this model, it would be most helpful to the user to have the necessary details of the numerical experimentation needed to arrive at eqn (9). It is worth pointing out that the model already seems to reveal the probable size effect in diagonal shear mode. It predicts that the ultimate shear capacity scales as $W^{-1/2}$ which agrees with the prediction of the Bažant–Kim model and also with the implied size effect in the concept of the brittleness number N_p.

The approach suggested by Gustafsson & Hillerborg is similar in principle (and in assumptions) to the Jenq–Shah model for diagonal shear mode of failure, but it can only be implemented numerically in a finite element program, which may restrict its use. Moreover, it is based on the notion of fracture energy G_F which is known to depend on the size of the beam as has been acknowledged even by its proponent [14]. It would seem that because of the intrinsic size dependency of G_F, the Gustafsson–Hillerborg model predicts that the ultimate shear capacity of longitudinally reinforced beams scales as $W^{-1/4}$, whereas all the other available models would suggest this variation to be $W^{-1/2}$. The scatter of the existing test data is so large that both scaling laws could be claimed to fit the data reasonably well, leaving the question of the scaling law still unresolved.

The Bažant–Kim model is the easiest of all the models to use in their present form. That is not surprising when it is recognized that it essentially scales the existing code formulae for diagonal shear failure to account for the size effect appropriate for the blunt crack growth in concrete. In this scaling procedure it assumes that the contributions from the composite action and the arch action scale in an identical manner. Whilst this may be acceptable for the arch-action contribution, it is yet to be demonstrated for the composite action contribution. The model moreover, assumes that the steel stress is constant over the entire shear span. This would seem to be a somewhat doubtful assumption in that it is tantamount to ignoring any bond-slip between the steel and concrete and to considering only the frictional part of the overall pull-out behaviour of the reinforcing bar. Nonetheless, this model gave the best prediction for the shear capacity of the test beam with $\rho = 0{\cdot}015$, when the shear span was measured from the location of the incipient diagonal tensile crack to the position of the applied concentrated load, but not from the support as required by the model. In the latter case the prediction is more conservative.

In conclusion, it may be helpful to delineate the most important questions that require the attention of the research community. From the foregoing discussion it is obvious that the contribution of the concrete matrix to the ultimate shear capacity is grossly underestimated because the major mechanism of retardation of a diagonal tensile crack under mixed-mode conditions, namely the frictional interlock between the crack faces is totally ignored. Likewise, it would seem that the pull-out behaviour of the reinforcing bar in the concrete matrix is not adequately represented by any of the available models. Finally, it appears that the question of the size dependency of the shear capacity is still far from resolved, judging by the differences in the scaling laws predicted by the models.

ACKNOWLEDGEMENT

Dr P. Nallathambi was instrumental in the successful execution of the test programme that was originally suggested by Professor S. P. Shah when he was a Visiting Professor at The University of Sydney.

REFERENCES

[1] Carpinteri, A., Stability of fracturing process in RC beams. *J. of Struct. Eng.* (ASCE) **110** (1984) 544–58.

[2] Carpinteri, A., Size-scale effects on the brittleness of concrete structures: dimensional analysis and snap-back instability. In *Fracture Mechanics: Application to Concrete,* ed. V. C. Li & Z. P. Bažant. American Concrete Inst., Detroit, SP-118, 1989, pp. 197–236.

[3] Bosco, C., Carpinteri, A. & Debernardi, P. G., Minimum reinforcement in high-strength concrete. *J. of Struct. Eng.* (ASCE), **116** (1990) 427–37.

[4] Jenq, Y.-S. & Shah, S. P., Shear resistance of reinforced concrete beams—a fracture mechanics approach. In *Fracture Mechanics: Application to Concrete,* ed. V. C. Li & Z. P. Bažant. American Concrete Inst., Detroit, SP-118, 1989, pp. 237–58.

[5] Bažant, Z. P. & Kim, J.-K., Size effect in shear failure of longitudinally reinforced beams. *ACI J.*, **81** (1984) 456–68.

[6] Gustafsson, P. J. & Hillerborg, A., Sensitivity in shear strength of longitudinally reinforced concrete beams to fracture energy of concrete. *ACI Struct. J.*, **85** (1988) 286–94.

[7] Hillerborg, A., Fracture mechanics and the concrete codes. In *Fracture Mechanics: Application to Concrete,* ed. V. C. Li & Z. P. Bažant. American Concrete Inst., Detroit, SP-118, 1989, pp. 157–70.

[8] Karihaloo, B. L. & Nallathambi, P., Notched beam test: mode I fracture toughness. In *Fracture Mechanics of Concrete: Test Method,* ed. S. P. Shah & A. Carpinteri. Chapman & Hall, London, 1991, pp. 1–86.

[9] Karihaloo, B. L. & Nallathambi, P., An improved effective crack model for the determination of fracture toughness of concrete. *Cem. & Conc. Res.,* **19** (1989) 603–10.

[10] Jenq, Y.-S. & Shah, S. P., Two-parameter fracture model for concrete. *J. Eng. Mech.* (ASCE), **111** (1985) 1227–41.

[11] Bažant, Z. P., Kim, J.-K. & Pfieffer, P. A., Determination of fracture properties from size effect tests. *J. of Struct. Eng.* (ASCE), **112** (1986) 289–307.

[12] Bažant, Z. P., Size effect in blunt fracture: concrete, rock, metal. *J. of Eng. Mech.* (ASCE), **110** (1984) 518–35.

[13] Hillerborg, A., Modéer, M. & Petersson, P. E., Analysis of crack formation and crack growth in concrete by means of fracture mechanics and finite elements. *Cem. & Conc. Res.,* **6** (1976) 773–82.

[14] Hillerborg, A., The theoretical basis of a method to determine the fracture energy G_F of concrete. *Matériaux et Constructions,* **18** (1985) 291–6.

[15] Shioya, T., Iguro, M., Nojiri, Y., Akiyama, H. & Okada, T., Shear strength of large reinforced concrete beams. In *Fracture Mechanics: Applications to Concrete,* ed. V. C. Li & Z. P. Bažant. American Concrete Inst., Detroit, SP-118, 1989 pp. 259–80.

Chapter 22

Reinforced Concrete Beam Behavior under Cyclic Loadings

ANDREA CARPINTERI

Istituto di Scienza e Tecnica delle Costruzioni, Università di Padova, 35131 Padova, Italy

ABSTRACT

If a reinforced concrete beam cross-section is subjected to a cyclic bending moment, the maximum value of which is greater than or equal to the bending moment of slippage or yielding of the reinforcement and lower than the concrete fracture bending moment, then a shake-down phenomenon due to steel plastic flow occurs. A Fracture Mechanics model is employed to analyze the above-mentioned phenomenon, which can be elastic, i.e. without energy dissipation, or plastic and dissipative. A numerical procedure follows the fatigue crack growth, and the energy dissipated during the cyclic loading and unloading processes is calculated from the hysteretic loops described in the moment–rotation diagram. Some experimental tests have been carried out in order to verify the reliability of the proposed theoretical model.

1. INTRODUCTION

When a reinforced concrete beam is subjected to cyclic loadings, it deteriorates progressively, its stiffness and loading capacity sensibly decreasing for different damage phenomena, like crushing and fracturing of concrete or pulling-out and yielding of the steel reinforcement.

Several models have been proposed to simulate the nonlinear

reinforced concrete beam behavior. For example, the dual component model, where each beam is replaced by an elastic element and an elasto-plastic element in parallel; the fiber model, where each cross-section is divided into many layers of fibers and the moment–curvature relationship is obtained from steel and concrete constitutive laws; the single component model, where each member is represented by an elastic beam element with inelastic springs at its two ends [1]. These models do not distinguish the contribution of each damage mechanism, but they are intended to represent the global phenomenon.

Other authors have studied the response of reinforced concrete structural elements, pointing out the hysteretic phenomena which can occur under cyclic loadings [2]–[8].

The Fracture Mechanics model proposed in Refs [9]–[11] analyzes the behavior of a reinforced concrete beam cross-section subjected to cyclic bending moment (Fig. 1), the maximum bending moment M being greater than or equal to the value M_P of slippage or yielding of the steel reinforcement [12] and lower than the unstable concrete fracture value M_F. It can be verified that an elastic shake-down due to steel plastic flow occurs for $M_P \leq M < M_{SD}$, with $M_{SD} = 2M_P$ = plastic shake-down bending moment; when $M_{SD} \leq M < M_F$, the phenomenon is plastic and dissipative, that is, the steel stress–strain curve shows hysteretic loops and energy dissipation.

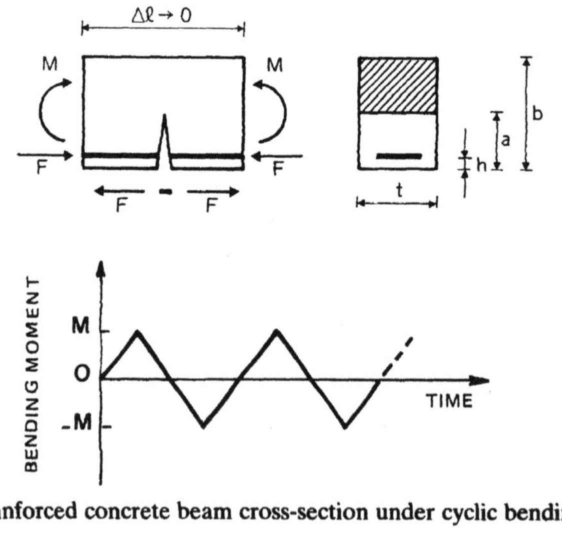

Fig. 1 Reinforced concrete beam cross-section under cyclic bending moment.

More precisely, a rectangular beam cross-section subjected to monotonic bending moment is firstly considered and a through-thickness edge crack is assumed to exist in the stretched part. The force transmitted by the reinforcement to the beam can be evaluated by means of a rotation congruence condition [12]. Applying Linear Elastic Fracture Mechanics, such a force increases linearly by increasing the applied bending moment, until the limit force of pulling-out or yielding of steel is reached. From this point onwards, a perfectly plastic behavior can be considered for the reinforcement. As a matter of fact, it is possible to show that even the slippage can be described by a rigid–plastic law [13] and the bond stress–slip relationship for monotonic loading in tension is almost identical to that in compression [14].

When the bending moment M_P of steel plastic flow has been exceeded, the cracked cross-section presents a linear hardening behavior until the unstable concrete fracture value M_F is reached.

If cyclic loading and unloading is examined [9]–[11], the maximum bending moment being greater than or equal to the value M_P, the phenomenon of shake-down can be studied. As already said, up to the value M_{SD} the shake-down is elastic, i.e. without energy dissipation; above this value the phenomenon becomes plastic and the energy absorbed can be calculated from the moment–rotation diagram.

Then the diagram ξ_{CR} against N_P can be obtained, ξ_{CR} being the critical relative crack depth, i.e. the relative crack depth $\xi = a/b$ for which $M_{SD} = M_F$. The parameter N_P is dependent on the mechanical and geometrical properties of the reinforced concrete beam cross-section [12] and is called the brittleness number. The phenomenon of plastic shake-down cannot occur for $\xi \geq \xi_{CR}$ because it is preceded by an unstable concrete fracture, M_F being lower than or equal to M_{SD}.

It is possible to verify that the critical relative crack depth decreases by increasing the dimensionless number N_P. Therefore, the plastic fatigue can occur only for very low values of the relative crack depth if the value N_P is large. Moreover, it can be remarked that the critical bending moment M_{CR} increases by increasing the dimensionless number N_P, M_{CR} being the value of the bending moments M_{SD} and M_F for $\xi = \xi_{CR}$.

A numerical procedure is developed which follows the fatigue crack growth and calculates the total energy dissipation. As a matter of fact, if an uncracked cross-section is subjected to a cyclic bending moment, fracturing of concrete can occur and the energy dissipated in the steel reinforcement can be computed for each loading cycle. When the crack depth increases, the hardening line of the moment–rotation diagram becomes more and more inclined and, therefore, the hysteretic loops shift toward the right-hand side. Finally, an unstable concrete fracture phenomenon occurs when the value ξ_F is reached, ξ_F being the relative crack depth for which the maximum value of the cyclic bending moment is equal to M_F.

Some experimental tests have been carried out [15], [16] in order to verify the reliability of the theoretical model proposed in Refs [9]–[11]. The beams have been subjected to cyclic loading by means of a four-point bending test.

Bending moment against rotation diagrams have been obtained for the centre-line cross-sections of the tested samples. On the basis of these diagrams, it can be remarked that the rigid-linear hardening behavior theoretically predicted is experimentally confirmed. Moreover, an elastic shake-down occurs for $M_P \leq M < M_{SD}$, while for $M_{SD} \leq M < M_F$ the shake-down is plastic and the hysteretic loops shift toward the right-hand side in the moment–rotation diagram.

2. BEHAVIOR OF RC BEAMS UNDER MONOTONIC LOADINGS

As is well-known, a cracked beam cross-section is equivalent to an elastic joint rotating under the action of the bending moment M^* and the axial tensile force F^* (Fig. 2) [17], [18]

$$\varphi = \lambda_{MM} M^* + \lambda_{MF} F^* = \text{cross-section rotation} \tag{1}$$

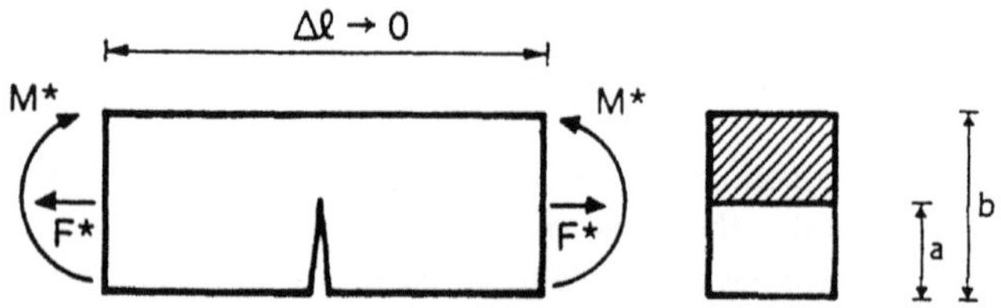

Fig. 2 Cracked beam cross-section.

where

$$\lambda_{MM} = \frac{2}{b^2 tE} \int_0^{\xi} Y_M^2(\xi)\, d\xi \tag{2}$$

$$\lambda_{MF} = \frac{2}{btE} \int_0^{\xi} Y_M(\xi) Y_F(\xi)\, d\xi \tag{3}$$

$$\xi = \frac{a}{b} = \text{relative crack depth} \tag{4}$$

$$Y_M(\xi) = 6(1{\cdot}99\xi^{1/2} - 2{\cdot}47\xi^{3/2} + 12{\cdot}97\xi^{5/2} - 23{\cdot}17\xi^{7/2} + 24{\cdot}80\xi^{9/2})$$

and

$$Y_F(\xi) = 1{\cdot}99\xi^{1/2} - 0{\cdot}41\xi^{3/2} + 18{\cdot}70\xi^{5/2} - 38{\cdot}48\xi^{7/2} + 53{\cdot}85\xi^{9/2}$$
$$\text{for } \xi \leq 0{\cdot}6 \text{ (from Ref. [19])} \tag{5}$$

E = Young's modulus of the material.

Moreover, the stress-intensity factor at the crack tip is equal to [19]

$$K_I = \frac{M^*}{b^{3/2}t} Y_M(\xi) + \frac{F^*}{b^{1/2}t} Y_F(\xi) \tag{6}$$

Consider a reinforced concrete beam cross-section subjected to a bending moment M (Fig. 3). A through-thickness edge crack with depth $a \geq h$ is assumed to exist in the stretched part, h being the steel cover. Therefore, the cross-section is globally subjected to the bending moment M (opening the crack) and to the eccentric axial force F (closing the crack) due to the statically undetermined reaction of the

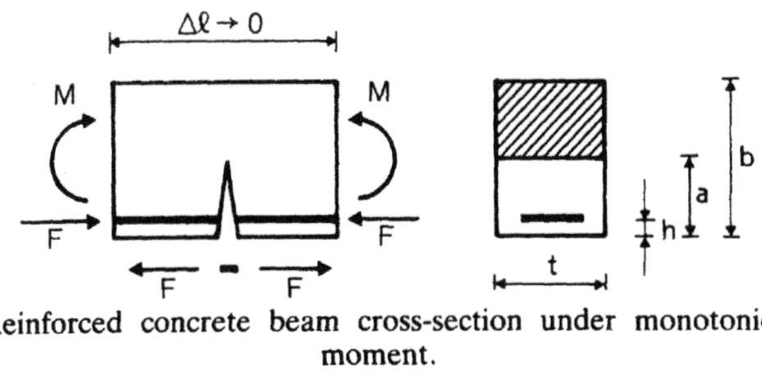

Fig. 3 Reinforced concrete beam cross-section under monotonic bending moment.

reinforcement. Consequently, the global actions on the cross-section are

$$M^* = M - F\left(\frac{b}{2} - h\right)$$
$$F^* = -F \tag{7}$$

Up to the moment when steel yields, the rotation is assumed to be equal to zero [12]

$$\varphi = \lambda_{MM}\left[M - F\left(\frac{b}{2} - h\right)\right] + \lambda_{MF}[-F] = 0 \tag{8}$$

where λ_{MM} and λ_{MF} are calculated from eqns (2) and (3), with E = Young's modulus of concrete. Applying this congruence condition, the force F transmitted by the reinforcement can be obtained as a function of the bending moment M

$$\frac{Fb}{M} = \frac{1}{\left(\frac{1}{2} - \frac{h}{b}\right) + r(\xi)} \tag{9}$$

with

$$r(\xi) = \frac{\int_0^{\xi} Y_M(\xi) Y_F(\xi)\, d\xi}{\int_0^{\xi} Y_M^2(\xi)\, d\xi} \tag{10}$$

Therefore, the bending moment of steel plastic flow M_P can be calculated from eqn (9)

$$M_P = F_P b\left[\frac{1}{2} - \frac{h}{b} + r(\xi)\right] \tag{11}$$

where F_P can indicate either the force of yielding $f_y A_s$ (f_y = steel yield strength; A_s = steel area) or the force of pulling-out, when the latter is lower than the former.

In conclusion, the cracked cross-section behavior is rigid-linear hardening. Thus, for $M \leq M_P$ the rotation φ is equal to zero, while for $M > M_P$ we have

$$\varphi = \lambda_{MM}\left[M - F_P\left(\frac{b}{2} - h\right)\right] + \lambda_{MF}[-F_P] \tag{12}$$

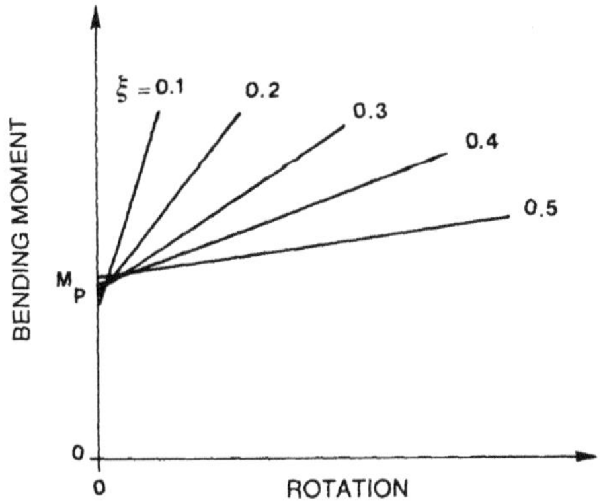

Fig. 4 Moment–rotation diagrams for different crack depths.

Equation (12) holds for $M < M_F$, where M_F is the unstable concrete fracture bending moment obtained by equalling eqn (6) to the concrete fracture toughness K_{IC}

$$M_F = \frac{K_{IC} b^{3/2} t}{Y_M(\xi)} + F_P b \left[\frac{1}{2} - \frac{h}{b} + \frac{Y_F(\xi)}{Y_M(\xi)} \right] \tag{13}$$

It can be remarked that, when the relative crack depth ξ increases, the bending moment of steel plastic flow M_P increases very slightly, while the hardening line slope λ_{MM}^{-1} decreases sharply (Fig. 4).

3. SHAKE-DOWN MODEL

As already said in Section 1, when a reinforced concrete beam cross-section is subjected to cyclic bending moment, a shake-down phenomenon due to steel plastic flow can occur [9].

Some important aspects of the above-mentioned phenomenon can be analyzed with reference to the simple system shown in Fig. 5 [20]. It consists of three elements with Young's modulus E, arranged in parallel and subjected to a force F: a central truss with cross-sectional area A and length l_I and two lateral trusses with equivalent area, corresponding to $A/2$, and length $l_{II} > l_I$. An elastic-perfectly plastic behavior of the material is assumed.

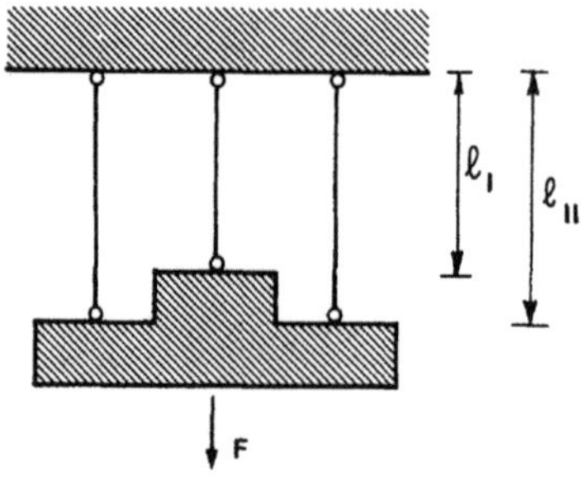

Fig. 5 Elementary shake-down model.

3.1 Monotonic Loadings

In elastic conditions, reactions X_I and X_{II} are equal to

$$X_I = Fl_{II}/(l_I + l_{II})$$
$$X_{II} = Fl_I/[2(l_I + l_{II})] \tag{14}$$

If F increases, the central element becomes plastic first, the related values of global force and displacement being given by

$$F_1 = \sigma_P A\left(1 + \frac{l_I}{l_{II}}\right)$$
$$\delta_1 = \sigma_P \frac{l_I}{E} \tag{15}$$

where σ_P is the yield stress. In the lateral elements, the yield condition occurs for

$$F_2 = 2\sigma_P A$$
$$\delta_2 = \sigma_P \frac{l_{II}}{E} \tag{16}$$

In conclusion, the behavior of the considered system under monotonic loadings is:

(a) linear elastic for $0 \leq \delta < \delta_1$. In this case, the relationship between F and δ is equal to

$$F = \frac{EA}{l_I l_{II}}(l_I + l_{II})\delta \tag{17}$$

(b) linear hardening for $\delta_1 \leq \delta < \delta_2$

$$F = \sigma_P A + \frac{EA}{l_{II}} \delta \tag{18}$$

(c) perfectly plastic for $\delta \geq \delta_2$

$$F = 2\sigma_P A \tag{19}$$

Dimensionless force–displacement diagrams for different values of the ratio l_I/l_{II} are shown in Fig. 6. For $l_I \rightarrow 0$, in the elastic stage the force is carried entirely by the central element, which becomes plastic for $F = \sigma_P A$. When $l_I = l_{II}$, on the other hand, the hardening stage is not present because all three elements become plastic simultaneously. It can be remarked that the straight line, to which the hardening portion of the diagram in Fig. 6 belongs, does not depend on the ratio l_I/l_{II}. As a matter of fact, when the central element has become plastic, its length is not taken into account in the analysis any more.

3.2 Cyclic Loadings

Consider now the response of the system in Fig. 5 to cyclic loading and unloading processes. In order to simplify the treatment, assume that:

(a) in each cycle the applied force is increased from zero to a maximum value F and then decreased to zero (Fig. 7);

(b) the lateral elements remain in the elastic field.

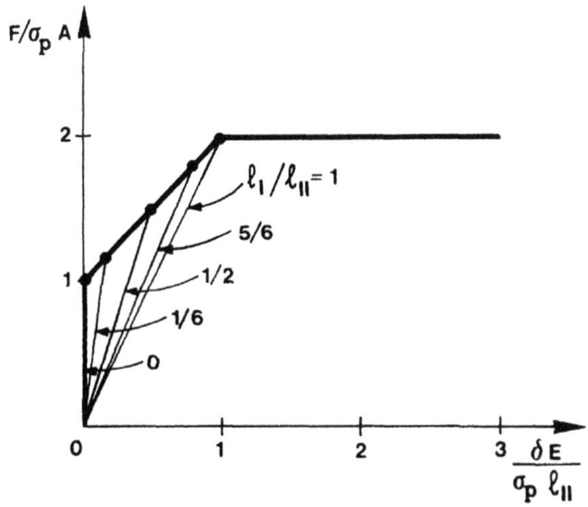

Fig. 6 Dimensionless force against displacement diagram.

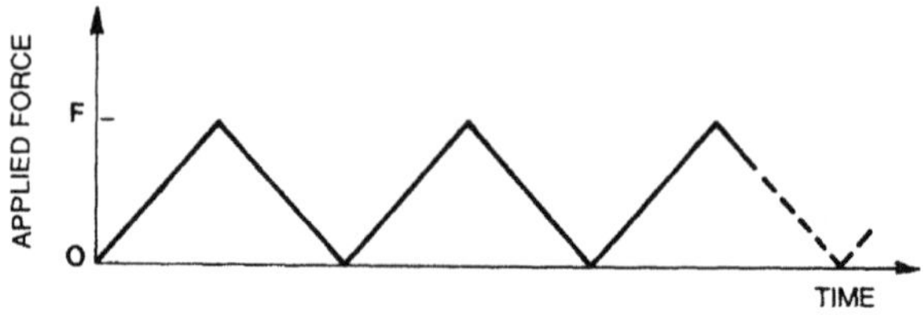

Fig. 7 Application pattern of force F.

Figure 8 shows a dimensionless force–displacement diagram obtained on the basis of the above-mentioned assumptions, in the case of $l_{\mathrm{I}}/l_{\mathrm{II}} = 1/3$. The system behaves as follows:

(1) for $0 \leq F < F_1$ (elastic condition), both loading and unloading occur along the segment 01;
(2) for $F_1 \leq F < F_{\mathrm{SD}}$, F_{SD} being a critical value called plastic shake-down force, the central element becomes plastic and, in the unloading phase, it is compressed by the lateral elements;
(3) for $F \geq F_{\mathrm{SD}}$, the central element yields even in compression.

In order to calculate the value F_{SD}, some considerations must be made. When the applied force reaches its maximum value F, the

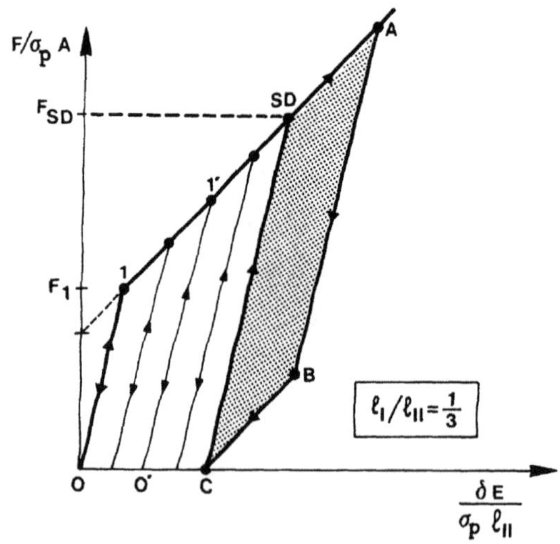

Fig. 8 Dimensionless force against displacement curves for loading–unloading processes.

equilibrium condition gives

$$X_{\mathrm{I,max}} + 2X_{\mathrm{II,max}} = F \tag{20}$$

whereas, if the applied force is equal to zero, the following equation holds

$$X_{\mathrm{I,min}} + 2X_{\mathrm{II,min}} = 0 \tag{21}$$

When the applied force is equal to F, the expansion of the central element is

$$\varepsilon_{\mathrm{I,max}} = \frac{l_{\mathrm{II}}}{l_{\mathrm{I}}}\,\varepsilon_{\mathrm{II,max}} \tag{22}$$

and, since the lateral elements are in the elastic field, we get

$$\varepsilon_{\mathrm{I,max}} = \frac{l_{\mathrm{II}}}{l_{\mathrm{I}}}\,\frac{X_{\mathrm{II,max}}}{EA/2} \tag{23}$$

When the applied force is equal to zero and inverse plasticization occurs, the expansion of the central element is

$$\varepsilon_{\mathrm{I,min}} = \varepsilon_{\mathrm{I,max}} - \frac{2\sigma_{\mathrm{P}}}{E} \tag{24}$$

and, since it is also

$$\varepsilon_{\mathrm{I,min}} = \frac{l_{\mathrm{II}}}{l_{\mathrm{I}}}\,\varepsilon_{\mathrm{II,min}}, \tag{25}$$

the following equation can be written

$$\frac{l_{\mathrm{II}}}{l_{\mathrm{I}}}\,\frac{X_{\mathrm{II,max}}}{EA/2} - \frac{2\sigma_{\mathrm{P}}}{E} = \frac{l_{\mathrm{II}}}{l_{\mathrm{I}}}\,\frac{X_{\mathrm{II,min}}}{EA/2} \tag{26}$$

Therefore, the reaction of the lateral elements under the maximum load can be obtained from eqn (26)

$$X_{\mathrm{II,max}} = X_{\mathrm{II,min}} + \sigma_{\mathrm{P}} A\,\frac{l_{\mathrm{I}}}{l_{\mathrm{II}}} \tag{27}$$

The assumption of inverse plasticization of the central element yields

$$X_{\mathrm{I,min}} = -\sigma_{\mathrm{P}} A \tag{28}$$

so that from eqn (21) we get

$$X_{\mathrm{II,min}} = \sigma_{\mathrm{P}} \frac{A}{2} \tag{29}$$

Moreover, since the central element is in the plastic condition under the maximum force, the following equation holds

$$X_{\mathrm{I,max}} = \sigma_{\mathrm{P}} A \tag{30}$$

Finally, recalling eqns (27), (29) and (30), the inverse plasticization load F_{SD} (plastic shake-down force) can be obtained from eqn (20)

$$F_{\mathrm{SD}} = \sigma_{\mathrm{P}} A + 2\left(\sigma_{\mathrm{P}} \frac{A}{2} + \sigma_{\mathrm{P}} A \frac{l_{\mathrm{I}}}{l_{\mathrm{II}}}\right) = 2\sigma_{\mathrm{P}} A\left(1 + \frac{l_{\mathrm{I}}}{l_{\mathrm{II}}}\right) = 2F_{\mathrm{I}} \tag{31}$$

that is to say, the inverse plasticization of the central element occurs only if $F \geq F_{\mathrm{SD}} = 2F_1$.

In conclusion, the behavior under cyclic loadings of the system in Fig. 5 can be summarized as follows (Fig. 8):

(1) For $0 \leq F < F_1$, the system is linear elastic and its representative point on the dimensionless force–displacement diagram oscillates along the segment 01.

(2) For $F_1 \leq F < F_{\mathrm{SD}}$, only the tensile plasticization of the central element occurs (loading phase). For subsequent unloading and reloading phases, the representative point oscillates elastically on the segment 0′1′ (elastic shake-down).

(3) For $F \geq F_{\mathrm{SD}} = 2F_1$, the yield condition for the central element occurs both in tension (loading phase) and in compression (unloading phase). The representative point of the system moves cyclically along the parallelogram SD–A–B–C (plastic shake-down). In particular, at the end of the first loading phase the point is in A; it reaches B elastically in the unloading phase, and then the plastic flow of the compressed central element starts. The flow stops at point C, when the force is equal to zero. The second loading cycle starts from C, it is elastic along C–SD and plastic along SD–A, etc. The energy dissipated in each cycle of hysteresis, due to the phenomenon of alternating plasticity, is equal to the area of the above-described parallelogram.

As already said, the study has been carried out with reference to only three elements and a cyclic loading pulsating from zero in order to simplify the treatment, but it could be easily extended to a larger number of elements and a more complex loading scheme (for example, alternate symmetrical, alternate non-symmetrical, etc.).

4. BEHAVIOR OF RC BEAMS UNDER CYCLIC LOADINGS

4.1 Plastic Shake-down Bending Moment

Consider the reinforced concrete beam cross-section subjected to cyclic bending moment (Fig. 9). The assumptions that the crack does not propagate and that the steel behavior is rigid–perfectly plastic are made.

If the cross-section is loaded with a maximum bending moment M greater than or equal to M_P (point A in Fig. 9) and then is unloaded (point B), a residual rotation due to the steel plastic flow remains and, therefore, the concrete compresses the steel segment with an unknown force F. The residual rotation at point B is equal to

$$\varphi = \lambda_{MM} F\left(\frac{b}{2} - h\right) + \lambda_{MF} F \tag{32}$$

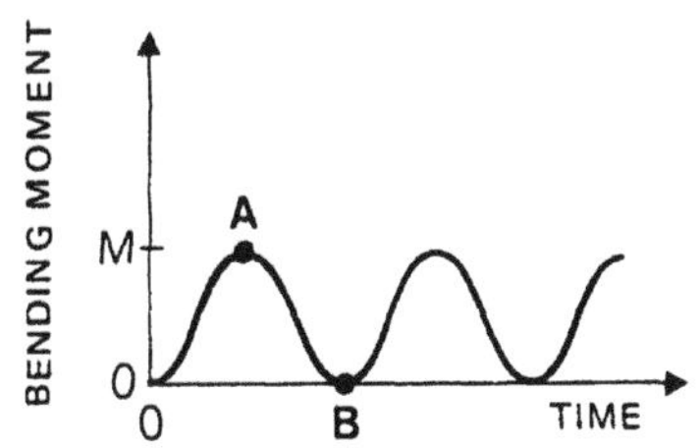

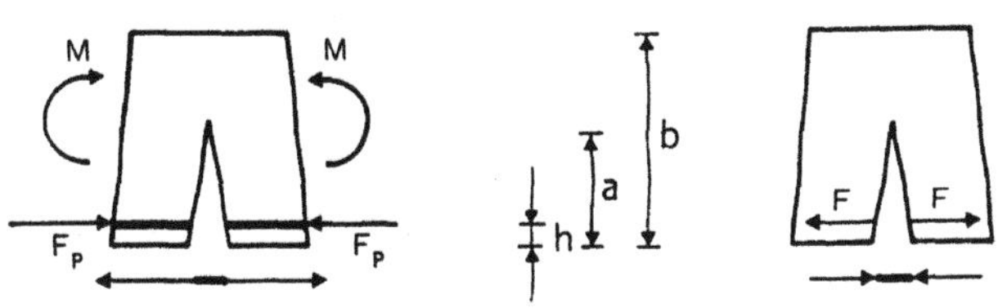

Fig. 9 Reinforced concrete beam cross-section. A, under maximum loading; B, after unloading.

Since the hypothesis of rigid–perfectly plastic steel has been assumed, this rotation coincides with the under-loading rotation at point A, given by eqn (12). Applying the above-mentioned congruence condition, the unknown steel compression F after unloading can be obtained as a function of the maximum moment M [9]

$$\frac{F}{F_P} = \frac{M}{M_P} - 1 \tag{33}$$

Consequently, the bending moment of plastic shake-down M_{SD} can be calculated from eqn (33) as the lowest maximum moment for which $F = F_P$, that is, for which the reinforcement yields even in compression

$$M_{SD} = 2M_P = 2F_P b\left[\frac{1}{2} - \frac{h}{b} + r(\xi)\right] \tag{34}$$

In conclusion, if the reinforced concrete cross-section is subjected to cyclic bending moment, the following fields of behavior can be defined (Fig. 10):

(1) elastic behavior for $0 \leq M < M_P$;
(2) elastic shake-down for $M_P \leq M < M_{SD}$;
(3) plastic shake-down for $M_{SD} \leq M < M_F$, M_F being given by eqn (13).

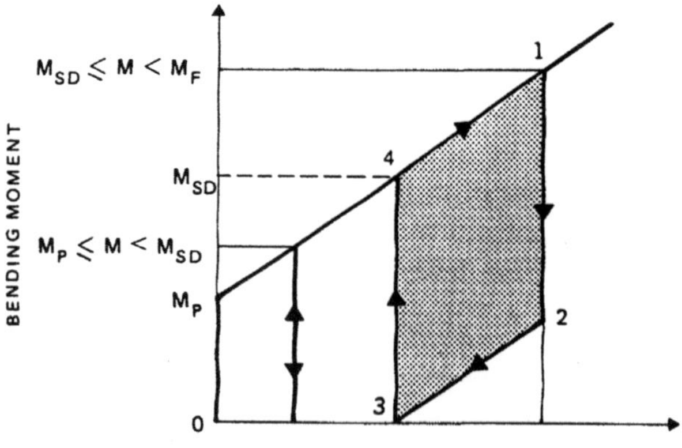

Fig. 10 Elastic shake-down (for $M_P \leq M < M_{SD}$) and plastic shake-down (for $M_{SD} \leq M < M_F$).

In case No. 3 ($M_{SD} \le M < M_F$), the moment–rotation diagram presents hysteretic loops and the energy dissipated in each cycle is equal to the area 1–2–3–4 (Fig. 10)

$$\frac{\text{work}}{\text{cycle}} = M_{SD}\lambda_{MM}(M - M_{SD}) \tag{35}$$

where M_{SD} and λ_{MM} are calculated from eqns (34) and (2), respectively.

It can be pointed out that the cyclic loading has been assumed to be unidirectional (Fig. 9), but analogous considerations could be made for the case of reversed cyclic loading.

4.2 Critical Crack Depth and Bending Moment

The plastic shake-down bending moment and the unstable concrete fracture bending moment can be written in dimensionless form

$$\tilde{M}_{SD} = \frac{M_{SD}}{K_{IC}b^{3/2}t} = 2N_P\left[\frac{1}{2} - \frac{h}{b} + r(\xi)\right] \tag{36}$$

$$\tilde{M}_F = \frac{M_F}{K_{IC}b^{3/2}t} = \frac{1}{Y_M(\xi)} + N_P\left[\frac{1}{2} - \frac{h}{b} + \frac{Y_F(\xi)}{Y_M(\xi)}\right] \tag{37}$$

where the brittleness number

$$N_P = \frac{F_P}{K_{IC}b^{1/2}t} = \frac{f_y b^{1/2}}{K_{IC}}\frac{A_s}{A} \tag{38}$$

is introduced, with $A = bt$ = cross-sectional area. As can be observed, the parameter N_P is a dimensionless number dependent on the mechanical and geometrical properties of the reinforced concrete beam cross-section.

The dimensionless bending moments $\tilde{M}_{SD}$ and $\tilde{M}_F$ against the relative crack depth are shown in Fig. 11 for $h/b = 0{\cdot}1$ and $N_P = 0{\cdot}3$ [10], [11]. It can be pointed out that $\tilde{M}_{SD}$ increases by increasing ξ, whereas $\tilde{M}_F$ decreases. The two curves intersect at a critical point, the coordinates of which are $\xi_{CR} = 0{\cdot}245$ and $\tilde{M}_{CR} = 0{\cdot}364$. The phenomenon of plastic shake-down cannot occur for $\xi \ge \xi_{CR}$, because it is preceded by the unstable concrete fracture, $\tilde{M}_F$ being lower than or equal to $\tilde{M}_{SD}$. For different h/b and N_P values it is possible to obtain $\tilde{M}_{SD}$ and $\tilde{M}_F$ curves qualitatively analogous to those in Fig. 11.

The diagram of the critical relative crack depth ξ_{CR} against the brittleness number N_P is shown in Fig. 12, varying the parameter h/b.

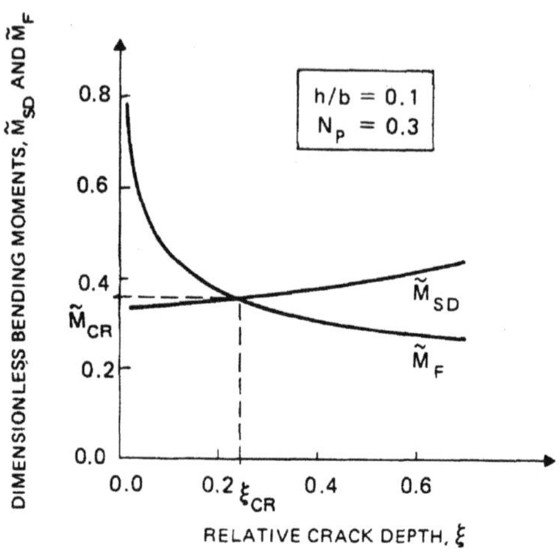

Fig. 11 Dimensionless bending moments $\tilde{M}_{SD}$ and $\tilde{M}_F$ against relative crack depth ξ, for $h/b = 0{\cdot}1$ and $N_P = 0{\cdot}3$.

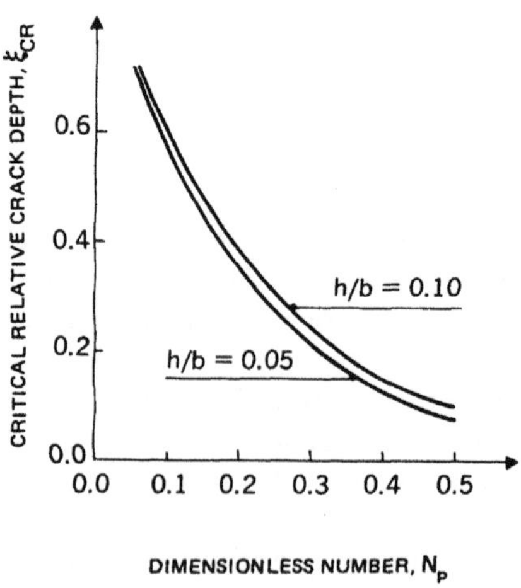

Fig. 12 Critical relative crack depth ξ_{CR} against brittleness number N_P, varying the parameter h/b.

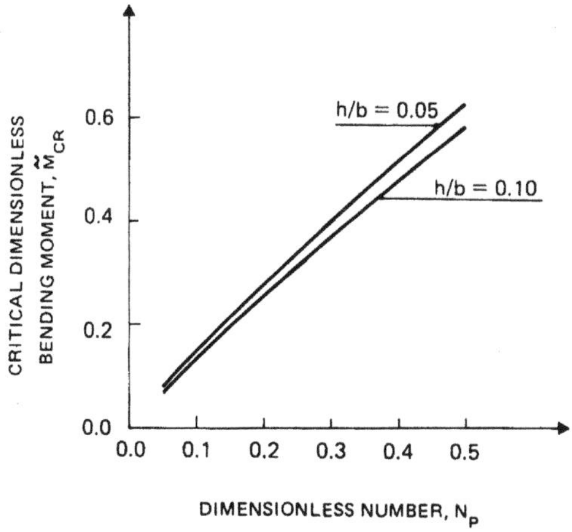

Fig. 13 Critical dimensionless bending moment $\tilde{M}_{CR}$ against brittleness number N_P, varying the parameter h/b.

It can be pointed out that ξ_{CR} decreases with increasing N_P. Therefore, the plastic shake-down can occur only for very low ξ if N_P is large. It can be seen from eqn (38) that the dimensionless number N_P is large if the cross-section depth b and/or the steel percentage A_s/A are large. The same result can be reached in the case of high steel yield strength and/or low concrete fracture toughness. Moreover, it can be noticed that the curve for $h/b = 0{\cdot}05$ is lower than that for $h/b = 0{\cdot}10$ (Fig. 12), even if the two curves are very close.

Eventually, it is possible to verify that the critical dimensionless bending moment $\tilde{M}_{CR}$ increases with increasing N_P (Fig. 13), the curve for $h/b = 0{\cdot}05$ being higher than that for $h/b = 0{\cdot}10$. However, the two curves are very close, as occurs for the curves ξ_{CR} against N_P in Fig. 12.

5. FATIGUE CRACK GROWTH AND ENERGY DISSIPATION

5.1 Theoretical Aspects

Consider now the possibility of crack propagation for the case of a reinforced concrete cross-section under cyclic bending moment [10], [11]. If the initial crack depth is assumed to be equal to ξ_1 and the cyclic loading is unidirectional, the slope of the hardening line in Fig. 14 is λ_{MM}^{-1}, λ_{MM} being obtained from eqn (2) by considering $\xi = \xi_1$. The energy dissipated in each cycle is equal to the area 1–2–3–4 and can be calculated from eqn (35) by assuming $\xi = \xi_1$.

When the relative crack depth increases up to ξ_2, the bending moment of steel plastic flow increases very slightly while the hardening line becomes more inclined (Fig. 14). The energy dissipated per cycle in the new configuration is equal to the area 5–6–7–8 and can be obtained from eqn (35), ξ being equal to ξ_2. Then, ξ can still increase and the energy dissipated in the steel reinforcement can be computed at each step, the hardening line being more and more inclined.

If the dimensionless maximum bending moment $\tilde{M}$ is greater than or equal to $\tilde{M}_{CR}$ (Fig. 15), the fatigue crack growth occurs from the initial relative crack depth ξ_I to the unstable fracture value ξ_F, the shake-down phenomenon being plastic during the whole crack propagation. Therefore, a hysteretic loop is described for each loading cycle and the total energy dissipation can be computed by adding all the values of dissipated energy per cycle (see eqn (35)) from ξ_I to ξ_F.

If $\tilde{M} < \tilde{M}_{CR}$ (Fig. 16), the fatigue crack growth occurs from ξ_I to ξ_F as in the previous case, but the shake-down is plastic only up to ξ_{SD}. Therefore, the energy dissipation is calculated from ξ_I to ξ_{SD} because the loop described in each loading cycle degenerates into a segment for $\xi \geq \xi_{SD}$.

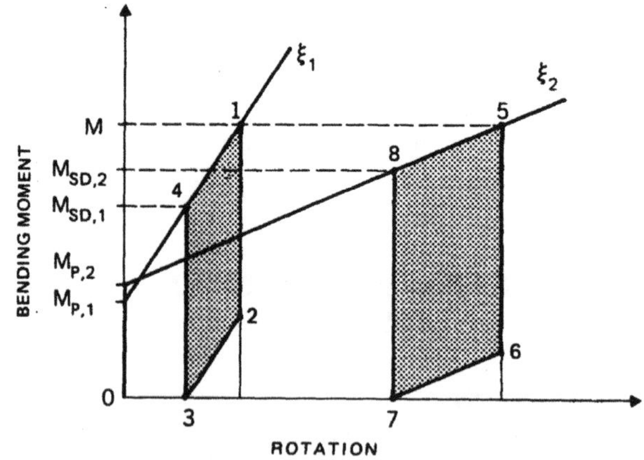

Fig. 14 Fatigue crack growth from ξ_1 to ξ_2 and hysteretic loops in moment–rotation diagram.

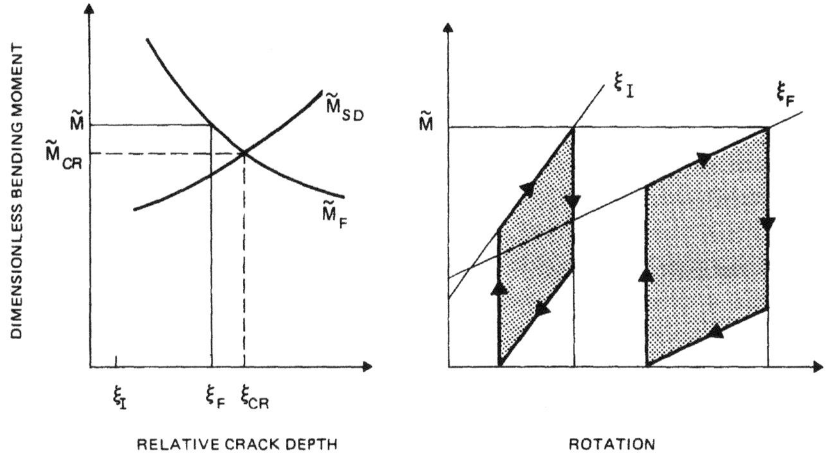

Fig. 15 Case $\tilde{M} \geq \tilde{M}_{CR}$: plastic shake-down from ξ_I to ξ_F.

The Paris–Erdogan law [21] can be utilized to follow the fatigue crack growth

$$da/dN = A(\Delta K_I)^m \tag{39}$$

where N = number of cycles, ΔK_I = stress-intensity factor range, A and m = material constants. Several authors have experimentally determined A and m for concrete [22]–[24]. In the next subsection, the

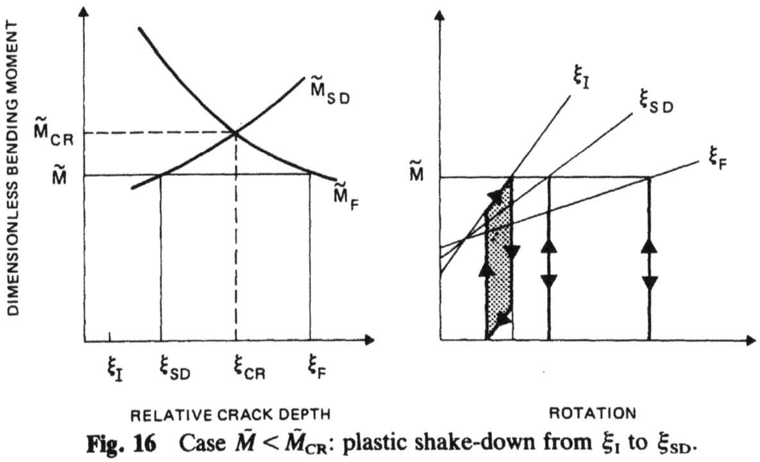

Fig. 16 Case $\tilde{M} < \tilde{M}_{CR}$: plastic shake-down from ξ_I to ξ_{SD}.

values proposed in Ref. [24] will be considered

$$A = 7{\cdot}71 \times 10^{-25}$$
$$m = 3{\cdot}12 \tag{40}$$

for crack propagation rate $\mathrm{d}a/\mathrm{d}N$ in m cycle^{-1} and stress-intensity factor range ΔK_{I} in N m$^{-3/2}$.

5.2 Numerical Examples

Consider a reinforced concrete beam cross-section with $t = 0{\cdot}2$ m, $b = 0{\cdot}3$ m, $a = h = 0{\cdot}03$ m, subjected to unidirectional cyclic loading. The concrete properties E and K_{IC} are assumed to be equal to $3{\cdot}21 \times 10^{10}$ N m^{-2} and $1{\cdot}75 \times 10^{6}$ N m$^{-3/2}$, respectively.

If the steel reinforcement consists of two bars with diameter $\phi = 10$ mm and yield strength $f_y = 373$ N mm^{-2}, the parameter N_P is equal to about 0·3. Different values N_P can be obtained, for instance, by varying the steel reinforcement. As an example, three values of N_P are considered: 0·1, 0·2 and 0·3. The critical bending moments $\tilde{M}_{\mathrm{CR}}$ related to these three values are 0·139, 0·255 and 0·364, respectively (see Fig. 13).

Figure 17 shows fatigue life predictions (continuous curves) against

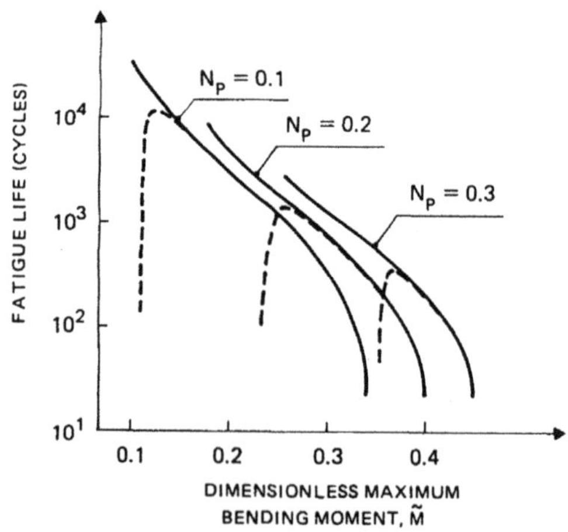

Fig. 17 Fatigue life (continuous curves) and number of hysteretic cycles (dashed curves) against dimensionless maximum bending moment $\tilde{M}$, for different numbers N_P, in the case $\xi_{\mathrm{I}} = 0{\cdot}1$.

dimensionless maximum bending moment $\tilde{M}$, the former being the number of loading cycles obtained by numerical integration of eqn (39) from the initial relative crack depth $\xi_I = 0{\cdot}1$ to the unstable fracture value ξ_F. It can be remarked that fatigue life obviously decreases with increasing $\tilde{M}$, while, for a certain value of $\tilde{M}$, it increases with increasing N_P.

The dashed branches in Fig. 17 represent the number of cycles for which the cross-sectional behavior is hysteretic. These curves coincide with the continuous ones for $\tilde{M} \geq \tilde{M}_{CR}$, according to what has been deduced in the previous subsection.

The energy dissipated per cycle during the fatigue crack growth is shown in Fig. 18 for three values of $\tilde{M}$. The total energy dissipated during the plastic shake-down can be obtained for each value of $\tilde{M}$: as a matter of fact, it is the area under the curve related to the value of $\tilde{M}$ is considered (Fig. 18).

The total dissipated energy against the dimensionless maximum bending moment is shown in Fig. 19. It can be remarked that the maximum value of energy dissipation decreases with increasing N_P, while the value of $\tilde{M}$ for which the above-mentioned maximum occurs increases with increasing N_P. For each value N_P considered in Fig. 19, the abscissa of the maximum is approximately equal to 0·16 for $N_P = 0{\cdot}1$, 0·28 for $N_P = 0{\cdot}2$ and 0·39 for $N_P = 0{\cdot}3$.

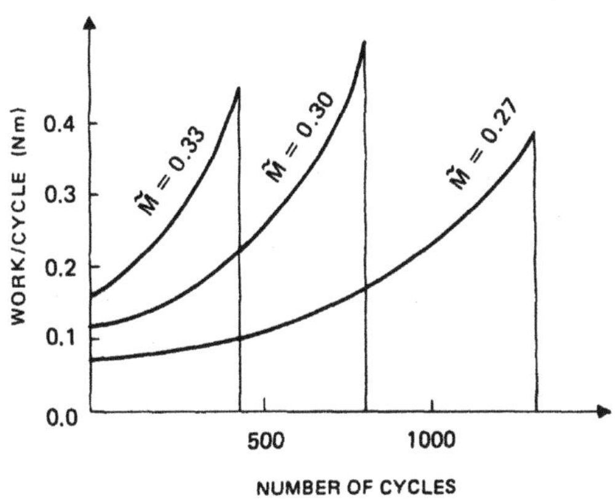

Fig. 18 Dissipated energy per cycle against number of cycles, for different values $\tilde{M}$, in the case $N_P = 0{\cdot}2$ and $\xi_I = 0{\cdot}1$.

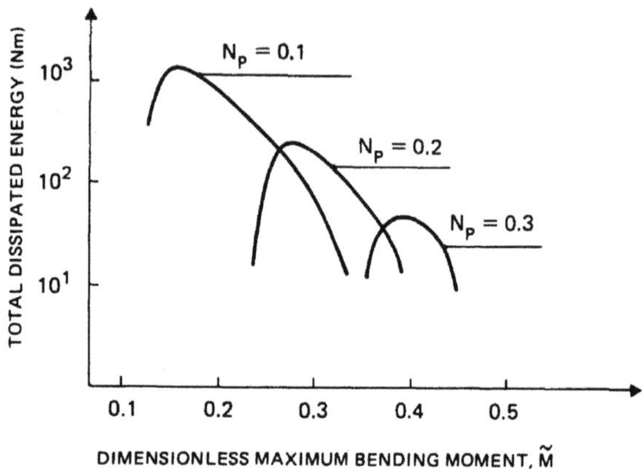

Fig. 19 Total dissipated energy against dimensionless maximum bending moment, for different numbers N_P, in the case $\xi_I = 0{\cdot}1$.

6. EXPERIMENTAL INVESTIGATION

An experimental investigation has been carried out to assess the reliability of the theoretical Fracture Mechanics model previously described. Some reinforced concrete beams have been tested under unidirectional cyclic loading to study the above-mentioned shake-down phenomena [15], [16].

6.1 Test Reinforced Beam Geometry

The test samples are reinforced concrete beams with $t = 20$ cm, $b = 30$ cm, $h = 3$ cm, total length $l = 220$ cm and span $s = 180$ cm (Fig. 20).

The longitudinal steel reinforcement consists of two bars (with diameter $\phi = 8$ mm) in the stretched part and two bars ($\phi = 8$ mm) for stirrup anchorage in the upper edge. The spacing of closed stirrups ($\phi = 6$ mm) is 10 cm near the external supports, while there is no transverse shear reinforcement in the central zone of the beams (Fig. 20).

A notch with depth $a = 3$ cm has been obtained in the stretched part of the center-line cross-section by means of either a disk saw 2-mm thick or a sheet-steel 1-mm thick embedded in the beam casting (Fig. 20). Therefore, the relative crack depth ξ is equal to $0{\cdot}1$.

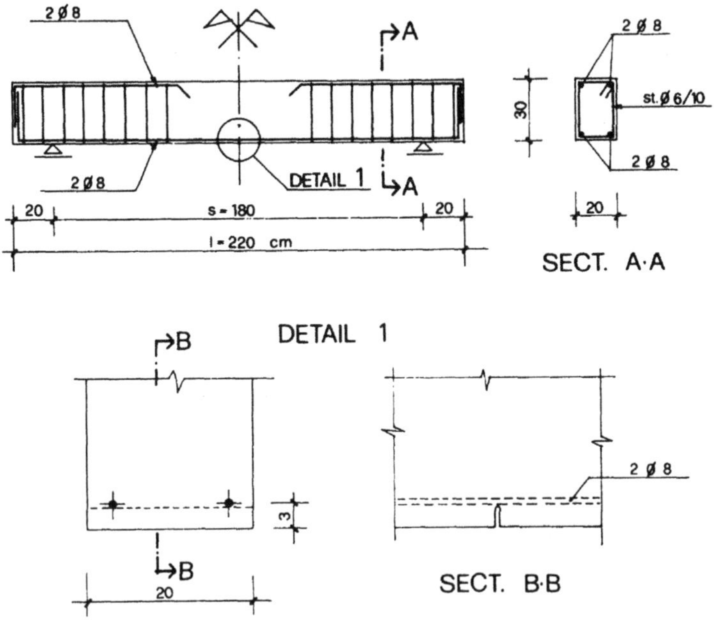

Fig. 20 Test reinforced beam geometry.

6.2 Material Properties

The reinforcement used is high bond steel Fe B 38 K with yield strength equal to 373 N mm^{-2}.

The concrete mixture is obtained from the following components: type 425 Portland cement (450 Kg m^{-3}), water (150 liter m^{-3}), well-rounded Brenta river gravel (sieve size 7–15 mm), crushed limestone sand (sieve size 0–3 mm) and a superplasticizer admixture (4·5 liter m^{-3}). The concrete was cast in metal formworks and subjected to either air curing or steam curing at atmospheric pressure. Then the beams were kept at 20°C and 60% relative humidity.

The following specimens were taken from the concrete casting to determine the material mechanical properties:

(a) cubic specimens with side 10 cm to evaluate the compressive strength f_{ck}. The average of the results obtained on four specimens is 63·77 N mm^{-2} for air curing and 68·38 N mm^{-2} for steam curing;

(b) prismatic specimens with base 10 cm × 10 cm to determine the longitudinal elastic modulus E. The mean value on three specimens is equal to 34·3 kN mm^{-2} for both curing conditions;
(c) specimens to check that the bond strength between steel and concrete (by means of the 'beam test'). The results show that steel reinforcement yielding occurs before slippage.

Then the concrete fracture energy G_F can be estimated in Nm^{-1} by means of an empirical equation proposed in Ref. [25]

$$G_F = \alpha (f_{ck})^{0\cdot 7} \tag{41}$$

where α is a coefficient dependent on the maximum aggregate size and f_{ck} is expressed in N mm^{-2}. Therefore, the critical value of the stress-intensity factor according to the equation $K_{IC} = (G_F E)^{1/2}$ results to be equal to 60·16 N mm$^{-3/2}$ for air curing and 61·65 N mm$^{-3/2}$ for steam curing.

6.3 Testing Setup

The beams were subjected to unidirectional cyclic loading by means of a four-point bending test (Figs 21 and 22). Each beam was loaded by applying two cyclic forces on the cross-sections at 1/3 and 2/3 of the span s. The loading was carried out by a hydraulic press and the load was recorded by a load-cell.

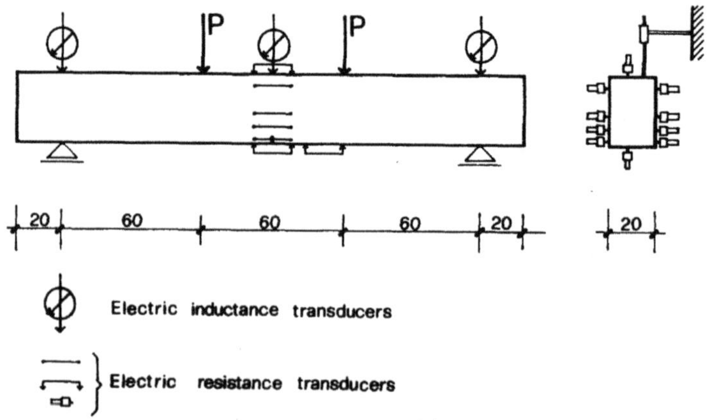

Fig. 21 Scheme of the unidirectional cyclic loading (four-point bending test).

Fig. 22 Testing apparatus.

Strain gages were used to measure strains, displacements and rotations (Figs 21–23):

(a) eleven electric resistance transducers with measurement base length = 100 mm, measurement field = ±2·5 mm and sensitivity = 2·5 μ, mounted near or astride the notch and the expected crack propagation trajectory;
(b) three electric inductance transducers with measurement field = ±5 mm and sensitivity = 25 μ.

6.4 Experimental Results

On the basis of the geometrical and mechanical properties of the test beams considered, the bending moment of plastic shake-down for $\xi = 0{\cdot}1$ is $M_{SD} = 2M_P = 13{\cdot}04$ kNm, from eqn (34), and the bending moment of unstable concrete fracture for the same relative crack depth is $M_F = 24{\cdot}42$ kNm for air curing and $M_F = 24{\cdot}88$ kNm for steam curing, from eqn (13).

Fig. 23 Transducers on one side of the test beam, in correspondence with the centre-line cross-section.

The center-line cross-section of each test beam was loaded by unidirectional cyclic bending moment with maximum value M equal to 9·81 kNm for five cycles and equal to 16·68 kNm for the other 20 cycles. Therefore, a phenomenon without energy dissipation was expected to occur in the first five cycles since the closed curve described in the moment–rotation diagram during each loading cycle degenerates into a segment when $M_P \leq M < M_{SD}$, while a dissipative phenomenon was theoretically predicted for the last 20 cycles because $M_{SD} \leq M < M_F$ (Fig. 10).

The bending moment versus rotation diagram is shown in Fig. 24 for the center-line cross-section of each tested sample. The samples are marked as follows:

(NO1) air curing; sheet-steel embedded in the beam casting and then removed before the testing;

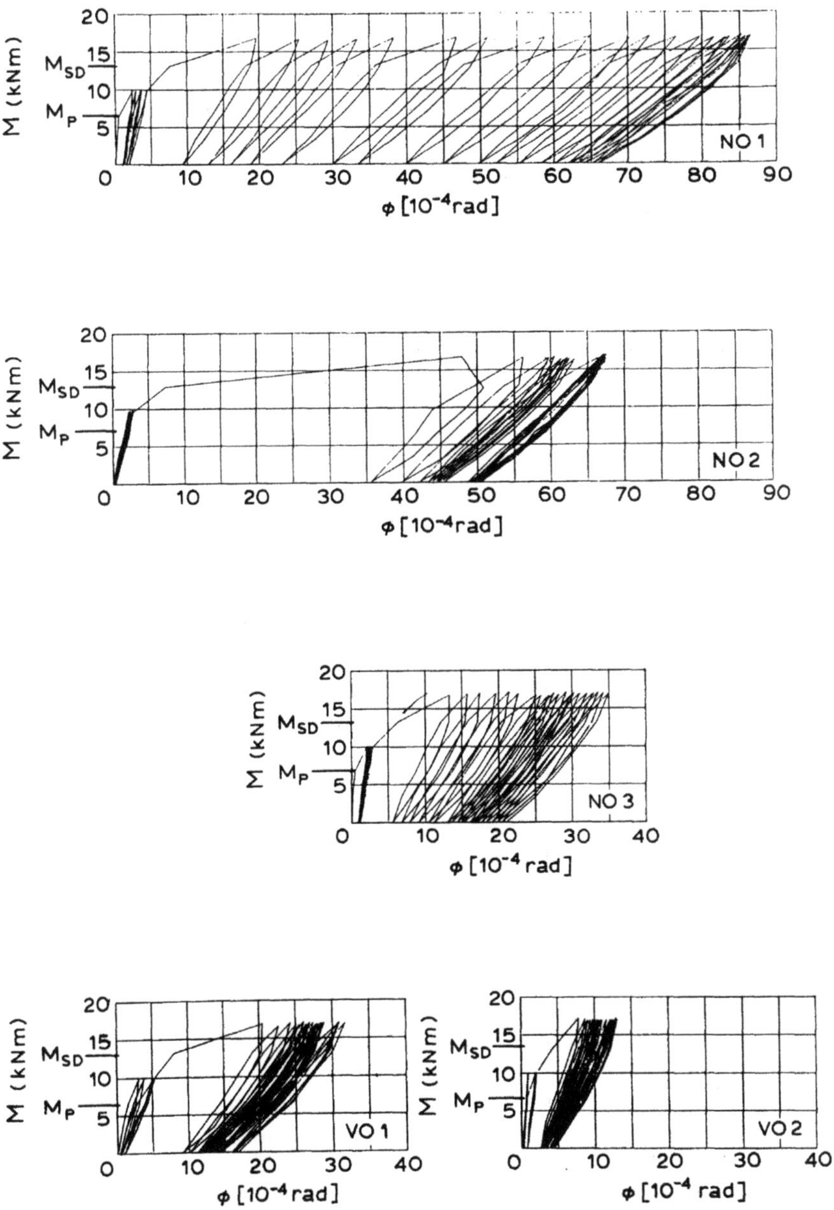

Fig. 24 Bending moment against rotation for five beams. The theoretical values M_P and M_{SD} are also reported.

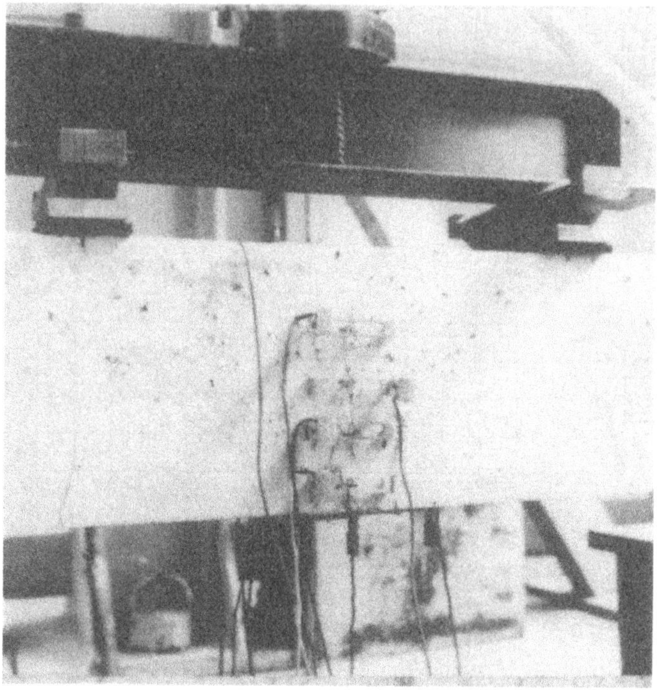

Fig. 25 Crack propagation during the cyclic loading for sample NO1.

(NO2) air curing; crack by means of the disk saw;
(NO3) air curing; sheet-steel embedded in the beam casting;
(VO1) steam curing; crack by means of the disk saw;
(VO2) steam curing; sheet-steel embedded in the beam casting.

The initial notch propagates during cyclic loading (Figs 25 and 26) and, on the basis of the recordings by the transducers mounted on both the beam sides, the final relative crack depth after the experimental testing appears to be about 0·7 in cases NO1 and NO2 (Fig. 27), 0·5 in cases NO3 and VO1, and 0·4 in case VO2. Therefore, according to the theoretical eqn (2), the ratio of the hardening slope λ_{MM}^{-1} for the generic beam tested to the hardening slope for NO1 is about unity in case NO2, 3·8 in cases NO3 and VO1, and 7·4 in case VO2. The experimental results in Fig. 24 confirm such predictions.

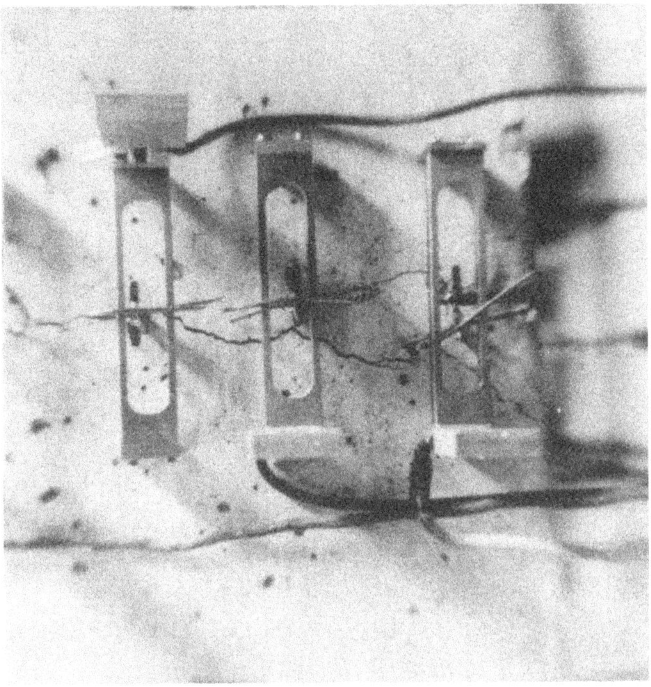

Fig. 26 Detail of the crack propagation in correspondence with the center-line cross-section (lower edge of sample NO1).

7. CONCLUSIONS

From the diagrams in Fig. 24, it can be remarked that the rigid-linear hardening behavior predicted by the proposed model [9]–[11] is experimentally confirmed, in relation to both the bending moment of steel plastic flow and the hardening line slope. Moreover, a phenomenon without energy dissipation (elastic shake-down) theoretically predicted for $M_P \leq M < M_{SD}$ is experimentally confirmed, that is to say, the closed curve described in the moment–rotation diagram during each loading cycle degenerates into a segment. On the contrary, a dissipative phenomenon (plastic shake-down) occurs for $M_{SD} \leq M < M_F$ and, when the relative crack depth increases, the hysteretic loops are observed to shift toward the right-hand side. As

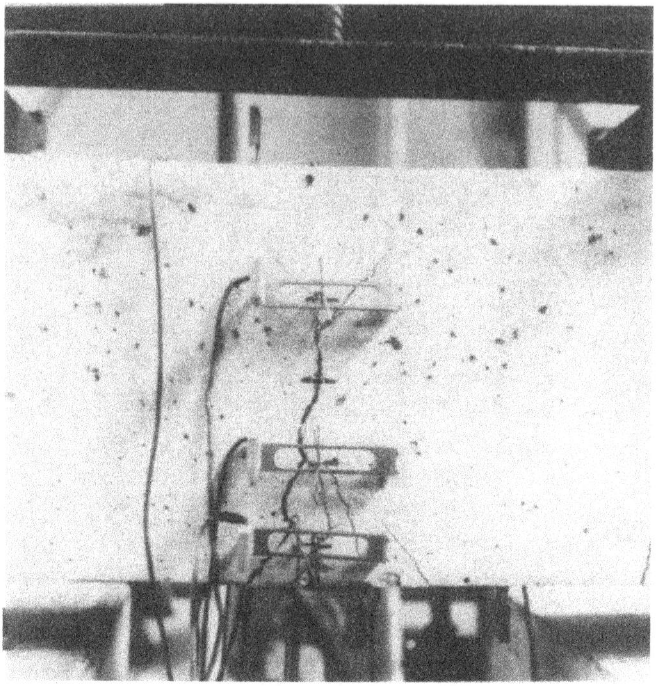

Fig. 27 Final relative crack depth after the experimental testing for sample NO1.

previously discussed, a numerical procedure can follow the fatigue crack growth under the considered cyclic loading and calculate the total energy dissipation during the hysteretic phenomenon.

ACKNOWLEDGEMENTS

The author gratefully acknowledges the research support for this work provided by the Italian Ministry for University and Technological and Scientific Research Ministry (MURST) and the Italian National Research Council (CNR).

REFERENCES

[1] Banon, H., Biggs, J. M. & Irvine, H. M., Seismic damage in reinforced concrete frames. *Journal of Structural Division* (ASCE), **107** (1981) 1713–29.

[2] Takeda, T., Sozen, M. A. & Nielsen, N. N., Reinforced concrete response to simulated earthquakes. *Journal of Structural Division* (ASCE), **96** (1970) 2557–73.

[3] Townsend, W. H. & Hanson, R. D., Reinforced concrete connection hysteresis loops. *ACI Publication,* **53** (1977) 351–70.

[4] Mehdi, S., Hysteresis models for reinforced concrete. *Journal of Structural Division* (ASCE), **108** (1982) 1077–86.

[5] Shah, S. P., Prediction of cumulative damage for concrete and reinforced concrete. *Materials and Structures* (RILEM), **17** (1984) 65–8.

[6] Filippou, F. C., Popov, E. P. & Bertero, V. V., Analytical studies of hysteretic behavior of R/C joints. *Journal of Structural Division* (ASCE), **112** (1986) 1605–22.

[7] Wang, M. L. & Shah, S. P., Reinforced concrete hysteresis model based on the damage concept. *Earthquake Engineering and Structural Dynamics,* **15** (1987) 993–1003.

[8] Shah, S. P., Wang, M. L. & Chung, L., Model concrete beam–column joints subjected to cyclic loading at two rates. *Materials and Structures* (RILEM), **20** (1987) 85–95.

[9] Carpinteri, Al. & Carpinteri, An., Hysteretic behavior of RC beams. *Journal of Structural Division* (ASCE), **110** (1984) 2073–84.

[10] Carpinteri, An., Fatigue crack growth in reinforced concrete. In *Proceedings of the International Conference on Micromechanics of Failure of Quasi-Brittle Materials,* ed. S. P. Shah, S. E. Swartz & M. L. Wang. Elsevier Applied Science Publishers, London, 1990, pp. 541–8.

[11] Carpinteri, An., Energy dissipation in R.C. beams under cyclic loadings. *Engineering Fracture Mechanics,* **39** (1991) 177–84.

[12] Carpinteri, Al., Stability of fracturing process in RC beams. *Journal of Structural Division* (ASCE), **110** (1984) 544–58.

[13] Giuriani, E., Experimental investigation on the bond-slip law of deformed bars in concrete. In *Proceedings of the IABSE Colloquium on Advanced Mechanics of Reinforced Concrete,* Technical University of Delft, Delft, 1981, pp. 121–42.

[14] Ciampi, V., Eligehausen, R., Bertero, V. & Popov, E., Analytical model for deformed bar bond under generalized excitations. In *Proceedings of the IABSE Colloquium on Advanced Mechanics of Reinforced Concrete,* Technical University of Delft, Delft, 1981, pp. 53–67.

[15] Carpinteri, An. & Zaupa, F., A Fracture Mechanics model for R.C. beams under repeated loadings. In *Proceedings of the International Conference on Recent Developments in the Fracture of Concrete and Rock,* ed. S. P. Shah, S. E. Swartz & B. Barr. Elsevier Applied Science Publishers, London, 1989, pp. 233–42.

[16] Carpinteri, An. & Zaupa, F., Behavior of R.C. beam cross-sections under unidirectional cyclic bending moments. In *Proceedings of the 8th Biennial European Conference on Fracture,* ed. D. Firrao. EMAS Publisher, London, 1990, pp. 828–33.

[17] Okamura, H., Watanabe, K. & Takano, T., Application of the compliance concept in Fracture Mechanics. *ASTM STP 536,* 1973, pp. 423–38.

[18] Okamura, H., Watanabe, K. & Takano, T., Deformation and strength of cracked member under bending moment and axial force. *Engineering Fracture Mechanics,* **7** (1975) 531–9.

[19] Tada, H., Paris, P. C. & Irwin, G. R., *The Stress Analysis of Cracks Handbook.* Del Research Corp., Hellertown, PA, USA, 1973.

[20] Carpinteri, A., 1991 (pers. comm.).

[21] Paris, P. C. & Erdogan, F., A critical analysis of crack propagation laws. *Journal of Basic Engineering* (Trans. ASME), **85D** (1963) 528–34.

[22] Mihashi, H. & Wittmann, F. H., Stochastic approach to study the influence of rate of loading on strength of concrete. *Heron,* **25** (1980) 54.

[23] Yam, A. S. T. & Mindess, S., The effects of fibre reinforcement on crack propagation in concrete. *International Journal Cement Composites and Lightweight Concrete,* **4** (1982) 83–93.

[24] Baluch, M. H., Qureshy, A. B. & Azad, A. K., Fatigue crack propagation in plain concrete. In *Proceedings of the International Conference on Fracture of Concrete and Rock,* ed. S. P. Shah & S. E. Swartz. Springer Verlag, 1989, pp. 80–87.

[25] CEB, Model Code 90, draft version, 1990.

Chapter 23

An Expert System Approach to Applying Fracture Mechanics to Reinforced Concrete

S. E. SWARTZ, Y.-C. KAN & K. K. HU
*Department of Civil Engineering,
Kansas State University,
Manhattan, Kansas 66506, USA*

ABSTRACT

After more than ten years of extensive—and intensive—research efforts world-wide investigating methods of fracture mechanics to explain localization, crack growth and fracture in plain and reinforced concrete structures, it is time to organize this material and present to the design engineering community practical approaches to its implementation. In this chapter an approach is presented, based on expert system concepts, for developing software which may be implemented on a personal computer. The domain areas include: tutorial, teaching aid, materials testing, and reinforced concrete design. One aspect of the expert knowledge used is the selectivity of the recommended procedures. This is a personal (expert) judgement which is necessary to provide reasonable limits to the scope of the program.

1. INTRODUCTION

The applications of methods of fracture mechanics to structural engineering are very diverse and include: life assessment of aircraft structural components, inspectability criteria for aircraft, inspectability criteria for bridges, failure evaluation (forensic engineering), design of metal structures, stability of rock masses, stability of dam structures, stability of bore holes, crack propagation in concrete, steel rebar/concrete matrix delamination, and failure mechanisms for cementitious materials.

In the area of application to cementitious materials it is stated that: 'Fracture researchers at present have no doubt that introduction of fracture mechanics into concrete design can bring about significant benefits. It will make it possible to achieve more uniform safety margins. This in turn will improve economy as well as structural reliability, and will make it possible to introduce new designs and utilize new concrete materials [1]. In the preface by Irwin to the text of *Fracture and Fatigue Control in Structures* [2], it is stated (with respect to metal structures)' 'Use of fracture mechanics analysis and data has explained many service fracture failures with a satisfactory degree of quantitative accuracy. By studying the possibilities for such fractures in advance, effective fracture control plans have developed.'

With respect to reinforced concrete structures, Elfgren concludes the report *Fracture Mechanics of Concrete Structures*: *From Theory to Applications* [3] with: 'The reasons for when and where cracks appear in concrete structures can be better understood with fracture mechanics. Not only static loads but fatigue loads and environmental loads attack the concrete and we need rational models to be able to predict what will happen so that we can counteract it with proper design measures. Here much can be achieved with the help of fracture mechanics.'

In the last ten years an 'explosion' of research effort has taken place in studying localization, crack growth/tension-softening, and fracture in plain and reinforced concrete structures within a framework of fracture mechanics based concepts. Swartz [4] lists almost 800 references and since then many hundreds more have appeared in the literature. Despite this activity by the research community, little evidence of this effort appears today in design practice.

The lack of acceptance, or indifference, by those engaged in the design of reinforced concrete structures may perhaps be attributed to one or more of the following:

—lack of consensus of researchers on basic issues;
—lack of standardized methods for obtaining fracture material parameters;
—limited evidence of practical implementation;
—limited (or no) understanding of the basic ideas.

The authors feel that the first two issues are rapidly being resolved—hence the need for workshops such as the present one on applications of fracture mechanics to reinforced concrete. Some evidence of practical implementation appears in the literature, particularly with respect to forensic engineering and/or repair. Two such reports are cited here—an investigation of the failure mechanism for the bridge at Schoharie Creek [5] and investigations of the cracking system at Kölnbrein Dam [6]. As was stated quite well by Hawkins (Sendai, Japan workshop, October 1988), the acceptance of fracture mechanics by the design community will first come through failure investigation, then through limited use on high risk-high cost projects, then through adaptation or revision of appropriate sections of national building codes.

However, even this sequence of events will not occur without a strong effort at educating present and future structural engineers. In the areas of metals this has been done now for many years through workshops and graduate level courses. Texts are available such as that by Rolfe & Barsom [2].

Such educational efforts need to be made to explain and promote the use of fracture mechanics in the design of concrete structures. The need becomes even more urgent when one considers the rapidly growing use of high-strength concrete (compressive strength >85 MPa) which ordinarily appears not to have a concomitant increase in fracture toughness and is a highly brittle material. The use of methods of fracture mechanics to characterize this material, e.g. 'brittleness number' [7] or 'characteristic length' [8], are especially valuable in providing caveats to the practitioner.

A possible mechanism to provide the design engineer with information on the use of fracture mechanics is through expert system based software. This could stand alone or be used in conjunction with undergraduate level or postgraduate short courses. The development of such a package is the subject of this paper.

2. EXPERT SYSTEMS

The use of knowledge based expert systems (ES) in many areas of engineering practice is developing very rapidly due to the availability of powerful computing equipment at modest cost and size (e.g. workstations) and the spreading knowledge of artificial intelligence (AI) concepts and their practical application to design.

In the area of structural engineering, a number of recent conferences and journals have appeared which are devoted to the use of automated structural analysis and design in general, and to developments in the use of AI techniques for conceptual, as well as, detailed structural design. See, for instance, Adeli [9], Adeli & Paek [10], [11], Kostem & Maher [12], Sriram *et al.* [13] and Maher [14].

2.1 What is an Expert System?

A definition given by Dym [15] is 'An expert system is a computer program that performs a task normally done by an expert or a consultant, and in so doing, it uses captured heuristic knowledge. In an expert system . . . the sequence of firing rules is determined within the program by an inference engine, and the rule premises may lead to multiple actions, or no action at all.'

Or, as stated by Krauthammer & Kohler [16], 'The so called 'expert systems' are applications of the heuristic search. The emphasis of an expert system is to develop a search procedure through a decision tree by making use of plausible inferences based on knowledge supplied by the user. The possible inferences are stored as rules in the program. These rules are similar to the rules people use to explain their decisions.'

The word 'heuristic' (akin to the word eureka) is defined as '. . . designating a method of education or computer programming in which the pupil or machine proceeds along empirical lines, using rules of thumb, to find solutions or answers.' (Webster's New World Dictionary, Second College Edition, Simon and Schuster, New York, 1984.) So, it may be considered that expert systems are computer programs designed to mimic the reasoning process of one or more experts in the field of application.

The basic components of an expert system are shown in Fig. 1 and include:

—input/output facilities to allow the user to communicate with the system and to create and use a data base for a specific situation;

—an inference engine that incorporates reasoning methods which act upon input data and knowledge from a knowledge base to solve the problem and also provide an explanation as to why (or why not) it arrived at a particular solution;

—a knowledge base which is developed from the expertise of one or more experts (not necessarily a programmer) which includes facts, opinions, rules (of-thumb), beliefs, 'gut feelings', which are unique to the expert(s) and the design domain;

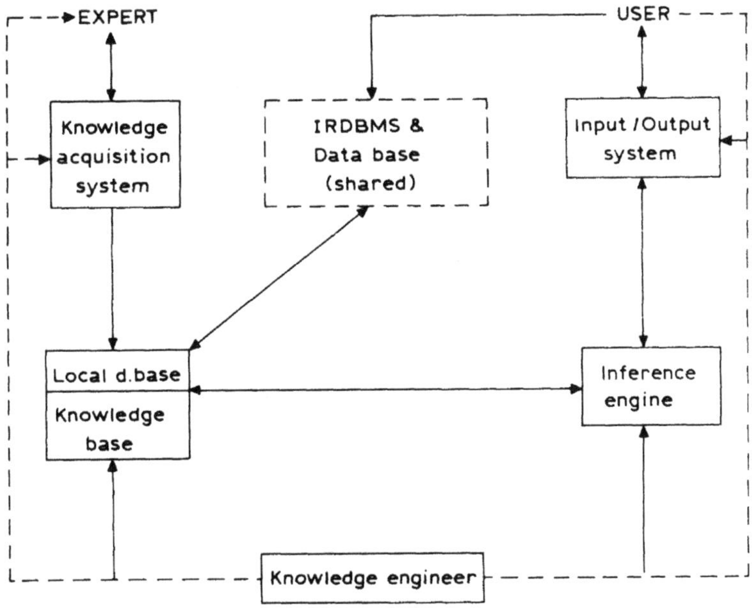

Fig. 1 Structure of expert systems.

—a knowledge-acquisition facility to allow the system to interface with the expert(s) and a knowledge engineer to build the knowledge base or to interface with other data bases, libraries, etc.

The knowledge engineer is the person who must develop and implement the model (software) which captures the expert's problem solving ability in the area of interest.

A useful publication summarizing expert systems and their utility in transportation engineering is given by Barnett *et al.* [17]. Some points which pertain to this section are given in the following.

1. Problems that are amenable to solution by expert systems are: diagnosis, interpretation/classification, prediction/forecasting, design, planning, and monitoring.
2. Expected benefits of the ES are given in terms of quantifiable aspects such as saving money, saving time, increasing productivity, enhancing a user's problem-solving ability and non-quantifiable aspects such as formalization of knowledge in a given problem domain and synthesizing the knowledge of a group of experts.

3. The need for an ES must be identified as well as an advocate, the end user, domain expert and knowledge engineer. The advocate defines the problem domain, expected benefits to a community of users and the experts. The end user provides a definition of the skill level that must be addressed, information on how problems in the field are solved, and advice on how the system must function—from the user viewpoint.
4. A crucial component in the expert system is the user interface. This is the means whereby the ES and user communicate in terms of input—via question and answer or menu format, output—graphics, tables, lists, etc., and explanation.

2.2 Expert Systems in Structural Engineering

A number of ES programs of varying complexity dealing with different aspects of structural engineering are given in the literature [14]. One of the first was SPERIL-I [18]. This performs structural assessment of existing structures which are subjected to earthquake excitation. The expertise consists of knowledge collected from experienced civil engineers and includes relations between factors such as structural damping, stiffness, creep, and buckling. This knowledge is represented as rules assessed throughout a forward chaining procedure.

Another example is SPERIL-II [19] which evaluates the general safety and damageability of an existing structure. The system analyzes inspection data and instrument records of structural responses during an earthquake, such as accelerations and displacements at certain locations in the structure. It then evaluates relevant safety characteristics such as interstory drift, stiffness and damping and the damageability of the structure's elements. The system's knowledge comes from case studies and is represented as rules of predicate logic.

These ES programs are quite large and complex. At a smaller scale, and capable of implementation on a personal computer, is MISS—A Microcomputer Integrated System for Structural Engineers [20]. This is an integrated, data based software system which combines finite element modeling with computer aided design to handle design of a multi-story reinforced concrete building. While not strictly an expert system, the key to its implementation on a microcomputer is its data management scheme with a central data bank which serves as a common denominator to all processing modules.

The ES program COLUMN [21] was written '... to explore the requirements of a computer program that, given the same information, will proportion reinforced concrete sections that are comparable to those produced by an experienced structural engineer.' The authors conclude that for proportioning and detailing reinforced concrete members, knowledge based expert systems can be written that will simulate the reasoning strategy used by design engineers, e.g. select a potential design, evaluate the design, and, if necessary, modify the design.

The recent availability of ES 'shells' reduces much of the programming effort. The shells typically provide a built-in user interface and inference engine so the programmer needs to provide only the model of some domain specific problem solving expertise. An example of an ES which is based on this concept is given by Krauthammer and Kohler [16] for the damage assessment of reinforced concrete structures after severe load applications (e.g. blast loads). An example of a decision tree for this problem is presented in Fig. 2. This approach is

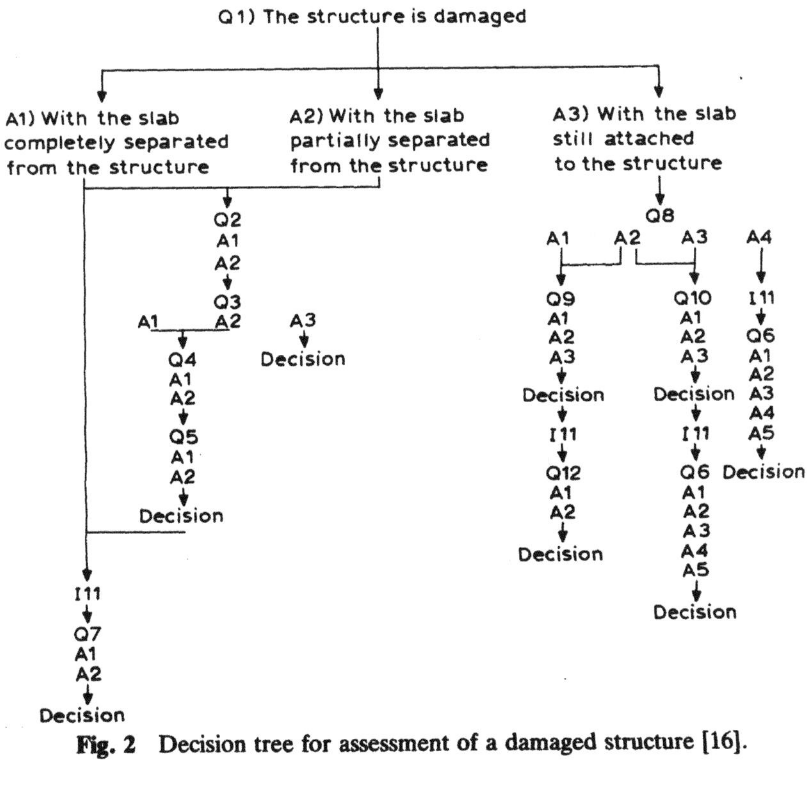

Fig. 2 Decision tree for assessment of a damaged structure [16].

an example of a *rule-based* knowledge representation scheme. The knowledge may consist of a collection of facts or condition-action (IF-THEN) rules or both. The inferencing technique that is used is deductive reasoning which consists of either: (1) deriving new facts from existing facts or rules—called 'forward chaining', or (2) proving that a given fact is true by combining selected facts in the knowledge base with selected rules—called 'backward chaining.' Note in Fig. 2 that multiple paths may result from a response to a question. This is of no concern for the system and it does not report any decision until all of the rules in the knowledge base have either been eliminated or verified.

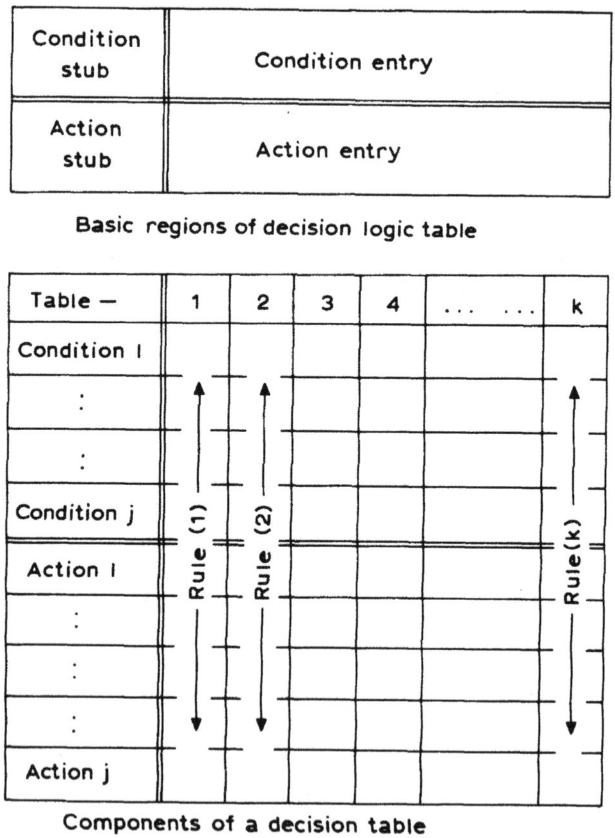

Fig. 3 Decision logic tables [24]: (a) basic regions of a decision logic table; (b) components of a decision table.

2.3 Decision Logic Tables

In a knowledge based expert system, it may be desirable to separate the knowledge from the inference process [22]. As an example, design standard knowledge could be represented separately from the process that applies this knowledge to solve design problems. The individual design standard knowledge requirements can be represented by using decision logic tables, or more simply, decision tables [23].

This approach was adopted by American Concrete Institute (ACI) Committee 118, Use of Computers, as a recommended format for the ACI Building Code [24]. An example of a generic decision table is shown in Fig. 3 where it is seen to consist of four major areas: condition stub, condition entry, action stub and action entry. The decision logic table format ensures that all logically possible situations are considered. (The first author was intimately involved in this exercise which did, in fact, cover the entire ACI 318-71 Building Code, and was intrigued to discover a number of logical combinations of conditions which led to no resulting rules or specifications! Thus,

U = Maximum (U_O, U_W, U_E)

where,

U = required load factor

U_O = (the minimum load factor) = $1{\cdot}4D + 1{\cdot}7L$

U_W = (the load factor for wind effects) given by Table 9.3.2

U_E = (the load factor for earthquake effects) given by Table 9.3.3

Table 9.3.2

W present ?	Y	Y	Y	N	Y	ELSE
D,W same sign ?	Y	Y	N	I	N	
D,L same sign ?	Y	N	Y	I	N	
$U_{W1} = 0{\cdot}75\,(1{\cdot}4D + 1{\cdot}7L + 1{\cdot}7W)$	X		X		X	
$U_{W2} = 0{\cdot}75\,(1{\cdot}4D + 1{\cdot}7W)$		X				
$U_{W3} = 0{\cdot}9D + 1{\cdot}3W$	X	X	X		X	
$U_W = 0$				X		
$U_W = \max(U_{W1}, U_{W3})$	X		X		X	
$U_W = \max(U_{W2}, U_{W3})$		X				
No provision for U_W						X

Table 9.3.3

Identical to Table 9.3.2, except substitute $1{\cdot}1E$ for W

Fig. 4 Decision table for Sections 9.3.1 through 9.3.3, ACI 318-71 [24].

this approach can also serve the purpose of discovering and removing 'holes' from the specification.)

The condition stub contains a list of all the conditions that must be considered for a given situation. The condition entry contains an array of yes (Y), or (N) or immaterial (I) responses in which each column represents a unique set of responses to the conditions. The action stub lists the appropriate actions for the given situation. The actions are listed in the order in which they are to be executed. The action entry contains indicators (X) and blank spaces. An indicator opposite an action specifies that the particular action is to be executed. Thus, each full column comprises a decision rule and represents one complete path of logic.

An example of a decision table which sets rules for overload factors (service load multipliers) when wind load is to be considered is presented in Fig. 4. The last decision rule is termed the 'ELSE' rule and is provided to indicate that sets of responses not presented in the condition entry represent impossible or disallowed events.

3. FRACTURE MECHANICS EXPERT SYSTEM

While the need to educate engineers with respect to methods of fracture mechanics and the applications to reinforced concrete structures may be accepted, the question still may be raised as to why should the expert system concept be employed. The authors have adopted this approach for the following reasons.

1. Software is portable, relatively inexpensive, and may be disseminated readily. At the present time, such software as may exist is in development and at a low level of resolution so workshops and educational seminars are needed as alternatives or supplements. Eventually, it is possible that such computer programs will be self sufficient.
2. Aspects of the software can be made to interact with present computer aided design systems in a logical way.
3. The most important reason is that the massive amount of information presently available has to be sorted, culled and distilled into a manageable form to be presented to the user. To do this requires selectivity on the part of the expert who supplies rules and procedures based on his experience and intuition. At

the same time, the ES approach allows for flexibility, including changes in the rules/procedures, based on additional experiences—successes or failures.

3.1 Scope of the Proposed System

The work described here has the following objectives:

1. Develop expert system software in the domain of the application of fracture mechanics to reinforced concrete structures which can be used as:
   —a stand-alone tutorial;
   —an aid to teaching (workshops, credit courses, etc.);
   —a guide to materials testing for concrete properties;
   —a dynamic (updatable) design aid for knowledgeable users, which may also interact with other data bases/software packages.
2. Develop appropriate knowledge and databases in:
   (a) Basic information—localization concepts, linear elastic fracture mechanics, nonlinear fracture mechanics (fictitious crack model, size effect law).
   (b) Materials testing—usual tests for properties of hardened concrete, tests to obtain fracture parameters, closed-loop testing, data reduction/interpretation.
   (c) Structural applications—failure analysis case studies, design, software availability and utilization. Current application areas include:
   —bar pull-out, bond
   —anchor bolts
   —shear
   —minimum reinforcement
   —fatigue
   —creep
   (d) Bibliography.
   (e) Heuristics—rules to derive the selection of appropriate procedures for given tasks.
3. Integrate the knowledge/data bases and heuristics with an inference engine and user interface to develop the ES using a commercially-available shell, VP-EXPERT [25].

The ES shell VP-EXPERT is a cost-effective software package which can be implemented on a personal computer (e.g. Zenith Z-154) as well as on larger machines. It has a rule-based inference engine that uses backward chaining for problem solving. It is compatible with, and uses, dBASE database files, Lotus 1-2-3 worksheet files and ASCII text files. It also has the ability to execute external DOS programs.

The organization of the major areas of the fracture mechanics program FRMECH is displayed in Fig. 5. The action stub and action entry sequence is selected on the assumption that use as a teaching aid will be independent of use for material testing or engineering. On the other hand, it may be necessary that the tutorial level is needed by all the others.

The development of this program is in progress and is expected to be a lengthy undertaking. It is being performed with succeedingly refined and denser levels of detail (resolution). What will be presented here is the most general level with the lowest resolution which will nevertheless display the organization of the program and the planned data bases.

Throughout the development, all phases will interact with a bibliog-

Tutorial ?	Y	N	N	N	Y	Y	Y	Y	N	E
Teaching aid ?	N	Y	N	N	Y	N	N	N	N	L
Material testing ?	N	N	Y	N	N	Y	N	Y	Y	S
Engineering design ?	N	N	N	Y	N	N	Y	Y	Y	E
Level I	X				X	X	X	X		
Level II		X			X					
Level III			X			X		X	X	
Level IV				X			X	X	X	
Exit										X

Level I = basic concepts, terminology, graphics
Level II = advanced concepts, teaching aids / graphics
Level III = material testing
Level IV = design applications

Fig. 5 Entry decision table for FRMECH.

raphy database. This has been developed and is called BIBLI [4]. At present the data set contains all known fracture mechanics references related to concrete up to the year 1985. Obviously, much more needs to be added! It was written in the framework of DBASE III which is a relational, database management system for 16-bit microcomputers. Its capabilities include 128 fields, 4000 bytes per record and up to one billion records per file with fast sorting, indexing and searching. At present it contains 783 citations.

3.2 FRMECH

Low level resolution flow diagrams for the tutorial overview, teaching aid overview, material testing overview, and engineering design

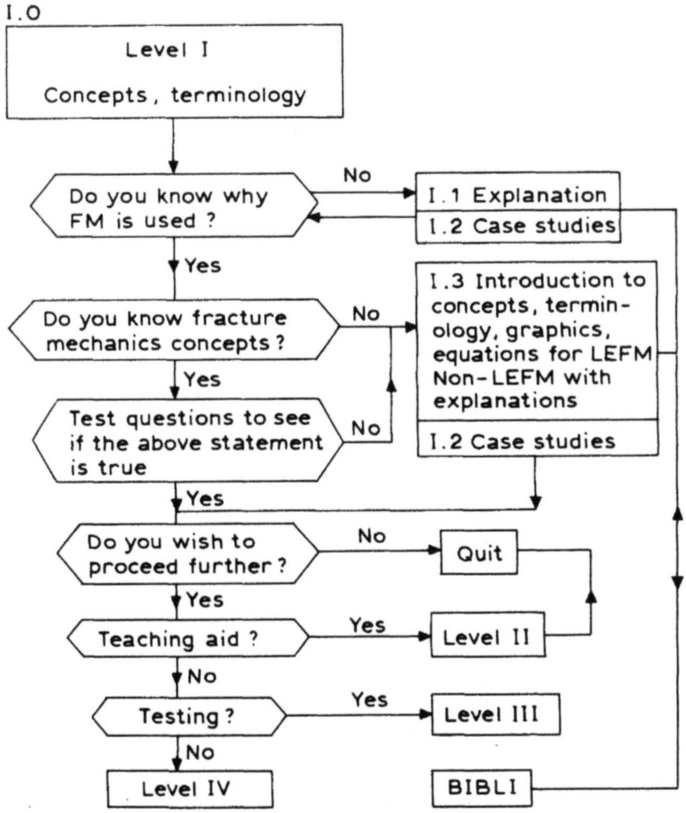

Fig. 6 Tutorial overview flow diagram.

overview are given in Figs 6–9. In Fig. 6, the case studies listed in I.2 are, at present, a non-linear, fracture mechanics, finite element analysis of the failure of Schoharie Creek bridge [5] and the cracking problems of Kölnbrein dam analyzed by Ingraffea and his co-workers [6].

With respect to testing methods, Fig. 8, the ones which will appear in this first version of the program are: RILEM proposal for determining fracture energy [26]; Bazant's size effect method to determine energy release [7]; Jenq and Shah two-parameter model [27]; the beam test method of Karihaloo and Nallathambi [28]; the beam test method of Go/Refai/Swartz [29]; and the wedge-loaded specimen [30]. A sequence of screens describing the RILEM method is given in Figs 10–15 and Bazant's method in Figs 16–23 [31].

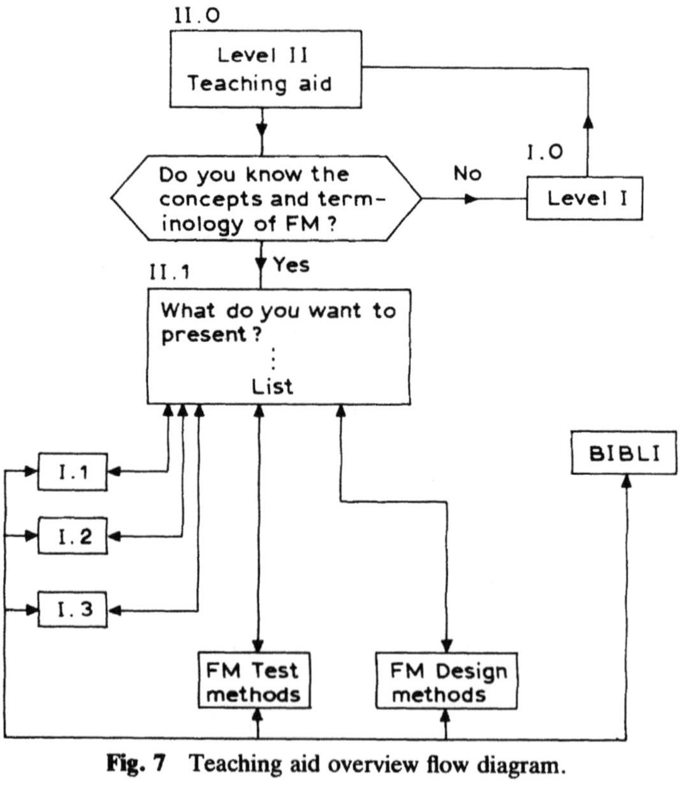

Fig. 7 Teaching aid overview flow diagram.

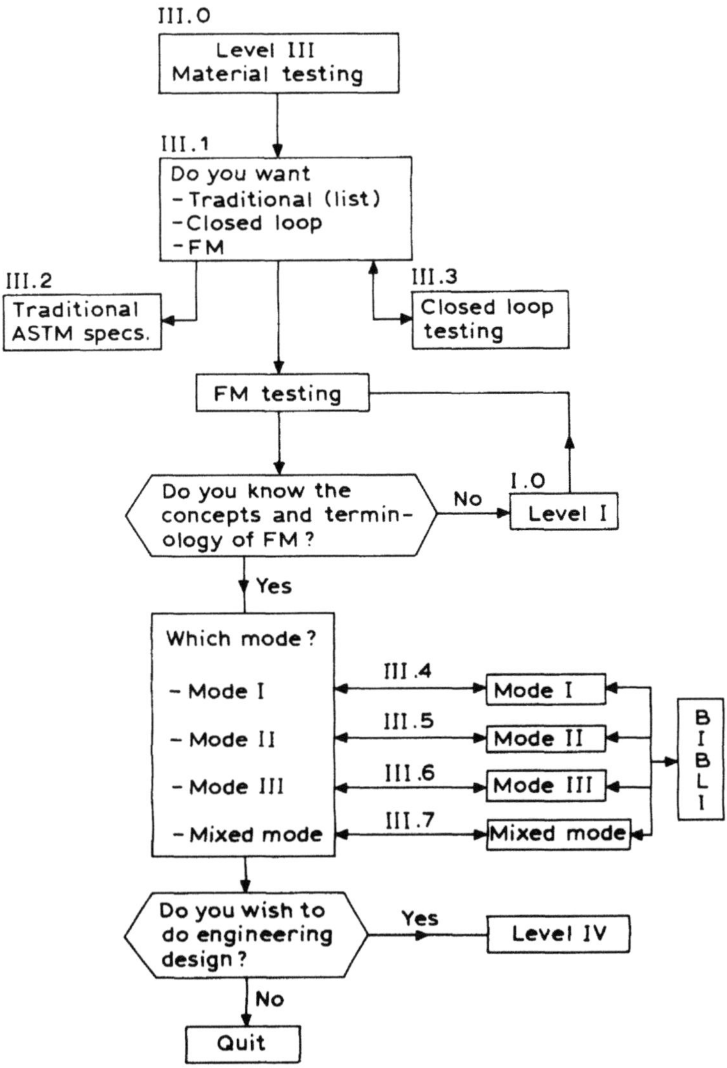

Fig. 8 Material testing overview flow diagram.

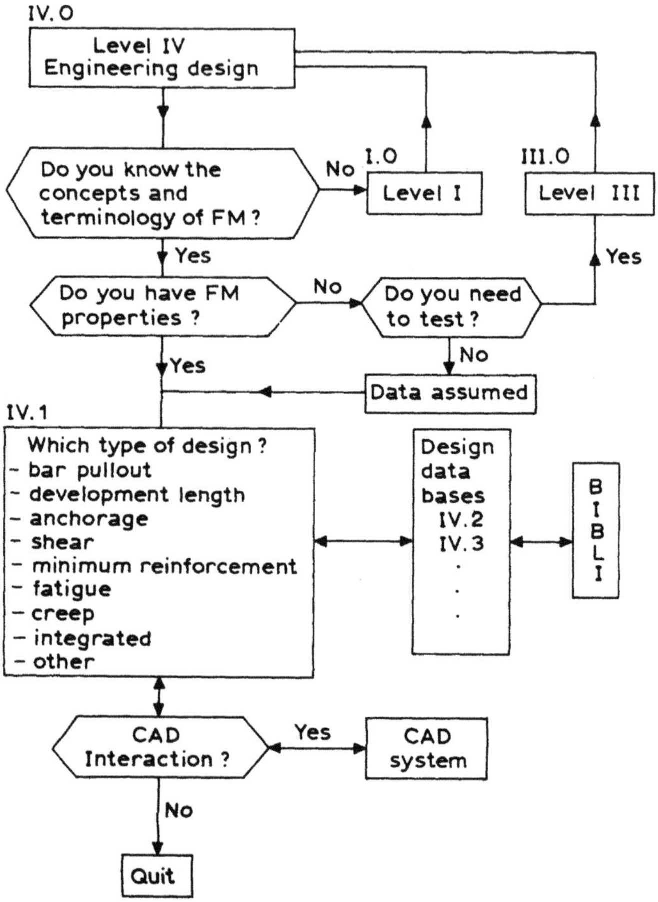

Fig. 9 Engineering design overview flow diagram.

Finally, the methods of application to reinforced concrete structures which are included in this first version of the program are:

—bar development (bend) [32]
—anchorage [33], [34]
—shear [8]
—minimum reinforcement [35]
—fatigue [36]–[40]
—creep [41]–[44]

RILEM METHOD TO DETERMINE
FRACTURE ENERGY (Hillerborg)

Background

Uses fictitious crack concept

The energy absorbed within the damage zone in a tensile test specimen is denoted by G_F and is the fracture energy per unit area of the fracture surface.

This is a non-LEFM method

Fig. 10 RILEM method, Screen 1.

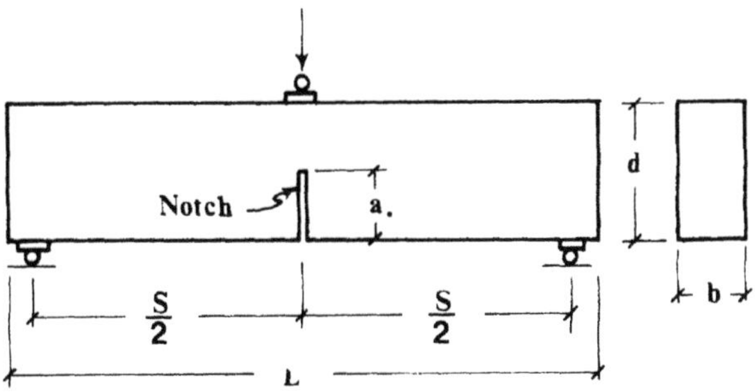

$a_0 = 05d \pm 5$ mm

Sizes of Specimens

Max. Agg. D_{max} mm	d mm	b mm	S mm	L mm
1–16	100 ± 5	100 ± 5	800 ± 5	840 ± 10
16.1–32	200 ± 5	100 ± 5	1130 ± 5	1190 ± 10
32.1–48	300 ± 5	150 ± 5	1385 ± 5	1450 ± 10
48.1–64	400 ± 5	200 ± 5	1600 ± 5	1640 ± 10

Fig. 11 RILEM method, Screen 2.

TESTING PROCEDURE

_A stiff testing machine or closed-loop machine (operating in deformation control) must be used to assure stable response in the test to failure.

_The deformation of the center of the beam shall be determined with regard to a line between two points on the beam above the supports. Accuracy: 0·01 mm.

_Load is applied at approximately a constant rate of deformation such that peak load is reached in 30–60 sec. Record load and displacement continuously.

Fig. 12 RILEM method, Screen 3.

4. CONCLUSIONS

The work presented here represents the start of a program to provide practical information to the engineering profession on the concepts of fracture mechanics and how these can be implemented in the design of reinforced concrete structures.

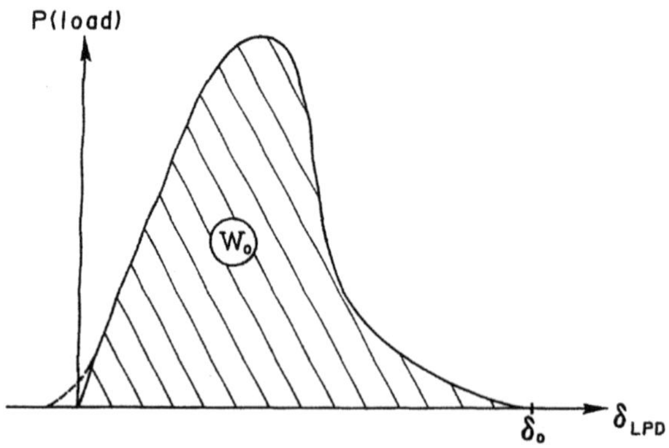

Fig. 13 RILEM method, Screen 4.

DATA EVALUATION

$$G_F = \frac{W_0 + mg\delta_0}{A_{lig}}$$

This is for downward load and displacement.

A_{lig} = projected area of fracture zone on a plane perpendicular to the beam axis;
= $(d - a_0)b$.

mg = weight of beam between supports plus carried fixtures.

For upward load and displacement Swartz and Yap showed that

$$G_F = \frac{W_0 - 1/2\ mg\delta_0}{A_{lig}}$$

Fig. 14 RILEM method, Screen 5.

DISCUSSION

Method is easy to apply.

Replicate tests should be made to determine average value and C.V. of G_F.

Numerous test data show that G_F is somewhat size dependent.

G_F is also crack or notch length dependent but by using $a_0 = 1/2d$ reasonable agreement with values from other methods is obtained.

Characteristic length is

$$l_{ch} = \frac{E_c G_F}{f_t^2}$$

Fig. 15 RILEM method, Screen 6.

BAZANT SIZE EFFECT METHOD FOR
DETERMINING FRACTURE ENERGY
OF CONCRETE

Background

The fracture energy G_f of a microscopically heterogeneous brittle material is the specific energy required for crack growth in an infinitely large specimen.

Size effect law

$$\sigma_N = Bf'_t\left[\left(1+\frac{d}{\lambda_0 d_a}\right)^r\right]^{-1/2r}$$

B, λ_0 = empirical constants; $r = 1$
d_a = maximum aggregate size
f'_t = strength parameter
$\sigma_N = P/bd$; b = specimen thickness
d = characteristic dimension

Fig. 16 Bazant method, Screen 1.

The size effect law is based on the fact that the nominal stress at failure of geometrically similar structures decreases with size.

Law obtained by dimensional analysis and

1. The energy release due to failure is a function of the length of the fracture;
2. It is also a function of the size of the process zone at the fracture front.

The obtained result, G_f is thus size and geometry independent.

Fig. 17 Bazant method, Screen 2.

SPECIMEN GEOMETRY

The method is valid for any geometry.
The proposal is directed to TPB beams.

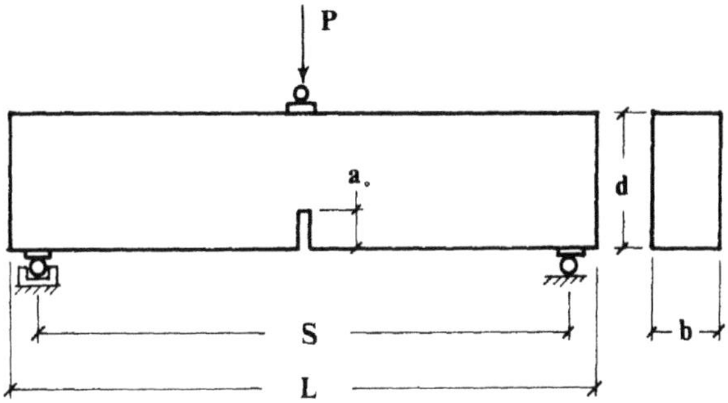

$$\frac{S}{d} \geq 2.5 \qquad 0.15 \leq \frac{a_0}{d} \leq 0.40$$

Notch width as small as possible
but at least $<0.5\ d_a$

$$b > 3d_a, \ d > 3d_a$$

$$\text{Min. } d \leq 5d_a, \ \text{Max. } d \geq 15\ d_a$$

$$\frac{\text{Max. } d}{\text{Min. } d} \geq 4, \quad \frac{d_{i+1}}{d_i} = \text{constant}$$

Fig. 18 Bazant method, Screen 3.

Size range should be as broad as feasible, e.g.

$\frac{d}{d_a} = 4, 8, 16$ is acceptable,

$\frac{d}{d_a} = 3, 6, 12, 24$ is better.

At least three identical specimens for each size.

For all sizes, keep b constant.

Fig. 19 Bazant method, Screen 4.

TESTING PROCEDURE

Use an ordinary uniaxial testing machine.

Load specimen at constant displacement rate.

Reach P_{max} in 1 to 10 min.

Record $P_j = P_{max}$ and d_j for each test.

Also record all pertinent geometry data, mix data, curing and storage conditions, temp. and humidity for each test.

Determine and record f'_c, E_c and the mean mass density of the concrete.

Fig. 20 Bazant method, Screen 5.

DATA EVALUATION

_assuming geometrically similar specimens
If $L_j \approx S_j$; $P_j^0 = P_j + 1/2 m_j g$
If L_j much greater than S_j,

$$P_j^0 = P_j + \frac{2S_j - L_j}{2S_j} m_j g$$

and $P_j^* = P_j^0$

$$\text{Define } Y_j = \left[\frac{bd_j}{P_j^*}\right]^2, \qquad X_j = d_j$$

and plot Y_j versus X_j.

Using linear regression determine slope A of the best fit line. Also check if the plot is approximately linear.

Statistics

Fig. 21 Bazant method, Screen 6.

$$G_f = \frac{g(\alpha_0)}{E_c A}$$

$$g(\alpha_0) = \left[\frac{S_m}{d_m}\right]^2 \pi \alpha_0 [1{\cdot}5F(\alpha_0)]^2$$

$\frac{S_m}{d_m}$ for mid-size specimen

$$\alpha_0 = \frac{a_0}{d}$$

$$F(\alpha_0) = F_4(\alpha_0) + \frac{(S_m/d_m) - 4}{4}[F_8(\alpha_0) - F_4(\alpha_0)]$$

$$F_4(\alpha_0) = 1{\cdot}090 - 1{\cdot}735\alpha_0 + 8{\cdot}2\alpha_0^2 - 14{\cdot}18\alpha_0^3 + 14{\cdot}57\alpha_0^4$$

$$F_8(\alpha_0) = 1{\cdot}107 - 1{\cdot}552\alpha_0 + 7{\cdot}71\alpha_0^2 - 13{\cdot}55\alpha_0^3 + 14{\cdot}25\alpha_0^4$$

Fig. 22 Bazant method, Screen 7.

DISCUSSION

Method is easy to apply but requires a minimum of 9 to 12 specimens.

The size of specimens can become fairly large, e.g. if $d_a = 1$ in. then least $d_{max} =$ 15 in. and least $S_{max} = 37{\cdot}5$ in. ($= 15 \times 2{\cdot}5$). If $S/d = 4$, a commonly used ratio, then the least $S_{max} = 60$ in.

Auxiliary calculations are needed to verify that the results are statistically reliable.

A measure of brittleness may be defined as $\beta = \frac{d}{\lambda_0 d_a}$. When

$\beta < 0{\cdot}1$ Plastic Limit Analysis

$0{\cdot}1 \leq \beta \leq 10$ Nonlinear Fracture Mechanics

$\beta > 10$ LEFM

Fig. 23 Bazant method, Screen 8.

The use of expert systems appears to offer a promising mechanism to develop computer software which may serve this education goal and which also may become a part of an integrated computer aided design system.

ACKNOWLEDGEMENTS

The work reported here was supported by the Center for Research in Computer Controlled Automation at Kansas State University. The bibliography data-base program BIBLI and associated data files were written by R. P. Bernhardt as an M.S. thesis project under the direction of the first author and was supported by the Naval Civil Engineering Laboratory Contract N62583/85 MT239. This support is gratefully acknowledged.

REFERENCES

[1] Bazant, Z. P. *et al.*, (members of ACI Committee 446, Fracture Mechanics). Fracture mechanics of concrete: concepts, models and determination of material properties. *Concrete International,* **12**(12) (1991) 67–70.

[2] Rolfe, S. T. & Barsom, J. M., *Fracture and Fatigue Control in Structures,* 1st edn. Prentice-Hall Inc., Englewood Cliffs, New Jersey, 1977.

[3] RILEM Committee TC 90-FMA, *Fracture Mechanics of Concrete Structures: From Theory to Applications,* ed. L. Elfgren. Chapman & Hall, London, 1989.

[4] Swartz, S. E., Applicability of fracture mechanics methodology to cracking and fracture of concrete. Final Report to the Naval Civil Engineering Laboratory, Contract N62583/85 MT239, November, 1985.

[5] Swenson, D. V. & Ingraffea, A. R., The collapse of Schoharie Creek bridge—a case study in concrete fracture mechanics. *International Journal of Fracture,* **51** (1991) 73–92.

[6] Linsbauer, H. N., Ingraffea, A. R., Rossmanith, H. P. & Wawrzynek, P. A., Simulation of cracking in large arch dam: part I, part II. *Journal of Structural Engineering* (ASCE), **115**(7) (1989) 1599–1615, 1616–30.

[7] Bazant, Z. P. & Pfeiffer, P. A., Determination of fracture energy from size effect and brittleness number. *ACI Materials Journal,* **84**(6) (1987) 463–80.

[8] Gustafsson, P. J. & Hillerborg, A., Sensitivity in shear strength of longitudinally reinforced concrete beams to fracture energy of concrete. *ACI Structural Journal,* **85**(3) (1988) 286–94.

[9] Adeli, H. (Ed.), *Expert Systems in Construction and Structural Engineering,* Chapman and Hall, London, 1988.

[10] Adeli, H. & Paek, Y., Computer-aided design of structures using LISP. *Computers and Structures*, **22**(6) (1986) 939–56.
[11] Adeli, H. & Paek, Y., Computer-aided analysis of structures in INTERLISP environment. *Computers and Structures*, **23**(3) (1986) 939–1007.
[12] Kostem, C. & Maher, M. (Eds) *Expert Systems in Civil Engineering*. American Society of Civil Engineers, New York, 1986.
[13] Sriram, D., Maher, M. L. & Fenves, S. J., Knowledge-based expert systems in structural design. *Computers and Structures*, **20**(1–3) (1985) 1–9.
[14] Maher, M. (Ed.) *Expert Systems for Civil Engineers*. American Society of Civil Engineers, New York, 1987.
[15] Dym, C. L., Expert systems: new approaches to computer aided engineering. Paper presented at the 25th Structures, Structural Dynamics and Materials Conference, Palm Springs, California 14–16 May 1984, pp. 99–115.
[16] Krauthammer, T. & Kohler, S., RC structures under severe loads—an expert system approach. In *Expert Systems in Civil Engineering*, ed. C. N. Kostem & M. L. Maher. American Society of Civil Engineers, New York, 1986, pp. 96–108.
[17] Barnett, D., Jackson, C. & Wentworth, J. A., Developing expert systems. Publication No. FHWA-TS-88-022, Federal Highway Administration, McLean, Virginia, USA, December, 1988.
[18] Ishizuka, M., Fu, K. S. & Yao, J. T. R., SPERIL-I: computer based structural damage assessment system. Report CE-STR-81-36, School of Engineering, Purdue University, Lafayette, 1981.
[19] Ogawa, H., Fu, K. S. & Yao, J. T. P., An expert system for damage assessment of existing structures. In *Proceedings of the First Conference on Artificial Intelligence Applications*. IEEE Computer Society, New York, USA, December, 1984.
[20] Kanok-Nukulchai, W., On a microcomputer integrated system for structural engineering practices. *Computers and Structures*, **23**(1) (1986) 33–7.
[21] Bezzina, A. S. & Simmonds, S. H., Knowledge-based expert systems in reinforced concrete design. *Concrete International*, **11**(2) (1989) 57–61.
[22] Garrett, J. H., Jr. & Fenves, S. J., Knowledge-based standard-independent member design. *Journal of Structural Engineering* (ASCE), **115**(6) (1989) 1396–1411.
[23] Fenves, S. J., Gaylord, E. H. & Goel, S. K., Decision table formulation of the AISC specification. *Civil Engineering Studies, Structural Research Series*, **347**, University of Illinois, Urbana, August 1969.
[24] Pyle, D., Fenves, S., Fisher, G., Goel, S., Ketchum, M., Noland, J., Palejs, A. & Swartz, S., Decision logic table format for building code requirements for reinforced concrete (ACI 318-71). *ACI Journal*, **70**(12) (1973) 788–92.
[25] VP-EXPERT, Version 2.0. Paperback Software International, 2830 Ninth St., Berkeley, CA 94710, 1988.
[26] RILEM Committee 50-FMC, Determination of the fracture energy of

mortar and concrete by means of three-point bend tests on notched beams. *Materials and Structures,* **18**(106) (1985) 43–8.

[27] Jenq, Y. S. & Shah, S. P., Two parameter fracture model for concrete. *Journal of Engineering Mechanics* (ASCE), **111**(10) (1985) 1227–41.

[28] Nallathambi, P. & Karihaloo, B. L., Determination of specimen-size independent fracture toughness of plain concrete. *Magazine of Concrete Research,* **38**(135) (1986) 67–76.

[29] Swartz, S. E. & Refai, T., Cracked surface revealed by dye and its utility in determining fracture parameters. In *Fracture Toughness and Fracture Energy,* ed. H. Mihashi, H. Takahashi & F. Wittmann. Balkema, Rotterdam, 1989, pp. 509–20.

[30] Rossi, P., Brühwiler, E., Jenq, Y.-S. & Chhuy, S., Fracture properties of concrete as determined by means of wedge splitting tests and tapered double cantilever beam tests. In *Fracture Mechanics Test Methods for Concrete,* eds S. P. Shah & A. Carpinteri. Chapman & Hall, London, 1991, pp. 87–128..

[31] Swartz, S. E., Proposals for practical testing of concrete to determine useful fracture properties. Paper presented at the 1990 Spring Convention of the American Concrete Institute, Toronto, Canada, 18–23 March 1990.

[32] Ingraffea, A. R., Gerstle, W. H., Gergely, P. & Saouma, V., Fracture mechanics of bond in reinforced concrete. *Journal of Structural Engineering* (ASCE), **110**(4) (1984) 871–90.

[33] Elfgren, L., Ohlsson, U. & Gylltoft, K., Anchor bolts analyzed with fracture mechanics. In *Fracture of Concrete and Rock,* ed. S. P. Shah & S. E. Swartz. Springer-Verlag, New York, 1989, pp. 269–75.

[34] Ohlsson, U. & Elfgren, L., Anchor bolts in concrete structures—two dimensional modeling. In *Fracture of Concrete and Rock: Recent Developments,* ed. S. P. Shah, S. E. Swartz & B. Barr. Elsevier Applied Science, London, 1989, pp. 747–53.

[35] Bosco, C., Carpinteri, A. & Debernardi, G., Minimum reinforcement in high strength concrete. *Journal of Structural Engineering* (ASCE), **116**(2) (1990) 427–37.

[36] Gylltoft, K., Fracture mechanics models for fatigue in concrete structures. PhD thesis, Lulea University of Technology, Lulea, Sweden, 1983.

[37] Swartz, S. E., Hu, K. K. & Jones, G. L., Compliance monitoring of crack growth in concrete. *Journal of the Engineering Mechanics Division* (ASCE), **104**(4) (1978) 789–800.

[38] Swartz, S. E., Huang, C.-M. J. & Hu, K. K., Crack growth and fracture in plain concrete–static versus fatigue loading. In *Fatigue of Concrete Structures,* ed. S. P. Shah. ACI Publication SP-75, American Concrete Institute, Detroit, 1982, pp. 47–69.

[39] Suresh, S., Fatigue crack growth in cementitious solids under cyclic compression: theory and experiments. In *Fracture of Concrete and Rock: Recent Developments,* ed. S. P. Shah, S. E. Swartz & B. Barr. Elsevier Applied Science, London, 1989, pp. 162–72.

[40] Hordijk, D. A. & Reinhardt, H. W., Growth of discrete cracks in concrete under fatigue loading. In *Toughening Mechanisms in Quasi-*

Brittle Materials, ed. S. P. Shah. Kluwer Academic Publishers, Dordrecht, 1991, pp. 541–54.

[41] Hu, X.-Z., Flaw analysis in time-dependent fracture for cementitious materials. In *Fracture of Concrete and Rock: Recent Developments*, ed. S. P. Shah, S. E. Swartz & B. Barr. Elsevier Applied Science, London, 1989, pp. 409–18.

[42] Bazant, Z. P. & Gettu, R., Determination of nonlinear fracture characteristics and time dependence from size effect. In *Fracture of Concrete and Rock: Recent Developments*, ed. S. P. Shah, S. E. Swartz, & B. Barr. Elsevier Applied Science, London, 1989, pp. 549–65.

[43] Liu, Z.-G., Swartz, S. E., Hu, K. K. & Kan, Y.-C., Time-dependent response and fracture of plain concrete beams. In *Fracture of Concrete and Rock: Recent Developments*, ed. S. P. Shah, S. E. Swartz & B. Barr. Elsevier Applied Science, London, 1989, pp. 577–86.

[44] Wittmann, F. H., Influence of time on crack formation and failure of concrete. In *Application of Fracture Mechanics to Cementitious Composites*, ed. S. P. Shah. NATO Advanced Research Workshop, Northwestern University, Evanston, 1984, pp. 443–64.

Index

www.ingramcontent.com/pod-product-compliance
Ingram Content Group UK Ltd.
Pitfield, Milton Keynes, MK11 3LW, UK
UKHW022005180926
475701UK00003B/205
9780367864613